T0138321

GEOGRAPHIES OF NINETEENTH-CENTURY SCIENCE

GEOGRAPHIES OF
NINETEENTH-CENTURY SCIENCE

Edited by
DAVID N. LIVINGSTONE
and CHARLES W. J. WITHERS

THE UNIVERSITY OF CHICAGO PRESS

CHICAGO AND LONDON

DAVID N. LIVINGSTONE is professor of geography and intellectual history at Queen's University, Belfast.

CHARLES W. J. WITHERS is professor of historical geography at the University of Edinburgh. They are the editors of *Geography and Revolution* and *Geography and Enlightenment*, both published by the University of Chicago Press.

The University of Chicago Press, Chicago 60637
The University of Chicago Press, Ltd., London
© 2011 by The University of Chicago
All rights reserved. Published 2011.
Printed in the United States of America

20 19 18 17 16 15 14 13 12 11   1 2 3 4 5

ISBN-13: 978-0-226-48726-7   (cloth)
ISBN-10: 0-226-48726-1   (cloth)

Library of Congress Cataloging-in-Publication Data

Geographies of nineteenth-century science / edited by David N. Livingstone and Charles W. J. Withers.
       p. cm.
   Includes bibliographical references and index.
   ISBN-13: 978-0-226-48726-7 (cloth : alk. paper)
   ISBN-10: 0-226-48726-1 (cloth : alk. paper)   1. Science—Great Britain—History—19th century.   I. Livingstone, David N., 1953–   II. Withers, Charles W. J.
   Q127.G4G46 2011
   509'.034—dc22   2010039367

CONTENTS

ACKNOWLEDGMENTS

This book is an exploration of the geographical dimensions of science in the nineteenth century. It is intended as a contribution to continuing discussions between geographers, historians of science, historians, and others about the importance of place and space to an understanding in historical context of what science was, why science took the shape it did in a particular place and not somewhere else, and why "where" questions are vital in explanation of science's "what," "how," "who," and "when." Not so long ago, perhaps, research into the historical geographies of science might have seemed, even for geographers, an obtuse field: after all, was not science supposed to be placeless? And so too for historians of science, whose subject of study, albeit of different periods and often in quite different ways, was about explaining the nature and content of science in its temporal dimensions and hardly at all about its geographical ones. But questions concerning the importance of space and place to an understanding of the making and communication of science—actually, of all intellectual work—have come to the fore in recent years. There has been, indeed, an evident "spatial turn" in the humanities and the social sciences. Scientists too—some of them anyway—recognize the socially constructed and situated nature of their authoritative claims over science. It is now possible to use terms such as the "geography of science," even the "historical geography of science," and for them to be understood as pertaining to a distinct field of inquiry.

Distinct, but not the preserve of any one set of practitioners. The book is a reflection of shared research interests in the geographical dimensions of science in the century in which, arguably, science—the sciences—became professional and popular, disciplined and discursively discrete, precisely institutionalized and widely instructive, as never before. The conference in which the chapters that make up this book had their first public airing was held in the Institute of Geography at the University of Edinburgh in July 2007. We are grateful to Becky Higgitt for her help as conference assistant over the four days of the meeting. As editors, we would particularly like to thank the two anonymous readers for the University of Chicago Press for their thoughtful, encouraging, and supportive comments, and our contributors for responding so well to suggestions for revisions to their spoken papers and earlier drafts.

Several funding bodies provided help toward the conference on which this book is based and, thus, to the volume itself. We are grateful to the British Academy, the British Society for the History of Science, the Royal Society of Edinburgh, the University of Edinburgh, and the Historical Geography Research Group of the Royal Geographical Society (incorporating the Institute of British Geographers). Funds from the latter allowed graduate students in particular to attend. Charles Withers owes a considerable debt to Professor Susan Manning, Dr. Pauline Phemister, and Anthea Taylor at the Institute for Advanced Studies in the Humanities of the University of Edinburgh for their support during a sabbatical fellowship that allowed for the revision of his chapter and for most of the joint editorial work. He would also like to thank the staff of the Institute for Advanced Studies at the University of Warwick for their warm welcome and assistance during his period there as visiting professor, which allowed further reflection on the project.

One of the great delights of undertaking this book has been the fact that we have been privileged to work again with Christie Henry, the epitome of the concerned, supportive, and engaging editor. We are greatly indebted to her and to her colleagues Abby Collier and Stephanie Hlywak at the University of Chicago Press. Nicolaas Rupke very graciously accepted our invitation to contribute an afterword to the collection, based on his closing remarks offered at the conference, and we are grateful to him for his words here and for his encouragement regarding the volume as a whole. Chiefly, our thanks go to our contributors, whose own endeavors speak eloquently of the rich insight that comes from thinking and working collaboratively and across disciplinary boundaries about the many questions concerning the geographical dimensions of science.

CHAPTER ONE

Thinking Geographically
about Nineteenth-Century Science

CHARLES W. J. WITHERS AND DAVID N. LIVINGSTONE

B ecause science is a spatially distributed entity, it makes sense to bring it—and its many dimensions—within the arc of geographical inspection. The list of scientific themes that might be scrutinized through geographical lenses is extensive. Scientific practices and procedures operate in different ways in different places, often in highly distinctive venues. Scientific knowledge is differentially spread across the surface of the earth, and moves from place to place through complex circulatory networks. Scientific institutions occupy distinct locations in different settings. Scientific theories are shaped by the prevailing political, economic, religious, and social conditions, as well as a host of other cultural norms in different geographical localities, and, not least in the natural and field sciences, they may bear the stamp of the environments within which they are constructed. The list could be elaborated *ad libitum.*

For this reason, various proposals have been put forward with regard to the organizational structures and conceptual taxonomies involved in understanding the geography of science. Scholars who have reviewed the field have offered a variety of taxonomic schema that differently order key questions having to do with the sites and social spaces of science's production and reception, the nature of its mobility across space and between different communities, and so on. For one, moves toward "the cultivation of a spatialised historiography of science," initially proposed under the headings of "the regionalisation of scientific style," the "political topography of scientific commitment" and the "social space of scientific sites," were later refined to advance "site," "circulation" and "region" as the principal organizational categories for understanding science's venues, movement, and variant areal expression.[1] For others, the shaping of scientific knowledge centered upon the themes of the territory, of working classes, pastoral privileges, metropolitan spaces, and research sites.[2] Yet others advocated a threefold approach to the geography of knowledge—a static geography of place, a kinematic geography of movement, and consideration of the dynamics of travel—or, simpler still, have advanced a twofold conceptualization of the places of science's making and reception (science *in situ*) and matters of science's movement across space (science in motion).[3]

Each of these classificatory concerns has been aimed not only at making sense of the spatial nature of science, but at the connections between such spatiality, the complex social practices that go to make up and to maintain scientific knowledge, and those matters of embodiment, credibility, and authority through which science is undertaken so as to be believed and acted upon.[4] In the end, each of these arrangements turn out to be, at best, rough approximations for thinking about two compelling sets of questions: the first having to do with the making and meaning of science *in place*, the second with science's movement *over space*. Why was science promoted *there* and not somewhere else? Why should science there have taken *that* form? How does science travel—within and between communities of practitioners, for example, or from "expert" to "lay" audience? How is science communicated over distance? There are, of course, many other spatially inflected questions. Addressing these and other spatial matters demands attention to much more than location. *Place*, to take just one term, has complex meanings.[5] As others have remarked, "Place is not mere background atmospherics, but provides for the very possibility of intellectual innovation."[6] So, too, *space* is not mere geographical extent, a surrogate for *territory* or *landscape*. As Henri Lefebvre has taught us, spaces are produced and are themselves productive of different social and material relationships. Spaces are therefore not to be thought of as "givens," as mere "containers" or "territories" inside and across which human activities simply take place. To the contrary, they are to be thought of as social productions, as constituted by social life in such a way that "material space" and "mental space" are elided. As Lefebvre styles it, his design is to elaborate an account of social life "which would analyse not things in space but space itself, with a view to uncovering the social relationships embedded in it." Because it signifies "dos and don'ts," space "is at once result and cause, product and producer."[7] And as others have framed it in relation to the broader "spatial turn" in the humanities and social sciences, thinking about the placed and spaced nature of science is vital to comprehension of the different meanings science has had for its actors, agents, artifacts, and audiences: "Geography matters, not for the simplistic and overly used reason that everything happens in space, but because *where* things happen is crucial to knowing *how* and *why* they happen."[8] Just how this is so for scientific spaces is critical to understanding how the geographies of science work at different times and in different arenas.

The possibilities for thinking geographically about science may, however, simply be too rich, too fertile, too intricate, to be reduced to the tidy mechanics of any easy classificatory system that will locate work under one or another of these general headings. But we insist nevertheless that thinking geographically matters. We do so because geography, like time and embodiment, *is* an essential thing: just as there is a rich history of science, so there is a rich geography of science. Explaining the places and spaces of and for science is as coherent and rich a project as locating it and explaining its temporal dimensions. Coming to terms with science's some*where* is as vital as surveying and explaining its some*time* and its some*bodies.*

Geography's disciplinary vocabulary and methods are instructive and informative in our concern with science's spatial dimensions. In addition to the language of *place* and *space*—albeit that they, with others, are not exclusively geographical terms—we may draw upon notions of *scale, hierarchy, distribution, location,* and, certainly, upon *maps* and *mapping* from the geographical lexicon. The mapping of scientific activity is undoubtedly a useful exercise in identifying where scientific pursuits take and took place and might profitably be surveyed at a range of different scales. Simply plotting where in Euclidian space various science-related enterprises are pursued can reveal significant patterns at global, regional, urban, and more local scales. But such cartographic ventures are, in a fundamental way, merely the prelude to the project of elucidating a comprehensive suite of scientific geographies. For any distributional map of scientific practices invites scrutiny of how those "truth spots"—as Thomas Gieryn characterizes them[9]—came into being, how they shape the knowledge that is produced in them, how they are occupied by human agents, and the relationship that they sustain between physical space and social space. If, then, as Gieryn further puts it, we must recognize that the laboratory is a significant "witness box" for science, we must also recognize that what have been called the "hallmarks of experimenting," namely, matters of reliability, claims to rationality and ideas of tolerance and accuracy, took the shape they did because of the ways in which built places materialized identities for the people, organizations, and practices they housed.[10] Nor should we neglect the fact that categories of scientific place, such as "the laboratory," or "the field" do not reduce to simple definition.[11]

Let us be clear, too, about what the geography of science is not, in our view at least. The idea of a geography of science has, from time to time, been taken to circumscribe the influence that the physical environment is purported to have exerted on scientific undertakings. With proximate roots in the environmentalist philosophy of Montesquieu and the more recent writings of figures like Ellsworth Huntington and Ellen Semple, this perspective has been championed in Harold Dorn's 1991 *Geography of Science.* This account centered on a Wittfogel-style narrative that attributed the development of science to the effects of those societies requiring techno-scientific initiative to deliver effective hydraulic management. In explaining the evolution of science, Dorn looked to "soil, climate, hydrology, and topographical relief, and to demographic fluctuations, latitude, and the differences between sown fields, steppe, and desert."[12] This form of ecological constructivism is problematic, trading as it does on large generalizations derived from *a priori* assumptions about the determining impact of the physical environment. But this should not be taken to mean that physical landscapes have *no* influence on the shape of scientific knowledge. As A. M. Keeling recently observed, the "emphasis on the cultural and social geographies of knowledge production, privileges the representations and practices of scientists, devoting rather less attention to the natural spaces and phenomena which they study and engage."[13] This raises critical questions about the role of agency. And it brings into view both Bruno Latour's project to disperse agency across the human *and* nonhuman worlds in networks of actants and Tim Ingold's efforts to recast the human agent as already inescapably embedded in the world and not abstractable from it.[14] Whatever the ontological status of these proposals, they open up new possibilities for thinking about the role of the physical environment in conditioning scientific knowledge and practice that do not collapse into naive ecological reductionism.[15]

Thinking geographically about science, however, does not reduce either to cartographic survey or to environmental causation. As the following chapters reveal in different ways, geography matters in manifold ways to making sense of the shape of science throughout the nineteenth century. For several reasons, as the chapters so ably show, the geographies of nineteenth-century science present a particularly fertile intellectual terrain for inquiries of this sort: science was being given disciplinary shape in this period as perhaps not before; science became in this period a public good with a variety of audiences and staging places; science's disciplinary emergence was evident in certain discursive procedures and methods that helped define "the field" in question; particular forms of dissemination, be they lectures, specialist

journals, or instrumental procedures, helped give science a public and professional credibility not readily enjoyed in earlier periods.[16] Presently, we
elucidate the structure we have devised to impose one form of coherence on
our authors' endeavors. Here, we prepare the ground for these more detailed
studies by addressing a sequence of geographical motifs that clarify, deepen,
and extend our understanding of the range, character, and performance of
science in the nineteenth century.

SCIENCE SPATIALIZED

What emerges from the essays presented in this volume is the remarkable range of sites in which scientific concerns have been engaged. All of
these are mappable locations, and each site is embedded in wider systems
of meaning, authority, and identity. Lightman's inspection of the places
of London science, to take just one example, reveals how different scientific sites operated different moral and epistemic economies. Some, like the
privileged venues patronized by figures like Joseph Banks, were aristocratic
in nature. Others, like those occupied by the British Association for the
Advancement of Science, were spaces where a gentlemanly reform of science manifested itself. Still others, such as those within London University,
presented themselves as sites of resistance to the utilitarian efforts of the
gentlemen of science and, in some places, of pressure for much more radical change in the practice of science. These sites and economies, and many
others, were thus not simply physical locations; they were, as Lightman
shows, symbolic urban places whose occupants were aligned for or against
aristocratic privilege, radical reform, or scientific naturalism. Taken at the
scale of a city as a whole, sites of scientific experimentation and of scientific
patronage appear within scientific institutions, including museums, and in
gentlemen's homes and offices.

Consider in these terms one recent survey of the connections between
science and the city that addresses what the editors term "an urban history
of science."[17] Rather than seeing the city as itself a "setting" for scientific
sites in any simple sense—as, say, a map of locations, a gazetteer of science's
urban places—Dierig, Lachmund, and Mendelsohn proposed attention to
four themes: the rise of urban expertise; science and the representation
of the city; places of knowledge and their urban context; and knowledge
from the street. Thinking about cities as sites of and for science might seem
to sanction imprecision; that is, with the change of scale away from the
extra-local making of science within given sites to bigger pictures, we lose
our grasp of science's conduct locally. This is not so. Rather, the change

from institutional site and practice to the city as both site and scale affords different ways of comparative thinking about the cartographies of science and, since scientific knowledge has to travel between and within city
sites and out from them to different audiences, it affords different perspectives upon the ways in which science moves within a city's interpretative
communities.

Just how particular kinds of symbolic transactions were implicated in
urban scientific sites comes through clearly in the essay by Samuel Alberti
on museums (chapter 3). Alberti's focus is on the way in which museum
space was used to convey credibility, authority, and expertise. Exploiting
the critical relationship between personal authority and material authority,
he shows the role played by physical objects—particularly the "type specimen"[18]—and their location within museum space in the negotiation of
cognitive authority. Museum space was never simply a container: as the
repository of a multitude of items, it transacted credibility.[19] Alberti's
analysis shows how museum space was routinely implicated in questions
rotating around identity, authority, and what is sometimes called object
epistemology. In museums, then, we see the possibilities for connecting in
particular ways what others have termed material autobiography and the
biographies of place.[20]

The city, then, is an important but variable arena for addressing the geographies of nineteenth-century science. If we were to adopt, for instance,
the theme of science and the representation of the city from within the
taxonomy above, and to do so with reference to the medical sciences,
then metropolitan science was both a form of knowledge and a matter of
technical sophistication. It required, as it did for contemporaries managing
city space, attention at city level: practices of urban sanitation that drew (or,
often, did not) upon revised etiologies of disease transmission, for example,
in which context science was applied to building design and the provision
of fresh water and seen as socially ameliorative, or the regulation of access
to botanical gardens and museums as educational and civic spaces. Cities,
nineteenth-century or otherwise, thus become a worthwhile scale of inquiry
within the geographies of science for, among other things, what is revealed
there of the social work that science does, how, and for whom.

But science was never just confined to metropolitan and civic sites and
geared solely to didactic display. Country houses were also sites of experimental testing.[21] And, as Donald Opitz shows (chapter 4), their influence
continued to manifest itself in the early history of genetics in Britain during the decades around 1900. Attending to why this distinctive geography

of experiment came to be cultivated and to the reasons behind William Bateson's enthusiasm for such a particular spatial formation as ideal for genetics research reinforces the value of thinking geographically about tests, trials, and experimentation and the particular nature of the performative sites wherein such things were conducted and evaluated.

Science is also made or, perhaps better, is made to disperse through institutional mobility, as the peripatetic activities of the British Association for the Advancement of Science illustrate. The circuit of industrial cities undertaken by the BAAS "Gentlemen of Science" and accompanying delegates was designed to transmit the benefits of scientific culture beyond the metropolitan core, even beyond the grand houses of the elite. The BAAS promoted what Withers calls "civic science"—science as a public good, a unifying, moral vision under the banner of scientific and political neutrality. But the particulars of this mission were moderated by the different urban and institutional contexts where the association convened. Here the oscillating currents connecting different geographical scales come into view and reveal something of how the local and the global are always mutually constituted.

Operating at a yet different interplay of scale—between the institutional and the continental—Sujit Sivasundaram likewise calls our attention to symbolic geography with Ceylon as his site of scrutiny (chapter 6).[22] As others have pointed out, islands provide a compelling focus of interdisciplinary study: as laboratory sites in biogeographical and natural history studies, as utopian and paradisial settings, or as sites of moral decline, for example, in literary studies.[23] Sivasundaram's concern is with the ways in which the island's twin topography and their long-standing iconic associations were constitutive of relations between British natural history and the natural knowledge that Kandyans cultivated around Buddhist temples. The low country and the high country, symbolized respectively by the coconut tree and the hill station, are used by Sivasundaram to challenge the presumption that the spaces of the colonial world were devoid of meaning and mere receptacles of an all-powerful, inward-moving European rationality. Indeed, it was quite otherwise, for in site after site, his excavation of layers of superimposed meaning discloses how Kandyan and British natural knowledge systems intermingled in complex and mutually transformative ways.

Places, of course, are very far from being just matters of physical location and symbolic meaning. They are also constitutive of social exchange, enabling or constraining activities that carried out within their confines. Finnegan's probing of what can be called the "speech spaces" of the lecture hall—"platform science" as he styles it—opens up new and fertile lines of inquiry into the reciprocal relationship between location and locution.[24] What lecturers felt free to say was circumscribed by the local customs and speech protocols of particular venues. Focusing on Edinburgh's Queen Street Hall, he demonstrates how its culture of oratory shaped the way in which scientific knowledge and its relationship to other spheres of life, religion in particular, was communicated. Speech spaces, of course, are not static entities, and the ways in which Queen Street Hall rhetoric evolved, not least to redraw the boundaries of what passed as scientific speech, are here exposed to close scrutiny. How these boundaries were policed, contested, and controlled all point to the reciprocal constitution of speech and space in the shifting culture of scientific communication. Here is a pointer to the need for further exploration into the aural geographies of scientific transmission at different times and places in order to grasp the dynamics of science's shaping in place and its circulation over space.

Speech is critically significant in explaining how scientific claims travel from person to person and from place to place, so it is not surprising that the role of conversation in the spread of scientific knowledge—both elite and popular—has been coming under increased scrutiny.[25] But science circulates from place to place over space in other forms too. Reputations also move between places and are differently constructed in different arenas. Isaac la Peyrère, the father of anthropological polygenism, was made to stand for different things at different times and places. In some, he was presented as a radical iconoclast undermining traditional authority. Elsewhere and at other times, he was seen as a visionary student of the human race who had been vilified at the hands of an autocratic church, or presented as a valiant defender of orthodoxy seeking for persuasive ways of reconciling scientific findings with traditional beliefs. La Peyrère's variegated reputational geography alerts us to the fact that theories mean different things, and are made to mean different things, in different locations. Darwinism, as David Livingstone points out, was likewise construed in significantly different ways in different cities.[26] Even within one regional setting, proponents of the same religious tradition could react very differently to his proposals, depending on local cultural politics.

The strong emotions that figures like La Peyrère and Darwin stirred up among people in different contexts alerts us to the role of the affective in scientific history. From time to time the contributors to this volume advertise elements of what might be called science's emotional geography. Diarmid Finnegan is suggestive of this in chapter 7, and Sally Kohlstedt's demonstration of how the American West was portrayed as spectacle (chapter 16) reminds us of the place of awe and wonder in scientific history.[27] Such sentiments—wonder, disbelief, incomprehension—have spatial components. Fear, for example, played a vital role in the story of the Victorian science of electricity. Whether electricity made a place safe or dangerous was crucial to how the public responded to electrical innovation. Champions of electrical technology thus felt the need to exhibit its safety in key public sites. As Graeme Gooday argues in chapter 9, the distribution of fearfulness and fearlessness of electricity in the late Victorian period meant that this new form of power had to prove itself by spatial testing. Nor was this a matter of theatrical demonstration alone. The practice of electrical safety had to be seen to be believed at other social, personal, and instrumental scales too as electricity's secure manipulation later underlay developments in physics, in engineering, and in the powering of cities and continents.[28]

The intimate connections between space and scientific testing, of course, have been recognized by scholars who have examined the culture of houses of experiment and laboratory life—the idea of a place's "factory stamp."[29] Gooday's focus on sites like the Crystal Palace and London's theatres significantly widens the scope of what passes as experimental space. So, too, does Smith's account (chapter 10) of how steamships and their builders were subjected to what, following Lefebvre, he calls "trial by space." His detailed description of the various time trials that Victorian steamships underwent, the technological innovations that were involved, and the debates surrounding how the trials were witnessed, warranted, and authorized, draws repeated attention to the unavoidable need for shipbuilders to "make their mark on space." The sciences that underlay the development of new marine technology depended upon the iterative performance of precision and different ideas of tolerance: the moral tolerance that underlay relationships between the operators and financiers, refined matters of instrumental tolerance as engines at sea were judged by their spatial efficiency, and questions of epistemic tolerance as marine engineers tried to make sense of the results of trials at sea.[30]

Textual circulation and institutional directives are not the only mechanisms implicated in science's mobility. The idea of scientific authoritativeness and the methods used to secure it may migrate from their point of origin—indeed, may not have a single point of origin—to numerous scientific destinations. This is a central element in the story that Lawrence Dritsas tells about expeditionary science (chapter 11). As he points out, the tensions between those who had direct experience of foreign lands and the claims of those engaged in what Murchison and his contemporaries termed "critical geography" by scrutinizing data at a distance revolved around how and where the authority of contemporary geographers had been attained. Woven into the geography of African exploration through expeditions were strains between indigenous testimony, metropolitan theory, expert analysis, and raw data. This was a distributed geography of credibility, production, and reception: produced, partly, in the field as specimens and reported tales and, also in part, upon returning home to be scrutinized. Expeditionary science could not be regarded as complete or agreed-upon until its claims had been subject to debate, its findings published and reviewed in different scientific journals, and its leaders feted, or not, by the public.[31]

Bringing space and place more centrally into our exploration of nineteenth-century science illuminates the ways in which talk and text, intellectual authority and credibility, and experimental test and trial operate in the generation and circulation of scientific knowledge and practice. It is the same with that other critical component of scientific culture—visualization. The visual has long been scrutinized by students of geography and of the history of science. Martin Rudwick drew attention to the way in which it had been ignored for too long in the latter context.[32] It has since then been the subject of numerous studies, not least by historians and geographers interested in the links between science, art and photography, and the authority of visual images.[33]

In this collection, visual matters are central to our inquiries in a variety of ways. The staging of nature as spectacle, the theatricality of electrical performance, and the romanticizing of highland wilderness in Ceylon are just a few of the ways in which the ocular assumes importance in this volume. And the geographies of the visual extend into other realms.[34] Anne Secord here directs our attention to the several ways in which the visual habits of students of botany were disciplined (chapter 12). What we might call the microspace of botanical publications served as guides to observation. The use of exemplary specimens helped develop "a practised eye." In the eighteenth century, scientific depiction presented "type" illustrations rather more than actual specimens: accuracy lay not in individuals but in archetypes. But in the face of new subject knowledge, disciplinary specialization, and shifts in the technologies of visualization, the search in the nineteenth century for what has been called "mechanical objectivity" demanded that variance in life forms was key, that actual specimens, even if irregular, be studied and displayed.[35] Her analysis thus brings geographical sensitivity to bear on observational practices in this period in numerous vital ways: it brings into focus the critical role of textual space in the honing of visual skills; it clarifies some key mechanisms in the circulation of scientific expertise; it demonstrates the intimate connections between local and regional collecting space and botanical guidebooks; and it renegotiates the boundaries between the so-called amateur and professional.

In whatever period, either traveling locally to observe and order plants or being on a continental expedition or voyage of global navigation and returning with new knowledge, specimens and reputation could count for little unless one's results made a yet further voyage—into print.[36] Behind and within volumes of print and reproduced images lie networks of exchange and correspondence. As Jonathan Topham shows here in his scrutiny of the flow of scientific knowledge (chapter 13), print has been of critical importance.[37] Several interlinking geographies are operating here. The circulatory networks through which print moved is one; another, the different ways in which key texts are appropriated in different settings. All of this renders problematic any assumption that scientific knowledge, either in its words or in its pictures, simply diffuses across the globe in any straightforward manner. Disruption of supply, translation between languages, selective reviewing of scientific literature, and local interpretations of meaning all point to the salience of textual geography and the study of the forms of its representation in the movement of scientific knowledge.

This is true too of the many guides to how to do the science in question and the quandaries such works raised about design and content in relation to different audiences.[38] The reciprocity between visual practices, local spaces, and textual guides was also significant in the ways in which popular natural history—among other things—was conveyed to scientifically untrained tourists in the nineteenth century. Here too guidebooks and handbooks directed and regulated the gaze of holiday-makers and day-trippers by recommending the scientific values of certain types of places, pointing them to the most popular geological sites, and instructing them in how best to view landscape vistas. Fyfe's account of this relatively neglected strand of nineteenth-century tourist culture shows just how tightly interwoven locality, textuality, and visualization were in popular encounters with science in this period. And if artistic illustration, photographic reproduction, and viewer discipline were all central to the development of nineteenth-century scientific culture, so, too, was cartographic representation.

Maps are spatial instruments. But as Simon Naylor reminds us (chapter 14), they are also a visual language, an important tool in the "visual technology" of the natural sciences.[39] Using the mapping of Cornwall's geology as a case in point, he shows how the cartographic enterprise can be read as a "form of territorial acquisition" on the part of mapping geologists and as the vehicle for the advancement of intellectual propriety-cum-property and of wider social agendas. The application of commonly used cartographic technologies to produce local maps does not obliterate local interests; rather, such technologies can become the vehicle through which power, politics, and personal authority may be channeled.

Kohlstedt's chapter in a sense brings us full circle—since, like those of Lightman and Alberti, hers is an account of the power of institutions to produce science and frame geographical understanding. Her chapter works dialogically between two spaces and scales—the museum space of the Smithsonian Institution and the expansive space of the American West. The symbolic role of the West in the American psyche emerged in significant part from its staging within the confines of the Smithsonian Institution, which defined its nature and heralded its significance. The museum thus operated not just as an exchange house for the trafficking of objects or as a site of accumulation for numerous artifacts; it also had a critical productive role in the construction of the space that was—and was thought to be—the American West itself. This account tellingly discloses how both spaces were inherently symbolic, how both remained unfixed and embodied meanings that were inherently unstable.

GEOGRAPHIES OF NINETEENTH-CENTURY SCIENCE

The detail afforded from the many studies of the ways of knowing and doing science in place and of the complex social and technical mechanisms influencing science's movement across space means that any typological classification is open to interpretation, always a flexible and contingent exercise rather than a once-and-for-all categorical enterprise.

In the chapters that follow, a remarkable range of geographical perspectives, analyses, and scales is mobilized by the authors to shed light on specific topics in place and over space concerning the geographies of nineteenth-century science. But neither we nor they presume completeness in our endeavors. The essays and the ideas worked with are intended to be more suggestive than comprehensive, illustrative rather than exhaustive. They are put forward in the hope that others may venture further into this conceptual territory and continue the task of exploring the spaces and places—the social, mental, material, performed, distributed, read, and debated geographies—of scientific enterprises.

In ordering the essays under three headings—"Sites and Scales," "Practices and Performances," "Guides and Audiences"—we have been concerned, naturally, to reflect the principal themes of the respective studies. But in so ordering them, we have also sought to build upon established taxonomies and to propose new and different ways of addressing science's geographical dimensions. To be sure, this threefold classification—in part at least—shares its concern in terms of practice with other discussions of the nature and definition of Victorian science.[40] It is also attentive to related work that has shown how science was then as perhaps never before in a competitive marketplace and how it had to perform in several ways to be noticed.[41] The popularity of science we disclose in different contexts here supplements recent collaborative efforts to think what, exactly, "popular" science was: less a single category in the nineteenth century, perhaps, than a level of science given a variety of meanings by the spaces in which it was undertaken and reacted to.[42] It may also be that the concern of the authors here with matters of practice in one way or another echoes earlier and wider disquiet over the dangers of the "localist turn"—the study of science in local context to the neglect of its wider significance—and, with others, a recognition of the importance of science's uneven transmission through various forms of embodied communicative action.[43] Nevertheless, science's practices and forms of communicative action are always grounded in particular settings, and questions regarding site, institutional organization, and social relationship in place will for that reason always continue to matter to an explanation of science's cognitive content and variable reception.

What we additionally suggest in pairing the geographical concepts of site and scale is that science's meanings and social purchase need to be considered as relational—not just as local but as national and even transnational—and that its meanings may change depending upon the scale at which we examine science.[44] As Kapil Raj has shown, it is through science's making in place and its circulation over different scales—the intra-institutional, the oceanic, the continental, even the global—that congruent ideas made far apart came together to constitute particular scientific practices such as botanical classification or topographical mapping.[45] Similarly, attention to scale and to notions of place as venue cannot easily be separated from matters of practice and of performance through which science was made there and received there or elsewhere: through speech, in experimental demonstration, by quiet contemplation of displayed phenomena, or in print and by reading. The idea of interpretative communities sharing certain procedures for making sense of what was read or heard is well known in this respect.[46] Likewise, the study of how and why science was put to work cannot easily neglect the purposes science served—to educate, to surprise, to act as a guide to intellectual conduct in a region, to test the claims of others—nor overlook the differences within these intentions or within so general a term as *audience*.

Our typology is fitting and novel, but we recognize that the chapters and the debates to which they refer embrace other organizational possibilities and that they rest upon the foundations provided by work within a field of inquiry now more than ever discernible as the geography of science. We are not for a moment suggesting that this threefold division—"Sites and Spaces," "Practices and Performance," "Guides and Audiences"—is the only way that this collection might have been ordered, any more than either we or our authors presume the essays to be definitive discussions of the topics under review. Elsewhere, others have faced similar issues in ordering the study of the geographies of discovery and exploration in the twentieth century.[47] Most of the essays embrace one or more of our ordering themes and so could easily have found an alternative place in our structure. Further commentaries follow in each of three sectional introductions. It will be clear upon reading these sectional introductions and the chapters to which they refer that other themes also connect the essays and cut against the grain of the ordering principles adopted: concerns with the cultures of scientific observation and with their mediation and transmission, for example, or with the importance of displaying, and with the practices of collecting and of printing. Science's texts matter not alone because of their astonishing variety but because of the ways in which science's reception and its practitioners' reputations did not straightforwardly follow from a work's publication or the giving of a well-turned lecture. Similarly, the connections between scientific credibility, socioscientific status, and the objects of science have been revealed in a variety of contexts: these, as well as the topics of circulation through space and the didactic purposes behind science, might each have merited sectional attention.

Geographies of Nineteenth-Century Science is thus a contribution to continuing interdisciplinary debates about the placed nature of science's making and reception, about the processes that were adopted to make scientific knowledge mobile for whom and with what consequence. Examining in these terms the cultural and political commodity that was science, or that was to become science, in the nineteenth century reveals that what has held to be science varied—by content over time, within institutions, at different scales, and for different audiences in different places. Science varied in its constitution and reputation in relation to place and personnel, in its nature by virtue of its different epistemic practices, in its audiences, and in the ways in which it was put to work. Because it did so in place and across space demands that its geographical dimensions should continue to be addressed.

NOTES

1. David N. Livingstone, "The Spaces of Knowledge: Contributions Towards a Historical Geography of Science," *Environment and Planning D: Society and Space* 13 (1995): 5–34; David N. Livingstone, *Putting Science in Its Place: Geographies of Scientific Knowledge* (Chicago: University of Chicago Press, 2003).

2. This is the framework adopted by Crosbie Smith and Jon Agar, eds., *Making Space for Science: Territorial Themes in the Shaping of Knowledge* (Basingstoke: Macmillan, 1998).

3. The three-part categorization is taken from Steven Harris, "Long Distance Corporations, Big Sciences, and the Geography of Knowledge," *Configurations* 6 (1998): 269–305. The two-part taxonomies are taken from Charles W. J. Withers, "The Geography of Scientific Knowledge," in *Göttingen and the Development of the Natural Sciences*, ed. Nicolaas A. Rupke (Göttingen: Wallstein, 2002): 9–18; and from Diarmid A. Finnegan, "The Spatial Turn: Geographical Approaches in the History of Science," *Journal of the History of Biology* 41 (2008): 369–88. For a further summary of work in the field, principally organized in relation to localisms and matters of practice, see Richard Powell, "Geographies of Science: Histories, Localities, Practices, Futures," *Progress in Human Geography* 31 (2007): 309–30.

4. Crosbie Smith and Jon Agar, "Introduction: Making Space for Science," in Smith and Agar, *Making Space for Science*, 2. They note that "such presuppositions have effectively become *sine qua non* to the history and sociology of knowledge, whether scientific, technological or medical"—to which we would add "the geography of [knowledge]."

5. For one summary, see Charles W. J. Withers, "Place and the 'Spatial Turn' in Geography and in History," *Journal for the History of Ideas* 70 (2009): 637–58.

6. Trevor J. Barnes, "Placing Ideas: Genus Loci, Heterotopia, and Geography's Quantitative Revolution," *Progress in Human Geography* 28 (2004): 568.

7. Henri Lefebvre, *The Production of Space*, trans. Donald Nicholson-Smith (Oxford: Blackwell, 1991), 89, 142.

8. Barney Warf and Santa Arias, "Introduction: The Reinsertion of Space in the Humanities and Social Sciences," in *The Spatial Turn: Interdisciplinary Perspectives*, ed. Barney Warf and Santa Arias (London: Routledge, 2009), 1.

9. Thomas F. Gieryn, "Three Truth Spots," *Journal of the History of the Behavioral Sciences* 38 (2002): 113–32; and Thomas F. Gieryn, "City as Truth-Spot: Laboratories and Field-Sites in Urban Studies," *Social Studies of Science* 36 (2006): 5–38.

10. See Thomas Gieryn, *Cultural Boundaries of Science: Credibility on the Line* (Chicago: University of Chicago Press, 1999), from which we take the notions of "witness box" and "truth spot." On the connections between buildings and practices and identities, see Thomas Gieryn, "Two Faces on Science: Building Identities for Molecular Biology and Biotechnology," in *Architecture of Science*, ed. Peter Galison and E. Thompson (Cambridge, MA: MIT Press, 1999): 423–55. On the hallmarks of experiment, see David Gooding, *Experiment and the Making of Meaning* (Dordrecht: Kluwer

Academic, 1999); and David Gooding, R. Pinch, and Simon Schaffer eds., *The Uses of Experiment* (Cambridge: Cambridge University Press, 1989).

11. For a discussion of laboratories within the geographies of science, see Powell, "Geographies of Science," 314–18, notably his attention to "laboratory ethnographies and ethnomethodologies"; and Robert E. Kohler, *Landscapes and Labscapes: Exploring the Lab-Field Border in Biology* (Chicago: University of Chicago Press, 2002).

12. Harold Dorn, *The Geography of Science* (Baltimore, MD: Johns Hopkins University Press, 1991), xi.

13. A. M. Keeling, "Charting Marine Pollution Science: Oceanography on Canada's Pacific Coast, 1938–1970," *Journal of Historical Geography* 33 (2007): 405.

14. See Bruno Latour, *Reassembling the Social: An Introduction to Actor-Network Theory* (Oxford: Oxford University Press, 2005); Tim Ingold, *The Perception of the Environment: Essays in Livelihood, Dwelling and Skill* (New York: Routledge, 2000).

15. In this context, we think, for example, of the importance accorded ecological reductionism in the work of Jared Diamond: see Jared Diamond, *Guns, Germs and Steel: The Fates of Human Societies* (London: Jonathan Cape, 1997); Jared Diamond, *Collapse: How Societies Choose to Fail or Survive* (London: Allen Lane, 2005).

16. On these matters, see for example, Martin Daunton, ed., *The Organisation of Knowledge in Victorian Britain* (Oxford: Oxford University Press, 2005); James Elwick, *Styles of Reasoning in the British Life Sciences: Shared Assumptions, 1820–1858* (London: Pickering and Chatto, 2007); John Pickstone, *Ways of Knowing: A New History of Science, Technology and Medicine* (Manchester: Manchester University Press, 2000).

17. S. Dierig, J. Lachmund, and A. J. Mendelsohn, "Introduction: Toward an Urban History of Science," in "Science and the City," special issue of *Osiris* 18 (2003): 1–19.

18. On type specimens, see Lorraine Daston, "Type Specimens and Scientific Memory," *Critical Inquiry* 31 (2004): 153–82.

19. See also Steven Conn, *Museums and American Intellectual Life, 1876–1926* (Chicago: University of Chicago Press, 1998); Lorraine Daston, "The Factual Sensibility," *Isis* 79 (1988): 452–70.

20. On the idea of "material autobiography," see Susan Pearce, *Collecting in Contemporary Practice* (London: Sage, 1998); and Samuel J. J. M. Alberti, *Nature and Culture: Objects, Disciplines and the Manchester Museum* (Manchester: Manchester University Press, 2009). On biographies of place as a "matter of practice" in sites, see Simon Naylor, "Introduction: Historical Geographies of Science—Places, Contexts, Cartographies," *British Journal for the History of Science* 38 (2005): 1–12.

21. Simon Schaffer, "Physics Laboratories and the Victorian Country House," in Smith and Agar, *Making Space for Science*, 149–80.

22. On islands as laboratories, see Beth Greenhough, "Imagining an Island Laboratory: Representing the Field in Geography and Science Studies," *Transactions of the Institute of British Geographers* 31 (2006): 224–37.

23. For one review of islands in these (and other) terms, see Klaus Dodds and Stephen A. Royle, "The Historical Geography of Islands. Introduction: Rethinking Islands," *Journal of Historical Geography* 29 (2003): 487–98, and the five papers that follow.

24. On, respectively, a national and a regional scale of reference, these issues concerning speech, scientific authority, and public witnessing have been discussed in

Diarmid A. Finnegan, *Natural History Societies and Civic Culture in Victorian Scotland* (London: Chatto and Windus, 2009); and Simon Naylor, *Regionalizing Science: Placing Knowledges in Victorian England* (London: Pickering and Chatto, 2010)—the latter a study of nineteenth-century Cornwall.

25. Alice Walters, "Conversation Pieces: Science and Polite Society in Eighteenth-Century England," *History of Science* 35 (1997): 121–54; Mary Hilton and Jill Shefrin, eds. *Educating the Child in Enlightenment Britain: Beliefs, Cultures, Practices* (Farnham: Ashgate, 2009); James A. Secord, "How Scientific Conversation Became Shop Talk," in *Science in the Marketplace: Nineteenth-Century Sites and Experiences*, ed. Aileen Fyfe and Bernard Lightman (Chicago: University of Chicago Press, 2007), 23–59.

26. David N. Livingstone, "Science, Text and Space: Thoughts on the Geography of Reading," *Transactions of the Institute of British Geographers* 35 (2005): 391–401.

27. On the nature and genealogy of "wonder" in science, see Lorraine Daston and Katharine Park, *Wonders and the Order of Nature, 1150–1750* (New York: Zone Books, 1998); Mary Baine Campbell, *Wonder and Science: Imagining Worlds in Early Modern Europe* (Ithaca, NY: Cornell University Press, 1999).

28. For one discussion of the power of electricity in shaping continental development, see Erik van der Vleuten and Arne Kaijser, eds. *Networking Europe: Transnational Infrastructures and the Shaping of Europe, 1850–2000* (Sagamore Beach, MA: Science History Publications, 2006).

29. See, for example, Steven Shapin, "The House of Experiment in Seventeenth-Century England," *Isis* 79 (1988): 373–404; Graeme Gooday, "The Premisses of Premises: Spatial Issues in the Historical Construction of Laboratory Credibility," in Smith and Agar, *Making Space for Science*, 216–45. We take this use of "factory stamp" in highlighting the interconnections between site, practice, and meaning from Peter Galison, *Einstein's Clocks, Poincaré's Maps* (London: Hodder and Stoughton, 2003).

30. We are in part prompted in this context by H. Otto Sibum, "Exploring the Margins of Precision," in *Instruments, Travel and Science: Itineraries of Precision from the Seventeenth to the Twentieth Century*, ed. Marie-Noëlle Bourguet, Christian Licoppe, and H. Otto Sibum (London: Routledge, 2003), 216–42.

31. For a discussion of nineteenth-century British scientific exploration in these terms, see Dane Kennedy, "British Exploration in the Nineteenth Century: A Historiographical Survey," *History Compass* 5/6 (2007): 1879–1900.

32. Martin Rudwick, "The Emergence of a Visual Language for Geological Science, 1760–1840," *History of Science* 14 (1976): 149–95; Martin Rudwick, *Scenes from Deep Time: Early Pictorial Representations of the Prehistoric World* (Chicago: University of Chicago Press, 2002).

33. See, for example, Martin Kemp, *The Science of Art: Optical Themes in Western Art from Brunelleschi to Seurat* (New Haven, CT: Yale University Press, 1990); Martin Kemp, *Visualizations: The Nature Book of Art and Science* (Oxford: Oxford University Press, 2000); Barbara Maria Stafford, *Artful Science: Enlightenment, Entertainment, and the Eclipse of Visual Education* (Cambridge, MA: MIT Press, 1994); Jennifer Tucker, "Photography as Witness, Detective, and Impostor: Visual Representation in Victorian Science," in *Victorian Science in Context*, ed. Bernard Lightman (Chicago: University of Chicago Press, 1997), 378–408; Jennifer Tucker, *Nature Exposed: Photography as*

Eyewitness in Victorian Science (Baltimore, MD: Johns Hopkins University Press, 2005); Denis Cosgrove, *Geography and Vision: Seeing, Imagining and Representing the World* (London: I. B. Tauris, 2008).

34. See Charles W. J. Withers, "Mapping the Niger, 1798–1832: Trust, Testimony and "Ocular Demonstration" in the Late Enlightenment," *Imago Mundi* 56 (2004): 170–93.

35. Lorraine Daston and Peter Galison, *Objectivity* (New York: Zed Books, 2007). On photography in capturing nature, see Jennifer Tucker, *Nature Exposed.*

36. On this point, see Marie-Noëlle Bourguet, "The Explorer," in *Enlightenment Portraits,* ed. Michel Vovelle (Chicago: University of Chicago Press, 1999), 257–315.

37. The literature on print culture is vast. Key works include the following: Roger Chartier, *The Cultural Uses of Print in Early Modern Europe* (Princeton, NJ: Princeton University Press, 1981); Adrian Johns, *The Nature of the Book: Print and Knowledge in the Making* (Chicago: University of Chicago Press, 1998); Peter Burke, ed., *New Perspectives on Historical Writing* (Cambridge: Cambridge University Press, 1991); Maria Frasca-Spada and Nicholas Jardine, eds., *Books and the Sciences in History* (Cambridge: Cambridge University Press, 2000); Jonathan R. Topham, "Scientific Publishing and the Reading of Science in Nineteenth-Century Britain: An Historiographical Survey and Guide to Sources," *Studies in History and Philosophy of Science* 31A (2000): 559–612.

38. For examples, see the essays in Frasca-Spada and Jardine, *Books and the Sciences in History;* Miles Ogborn, *Indian Ink: Script and Print in the Making of the English East India Company* (Chicago: University of Chicago Press, 2007); and Ann B. Shteir and Bernard Lightman, eds., *Figuring It Out: Science, Gender and Visual Culture* (Hanover, NH: University Press of New England, 2006).

39. Jane Camerini, "Evolution, Biogeography, and Maps: An Early History of Wallace's Line," *Isis* 84 (1993): 705.

40. The twenty essays making up Lightman, *Victorian Science in Context* were organized under three headings: "Defining Knowledge," "Ordering Nature," and "Practicing Science."

41. Fyfe and Lightman, *Science in the Marketplace;* Bernard Lightman, *Victorian Popularizers of Science: Designing Nature for New Audiences* (Chicago: University of Chicago Press, 2007). On the emergent organization of different sciences in the nineteenth century, see the essays in Daunton, *The Organisation of Knowledge in Victorian Britain.*

42. For a discussion on "popular science," see Jonathan Topham, "Introduction," *Isis* 100 (2009): 310–18, and the set of papers making up the "Focus" section there.

43. On these points, see Harris, "Long Distance Corporations," and James A. Secord, "Knowledge in Transit," *Isis* 95 (2004): 654–72.

44. This is to subscribe to the local and transnational agenda posited by Lewis Pyenson, for example, but to allow that national frames of reference may continue to be relevant in explanations of science's geographies, and, like Raj and others, that we continue to consider the mediated *and situated* connections between forms of knowledge in local, national, and global context: see Lewis Pyenson, "An End to National Science: The Meaning and Extension of Local Knowledge," *History of Science* 40 (2002): 251–90. On the important role of mediation and of cultural intermediaries in the sciences and literatures of Western exploration, see Simon Schaffer, Lissa Roberts, Kapil Raj, and James

Delbourgo, eds., *The Brokered World: Go-Betweens and Global Intelligence*, 1770–1820 (Sagamore Beach, MA: Science History Publications, 2009).

45. Kapil Raj, *Relocating Modern Science: Circulation and the Construction of Knowledge in South Asia and Europe*, 1650–1900 (Basingstoke: Macmillan, 2006).

46. Stanley Fish, *Is There a Text in This Class?: The Authority of Interpretive Communities* (Cambridge, MA: Harvard University Press, 2003).

47. See James R. Ryan and Simon Naylor, "Exploration and the Twentieth Century," in *New Spaces of Exploration: Geographies of Discovery in the Twentieth Century*, ed. James R. Ryan and Simon Naylor (London: I. B. Tauris, 2010), 1–22.

PART ONE

Sites and Scales

The essays in this first section are centrally concerned with different scientific sites—not simply as locations but as social spaces and epistemic venues—and with addressing how scientific ideas change their meaning when framed in regard to different geographical scales.

Bernard Lightman offers a survey of several sites for science in London. As he shows, to plot different spaces of knowing in London's scientific world is to plot different ways of knowing and, in order to explain why different sites and ways had the authority and the longevity they did, to reveal the networks of political influence that underlay, for example, Huxley's creation of his privileged sites in South Kensington or, earlier, Banks's center of calculative authority in Soho Square. Lightman's metropolitan cartography of science might, as he suggests, be expanded beyond London, beyond Britain. Yet it provides, too, a compelling case for the citywide study of science's geographies.

The museum is a fitting site in which to look within the city at science's making because of the particular relationships there between practitioners, institution, and things. Samuel Alberti argues that museums achieved their privileged status as places and sites of knowledge from a combination of practitioners' curatorial credibility, authority, and expertise. On the one hand, the object-based epistemology of the material cultures of botany, geology, zoology depended upon museological practices undertaken there—collecting, displaying, cataloguing, educating, and so on. But they did not straightforwardly do so as part of a move toward permanency of display site or of taxonomy since, at different times, new knowledge and different purposes could reposition the material objects in more ways than one. On the other hand, the museum derived its authority as a place in which such knowledge was situated—sited, sighted, and cited—because of the status of the people doing the work there: social space, museum practice, and physical space interlocked and provided mutually supportive means to secure knowledge claims. While publication lent permanency (of a sort) to specimens, and specimens lent status to collections, what was particularly crucial was the relationship between curatorial authority and the collections.

In the distinctive experimental geography of "country-house science" as Donald Opitz calls genetics research in early twentieth-century Cambridge, experimental space and social status were closely aligned. This was also, initially, a distributed geography. Further refinement of the science required its concentration into a single experimental station, for the centralizing of science was crucial to its conduct, and remained so at least for twenty years before the design of Whittingehame Lodge hindered the very science it was built to contain. The nature of the science of genetics in Cambridge was determined not so much by the aristocratic authority of its practitioners as from the replication at a smaller scale of distinctive and separate experimental spaces for doing such science beyond the university. The fact that later changes in the nature of that science required a different experimental geography, that new laboratories followed new procedural methods—yet different architectures of and for science—serves only to strengthen the connections that Opitz discloses between particular sites and genetic science's "factory stamp."

However much there was a resident culture of scientific inquiry in towns and in some country houses, the towns and cities of nineteenth-century Britain and Ireland were for the British Association for the Advancement of Science also temporary venues for associational civic science, in the form of its week-long annual meeting. Charles Withers shows how a national policy of being provincial was given local inflection through different practices in demonstrating and displaying the scientific credentials of the host venue and the status of the science. What is also disclosed, with reference to geography's status as a science and a section within the BAAS, is that public audiences much preferred listening to accounts of exploratory endeavor rather than the declarations of geography's emergent credentials as a science that characterized the section's presidential addresses. Within BAAS meetings, and within different subjects, as well as across national geographies with local civic expression, what geography and science were held to be varied by site and by audience.

For Sujit Sivasundaram, the scale of inquiry is the island of Ceylon, and includes the ways in which nineteenth-century networks of colonial understanding about that island could not be separated from questions about the physical topography of the island itself. Here, differences within the island mattered to the sorts of natural historical knowledge that was secured: be-ing "up country" or on the coastal belt—the divide between these topog-raphies often signaled by the presence/absence of coconut trees—mattered to what was known. The hill station at Nuwara Eliya in the center of the island was at once an experimental station, a site of European recupera-tion from the rigors of the tropics, and, in helping shape larger narratives, it provided for the British a productive venue for those dominant motifs of colonial development that traveled well beyond the island.

CHAPTER TWO

Refashioning the Spaces of London Science: Elite Epistemes in the Nineteenth Century

BERNARD LIGHTMAN

A writer for *All the Year Round*, the *World*, and other papers, Bernard Becker was also on the staff of the *Daily News*. During the winter of 1878–79 he was sent by the *Daily News* to investigate the distress of the manufacturing populations of Sheffield and Manchester. Two years later he studied and wrote about the conditions in Ireland. His *Disturbed Ireland: Letters Written during the Winter of 1880–81* attracted considerable attention and was discussed in the House of Commons.[1] In an earlier foray into investigative journalism in 1874, Becker ventured into the scientific spaces of London. In the preface to his book *Scientific London*, Becker explained why he had seen this as an important mission—these spaces were relatively uncharted territory. "I was astonished," he recalled, "to find that the written records" of the deeds of the learned societies "were few and far between." Though his mission was not as dangerous as exploring deepest darkest Africa or as perilous as exposing the poverty in the slums of the east end of London, Becker presented himself as a traveler in search of science in order to capitalize on the popular travel narrative genre. At the Royal Society at Burlington House in London, he was impressed by the "thoughtful-looking, grey-haired men" who attended the meeting and by the symbols of royalty in the room. By contrast, when he visited the Chemical Society, also at Burlington House, he was shocked by "the extreme youth of many among the audience," some of whom turned out to be fellows of the society. On Friday evenings in the densely packed "elegant throng" of the lecture hall of the fashionable Royal Institution on Albemarle Street, he encountered "lions of the first magnitude," who roared "by turns."[2]

Like Becker, I too am a traveler in search of science. In this chapter I will be "visiting" various sites of nineteenth-century science, accompanied, hopefully, with a more sophisticated historiographical framework than

Becker's. There is substantial scholarship to draw on now to help guide our exploration of the spatial nature of science. It is no longer necessary to prove that place mattered—this has been amply demonstrated beyond a doubt. We can turn our attention to *how* space mattered. We can examine the dynamic dialectic of the spatial and the social, or, as Livingstone has neatly put it, the "multi-layered mosaic of social spaces which are both cause and consequence of human agency."[3] To provide some signposts for those mapping the myriad sites of nineteenth-century science, I offer a synthesis of the scholarship to date. Though parts of the map may be familiar, as a whole it contains some new features. Of course it is not a complete map—many details are left out, and parts of it may need to be redrafted. I concentrate on one national context, Britain, so that I can capture in some depth the dynamic shifts in the spatial configurations of science during the nineteenth century. The map will be London-centered, since I focus on the most influential spaces. This paper is not just about the power of place. It also concerns places of power and the epistemic systems produced therein.

In his stimulating *Ways of Knowing*, the historian of medicine John Pickstone has laid out a much more ambitious "big picture." Applying a Foucauldian analysis to science, technology, and medicine from the Renaissance to the present, he explores the different "epistemes," or ways of knowing, fundamental to each period of the modern world. Much of modern science, technology, and medicine, he argues, can be understood in terms of three ways of knowing and their interactions: natural history, analysis, and experimentalism.[4] Like Pickstone, I want to identify broad epistemic patterns across disciplines and to see how they change over the course of time. Pickstone's periodization for his epistemes in the nineteenth century corresponds roughly to my own. He usefully explores how ways of knowing are embodied in the workings of institutions. His approach therefore is relevant for mapping the geography of nineteenth-century science. By aiming to place his discussion of science within a global context, he can only treat Britain in very broad brushstrokes, however, and he does not put the theme of space and place at the forefront of his analysis. We need a more detailed and nuanced treatment of the geographies of nineteenth-century British science, which, in terms of scale, lies somewhere between the global and the local.[5]

During the 1970s and up to the middle of the 1980s, historians tried to locate science within the larger story of the transformation of Britain over the course of the nineteenth century from an agrarian, rural, aristocratic, and Christian society to an industrial, urban, middle-class, and secular society. The emphasis was on the shift within the scientific elite from the gentlemen of science to the professionals, and on the change in the epistemological foundations of scientific theory from the natural theology of the first half of the century to the scientific naturalism of the second half of the century. The move in the mid-1980s to study scientific practice led to more sophisticated work on elite science and its theory-making.[6] Newer scholarship since the 1990s has overlaid the attention paid to elite theory and practice with an increased awareness of marginalized and subordinate scientific figures—the women, workers, invisible assistants, and popularizers who were not members of the elite. Although the issue of place had lurked in all of this scholarship since the 1970s, until quite recently the emphasis has remained on people, their theories, and their practices. What happens if we highlight, even more, the role of space? If we look at the development of British science through biographies of the important sites of the nineteenth century? If scientific sites become agents, or at least coactors, with people working within them? Let us see if any of the familiar features of the topography need to be redrawn if we try to organize the map around the distinctive sites of nineteenth-century British science.

SITES OF GENTLEMANLY AND UTILITARIAN SCIENCE

If Becker had been alive in the first few decades of the nineteenth century and had visited the "truth-spots" of the period—the privileged spaces of science from which emanated discourses of immense power—where would he have gone?[7] Though vibrant scientific cultures existed throughout Britain, he would have found himself, again, in London, at the Royal Institution and at the Royal Society. He may have also had to make the trip to Kew Gardens, though he might have saved himself the trouble of going to all three places by paying a visit to just one location: the study of the botanist and scientific statesman Joseph Banks. It was from this study, a key center of calculation, that these three privileged sites were run, and they formed the heart of a vast scientific network that Banks had constructed.[8] All three were to play a significant role throughout the nineteenth century, but at that point they were spaces of the landed aristocracy and the upper class where the importance of agricultural science was emphasized. At the end of the eighteenth century both George III and the Tories saw agricultural improvement as the path to national prosperity. After Banks was appointed by George III in 1772 to reorganize Kew, it became "a great botanical exchange house for the empire." Under Banks the collections at the Royal Garden expanded, and botanists trained there worked on Admiralty vessels, for the English East India Company, and in colonial gardens around the world.[9] The

Royal Society, which Banks presided over from 1778 until his death in 1820, was devoted to natural history, the study of antiquities, and agricultural improvement.[10] Banks opposed on principle the founding of metropolitan specialist societies because he believed that they undermined the power of the Royal Society. The Royal Institution was first proposed at a meeting in 1799 at Banks's Soho Square house with the goal of applying science to the needs of the nation and especially to agriculture.[11] Banks ensured from the start that there were strong links between the Royal Institution and the Royal Society.[12]

After Banks's death in 1820, scientific sites slowly began to shed facets of their aristocratic nature. Not that aristocrat sites disappeared entirely. In 1844 the Earl of Rosse (1800–1867) built, with his own money, the largest telescope in the world on the grounds of his castle in Ireland. Nicknamed the Leviathan of Parsonstown, Rosse trained his telescope on the Orion nebula in order to discredit the nebular hypothesis and the religious and political radicalism associated with it.[13] But from about 1820 to 1850, old sites were modified, and new sites arose as the geography of British science was subtly reconfigured. Whereas Banks and his supporters had exploited and reinforced relations of genteel patronage and obligation, reformers wanted to alter the politics of science.[14]

The reformers came mainly from two groups, the gentlemen of science and the Utilitarians. Relatively few of the gentlemen of science were actually genteel by birth. They were committed to the serious pursuit of knowledge as a vocation, but not for pay.[15] They pushed for the advancement of science in such new spaces as the British Association for the Advancement of Science. Many of the gentlemen of science were educated at Anglican Cambridge, and they belonged to the Cambridge Network, a loose convergence of scientists, historians, dons, and other scholars. George Airy, Charles Babbage, John Herschel, George Peacock, and William Whewell were key members of the Network. They were the "young Turks" of their era, as they attempted from 1815 to 1830 to break into the metropolitan scientific scene. They worked to modernize the study of mathematics at Cambridge by converting the university from Newtonian mathematics to the new French methods of analysis. They inaugurated a new era in British physical science and moved to undermine the emphasis on natural history, especially the agricultural sciences. Other Cambridge scientists, such as Adam Sedgwick and William Hopkins, were part of the Network. Scientific men of the armed forces formed alliances with members of the Network, including Edward Sabine, an expert on geodesy and terrestrial magnetism, and Francis Beaufort, appointed as hydrographer to the Navy in 1829. The

Network promoted the modernization and internationalization of English science, scholarship, and religious thought.[16]

The gentlemen of science were joined by the Utilitarians in the push to reform aristocratic spaces of science. A general rise in professional consciousness in the 1820s affected law, medicine, education, and government administration. Composed largely of members of the professional middle class, the Utilitarians were interested in social reform. Science was neither polite knowledge for a hereditary elite, nor merely useful knowledge for agricultural improvers. Science was seen as a professional tool to be used to create a body of knowledge useful to the reformers in government and in the professions. It provided the basis for the professionals' expertise and the key to an organized, efficiently administered, and smoothly functioning society. This vision of science was embodied in the founding in 1826 of London University. It was set up as a secular institution modeled on the universities of Berlin and Bonn, and, unlike Cambridge and Oxford, it opened up its doors to non-Anglicans.[17] Utilitarian conceptions also pervaded the spaces of the Royal Institution, through the chemist William Thomas Brande, the Statistical Department of the Board of Trade (founded in 1832), and the Geological Survey (established in 1835).[18]

The three sites that Banks had controlled were not immune to the winds of change, and the reformers refashioned them. We are used to thinking about self-fashioning by individuals, but we should consider also how spaces are refashioned. At the Royal Institution, Michael Faraday (1791–1867) had become an important figure by the end of the 1820s. A friend, though not a member, of the Cambridge Network, he was the contemporary scientist most admired by the Network.[19] He had been a bookbinder's apprentice before he was hired as Davy's assistant. Faraday and the Royal Institution were well suited to each other. The Royal Institution provided him with both a laboratory in which to undertake his research and an audience to listen to his lectures about science[20] (fig. 2.1). Faraday played a role in the shift in emphasis at the Royal Institution from agricultural improvement to natural philosophy and the value of pure research.[21] The establishment by Faraday in 1825 of the very successful Friday Evening Discourses gave the Royal Institution an even greater public presence. When he was offered the chair of chemistry at the University of London in 1827, Faraday declined, explaining that despite being unsatisfied with his salary at the Royal Institution he possessed "the kind feelings and good will of its Authorities and members [and] all the privileges it can grant or I require." By 1836 he could not imagine himself to be anywhere else. That year he told one correspondent, "I have been here so long (three and twenty years) attached to

Fig. 2.1. *Michael Faraday (1791–1867) in his basement Laboratory*, 1852 (w/c on paper),
by Harriet Jane Moore (1801–84), The Royal Institution, London,
UK/ The Bridgeman Art Library.

the Royal Institution that I feel as if I were a limpet on a rock and that any
chance which might knock me from my position would leave me but little
hopes of attaching myself any where again."[22] Ever loyal to the Royal Institution, Faraday remained there for the rest of his life.

Although Faraday represented the public face of the Royal Institution,
his emphasis on research was not the only factor shaping this important scientific site in the second quarter of the nineteenth century. More and more
members of the professional middle class, especially physicians and barris-
ters, began to participate in the activities of the Royal Institution. The Utilitarian component of Royal Institution governorship rose steadily after 1811.
From 1826 to 1831 it became dominant, displacing but not eliminating the
aristocratic landowners. The chemist William Thomas Brande (1788–1866),
who led the Royal Institution prior to Faraday, from 1813 to 1831, embodied

Utilitarian ideals. Under Brande, the Royal Institution undertook a series of activities that gave it the reputation of being a metropolitan powerhouse for the scientific management of social problems. The Royal Institution produced textbooks and a journal, and became involved in consultancy, commercial analysis, and legal and government testimony. Faraday may have preferred to devote his time to his electromagnetic researches, but even he was drawn into undertaking demanding technical research long after Brande was gone, including the testing of lighthouse arrangements, checking gas works, and analysis of many different substances. The Utilitarian governors were just as influential as Faraday in moving the Royal Institution away from its emphasis on agricultural improvement.[23] Though the Utilitarians and the gentlemen of science were distinct groups, they shared, in addition to their reformist inclinations, a common episteme: analysis. In this way of knowing, scientists followed the French in conceiving of chemicals, medical bodies, animals, and plants as being composed of various elements, tissues, or organs compounded in different ways; or followed the Germans in emphasizing the analysis of form.[24] They established analysis as the most credible epistemological approach for doing science.

Kew was also undergoing a metamorphosis from royal to public garden. After Crown patronage had declined in the twenty years following Banks's death, Kew was seen by critics as a poorly managed royal monopoly that failed to serve the public or the interests of science. Transferred to the British government in 1840, Kew was a public institution when William Hooker (1785–1865) took charge in 1841. Before taking up his post at Kew, Hooker had been Regius Professor of Botany at Glasgow University. He had been appointed to that position in 1820 with Banks's recommendation. But he was gradually drawn into the orbit of the gentlemen of science. He was appointed to the British Association for the Advancement of Science's first Council in 1832, and he played a major role in the organization of the association's annual meeting at Glasgow in 1840.[25] After moving to Kew, Hooker strived to transform it into a center for scientific research as well as a place for the amusement and edification of the nation. Open every weekday from 1:00 to 6:00 P.M. to the public, the number of visitors rapidly increased from 9,174 in 1841 to 179,627 in 1850.[26] Similar developments took place at the Zoological Society of London's Regent's Park Zoo, which, after a serious financial crisis in 1846, was forced to abolish admissions restrictions. By 1850, attendance had increased from the 168,895 visitors in the previous year, to 360,402 visitors, due largely to the introduction of a hippopotamus. By exhibiting exotic animals from the ends of the earth, it became a national public institution illustrative of Britain's economic and imperial power. It could also claim to serve a useful purpose in its contribution to the commercial potential of acclimatization.[27]

Banks's Royal Society was viewed by the gentlemen of science as being tainted by corruption and the cultivation of the nobility. They were opposed to its domination by men who were ignorant of science. Led by John Herschel in 1830, reformers failed to shift the balance of power in the Royal Society from aristocrats and wealthy members to men of science. The lack of significant reform was still an issue more than ten years later. Faraday, who sympathized with the reform group in the Royal Society, wrote to a correspondent in 1843 that he had not attended a meeting in several years since "I do not like the present constitution of it, and want to restrict it to scientific men. As these opinions are not acceptable, I have withdrawn from any management in it."[28] But by 1848 traditional loyalties to the Crown and Church were replaced by new contractual allegiances based on service to knowledge and utility to the state. Alterations in the statutes of the Royal Society transformed it from an absolute to a constitutional monarchy.[29]

While older aristocratic sites were being refashioned, the gentlemen of science opened up new scientific spaces during this period. The British Association for the Advancement of Science was created in 1831 as a peripatetic organization. Initially provincial in origin, by 1833 Cambridge and Oxford academics began to dominate the London-based Council, particularly liberal Anglicans from Cambridge's Trinity College. The Cambridge Network was central to the BAAS. William Whewell, Adam Sedgwick, George Peacock, and George Biddell Airy were intimately involved in its intellectual and managerial policies. They were attracted to it, in part, because of the failed revolt in the Royal Society. Conservative reformers, they were opposed to the political claims of both die-hard Tories and aggressive democrats. Their liberal theology and their science were connected to their political position. Embracing natural theology, they pointed to a divine order behind both nature and society, and to the role of science as a neutral means for obtaining desirable ends. This was a message tailored for a British nation undergoing the stress of the 1830s and 1840s, turbulent decades when heated debates over reform brought the country to the brink of revolution.[30]

But the gentlemen of science needed the support of other groups to make the BAAS a success. The advantages of aristocratic patronage were understood—alliances with key aristocrats allowed them to develop their power base. Aristocrats were wooed to serve as presidents up to 1841 while the association was still establishing itself. But agricultural science was never given its own section because the gentlemen of science did not want the as-

Fig. 2.2. The Hunterian Museum in 1842, drawn by Thomas Hosmer Shepherd.
© Copyright 2005 by The Hunterian Museum at The Royal College of
Surgeons of England.

sociation to appear as a vehicle for the landed interest. Instead, the emphasis was on the physical sciences. Moreover the BAAS was closely connected
to the metropolitan specialist societies that Banks had opposed, such as the
Geological Society (founded 1807) and the Astronomical Society (1820). The
London gentry, liberal professionals whose mental cultivation allowed them
to mingle with the aristocracy, was another group that the gentlemen of

science attracted to the association. The gentlemen of science also attempted to draw on the interest in science in the newer commercial and manufacturing towns and courted the leaders of provincial societies.[31] Like the Royal Institution and Kew Gardens, the BAAS reached out to the public.

The rapid growth of museums was another signal that scientific spaces were in the process of being reconfigured.[32] Rather than the laboratory, the museum was the central institution of Victorian science.[33] Museum architecture and the organization of exhibits often embodied an epistemological perspective. The Museum of Practical Geology on Jermyn Street in London (opened in 1851) brought together and displayed materials gathered by the Geological Survey. It defined nature as useful by teaching viewers to understand how natural resources were transformed into commercial products.[34] At the Hunterian Museum at the Royal College of Surgeons, Lincoln's Inn Fields, in London, the prominence of vertebrate paleontology illustrated the scientific epistemology of natural theology (fig. 2.2). Richard Owen, who worked at the Hunterian Museum from 1827 to 1856, was connected to the gentlemen of science through Oxford geologist William Buckland, whom he had met in the early thirties. Buckland became his close friend and patron. Sedgwick and Whewell also served as his patrons. Owen was one of most active supporters of the movement to expand museum collections and to turn them to educational and research purposes.[35] Among the life scientists, he was the most dominant adherent to the style of analysis.[36] Though the gentlemen of science and the Utilitarians were in favor of reforming the older, privileged spaces of Banks's aristocratic world, they were not opposed in principle to the existence of privileged sites in science. They conferred a privileged status to the scientific sites in which they worked, whether they were new, like the museum, or refashioned sites like the Royal Institution or Kew Gardens.

SITES OF RESISTANCE TO GENTLEMANLY
AND UTILITARIAN SCIENCE

But not all sites were scenes of gentlemanly and Utilitarian science in the second quarter of the nineteenth century. For some, the reformist inclinations of gentlemen and Utilitarians did not go nearly far enough. The founding of the secular London University provided a hospitable site on Gower Street for Scottish anatomists enamored with French evolutionary theory. From this base of operations, they could use radical Lamarckianism to challenge the Tory-Anglican establishment and argue for the reform of privileged aristocratic institutions. But they were also critical of the gentlemen of science, who appeared to them to be too conciliatory in their push for reform and too dependent on natural theology as a framework for their science. Robert Grant, who held the chair of comparative anatomy from 1827, clashed with the gentlemen of science in the Zoological Society, the Royal Society, and the Geological Society.[37] London University was not the only new educational institution founded at this time. A number of new "private" or nonhospital medical schools were founded after the late 1820s with close ties to Nonconformism. Disadvantaged both socially and in the medical world, medical Dissenters were receptive to the new Continental anatomies established at London University.[38]

Whig MP, lawyer, and educationalist Henry Brougham had originally supported the founding of London University in the hopes that it would crush bigotry and intolerance. But the Scottish medical professors eventually hired were more radical than he could have anticipated.[39] Brougham, who was on friendly terms with the gentlemen of science, was involved in the creation of new public sites designed to counter the impact of radical scientific ideas even before London University was established: the mechanics' institutes and the Society for the Diffusion of Useful Knowledge. In 1823 a manifesto was published in the *Mechanic's Magazine* calling for the establishment of a London Mechanics' Institution. Brougham, George Birkbeck, and James Mill were among the prominent backers of the proposal. The institution formally opened in 1824 and had as its goal the scientific instruction of the artisan. This marked the origins of the mechanics' institute movement in England. Brougham's *Practical Observations upon the Education of the People* (1825) helped to spark the rapid growth of these new scientific spaces. In 1826, 101 institutions existed in Britain.[40] By 1851 there were more than 700 institutions with more than 120,000 members.[41] They offered libraries, evening classes, and lectures. The institutes were run for the working classes, but not by them. Members of the middle classes, such as medical men, lawyers, journalists, and bankers, founded the majority of them. They believed that rational recreation based on scientific knowledge, which revealed the design in nature, would lead to moral improvement and counter the impact of radical scientific ideas.[42]

Brougham was more directly involved in the Society for the Diffusion of Useful Knowledge, founded in 1826. Here again, the aim was to undermine political radicalism with rational information. This time, however, the medium was the printed word. Inexpensive volumes in Brougham's Library of Useful Knowledge appeared every two weeks and covered all of the sciences. Other publishers, such as John Murray, in his Family Library, and Longman, in his Cabinet Cyclopaedia, followed Brougham in publishing cheap science books in series form.[43] These publishing experiments in the twenties and thirties were crucial in establishing that print culture could provide a public scientific space that was both commercially viable and intellectually valuable. They constituted the beginnings of that vast communications revolution that would produce affordable scientific reading for the public in the second half of the century. Through the mechanics' institutes and the publication of cheap books on science, sites of resistance to gentlemanly and Utilitarian science could be kept in check.

AFTER 1850: SITES OF SCIENTIFIC NATURALISM

As a new generation of practitioners arrived on the scene at the midpoint of the century, the sites of science were reconfigured once again. Their aim was to make scientific spaces hospitable to their goals, which included the secularization of nature, the professionalization of their discipline, and the promotion of expertise. Some of the newcomers were shaped by their experiences at sites of resistance to gentlemanly science, such as the dissenting medical academies. Others came out of the refashioned aristocratic sites, such as Kew Gardens, or from one of the new sites created in the name of public utility. When in 1874 Barber visited the key scientific sites of his time, he encountered Thomas Henry Huxley (1825–1895) lecturing to workers at the Royal School of Mines, John Tyndall (1820–1893) demonstrating experiments to a fashionable audience at the Royal Institution, and Joseph Dalton Hooker (1817–1911) sitting in the president's chair at the Royal Society. All three sites were located in London. Huxley, Hooker, and Tyndall were quite sensitive to the power of place. From the point of view of these three influential representatives of scientific naturalism, London was the focal point of British science. When in 1854 it seemed possible that Huxley would be offered a professorship at Edinburgh, he wrote to his sister Lizzie that he did not want to leave London. "It is *the* place," he told her, "the centre of the world."[44]

Huxley, Tyndall, and Hooker recognized that to accomplish their goals they needed to refashion many of the sites of gentlemanly science and that it was imperative to create new scientific spaces in which they could discover and communicate knowledge. They had more in common with the Utilitarians, but even the Royal Institution had to be refashioned to fit their needs. In order to reform aristocratic sites, the gentlemen of science and the Utilitarians had argued for a more public dimension to many scientific spaces. They wanted to distance themselves from aristocratic privilege, yet they wanted to protect certain privileged sites of knowledge as spaces where they could create forms of knowledge that undermined radicalism. If the gentlemen of science and the Utilitarians lived in an age of reform, the scientific naturalists lived in a period when the democratic impulse was even more powerful. In pursuing the goal of professionalizing science the scientific naturalists were faced with seemingly contradictory goals: on the one hand, emphasizing the privileged status of spaces of expertise, while on the other cultivating public spaces to win the support of the masses. The Reform Act of 1867 expanded the franchise, and now many members of the working class had the power to vote in politicians who would determine the degree of state support for science. One space of expertise was particularly important to the scientific naturalists—the laboratory. Their emphasis on this space was in keeping with the way of knowing referred to by Pickstone as "experimentalism." Pickstone argues that the creation of systematic experimentation took place around the middle of the nineteenth century and that it constituted a distinctive episteme. He points out that it was institutionalized in the Royal Institution in London, and it was systematically built up in universities oriented toward research, particularly in Germany. Pickstone explicitly identifies Huxley and Tyndall as experimentalists.[45] Three sites of scientific naturalism, Kew Gardens, the Royal Institution, and the South Kensington biological laboratories, illustrate the complicated spatial dynamics of science in the latter half of the nineteenth century. In the laboratories located at each of these privileged sites, scientific naturalists used experiment as a means for constructing an evolutionary, secular world.

During his tenure as director, William Hooker had emphasized Kew's role as a public garden in order to obtain government funding for new lakes and plant houses. He also tried to strengthen Kew's research facilities, however, and by the end of the 1850s, he had overseen the completion of a herbarium and a library. His son, Joseph, succeeded him as director of Kew in 1865, after having served for ten years as deputy director. Under Joseph, a fundamental change took place in Kew's identity as an institution (fig. 2.3). By 1872 Joseph was presenting the gardens as principally committed to pure research and the imperial economy.[46] Kew, he argued, could best serve the public interest by becoming both a center for botanical research as well as a public garden. He continued to allow only serious botanical students and artists into the garden in the morning and resisted all attempts to extend the garden's opening hours for the general public.[47]

Joseph Hooker's refashioning of Kew into a research and imperial space was essential for responding to the machinations of Richard Owen, who was notoriously hostile to Kew, Huxley, and the other scientific naturalists. When Acton Smee Ayrton, who was first commissioner of works in

Fig. 2.3. Kew Gardens. From the private collection of Professor Janet Browne.

Gladstone's government, was engaged in a bitter dispute with Hooker from
1870 to 1872, he turned to Owen to strengthen his hand. Ayrton wanted to
turn Kew into a public recreation park, and he asked Owen to write an official report. Owen recommended the transfer of the Kew collections to the
British Museum, where they would come under his jurisdiction. Hooker's
allies helped him to preserve Kew as a site where research, as defined by the
scientific naturalist, continued to be one of its defining activities.[48] Huxley
saw Ayrton's attack on Hooker as nothing less than an assault on science
itself.[49] Tyndall drafted an appeal to Gladstone, signed by Darwin, Huxley,
and others, objecting to Ayrton's actions. The publication of the appeal led
to a discussion in both Houses of Parliament. In 1872 Tyndall wrote a letter
to Lord Derby, Hooker's chief supporter in the House of Lords, in which he
asserted that Kew was "of the very highest importance to botanical science
and to the application of that science in India and the Colonies."[50] After
winning the controversy with Ayrton, Hooker persuaded his friend Thomas
Jodrell Phillips-Jodrell to build and equip a botanical laboratory. The Jodrell
Laboratory opened its doors in 1876, designed for research in plant physiology, palaeobotany, anatomy, cytology, and other branches of botany requiring controlled laboratory experiments.[51] The scientific empire of Kew

reached its zenith under William Thiselton-Dyer, Joseph Hooker's son-in-law, who became director in 1885. Thiselton-Dyer had been a demonstrator in Huxley's lab at South Kensington in the early seventies. Championing the study of botanical physiology and laboratory research, Thiselton-Dyer oversaw the formation of a system of colonial botanical establishments.[52]

At the Royal Institution, the mantle of leadership was in the process of passing from Faraday to Tyndall. In the early fifties, when both Huxley and Tyndall were searching for permanent scientific positions in London, Huxley advised Tyndall that the Royal Institution was where he "ought to be—looking to Faraday's place." Both of them understood that working at the Royal Institution gave Tyndall the prestige needed to push forward the agenda of scientific naturalism. "What they want," Huxley wrote to Tyndall on February 25, 1853, "and what you have, are *clear powers of exposition*—so clear that people may think they understand even if they don't. That is the secret of Faraday's success."[53] The Royal Institution was the primary space in which Tyndall practiced his science from his appointment as professor of natural philosophy in 1853 until he retired in 1887. As a result, he became closely identified with the Royal Institution. Shortly after his death, an obituary notice in the *Standard* stated that "Tyndall's scientific career was co-extensive with his connection with the building in Albemarle Street."[54] Before he arrived at the Royal Institution, he had worked as a draughtsman and practical surveyor in spaces with links to reformers who championed the utility of science, such as the Ordnance Surveys of Ireland and England. He spent some of his free time in local mechanics' institutes.[55] Determined to receive formal training, Tyndall completed a Ph.D. in chemistry at the University of Marburg in 1849. In the Royal Institution's well-equipped laboratory, still a rarity in mid-century Britain, Tyndall continued Faraday's emphasis on original research. But the public face of the Royal Institution, especially the opportunities for extensive public lecturing, also allowed Tyndall to develop his performative skills and to campaign for the appreciation of science as a means of culture.[56] Since Faraday had acted as Tyndall's mentor in his early years at the Royal Institution, he was restrained in his advocacy of the agenda of scientific naturalism within its walls. Tyndall's most aggressive challenges to the authority of Christian clerics were presented in other rhetorical spaces, such as in his presidential address to the British Association for the Advancement of Science in Belfast in 1874.

In his sixty-foot biological laboratory in the Science Schools Building in South Kensington, Huxley was free to teach his students to view nature through secular eyes. But Huxley moved through a series of scientific spaces before he had the opportunity to create his lab, and they each left their mark on him. At Sydenham College, a cut-price private medical school in London, Huxley was trained in medicine in the early 1840s. Here he was exposed to dissenting teachers engaged in a war with elite surgeons who upheld the medical establishment. After his voyage (1846–1851) on *HMS Rattlesnake* as assistant surgeon, in 1854 Huxley was appointed as lecturer at the Royal School of Mines and as paleontologist to the Geological Survey. Huxley transformed the Royal School of Mines into the leading institution for applied science education. But Jermyn Street lacked the laboratory facilities Huxley needed—it was dominated by the geology museum. Without a laboratory, Huxley could only show an experiment or dissection during his lectures.[57]

Like Tyndall, Huxley sought to present the laboratory as the privileged scientific site. Huxley's lobbying paid off when the four-floor Science Schools Building was constructed by the government in 1871. The top floor housed a laboratory based on the Berlin and Bonn models. Huxley's laboratory was designed to be an authoritative space in which to encounter "Nature." Each student was instructed in scientific method, to verify every fact in the lab, and to see the natural world in wholly secular and evolutionary terms. Joined by a team of lab assistants, Huxley could train the science teachers who would return to the factory towns. On June 4, 1872, Huxley wrote to Tyndall about his "new system of teaching which, if I mistake not, will grow into a big thing and bear great fruit." The South Kensington laboratory became a model of effective practical teaching in botany and physiology that was exported to other academic institutions in the 1870s and 1880s by Huxley's former demonstrators after they left South Kensington. Huxley's new site signaled the shift from museum display to the lab as the new knowledge-manufacturing space for science.[58] Although Huxley touted the privileged space of the lab, where access was restricted to the trained expert, he did not ignore the public spaces of science. Like Tyndall he was a gifted lecturer who spoke in a variety of venues. Huxley also frequently made use of the space of print culture to present his case for scientific naturalism. As Darwin's bulldog, he used the controversies surrounding evolution to present his case for professional science in books and periodicals such as *Nature* or *Nineteenth Century*.

CONTESTED SPACES AND SITES OF RESISTANCE
TO SCIENTIFIC NATURALISM

Although scientific naturalists successfully refashioned sites in line with their agenda, such as the Royal Institution and Kew Gardens, and although they created new spaces to further their goals, such as Huxley's South Kensington laboratory, vast tracts of the scientific landscape were not under their control in the second half of the nineteenth century. The scientific naturalists held positions of power in the Royal Society; nevertheless a form of aristocratic science survived in other sites.[59] Aristocrats, such as Lord Salisbury, William Siemens, William Armstrong, and Lord Rayleigh, built laboratories on their landed estates. The country-house laboratory was their privileged "truth-spot," where they created knowledge congruent with their social and religious beliefs. In these aristocratic spaces outside London, they examined gases, spectra, photographs, and the ether, while also enquiring into spiritualist phenomena.[60] Salisbury and Rayleigh were part of an extensive aristocratic family network comprised of the Balfours of Whittingehame, the Gascoyne-Cecils of Hatfield, the Strutts of Terling, the Sidgwicks of Hillside, the Campbells of Inveraray, and the Parsons of Birr. From these families came some of the most eminent men and women of science in the second half of the century. In addition to the physicist Lord Rayleigh (1842–1919), the group included among its ranks the zoologist Francis Maitland Balfour (1851–1882), the astronomer William Parsons, and the mathematician Eleanor Mildred Sidgwick (1845–1936).[61]

The science practiced in these aristocratic country houses was informed by a moderate evangelicalism based on the belief that God's providence worked through physical laws that were a manifestation of a permanent, moral law. Members of this network were sympathetic towards spiritualism, champions of amateur practice, and skeptical about natural selection. They held to a worldview that reconciled science and religion while offering an alternative to the conception of science championed by scientific naturalism. Their science was embedded in aristocratic and Christian values. A number of these individuals had a tremendous impact on other sites, such as the British Association for the Advancement of Science and the University of Cambridge, where they brought with them the traditions of country-house science. As presidents of the British Association, Rayleigh (1884), Salisbury (1894), and Arthur Balfour (1904) delivered addresses in which they defended theism and criticized scientific naturalism. At Cambridge, Francis Balfour and Rayleigh were active participants in an informal aristocratic network coalescing at Cambridge in the late 1870s. Before his early death in 1882, Balfour became a central figure in the developing Cambridge school of embryology. In 1879, Rayleigh was elected the second professor of experimental physics at the Cavendish Laboratory at Cambridge, a position he held until 1884. Aristocrats and their country-house science continued to play significant roles in the second half of the nineteenth

century, challenging the authority of scientific naturalists and their favored
episteme.[62]

The museum, one of the key spaces opened up by the gentlemen of science, could provide a site for resisting the aims of scientific naturalists. The
Oxford University Museum, completed in 1860, embedded the principles
of the natural theology tradition in its architecture. The interior, which
resembled a cloister, was designed to emphasize the harmonious relationship between science and religion.[63] The Natural History Museum in South
Kensington, which opened in 1881, was built along similar lines (fig. 2.4).
Owen, who had moved from the Hunterian Museum to the British Museum in 1856 to become superintendent of the natural history collections,
had been pushing since 1859 for a separate but unified museum of natural
history. Huxley opposed this plan. Not only would it have concentrated all
collections in Owen's hands, it would also have increased the status of the
museum as a site of ongoing research. Huxley believed that the laboratory
should be considered the most important space for the practice and teaching
of science.[64] In opposition to Owen, he proposed that the botanical collections go to Kew (to benefit Hooker) and to the Museum of Practical Geology, where he was then located.[65] Huxley and Owen also disagreed on the
internal organization of museums. Huxley advocated a more rigid physical
separation between the spaces for the public and those for the practitioner,
true to his emphasis on scientific expertise. Only a few representative specimens would be shown to the public.[66] Owen believed that the purpose of
a natural history museum was to show the greatness of God in the variety
of nature. All specimens would be on display, though some galleries were
intended primarily for practitioners and naturalists. Due to his friendship
with liberal politicians, Owen was able to overcome Huxley's opposition,
and the museum that was built stood as a physical embodiment of early, not
late Victorian, science, as it evoked natural theological themes. Huxley may
have had the last laugh. Although Owen had seen his long-standing dream
of a separate, national museum of natural history come to fruition before he
retired in 1883, his successor was William Henry Flower, Huxley's friend
and an exponent of Huxley's notion that public spaces should be segregated
from those reserved for practitioners.[67]

Spaces such as the laboratory and print culture, which scientific naturalists entered to establish a privileged site in which to produce and teach
knowledge or a site in which to communicate that knowledge to the public,
were not fully under their control. North British physicists, hostile toward
the agenda of scientific naturalists, also established important laboratories.
Here they developed a distinctive, new energy physics designed to be in

Fig. 2.4. The central hall of the Natural History Museum. Courtesy of the Natural History Museum Archives, reference number PH/172/12.

harmony with Christian belief.[68] Kelvin, who had set up a precision measurement laboratory at Glasgow University in the early 1850s, well before Huxley's South Kensington facility was built, had had a major impact on the development of laboratory science in Britain, and he was no scientific naturalist. Another North British physicist, the evangelical James Clerk Maxwell, introduced experimental physics into the mathematical and moral culture of Cambridge when he was appointed to the new Cavendish chair at Cambridge in 1871. The laboratories at Glasgow and Cambridge, like Tyndall's lab at the Royal Institution, emphasized research. Huxley's lab was more in line with the teaching laboratories established in this period.[69] The proliferation of laboratories of different kinds, some in spaces that were not congenial to scientific naturalism, complicated Huxley and Tyndall's stress on the laboratory as a site for guaranteeing their particular epistemic position.

Similarly, the spaces of print culture were not always so accommodating to the goals of scientific naturalists. Some publishers, like Macmillan, who published *Nature* and Huxley's Science Primers series, and some editors, like James Knowles, who welcomed Huxley's contributions in his journal *Nineteenth Century*, liked to work with practitioners.[70] But others, such as the newspaper editor W. T. Stead, a proponent of the new journalism, and publishers such as George Routledge and Thomas Jarrold, did not believe that the men of science were necessarily the best communicators for a popular audience. Routledge and Jarrold hired Anglican clergymen like John George Wood and Ebenezer Brewer, neither of them practitioners, to write their science books for the public. Both popularizers were extraordinarily successful. The sales of Wood's *Common Objects of the Country* (1858) and Brewer's *Guide to the Scientific Knowledge of Things Familiar* (1847) both exceeded the sales of Darwin's *Origin of Species* (1859). There was a whole contingent of popularizers of science like Wood and Brewer, who, though not as successful in terms of sales, produced a steady stream of books containing theologies of nature that challenged the scientific naturalists' secularized perspective, their emphasis on expertise, and their privileging of the laboratory.[71] All in all, a formidable array of sites of resistance to scientific naturalism was in existence in the second half of the century.

SCIENTIFIC ELITES AND CHANGING SPATIAL CONFIGURATIONS

Thinking spatially about science over the entire course of the nineteenth century gives us a renewed appreciation for how places can become sites of contention as the composition of the scientific elite changes, as the role of that elite is altered, and as new groups attempt to force their way into the charmed circle of power. Behind the formidable and seemingly solid walls of the buildings that house scientific sites, we discover malleable spaces. Sites are refashioned several times over the century to fit the needs of various groups or individuals. New sites are also created to serve the needs of new generations. And most significantly, we have found, paradoxically, that as some scientific sites became more and more accessible to the public, some were more carefully controlled by elite scientists. In 1800 the most important sites of science were under the thumb of one man, Joseph Banks, who reigned over these aristocratic spaces like an absolutist monarch. After his death in 1820, the gentlemen of science and the Utilitarians constructed sites in which they shared power among themselves while granting the public some access to scientific spaces such as Kew, the museum, and the Royal Institution. The agenda of scientific naturalism, which emphasized training, expertise, and laboratory research, led to an even greater split between the public and professional spaces of science. Huxley and his allies attempted to maintain a tight control over the spaces of elite science in order to expel the gentlemen of science and undermine the credibility of their natural theology. But they were not completely successful in controlling the laboratory. Nor were they effective in organizing the public spaces of science to suit their goals. The large space previously assigned to scientific naturalists on our maps of British science in the second half of the century is most in need of correction in light of this overview of the terrain.

At best, I have provided a provisional map of one sector of the geographies of nineteenth-century science, a rough "lay of the land." It needs to be expanded to more fully take into account the British spaces outside London. It needs to be enlarged to include scientific sites outside Britain, as metropolitan science was inextricably connected to, and even shaped by, the colonies.[72] And, in addition to increasing the scale to the global level, it needs to be extended chronologically, so that it can be contrasted to and aligned with the "big picture" drawn by Pickstone. But perceptive readers will already have recognized that I have neglected a crucial dimension of science by focusing primarily on scientific sites as coactors with the people who worked within them. If scientific objects could be added to the discussion, we would have a much more comprehensive map. As Pickstone observes, each way of knowing creates, or construes, particular kinds of "objects."[73] Ideally, we want to construct a historical map depicting the geographies of science that incorporates objects as well as epistemes. As scholars continue to explore how place was reconfigured and sites refashioned, and as they take into account all of these factors, that map will be drawn and redrawn repeatedly.

NOTES

I am indebted to Ruth Barton, Jamie Elwick, and Graeme Gooday for their helpful suggestions.

1. John Foster Kirk, "Becker, Bernard Henry," in *A Supplement to Allibone's Critical Dictionary of English Literature and British and American Authors* (Philadelphia: J. B. Lippincott, 1899), 1:118; Frederic Boase, "Becker, Bernard Henry," in *Modern English Biography* (Truro: Netherton and Worth, 1905), 4:333.

2. Bernard H. Becker, *Scientific London* (London: Frank Cass, 1968), v, 24, 46, 48, 146–47.

3. David N. Livingstone, "Science, Site and Speech: Scientific Knowledge and the Spaces of Rhetoric," *History of the Human Sciences* 20 (2007): 73.

4. John V. Pickstone, *Ways of Knowing: A New History of Science, Technology and Medicine* (Manchester: Manchester University Press, 2000); John V. Pickstone, "Working Knowledges before and after circa 1800: Practices and Disciplines in the History of Science, Technology, and Medicine," *Isis* 98 (2007): 489–516.

5. James Elwick has perceptively applied Pickstone's approach to an analysis of the British life sciences between 1820 and 1858. See James Elwick, *Styles of Reasoning in the British Life Sciences: Shared Assumptions, 1820–1858* (London: Pickering and Chatto, 2007).

6. James A. Secord, "Knowledge in Transit," *Isis* 95 (2004): 658.

7. Thomas F. Gieryn, "Three Truth-Spots," *Journal of the History of the Behavioral Sciences* 38 (2002): 113; Livingstone, "Science, Site and Speech," 73.

8. David Philip Miller, "Joseph Banks, Empire and 'Centers of Calculation' in Late Hanoverian London," in *Visions of Empire: Voyages, Botany, and Representations of Nature*, ed. David Philip Miller and Peter Hanns Reill (Cambridge: Cambridge University Press, 1996), 21–37.

9. Richard Drayton, *Nature's Government: Science, Imperial Britain, and the "Improvement" of the World* (New Haven, CT: Yale University Press, 2000), 101, 108, 125.

10. Iwan Morus, Simon Schaffer, and Jim Secord, "Scientific London," in *London— World City, 1800–1840*, ed. Celina Fox (New Haven, CT: Yale University Press, 1992), 130.

11. Drayton, *Nature's Government*, 88; Frank A. J. L. James and Anthony Peers, "Constructing Space for Science at the Royal Institution for Great Britain," *Physics in Perspective* 9 (2007): 141. The Royal Institution was influential as a model to be imitated by other institutions, both within London (e.g., the London Institution, founded in 1805; the Surrey Institution, established in 1808; and the Russell Institution, created in 1808) and outside London in Cornwall, South Wales, Liverpool, and Manchester. See Frank A. J. L. James, "Introduction," in *"The Common Purposes of Life": Science and Society at the Royal Institution of Great Britain*, ed. Frank A. J. L. James (Aldershot: Ashgate, 2002), 2; Morus, Schaffer, and Secord, "Scientific London," 131.

12. David Knight, "Establishing the Royal Institution: Rumford, Banks and Davy," in *"The Common Purposes of Life,"* ed. Frank A. J. L. James (Aldershot: Ashgate, 2002), 108.

13. Simon Schaffer, "The Leviathan of Parsonstown: Literary Technology and Scientific Representation," in *Inscribing Science: Scientific Texts and the Materiality of Communication*, ed. Timothy Lenoir and Hans Ulrich Gumbrecht (Stanford, CA: Stanford University Press, 1998), 182–222.

14. Morus, Schaffer, and Secord, "Scientific London," 135.

15. James A. Secord, *Victorian Sensation: The Extraordinary Publication, Reception, and Secret Authorship of "Vestiges of the Natural History of Creation"* (Chicago: University of Chicago Press, 2000), 403–4.

16. Susan Faye Cannon, *Science in Culture: The Early Victorian Period* (New York: Dawson and Science History Publications, 1978), 30, 33–35, 39–40, 44, 63; David Philip Miller, "The Revival of the Physical Sciences in Britain, 1815–1840," *Osiris* 2 (1986): 110.

17. Morris Berman, *Social Change and Scientific Organization: The Royal Institution, 1799–1844* (Ithaca, NY: Cornell University Press, 1978), 110, 113–23; Peter Alter, *The Reluctant Patron: Science and the State in Britain, 1850–1920* (Oxford: Berg, 1987), 25; Adrian Desmond, *The Politics of Evolution: Morphology, Medicine, and Reform in Radical London* (Chicago: University of Chicago Press, 1989), 26, 33–41.

18. Desmond, *Politics of Evolution*, 28.

19. Cannon, *Science in Culture*, 31, 58.

20. Frank A. J. L. James, ed., *The Correspondence of Michael Faraday* (London: Institution of Electrical Engineers, 1993), 2:xxvi.

21. Sophie Forgan, "Faraday—From Servant to Savant: The Institutional Context," in *Faraday Rediscovered: Essays on the Life and Work of Michael Faraday*, ed. David Gooding and Frank A. J. L. James (Basingstoke: Macmillan, 1985), 64.

22. James, *Correspondence of Michael Faraday*, 1:442; 2:386.

23. Berman, *Social Change and Scientific Organization*, 101–67.

24. Pickstone labels the period from 1780 to 1850 the "Age of Analysis," and he asserts that analysis replaced natural history as the dominant force across the sciences in the early nineteenth century. He sees the analytical sciences as products of major institutional novelties, such as French professional schools, hospitals, and museums, German reformed universities, and the British Industrial Revolution. He stresses the relation between analysis and professional education and consultancy; see Pickstone, *Ways of Knowing*, 12, 83–85. Pickstone's broad brushstrokes obscure the differences between Britain and the rest of Europe. Not only did British science lag behind in its acceptance of analysis, but, in the case of the gentlemen of science, analysis could be wed to natural theology. The important role of natural theology in British science is not a part of Pickstone's story.

25. Jack Morrell and Arnold Thackray, *Gentlemen of Science: Early Years of the British Association for the Advancement of Science* (Oxford: Clarendon Press, 1981), 211, 215.

26. Drayton, *Nature's Government*, 168, 180, 184, 188.

27. Harriet Ritvo, *The Animal Estate: The English and Other Creatures in the Victorian Age* (London: Harvard University Press, 1987), 217, 230, 239.

28. James, *Correspondence of Michael Faraday*, 3:126.

29. David Philip Miller, "Between Hostile Camps: Sir Humphry Davy's Presidency of the Royal Society of London, 1820–1827," *British Journal for the History of Science* 16 (March 1983): 18; Roy MacLeod, "Whigs and Savants: Reflections on the Reform Movement in the Royal Society, 1830–48," in *Metropolis and Province: Science in British Culture, 1780–1850*, ed Ian Inkster and Jack Morell (Philadelphia: University of Pennsylvania Press, 1983), 55–90.

30. MacLeod, "Whigs and Savants," 66; Morrell and Thackray, *Gentlemen of Science*, 21–25, 31–33.

31. Morrell and Thackray, *Gentlemen of Science*, 109–10, 117–19, 124, 127, 283; Simon Naylor, "The Field, the Museum and the Lecture Hall: The Space of Natural History in Victorian Cornwall," *Transactions of the Institute of British Geographers* 27 (2002): 494–513.

32. Pickstone argues that museums, which emerged during the episteme of natural history, were reoriented in the early nineteenth century along analytical lines. (Pickstone, *Ways of Knowing*, 131.) But since British natural theology tradition is not an essential part of his story, Pickstone does not deal with the crucial link between it and important Victorian museums.

33. James A. Secord, "Introduction," in *Vestiges of the Natural History of Creation and Other Evolutionary Writings*, by Robert Chambers (Chicago: University of Chicago Press, 1994), xii.

34. Carla Yanni, *Nature's Museums: Victorian Science and the Architecture of Display* (New York: Princeton Architectural Press, 2005), 52, 58. The Edinburgh Museum of Science and Art, completed in 1889, adopted a similar emphasis. See Yanni, *Nature's Museums*, 91–106.

35. Nicolaas A. Rupke, *Richard Owen: Naturalist* (New Haven, CT: Yale University Press, 1994), 13, 59, 61.

36. Elwick, *Styles of Reasoning in the British Life Sciences*, 8.

37. Desmond, *Politics of Evolution*, 134.

38. Ibid., 152–53.

39. Ibid., 25.

40. Guy Stuart Kitteringham, "Studies in the Popularisation of Science in England, 1800–30" (PhD diss., University of Kent at Canterbury, 1981), 313–14.

41. Steven Shapin and Barry Barnes, "Science, Nature and Control: Interpreting Mechanics' Institutes," *Social Studies of Science* 7 (1977): 33.

42. Ibid., 35–36; Kitteringham, "Studies in the Popularisation of Science in England," 313–14.

43. Secord, *Victorian Sensation*, 48.

44. Leonard Huxley, *Life and Letters of Thomas H. Huxley* (New York: D. Appleton, 1902), 1:129.

45. Pickstone, *Ways of Knowing*, 13, 30, 145, 150.

46. Drayton, *Nature's Government*, 219.

47. Jim Endersby, "Hooker, Joseph Dalton," in *Dictionary of Nineteenth-Century British Scientists*, ed. Bernard Lightman (Bristol: Thoemmes Continuum Press, 2004), 2:999.

48. Leonard Huxley, *Life and Letters of Sir Joseph Dalton Hooker* (London: John Murray, 1918), 2:174; W. B. Turrill, *Joseph Dalton Hooker: Botanist, Explorer, and Administrator* (London: Thomas Nelson and Sons, 1963), 90, 123.

49. Huxley, *Life and Letters of Sir Joseph Dalton Hooker*, 2:165.

50. A. S. Eve and C. H. Creasey, *Life and Work of John Tyndall* (London: Macmillan, 1945), 166.

51. Turrill, *Joseph Dalton Hooker*, 136.

52. Drayton, *Nature's Government*, 253, 267.

53. Leonard Huxley, *Life and Letters of Thomas H. Huxley*, 1:124.

54. Eve and Creasey, *Life and Work of John Tyndall*, 284.

55. Joe Burchfield, "Tyndall, John," in *Dictionary of Nineteenth-Century British Scientists*, ed. Bernard Lightman (Bristol: Thoemmes Continuum Press, 2004), 4:2053.

56. J. D. Burchfield, "John Tyndall at the Royal Institution," in *"The Common Purposes of Life,"* ed. Frank A. J. L. James (Aldershot: Ashgate), 150–51, 157.

57. Adrian Desmond, *Huxley: From Devil's Disciple to Evolution's High Priest* (Reading, MA: Addison-Wesley, 1997), 15–16, 354, 394.

58. Desmond, *Huxley*, 395; Graeme Gooday, " 'Nature' in the Laboratory: Domestication and Discipline with the Microscope in Victorian Life Science," *British Journal for the History of Science* 24 (1991): 333–40; Leonard Huxley, *Life and Letters of Thomas H. Huxley*, 1:408.

59. A number of the most influential scientific naturalists formed the X-Club in 1864 to coordinate their efforts to reform Victorian science in accordance with their agenda. The nine members were Huxley, Hooker, Tyndall, Spencer, George Busk, Edward Frankland, Thomas Archer Hirst, John Lubbock, and William Spottiswoode. All of the X-Club members, with the exception of Spencer, were Fellows of the Royal Society. Hooker was president of the Royal Society from 1873 to 1878. During his term he was involved in a number of reforms that further restricted the privileged, nonscientific element in the society. (See Turrill, *Joseph Dalton Hooker*, 164; Leonard Huxley, *Life and Letters of Sir Joseph Dalton Hooker*, 2:132–35.) Since Huxley was secretary to the Royal Society from 1871 to 1880, when his term overlapped with Hooker's tenure as president, the scientific naturalists were virtually in control of the society for most of the 1870s. From 1883 to 1885, Huxley also served as president. (See Leonard Huxley, *Life and Letters of Thomas H. Huxley*, 1:383.) The X-Club dominated the Royal Society in the 1870s and the first half of the 1880s: see Ruth Barton, "'An Influential Set of Chaps': The X-Club and Royal Society Politics, 1864–65," *British Journal for the History of Science* 23 (1990: 53–81.

60. Simon Schaffer, "Physics Laboratories and the Victorian Country House," in *Making Space for Science: Territorial Themes in the Shaping of Knowledge*, ed. Crosbie Smith and Jon Agar (Basingstoke: Macmillan, 1998), 172–77.

61. Donald Luke Opitz, "Aristocrats and Professionals: Country-House Science in Late-Victorian Britain" (PhD diss., University of Minnesota, 2004), 7–10.

62. Ibid., 33, 101, 109, 256–57; Donald L. Opitz, "'This House Is a Temple of Research': Country-House Centres for Late Victorian Science," in *Repositioning Victorian Sciences: Shifting Centres in Nineteenth-Century Scientific Thinking*, ed. David Clifford, Elisabeth Wadge, Alex Warwick, and Martin Willis (London: Anthem Press, 2006), 143–53.

63. Yanni, *Nature's Museums*, 80–84.

64. Paul White, *Thomas Huxley: Making the "Man of Science"* (Cambridge: Cambridge University Press, 2003), 34–35, 56–57, 65.

65. Rupke, *Richard Owen*, 97–100.

66. Huxley actually believed that each discipline required three museums: popular, scientific, and economic. See Leonard Huxley, *Life and Letters of Thomas H. Huxley*, 1:145.

67. Yanni, *Nature's Museums*, 112–13, 129–33, 146.

68. Crosbie Smith, *The Science of Energy: A Cultural History of Energy Physics in Victorian Britain* (Chicago: University of Chicago Press, 1998).

69. Graeme Gooday, "Precision Measurement and the Genesis of Physics Teaching Laboratories in Victorian Britain," *British Journal for the History of Science* 23 (1990): 25–51.

70. Even *Nature* could become an unfriendly space for scientific naturalists. See Ruth Barton, "Scientific Authority and Scientific Controversy in *Nature*: North Britain against the X Club," in *Culture and Science in the Nineteenth-Century Media*, ed. Louise Henson, Geoffrey Cantor, Gowan Dawson, Richard Noakes, Sally Shuttleworth, and Jonathan R. Topham (Aldershot: Ashgate, 2004), 223–35.

71. Bernard Lightman, *Victorian Popularizers of Science: Designing Nature for New Audiences* (Chicago: University of Chicago Press, 2007).

72. Bruce Hunt, "Doing Science in a Global Empire: Cable Telegraphy and Electrical Physics in Victorian Britain," in *Victorian Science in Context,* ed. Bernard Lightman (Chicago: University of Chicago Press, 1997), 312–33; Sujit Sivasundaram, *Nature and the Godly Empire: Science and Evangelical Mission in the Pacific, 1795–1850* (Cambridge: Cambridge University Press, 2005); Fa-Ti Fan, *British Naturalists in Qing China: Science, Empire, and Cultural Encounter* (Cambridge, MA: Harvard University Press, 2004).

73. Pickstone, "Working Knowledges," 494. See also Lorraine Daston, "Introduction: The Coming into Being of Scientific Objects," in *Biographies of Scientific Objects,* ed. Lorraine Daston (Chicago: University of Chicago Press, 2000), 5.

CHAPTER THREE

The Status of Museums:
Authority, Identity, and Material Culture

SAMUEL J. M. M. ALBERTI

"Taxonomic scientists are often referred to by their specialty," observed
former Natural History Museum curator Richard Fortey in his recent
paleontological memoir. "Thus 'bat man' would be an expert on bats, 'worm
man' on worms, and an anthropologist would naturally be a 'man man.' I
suppose I was known as 'trilobite man' even though it sounds like a crea-
ture from a horror film."[1] A curator's identity and authority is unusually
closely bound with his or her object(s) of study. This is because the museum
as a place for scientific knowledge is characterized by a particular relationship between practitioner, institution, and things. In this chapter I seek to
determine when, where, and how this relationship was formed, concluding
that this configuration gave museums particular credibility at a particular
moment. As several other chapters here will illustrate, science's making is
dependent upon the relationships between place and practice. Museums, no
less than other sites, were authoritative nodes in a wider system of material "knowledge in transit," as objects and the knowledge made with them
passed between collectors, curators, and audiences.[2]

Museums have their roots in the cabinets of curiosity, gardens, and
shows of Renaissance Europe. They are more plentiful than ever today, but
do not hold the same esteem compared to other scientific venues as they
have in the past. At what point, then, did the museum reach its apogee in status as a site for the production and consumption of natural knowledge? Historians of museums have shied away from specificity in this regard because
of the bewildering variety in museum sites and practices, even within specific national contexts. This chapter proposes, for Britain at least, that this
"moment" began in the middle of the nineteenth century and would last for
around eight decades. It is not my purpose to identify a nostalgic Golden Age
of museums, nor to assert that for any or all of this period the museum had

more authority than other sites for the practice of science. Rather, I explore the reasons why late nineteenth- and early twentieth-century museums had more influence than they had at any other time in their own history. Such a study involves not only comparison with the years before and after this qualitatively and quantitatively distinct period, but also contrast with other sites and spaces. For Victorian museums were closely linked to a range of other institutions, respectable or otherwise, in what Tony Bennett has help-fully dubbed the "exhibitionary complex."[3] Against "approved" venues for science—the field, laboratory, garden, lecture hall, and hospital ward—the museum needed a distinct offer. Some less savory sites, meanwhile, were too close for comfort for those curators striving for respectability: the cir-cus, menagerie, freak show, taxidermist's shop, or commercial zoo, which displayed the same objects as museums, often in similar ways. These hazy distinctions needed to be clarified.

In identifying what rendered the museum distinct from these poten-tially similar sites, this chapter is concerned with the *status* of the museum, which I take to be a combination of credibility, authority, and expertise. For as Graeme Gooday observes, the authority of scientific experts was not simply a given—it did not necessarily map onto elites or those educated in a particular way.[4] It was heavily dependent on context: the authority of the museum did not stem solely from personal expertise, but relied also on the credibility afforded to individuals in a particular venue, which depended on a distinct relationship between person, place, and thing. Credibility, then, was built on trust, character, and patronage within particular localities and spaces.[5] Curators spoke with authority because of *where* they worked. Fol-lowing David Livingstone, this chapter is therefore concerned with the in-terplay between location and locution.[6]

In looking in detail at the relationship between personal authority and site, this analysis informs and is informed by studies of the grand architec-ture of museum buildings, the professions who worked in them, and their visitors.[7] For the construction of authority was not a one-sided process, and those within museums depended on those outside to shore up, reinforce, and witness their credibility. But here I concentrate on the practitioners' role in this process by examining where and how they deployed ostensi-bly "natural objects"—the material culture of geology, botany, zoology, and anatomy.[8] The structure of the argument that follows reflects my concern with where, who, and what made museum science credible. Discussing first the role of the curator and then the importance of material culture allows us to think about the particular relationship between curator and object in the museum site.

My account of this relationship begins in the middle of the nineteenth century, a key moment for museums in Britain. The Great Exhibition of 1851 brought together the display techniques of the cabinet of curiosities, art galleries, arcades, and mechanics' institutes, and gave rise to an exhibitionary complex on a new scale: massed material culture on display to massed publics.[9] "Having achieved its institutional apotheosis in the Crystal Palace," claims the art historian Donald Preziosi, "the museum's history came to an end."[10] On the contrary, I argue here, the Great Exhibition was just the beginning.

The midcentury also marked the genesis of a significant shift in museum governance in Britain. Following the Act for Encouraging the Establishment of Museums in Large Towns in 1845, successive legislation first allowed local governments to raise taxes for free libraries and museums, then to shore up their funding and position within the "municipalization" of civic culture, culminating in the Public Libraries Act of 1919.[11] This was not an immediate seismic shift but a gradual take-up across the provinces that marked out the parameters of the period in which museums occupied an especially hegemonic place in civic culture. By 1850, Sunderland, Salford, and Ipswich had already taken advantage of the first Act. Large towns such as Sheffield followed suit in the 1870s and 1880s, and still in the 1920s cities such as Leeds were initiating the legislation. By the turn of the century, there were 150 natural history museums in Germany, twice that number in France, and 250 in Britain, a fivefold increase from midcentury.[12] The exponential increase in British public museums in the second half of the nineteenth century was for the most part due to municipal absorption of private and associational collections rather than the creation of new museums. The shift in governance served to increase the number of objects therein (as more donations arrived from a wider social constituency) and to render collections more visible and the spaces they occupied more authoritative. But public museums were not the only museums in this period—institutional teaching collections, associational museums, and even some private cabinets were also credible, authoritative venues.[13]

As well as innovations in display and governance, later nineteenth-century museums were fueled by the colonial enterprise during the "payday of empire."[14] Especially from 1880, the character and size of the British imperial project played a significant part in museum growth and foundation, and the peak of colonialism matches the height of the credibility of the museum, from the "scramble for Africa" to the fall of the Raj.[15] In studying the interrelationship between museums and empire, then, we should be careful not to concentrate exclusively on the nineteenth century; rather, we must extend our analyses to encompass the peak of empire in the interwar period. Accordingly, in closing, this chapter briefly assesses the fate of museums in the twentieth century.

PERSONAL AUTHORITY—THE CURATORS

Despite the importance of the Great Exhibition, our first point of call is not with Henry Cole in South Kensington, but with Richard Owen in Westminster.[16] Owen, at the time conservator of the Hunterian Museum at the Royal College of Surgeons (fig. 3.1), was called as a witness before the commission appointed to inquire into the constitution and government of the British Museum, which met in the late 1840s at the behest of Francis Egerton, Earl of Ellesmere, and reported in 1850. Owen defended not only the natural history collections (which he would later argue should be removed from Bloomsbury altogether), but also the authority of their custodians:

> If due care has been had in the choice of the Curator, and due opportunities given him, as the Curator of a public collection, to show before the public his capacity, and not only his knowledge but his judgement, . . . I cannot but think that a Curator being chosen upon those principles, must be himself responsible for his mode of naming or his mode of doing any other thing by which his collection is made available for the diffusion of knowledge, and that *any interference with such practical details of duty by a superior Board would be inadvisable.*[17]

As was his wont, Owen impressed the commission. Although more concerned with the question of the removal of the natural history collections, the subsequent Select Committee also detailed "the qualities necessary to constitute a good curator," namely, "Patient research, constant attention to details, care in the compilation of catalogues, taste and skill in arrangement [and] capacity of administration."[18] Even Owen's intractable foe Thomas Henry Huxley concurred, telling the Select Committee that "the proper performance of the duties of the curator requires very special faculties; in order to keep any collection of natural objects in thoroughly good order you require a great deal of special knowledge; and to do it well you must be gifted by nature with a peculiarly quick eye and ready appreciation of differences and resemblances."[19]

What was so pressing that even Owen and Huxley agreed? The Ellesmere Commission had exposed a perceived lack of status and authority on the part of the British Museum curators, and their need, or at least desire,

NEW MUSEUM OF THE ROYAL COLLEGE OF SURGEONS, LINCOLN'S-INN FIELDS.—(SEE NEXT PAGE)

Fig. 3.1. The Royal College of Surgeons Museum, from the *Illustrated London News*, 20 May 1854. Wellcome Library, London.

to act upon their own expertise in the care, arrangement, and acquisition of collections, loosening the control of the trustees.[20] Owen was required to live on the premises, as a caretaker would, and across Britain, collections were in the care of various boards, committees, honorary curators, and learned councils, while many of the curators, conservators, preparators, or keepers who worked on the collections were given little more autonomy than servants.

But in the second half of the century, the control of the collections began to shift from remote honoraries toward those who worked on site. The 1853 Select Committee on the National Gallery, for example, addressed the relationship between curators and the trustees.[21] In 1855 the gallery's trust was reduced in power and size (partly by abolishing *ex officio* trustees) relative to the new post of director. The public museums established after the 1845 Act dispensed with trustees altogether, as did the Museum of Economic Geology and the South Kensington Museum (whose directors answered directly to government). Owen advocated this model of governance, thereby seeking not only to establish the independence of the natural history collections, but also to reinforce the role of keepers more generally. His efforts to elevate the duties (and salary) of the curator from menial to skilled were

part of a wider conflict with the Museum Committee at the Royal College of Surgeons, a battle that ended in his bitter resignation. Ultimately, of
course, Owen was successful in his bid to elevate his own status, at least,
and went on to be the first director of the British Museum (Natural History)
in its own dedicated building. More generally, the Ellesmere Commission
and the National Gallery select committees marked the beginning of the
emergence of the independent, expert curator *qua* curator: not necessarily professional (many were still unpaid), but nevertheless a credible practitioner in a particular site.

Humbler curators faced similar challenges. In provincial philosophical
society museums, the salaried curator was merely an assistant to a gaggle
of honorary curators, each ostensibly responsible for a branch of the collection.[22] These five or ten honoraries were elected by ballot or council appointment not only according to their interests, but also in light of how
likely they were to donate valuable or interesting specimens. The qualifications and duties of the salaried staff varied enormously, but in the second half
of the century we can discern a gradual change in their standing. During his
forty years in post as assistant curator and librarian at the Sheffield Literary
and Philosophical Society Museum, the poet and writer John Holland established himself as the foremost authority on local culture and floriculture.[23]
According to the president of the society, rather than serving it, Holland
"might almost have been called the Society itself."[24]

It became more common for museum positions to be taken up by those
with medical training, replacing those from an artisanal background. Before
the midcentury, the widely variant skills and attitudes of paid staff had been
problematic. "If, before my acceptance of the curatorship of the Manchester
Museum I had known all I subsequently learned," wrote William Crawford
Williamson, "I should certainly have shrunk from taking the step."[25] Hired
as chief curator, general manager, and superintendent in 1835 at the tender
age of nineteen, Williamson was the son of John Williamson, curator of the
Scarborough Museum, and had come to Manchester to serve his apprenticeship at the Manchester Royal Infirmary. His complaints stemmed from a
clash with the museum's conservator, Timothy Harrop, a weaver-turned-
taxidermist. Williamson considered Harrop to be "wholly ignorant of every
branch of science except taxidermy . . . probably the worst bird-stuffer in
Europe."[26] (Harrop was so concerned with appearance over all else that he
allegedly installed only one glass eye in many of the birds, on the side that
faced the audience.) Harrop in turn was furious at Williamson's demands for
a larger salary, especially given that he had trained the young upstart's father in the art of taxidermy years earlier. Williamson eventually resigned in

disgust in 1838 and left the region, returning in 1851 as professor of natural history at Owens College, which later absorbed the Manchester Museum. By this time, his authority was no longer in dispute. His successors at the museum, who eventually dispatched Harrop, were likewise medically trained, and gradually transformed the curator's position from glorified janitor to expert naturalist.

Similar struggles were evident up and down the country. In Newcastle upon Tyne, the taxidermist-curator in the early century had irons in a number of fires, offering his services to other naturalists in addition to his museum duties in order to supplement his income. It was not unusual for such practitioners to have their own (considerable) collections on the side, and the boundaries between these and the formal collections, between museum and shop, could easily become blurred.[27] William King, the Newcastle curator from 1840, was dismissed after seven year's service for running a fossil trade.[28] Not only did he fail to complete a catalogue of the society's collections, he also kept his own stock in the museum and supposedly profited from sales of specimens from the museum. Furious at his dismissal, King refused to give up the keys to the museum. His replacement also left under a cloud, and only in 1853 did the society that ran the museum find a successful appointee in Joseph Wright, who was to serve for half a century.[29]

The museum's credibility as a space, then, arose from differentiating expert curators from dabbling trustees on the one hand and lowly craftsmen on the other. They were also to be distinguished from avaricious showmen. Even more than the natural history museum required distancing from the taxidermist's shop, formal teaching collections in hospitals and colleges needed to be aloof from the proliferating freak shows and commercial anatomy museums. As historians of medicine have demonstrated, in the mid–nineteenth century there was a marked shift in attitude in the medical press toward medical exhibitioners such as Joseph Kahn and J. W. Reimers.[30] In 1851, The Lancet considered Joseph Kahn's private anatomical museum "a splendid scientific collection," and yet a mere five years later the same show was viewed with "unqualified disgust and condemnation" as it was capable of inflicting "monstrous evils to society," whereas those collections "attached to the hospitals of this metropolis far surpass[ed] in real interest and instructiveness anything that a private speculator can hope to bring together."[31] Commercial anatomy shows proliferated in the 1850s and 1860s, to be met with condemnation from the medical profession: some were then prosecuted (using for example the Obscene Publications Act), and they had all but disappeared by the 1870s. And yet, Kahn's and other commercial shows had displayed very similar objects, often in ways similar to the methods of respectable hospital museums, and they had been hugely popular. They were censured not by the public but by medical curators in hospitals and colleges, who were shoring up their credibility, claiming authority over the dead body.

Just as complex as the boundaries between orthodox and other sites for display (and sale) were the distinctions between the museum and other acceptable spaces, such as the lecture hall and library. Taxidermists and pre-parators were gradually rendered invisible technicians working for (rather than as) curators, but the relationship between museum practitioners and lecturers was more complicated. The latter included not only populist rabble-rousers but also elite educators in the ancient universities and emerging university colleges.[32] Curators were keen to disaggregate lecturing from their duties: John Edward Gray, long-standing zoologist at the British Museum, argued "that the curator's time is fully occupied with his duties as curator; and that a lecturer is a distinct person from the curator."[33] Nevertheless, there was considerable interaction and overlap between curators and profes-sors. As a mark of their increasing authority, many curators held twin posts in museums and universities, or curated the large collections gathered by colleges.[34] The biologist Louis Miall was both curator at the Leeds Philo-sophical Society and Professor of Biology at the Yorkshire College of Sci-ence.[35] Chairs of natural history or of anatomy often came with curatorial duties, as for example in the Scottish universities.[36] The Royal College of Surgeons had prevented Richard Owen from taking up the Fullerian Chair in physiology at the Royal Institution in 1837, but he finally did so in 1858, after he had moved to the British Museum (Natural History).[37]

Many curators also acted as librarians, and as we have seen, museum and library benefited from the same acts of Parliament. But it gradually became more important to distinguish between spaces for print and of ma-terial culture. Curators were not to be outdone by the formation of the Li-braries Association in 1877, and met to form a Museums Association in 1888.[38] (The American Association of Museums, founded in 1906, lagged even further behind the American Library Association, established thirty years before). The association, and especially its journal, aimed to give curators a distinct vocational identity and to standardize techniques and practices such as labeling and preservation. Nevertheless, the foundation of the Museums Association did not yet signal the emergence of a distinct professional community of curators but an alliance of a number of differ-ent subject-specialist vocations (including many amateurs). For even those curators with responsibility for large and diverse collections tended to have specialist areas within them. As the free museum and library campaigner Thomas Greenwood argued,

> Curators do not profess to know everything about a Museum; the man who says he does should be immediately pensioned. One of the drawbacks in Museum work is that owing to so many special departments there is at times a pardonable bewilderment. If one finds, or has given to him, an odd-looking coin, such as he has never seen before, or has sent him a conical looking beetle, or a bird, and does not know the name and value of either, he forthwith darts off to the nearest curator and asks what the coin is worth, or to what class the beetle belongs, or what name is the bird. Anything from an old Roman nail to the skeleton of an elephant comes within the category, and it cannot be expected that the Curator shall be able straight off to give chapter and verse for, and a brief description on, whatever is placed before him. Furthermore, a walking encyclopædia at £120 a-year is a commodity which in these days of education ought not to be expected.[39]

As might be expected of a group campaigning to elevate its status and authority, this issue of inadequate remuneration was a constant refrain. Consider the following claim by Owen's successor at both the Hunterian and the British Museum (Natural History), William Henry Flower:

> What a museum really depends on for its success and usefulness is not its building, not its cases, not even its specimens, but its curator. He and his staff arc the life and soul of the institution, upon whom its value depends; and yet in many—I may say most of our museums—they are the last to be thought of. The care, the preservation, the naming of the specimens are either left to voluntary effort—excellent often for special collections and for a limited time, but never to be depended on as a permanent arrangement—or a grievously undersalaried and consequently uneducated official is expected to keep in order, to clean, dust, arrange, name and display in a manner which will contribute to the advancement of scientific knowledge, collections ranging in extent over almost every branch of human learning, from the contents of an ancient British barrow to the latest discovered bird of paradise from New Guinea.[40]

Greenwood and Flower alike were concerned with cementing the credibility of museums by "the elevation of the position and acquirements of those who have the care of them."[41] This endeavor was based, I argue, on the particular role of material culture in the museum, and especially on the relationship between curators and their collections.

MATERIAL AUTHORITY—THE COLLECTIONS

Stephen Conn has identified the half-century from the U.S. centenary in 1876 as the height of what he terms the "object-based epistemology" in natural history, anthropology, commerce, history, and art.[42] Museums and galleries were invaluable visual educational tools in a polyglot nation. In late Victorian Britain, too, as Amiria Henare argues, there was evident a distinct epistemology of artifacts, lost with the linguistic turn after the First World War.[43] Although the tension between object and text may not have been as stark as Henare claims, it is clear that particular stock was placed in material culture in this period, and especially material culture located in formal collections. The cachet of museums as spaces contributed to the cultural stature of objects, and vice versa. As Flower claimed, "One of the most potent means of registering facts, and making them available for future study and reference, is to be found in actual collections of tangible objects."[44]

Part of their appeal stemmed from the sheer quantity of things. Robert Kohler and Susan Pearce have both identified the decades around 1900 as the quantitative peak of collecting in the sciences and the arts.[45] Material flowed to museums in staggering volumes, and it seemed that every town could have an encyclopedia writ large in things. And more objects meant more identification. Taxonomy thrived in this era, and lumpers and splitters alike classified species to their hearts' content.[46] New species were identified in unprecedented quantities, in museums and with museum specimens. Already in 1855, that most vigorous of taxonomic crusaders, the ornithologist Hugh Strickland, had condemned the synonymic confusion generated by those who worked away from metropolitan collections.[47] Strickland advocated centralized, public research collections, to which other naturalists had to travel. This leant even more credibility to collections, especially those in national or large university museums, such as the Manchester Museum, where the curator, William Boyd Dawkins, advocated formal representative series rather than the miscellany of the local society cabinet.[48] Publicly owned collections in their grand spaces held more prestige than private cabinets, and dead specimens more credibility than those in gardens or the field.

Comparative anatomy also continued to flourish. After all, the larger the collection on which to base comparison, the more credible the conclusion, and to legitimate their practices, museum-based scholars from Richard

Owen in South Kensington to Louis Miall in Leeds explicitly played on the authority of Cuvier, who himself had begun the struggle to wrest hegemony from the field naturalist.[49] Miall, for example, spent three years dissecting an elephant in a makeshift shed behind his museum, carving it up for the collection and periodically injecting it with preservatives to prevent putrefaction.[50] Only this kind of site and this stretch of time could give such intimate, discriminating knowledge of the material culture of nature, from which painstaking labor curators claimed moral authority.[51] Few had the resources or the opportunity to take on such a mammoth task: at the other end of the scale, many naturalists could (and did) preserve plants on sheets or pin insects in the humblest of settings. But it was those who were doing so within the legitimating space of the museum, with its vast collection, who did so with most authority.

Furthermore, as the nineteenth century drew to a close, ecological studies in both zoology and botany relied not only upon intense fieldwork, but on new generations of collected and stored specimens.[52] Such detailed ecological studies were essential for stocking the habitat diorama, an innovative museum display technique in the period.[53] Other object-based endeavors, including ethnology and archaeology, also thrived in this era.[54] Like taxonomy and comparative anatomy, these were established enterprises that gained new status in the decades around 1900. Such scientific practices could only be undertaken in, and were characteristic of, the museum space, the only place in which a permanent, indelible record of nature could be seen in its full glory—the key site for the Victorian "panoramic philosophy."[55]

The clearest exemplification of the authority of the object was in the consolidation of the unique type specimen in natural history. As Paul Farber and Lorraine Daston have argued, the earlier practice of basing species descriptions on a range of specimens, living and dead, gave way from the middle of the nineteenth century to the isolation of a single, unique type—the holotype.[56] The name of a species inhered not to a living specimen in the field nor to a conglomerate of examples, but to a single, purportedly permanent individual. The classification of a species depended on a concrete thing in a particular place. And that place was the museum, which "by edict serve[d] as the last court of appeal in all questions and disputes about species definition, membership, and names."[57] In the later nineteenth century, public museums emerged as the repositories of choice for voucher specimens generally and types in particular. As one American taxonomist wrote in 1897,

All naturalists concede that type specimens constitute the most important material in a museum of natural history. The true appreciation

of this fact, however, is of recent date, and is shown in the numerous
lately published catalogues of types possessed by different museums.
The greater number of these publications have appeared in England
and America. This just valuation of type material in recent years has
come about through the work of specialists in their efforts to monograph
groups of organisms. . . . It is upon the type material that the entities of
nature history and its taxonomy rest. It is therefore of the greatest importance to learn the whereabouts of types.[58]

Type specimens were singled out in this period for particular care and identification in collections. The training and skill required to identify, isolate,
and maintain type specimens was the domain of museum-based practitioners. Debates among taxonomists leading up to the international codes on
botanical nomenclature (from 1867) and the International Commission on
Zoological Nomenclature (in 1895) enshrined type specimens as taxonomic
icons, and, I would argue, cemented the authority of collections and the
museum space.[59]

The crucial moment in the singularization of the "type" was its first
publication. As Daston wryly comments, "The phrase 'author of the species' sounds faintly blasphemous,"—and yet this was the claim made by
taxonomists, based on their relationship with the specimen.[60] Species may
not have been created in museums, but they were authored. Museum practitioners from geologists to archaeologists do not publish *on* or *about* a
specimen or a piece, they *publish the specimen*. To do so is to imbue it with
a particular permanence, which in turns lends the collection itself kudos.
With every specimen published (by a curator or anyone else), the prestige
of that collection grows, and the practice of measuring the importance of
a museum by the number of type and figured specimens in its collection
emerged in the late nineteenth century. They were widely publicized not
only in scientific periodicals but in printed catalogues that became ever
more detailed, ever more elaborate, ever larger. They were sent out far and
wide, acting as textbooks in their own right and ambassadors for the collection. Museum and print spaces were mutually constitutive, and these
mighty tomes shored up the authority of the collection.

But objects in sites other than museums were the basis for publication.
Other groups, professional and amateur, had authoritative identities in this
era. Other venues had credibility, and material culture had status elsewhere.
What was particular about the museum was the relationship between person and thing in this space. Historians and psychologists of collecting have
explored how the identity of private collectors becomes wrapped up in their

collections—what Susan Pearce has termed "material autobiography"—and yet such analyses have rarely been applied to institutional collections.[61] Just as private collections reflected the identities and interests of their collectors—and became part of "the extended self"—so the character, strengths, and biases of museums reflected not only their founders, but also subsequent curators.[62] This is especially striking, given the stark contrast one might expect between partial personal collections and ostensibly objective, rational museums.

Just as the identities of those who exchange things become "entangled" in the objects themselves, so too curators become wrapped up in their objects—and vice versa.[63] The routes taken by objects on their way to museums, and the people through whose hands they passed are well studied.[64] But objects' "careers" do not end once they arrive in the collection: most will then spend far more time in the company of the curator than the manufacturer, collector, or donor. We should not then be surprised at how attached museum practitioners were (and are) to the objects in the care. Despite the careful boundaries constructed between private collections and public museums, they referred to "my" specimens and "my" collection—which was why it was so important to distinguish between the zoological museum and the taxidermist's shop, as discussed above. To reiterate Flower's point, curators were "the life and soul of the institution," forming an intense association and proprietorship.[65] Over their long tenures, curators built up formidable textual and oral familiarity with the specimens, and their credibility relied on this memory of a collection.

Cabinets of curiosity had been "memory theatres," and so too modern museums were intended to act as memory stores for nations, although museologists are beginning to challenge the commonplace assumption that the museum acts as a collective memory.[66] As yet unacknowledged is the particular role of the curator's memory in the museum. The status of the museum lay not in its function in societal commemoration; rather, it was dependent on the intense association between keeper and things kept. Whereas museums had previously displayed most of what they had (or retained duplicates in cabinets underneath the exhibits), toward the end of the nineteenth century the massive influx of specimens necessitated storage facilities away from the galleries, in basements, attics, or even in other buildings. Advocates of the "new museum" idea agitated for distinct research collections, away from the galleries, only accessible to a small elite.[67] Most visitors were unaware they were seeing only the tip of the iceberg in the collections on display: far from acting as a storehouse of memories, museums would have been engines of forgetting, had it not been for the gatekeeper

role played by the curators. Their status, particular to museum practitioners, rested on their close association with the countless objects not only in
the vitrines but also in the darkest stores. Too often, scholarly analyses of
museums concentrate only on exhibitionary aspects, ignoring their cumula-
tive character, which stemmed from the countless stored specimens.

The relationship between those who worked in the museum and the
objects they cared for extended beyond knowing what was there (and where
to find it). In slowly adding, preserving, and adapting the collection, each
curator left an indelible mark upon objects and the museum, from the design of a new gallery to the angle of the butterfly pins. Keepers' identities
were inscribed in the collections and the buildings, and they bequeathed
their successors innumerable traces of themselves, for better or for worse
(woe betide the curator who could not read his predecessor's handwriting).
The potential tribulations of the transfer from one curator to the next were
avoided if the outgoing keeper could pass on his skills and tacit understanding of the collections directly to his successor during a period of apprenticeship—the assistant succeeding the keeper. In many cases, curation became
a generational vocation. Dynasties of keepers included the Tradescants in
Lambeth, the Grays at the British Museum, and the Geoffroy Saint-Hilaires
in Paris.[68] At the Manchester Museum, the zoologist Robert Standen, appointed in 1896, helped to secure a post as printer for his daughter Alicia
(a keen naturalist), who undertook cataloguing and collection care in addition to her formal duties. She then married the museum's paleontologist,
Wilfrid Jackson. After Standen's wife died, the widower came to live with
the young couple, and the curators lived and worked together for nearly
two decades. Jackson would eventually retire after thirty-eight years at the
museum. As their daughter remembered of her father and grandfather, the
museum was "their whole life and their whole hobby."[69] This did not always end so well, however. Only after the firm encouragement of a special
subcommittee appointed to report on the dilapidated state of the collections
in his care did their entomological colleague J. Ray Hardy finally depart
the Museum in 1918, aged seventy-four. "I feel . . . torn away," he wrote in
resignation.[70]

In happier times, curators like Hardy wielded their prolonged contact
and intense knowledge of the collections to establish their authority. They
relied on the exclusive association between themselves and their objects in
the particular space of the museum to counter the valor and effort of the
field-worker or the technology of the laboratory scientist. They became
gatekeepers: to gain access to the vast reserve collections, especially type
specimens, visiting researchers needed to go through (or rather, with) their

curators. Taxonomists needed not only firsthand experience of the object, but also the tacit knowledge of the keeper.

Other visitors also served to shore up the authority of the curators as larger and more diverse audiences visited museums in the later nineteenth century and developed trust in museums and public institutions in a particular way. The vocational (if not professional) identities of curators were based on their relationships with these audiences as well as with material culture. The significance of the museum within the geographies of nineteenth-century science rested on this distinct, object-based, and mutually constitutive construction of expertise between practitioner and visitor in particular spaces. And those spaces, the "cathedrals of science" erected in the decades around 1900, were designed to lend the exhibits (and stores) within them moral authority. The object-person relationship contributed to the production of the site, which then fed back into the credibility of the practices enacted there.

Curators' authority depended on a particular space at a particular time. Looking beyond the Victorian era, the credibility of the museum as a site for the production of science survived well past the Great War. New generic training courses planted the seeds of a coherent professional identity for curators; large museums, including the British Museum (Natural History) and the American Museum of Natural History remained important sites for research; imperial endeavors such as economic entomology relied on collections; and an influx of Carnegie funding prompted something of a Renaissance in U.K. museums in the 1920s and 1930s.[71] The "moment" I have sought to recover, the apogee of museological hegemony, was nearly a century long. It was not, however, infinite. Even allowing for considerable variation across place and discipline, it is clear that before the mid-twentieth century the status of the museum as a site for the production of knowledge had fundamentally shifted. This change was linked to the fates of museological disciplines, as natural history lost credibility (many curators began to refer to their collections as "natural science"), and field-based ethnography was more favored than collection-centered ethnology.[72] Given the chronological parameters of the present volume, this is not the place to discuss this era and the reasons for this shift in detail: suffice it to note that any explanation must take into account the emergence of "big science" and of the mass media, and that this did not mean that museums declined in popularity with their audiences (far from it), but that curators' credibility shifted in comparison with that of other scientists on the one hand, and with exhibit makers on the other.[73]

This chapter began in 1850 with the most prominent natural history curator in England, and I end it with possibly the lowliest. Bruce Frederic Cummings worked as a second-class assistant in the British Museum (Natural History) insect room from 1912. Not long afterwards, he wrote, "I am quite disenchanted of Zoology. I work—God save the mark—in the Insect Room!" Instead of becoming a "first-rate zoologist," he considered himself to be in the "very low hole of economic entomology":

> The B.M. is a ghastly hole. They will give me none of the apparatus I require. If you ask the Trustees for a thousand pounds for the propagation of the Gospel in foreign parts they will say, "Yes." If you ask for twenty pounds for a new microscope they say, "No, but we'll cut off your nose with a big pair of scissors." . . . I who had been dissecting for dear life up and down the whole Animal Kingdom in a poorly equipped attic laboratory at home, with no adequate instruments, was bitterly disappointed to find still less provision made even in a so-called Scientific Institution so grandly styled the British Museum (N.H.).[74]

Ill-health forced him to resign in 1917, giving him time, under the pseudonym W. N. P. Barbellion, to publish these semifictional musings in *The Journal of a Disappointed Man* (1919). In a review, *Nature* "regretted to see another promising entomologist pinned for life to systematic entomology" in the museum's "cramping environment."[75] The huge collections that had been the basis of their status now denied museums the physical or conceptual flexibility to adapt to the changing intellectual climate.

NOTES

I am grateful to Richard Bellon, Fay Bound Alberti, Graeme Gooday, and Helen Rees Leahy for their advice; and to David Livingstone and Charles Withers for the invitation to participate in the conference on which this volume is based. The chapter was drafted during Wellcome Trust research leave.

1. Richard Fortey, *Dry Store Room No. 1: The Secret Life of the Natural History Museum* (London: Harper, 2008), 35. This practice is slowly disappearing now, as the relationship between expertise and museums and the gender balance of staff change.

2. On knowledge in transit, see James A. Secord, "Knowledge in Transit," *Isis* 95 (2004): 654–72.

3. Tony Bennett, "The Exhibitionary Complex," *New Formations* 4 (1988): 73–102;

Tony Bennett, *The Birth of the Museum: History, Theory, Politics* (London: Routledge, 1995); Tony Bennett, *Pasts Beyond Memory: Evolution, Museums, Colonialism* (London: Routledge, 2004).

4. Graeme J. N. Gooday, "Liars, Experts and Authorities," *History of Science* 46 (2008): 431–56.

5. On moral authority and scientific credibility, see Richard Bellon, "Inspiration in the Harness of Daily Labor: Darwin, Botany and the Triumph of Evolution, 1859–1868," *Isis* 102 (forthcoming, 2011).

6. David N. Livingstone, *Putting Science in Its Place: Geographies of Scientific Knowledge* (Chicago: University of Chicago Press, 2003).

7. On museum architecture, see Sophie Forgan, "Building the Museum: Knowledge, Conflict, and the Power of Place," *Isis* 96 (2005): 572–85; Michaela Giebelhausen, ed., *The Architecture of the Museum: Symbolic Structures, Urban Contexts* (Manchester: Manchester University Press, 2003); Suzanne MacLeod, ed., *Reshaping Museum Space: Architecture, Design, Exhibitions* (London: Routledge, 2005); Carla Yanni, *Nature's Museums: Victorian Science and the Architecture of Display* (London: Athlone, 1999; repr., New York: Princeton Architectural Press, 2005). On the construction of professional communities in and around museums, see for example Samuel J. M. M. Alberti, *Nature and Culture: Objects, Disciplines and the Manchester Museum* (Manchester: Manchester University Press, 2009); J. Lynne Teather, "The Museum Keepers: The Museums Association and the Growth of Museum Professionalism," *Museum Management and Curatorship* 9 (1990): 25–41. On museum visitors, see Samuel J. M. M. Alberti, "The Museum Affect: Visiting Collections of Anatomy and Natural History," in *Science in the Marketplace: Nineteenth-Century Sites and Experiences*, ed. Aileen Fyfe and Bernard Lightman (Chicago: University of Chicago Press, 2007), 371–403; Clare Haynes, "A 'Natural' Exhibitioner: Sir Ashton Lever and His *Holosphusikon*," *British Journal for Eighteenth-Century Studies* 24 (2001): 1–14; Kenneth Hudson, *A Social History of Museums: What the Visitors Thought* (London: Macmillan, 1975).

8. Many of the same arguments can also be applied to ethnology and archaeology, both part of an extended natural history in this period. See John V. Pickstone, *Ways of Knowing: A New History of Science, Technology and Medicine* (Manchester: Manchester University Press, 2000).

9. Richard D. Altick, *The Shows of London: A Panoramic History of Exhibitions, 1600–1862* (Cambridge, MA: Belknap, 1978); Jeffrey A. Auerbach, *The Great Exhibition of 1851: A Nation on Display* (New Haven, CT: Yale University Press, 1999); Richard Bellon, "Science at the Crystal Focus of the World," in Fyfe and Lightman, *Science in the Marketplace*, 301–35; Robert Brain, *Going to the Fair: Readings in the Culture of Nineteenth-Century Exhibitions* (Cambridge: Whipple Museum of the History of Science, 1993); Robert W. Rydell, *All the World's a Fair: Visions of Empire at American International Expositions, 1876–1916* (Chicago: University of Chicago Press, 1984).

10. Donald Preziosi, "Philosophy and the Ends of the Museum," in *Museum Philosophy for the Twenty-First Century*, ed. Hugh H. Genoways (Lanham, MD: Altamira, 2006), 77.

11. House of Commons, "A Bill [as Amended by the Committee] to Enable Town Councils to Establish Museums of Art in Corporate Towns," *Sessional Papers* 223, 16

April 1845, 4:441. The Museum Act(s) and the municipalization of provincial culture are discussed in Yun Shun Susie Chung, "John Britton (1771–1857)—A Source for the Exploration of the Foundations of County Archaeological Society Museums," *Journal of the History of Collections* 15 (2003): 113–25; Elizabeth Frostick, "Museums in Education: A Neglected Role?" *Museums Journal* 85 (1985): 67–74; Thomas Greenwood, *Museums and Art Galleries* (London: Simpkin, Marshall, 1888); Simon Gunn, *The Public Culture of the Victorian Middle Class: Ritual and Authority in the English Industrial City, 1840–1914* (Manchester: Manchester University Press, 2000); Kate Hill, *Culture and Class in English Public Museums, 1850–1914* (Aldershot: Ashgate, 2005); Elijah Howarth, "On Some Recent Museum Legislation," *Report of the Proceedings of the Museums Association* 2 (1891): 121–24; Elijah Howarth, "Library and Museum Legislation," *Report of the Proceedings of the Museums Association* 3 (1892): 87–95.

12. Bennett, *Birth of the Museum;* Lewis Pyenson and Susan Sheets-Pyenson, *Servants of Nature: A History of Scientific Institutions, Enterprises, and Sensibilities* (London: HarperCollins, 1999).

13. Samuel J. M. M. Alberti, "Placing Nature: Natural History Collections and Their Owners in Nineteenth-Century Provincial England," *British Journal for the History of Science* 35 (2002): 291–311.

14. Susan M. Pearce, *On Collecting: An Investigation into Collecting in the European Tradition* (London and New York: Routledge, 1995).

15. On museums and empire, see Tim Barringer and Tom Flynn, eds., *Colonialism and the Object: Empire, Material Culture, and the Museum* (London: Routledge, 1998); Bennett, *Pasts Beyond Memory;* Annie E. Coombes, *Reinventing Africa: Museums, Material Culture and Popular Imagination in Late Victorian and Edwardian England* (New Haven, CT: Yale University Press, 1994); Claire Loughney, "Colonialism and the Development of the English Provincial Museum, 1823–1914" (PhD diss., Newcastle University, 2005); John M. MacKenzie, *Museums and Empire: Natural History, Human Cultures and Colonial Identities* (Manchester: Manchester University Press, 2009); Susan Sheets-Pyenson, *Cathedrals of Science: The Development of Colonial Natural History Museums during the Late Nineteenth Century* (Kingston, Ontario: McGill-Queen's University Press, 1988); Anthony A. Shelton, "Museum Ethnography: An Imperial Science," in *Cultural Encounters: Representing "Otherness,"* ed. Elizabeth Hallam and Brian Street (London: Routledge, 2000), 155–93.

16. Nicolaas A. Rupke, *Richard Owen: Biology without Darwin,* 2nd ed. (Chicago: University of Chicago Press, 2009); William Thomas Stearn, *The Natural History Museum at South Kensington: A History of the British Museum (Natural History), 1753–1980* (London: Heinemann, 1981).

17. Royal Commission to Inquire into the Constitution and Government of the British Museum, *Report, Minutes of Evidence, Index; Appendix,* Parliamentary Papers, House of Commons, sess. 1850, vol. 24, no. 1170, para. 2645; emphasis added.

18. Select Committee on the British Museum, *Report, Proceedings, Minutes of Evidence, Appendix, Index,* Parliamentary Papers, House of Commons, sess. 1860, vol. 16, no. 540, xiv.

19. Ibid., para. 91.

20. Arthur MacGregor, *Curiosity and Enlightenment: Collectors and Collections*

from the Sixteenth to the Nineteenth Century (New Haven, CT: Yale University Press, 2007); Stephanie Moser, *Wondrous Curiosities: Ancient Egypt at the British Museum* (Chicago: University of Chicago Press, 2006).

21. Nicholas M. Pearson, *The State and the Visual Arts: A Discussion of State Intervention in the Visual Arts in Britain, 1760–1981* (Milton Keynes: Open University Press, 1982); Nick Prior, *Museums and Modernity: Art Galleries and the Making of Modern Culture* (Oxford: Berg, 2002); Christopher Whitehead, *The Public Art Museum in Nineteenth-Century Britain: The Development of the National Gallery* (Aldershot: Ashgate, 2005).

22. Alberti, "Placing Nature"; Simon J. Knell, *The Culture of English Geology, 1815–1851: A Science Revealed Through Its Collecting* (Aldershot: Ashgate, 2000).

23. Holland (1794–1872) curated the collection from 1833 until his death and edited the *Sheffield Iris*. See John Holland, *The Tour of the Don* (London: Groombridge, 1837); William Smith Porter, *Sheffield Literary and Philosophical Society: A Centenary Retrospect* (Sheffield: Northend, 1922).

24. Henry Clifton Sorby cited in William Hudson, *The Life of John Holland of Sheffield Park* (London: Longmans Green, 1874), 255.

25. William Crawford Williamson, *Reminiscences of a Yorkshire Naturalist* (London: Redway, 1896), 59.

26. Williamson, *Reminiscences*, 60.

27. *A Catalogue of the Valuable Collection of Natural History Belonging to the Late Mr Richard Rutledge Wingate* (Newcastle: Blackwell, 1859).

28. Natural History Society of Northumberland, Durham, and Newcastle upon Tyne, *Report* (Newcastle upon Tyne: private circulation, 1847).

29. E. Leonard Gill, *The Hancock Museum and Its History* (Newcastle: Natural History Society, 1908); T. Russell Goddard, *History of the Natural History Society of Northumberland, Durham and Newcastle Upon Tyne, 1829–1929* (Newcastle upon Tyne: Reid, 1929).

30 Alberti, "The Museum Affect"; Alan W. Bates, "'Indecent and Demoralising Representations': Public Anatomy Museums in Mid Victorian England," *Medical History* 52 (2008): 1–22; Maritha Rene Burmeister, "Popular Anatomical Museums in Nineteenth-Century England" (PhD diss., Rutgers University, 2000); Helen MacDonald, *Human Remains: Dissection and Its Histories* (London: Yale University Press, 2006); Michael Sappol, *A Traffic of Dead Bodies: Anatomy and Embodied Social Identity in Nineteenth-Century America* (Princeton, NJ: Princeton University Press, 2002).

31. "Dr. Kahn's Anatomical Museum," *Lancet*, 26 April 1851, 474; "An Obscene Exhibition," *Lancet*, 5 April 1856, 376–77.

32. Martin J. Daunton, ed., *The Organisation of Knowledge in Victorian Britain* (Oxford: Oxford University Press, 2005); Bernard V. Lightman, *Victorian Popularisers of Science* (Chicago: University of Chicago Press, 2007).

33. Select Committee on the British Museum, *Report*, para. 792.

34. Samuel J. M. M. Alberti, "Civic Cultures and Civic Colleges in Victorian England," in Daunton, *Organisation of Knowledge*, , 337–56. On professors and curators in an earlier period, see Richard W. Burkhardt, "The Leopard in the Garden: Life in Close Quarters at the Muséum D'histoire Naturelle," *Isis* 98 (2007): 675–94.

35. Samuel J. M. M. Alberti, "Amateurs and Professionals in One County: Biology and Natural History in Late Victorian Yorkshire," *Journal of the History of Biology* 34 (2001), 115–47; R. A. Baker and J. M. Edmonds, "Louis Compton Miall (1842–1921)—The Origins and Development of Biology at the University of Leeds," *The Linnean* 14 (1998): 40–48.

36. Elizabeth Hallam, *Anatomy Museum: Death and the Body Displayed* (London: Reaktion, 2011); Lawrence Keppie, *William Hunter and the Hunterian Museum in Glasgow, 1807–2007* (Edinburgh: Edinburgh University Press, 2007); Geoffrey N. Swinney, "Who Runs the Museum? Curatorial Conflict in a National Collection," *Museum Management and Curatorship* 17 (1998): 295–301.

37. Rupke, *Richard Owen.*

38. Geoffrey Lewis, *For Instruction and Recreation: A Centenary History of the Museums Association* (London: Quiller, 1989); Teather, "Museum Keepers."

39. Greenwood, *Museums and Art Galleries*, 362–63.

40. William Henry Flower, *Essays on Museums and Other Subjects Connected with Natural History* (London: Macmillan, 1898), 12.

41. Flower, *Essays on Museums*, 35.

42. Steven Conn, *Museums and American Intellectual Life, 1876–1926* (Chicago: University of Chicago Press, 1998).

43. Amiria Henare, *Museums, Anthropology and Imperial Exchange* (Cambridge: Cambridge University Press, 2005).

44. Flower, *Essays on Museums*, 253.

45. Robert E. Kohler, *All Creatures: Naturalists, Collectors, and Biodiversity, 1850–1950* (Princeton, NJ: Princeton University Press, 2006); Pearce, *On Collecting.*

46. Splitters liked to find new species in every variation; lumpers argued for fewer, more broadly defined species. On taxonomy, see, for example, Paul Lawrence Farber, *Finding Order in Nature: Naturalist Tradition from Linnaeus to E. O. Wilson* (Baltimore, MD: Johns Hopkins University Press, 2000); Harriet Ritvo, *The Platypus and the Mermaid and Other Figments of the Classifying Imagination* (Cambridge, MA: Harvard University Press, 1997).

47. Lorraine Daston, "Type Specimens and Scientific Memory," *Critical Inquiry* 31 (2004): 153–82; Gordon R. McOuat, "Species, Rules and Meaning: The Politics of Language and the Ends of Definitions in Nineteenth-Century Natural History," *Studies in History and Philosophy of Science* 27 (1996): 473–519; Hugh Edwin Strickland, *Ornithological Synonyms*, ed. Mrs. Hugh Edwin Strickland and William Jardine (London: Van Voorst, 1855).

48. Alberti, *Nature and Culture.*

49. Dorinda Outram, "New Spaces in Natural History," in *Cultures of Natural History*, ed. Nicholas Jardine, James A. Secord, and Emma C. Spary (Cambridge: Cambridge University Press, 1996), 249–65.

50. Louis Compton Miall and F. Greenwood, "Anatomy of the Indian Elephant," *Journal of Anatomy and Physiology* 12 (1878): 261–87.

51. Bellon, "Inspiration in the Harness of Daily Labor"; cf. Fiona Candlin, *Art, Museums and Touch* (Manchester: Manchester University Press, 2010).

52. Peter Bowler, *The Fontana History of the Environmental Sciences* (London:

Fontana, 1992); John Sheail, *Seventy-Five Years in Ecology: The British Ecological Society* (Oxford: Blackwell, 1987); Donald Worster, *Nature's Economy: The Roots of Ecology* (San Francisco: Sierra Club, 1977).

53. Lynn K. Nyhart, "Science, Art, and Authenticity in Natural History Displays," in *Models: The Third Dimension of Science*, ed. Soraya De Chadarevian and Nick Hopwood (Stanford, CA: Stanford University Press, 2004), 307–35; Karen Wonders, *Habitat Dioramas: Illusions of Wilderness in Museums of Natural History* (Uppsala: Almqvist and Wiksell, 1993).

54. Bennett, *Pasts Beyond Memory*; Coombes, *Reinventing Africa*; Chris Gosden, *Anthropology and Archaeology: A Changing Relationship* (London: Routledge, 1999); Henare, *Museums, Anthropology and Imperial Exchange*; Fred R. Myers, ed., *The Empire of Things: Regimes of Value and Material Culture* (Oxford: Currey, 2001); George W. Stocking, ed., *Objects and Others: Essays on Museums and Material Culture* (Madison: University of Wisconsin Press, 1985); A. Bowdoin van Riper, *Men among the Mammoths: Victorian Science and the Discovery of Human Prehistory* (Chicago: University of Chicago Press, 1993).

55. Bellon, "Science at the Crystal Focus of the World."

56. Daston, "Type Specimens." See also Kristin Johnson, "Type-Specimens of Birds as Sources for the History of Ornithology," *Journal of the History of Collections* 17 (2005): 173–88; Phil Rainbow and Roger J. Lincoln, *Specimens: The Spirit of Zoology* (London: Natural History Museum, 2003).

57. Daston, "Type Specimens," 158.

58. Charles Schuchert, "What Is a Type in Natural History?" *Science* 5 (1897): 636.

59. Richard V. Melville, *Towards Stability in the Names of Animals: A History of the International Commission on Zoological Nomenclature, 1895–1995* (London: International Trust for Zoological Nomenclature, 1995).

60. Daston, "Type Specimens," 162. On collections and publications, see Knell, *Culture of English Geology*.

61. Pearce, *On Collecting*, 272. See also Russell W. Belk, *Collecting in a Consumer Society* (London: Routledge, 1995); John Elsner and Roger Cardinal, eds., *The Cultures of Collecting* (London: Reaktion, 1994); Werner Muensterberger, *Collecting: An Unruly Passion. Psychological Perspectives* (Princeton, NJ: Princeton University Press, 1994); Susan M. Pearce, *Collecting in Contemporary Practice* (London: Sage, 1997).

62. Russell W. Belk, "Possessions and the Extended Self," *Journal of Consumer Research* 15 (1988): 139.

63. Nicholas Thomas, *Entangled Objects: Exchange, Material Culture, and Colonialism in the Pacific* (Cambridge, MA.: Harvard University Press, 1991); Annette B. Weiner, *Inalienable Possessions: The Paradox of Keeping-While-Giving* (Berkeley and Los Angeles: University of California Press, 1992).

64. See, for example, Chris Gosden and Frances Larson with Alison Petch, *Knowing Things: Exploring the Collections at the Pitt Rivers Museum, 1884–1945* (Oxford: Oxford University Press, 2007); Henare, *Museums, Anthropology and Imperial Exchange*; Knell, *Culture of English Geology*; Pearce, *On Collecting*; Hugh Torrens, "Mary Anning (1799–1847) of Lyme: 'The Greatest Fossilist the World Ever Knew,'" *British Journal for the History of Science* 28 (1995): 257–84.

65. Flower, *Essays on Museums*, 12.

66. Paul Connerton, *How Societies Remember* (Cambridge: Cambridge University Press, 1989); Susan A. Crane, ed., *Museums and Memory* (Stanford, CA: Stanford University Press, 2000); David Gross, *Lost Time: On Remembering and Forgetting in Late Modern Culture* (Amherst: University of Massachusetts Press, 2000); Maurice Halbwachs, *On Collective Memory*, trans. Lewis A. Coser (Chicago: University of Chicago Press, 1992); Gaynor Kavanagh, *Dream Spaces: Memory and the Museum* (Leicester: Leicester University Press, 2000); Nick Merriman, "Museum Collections and Sustainability," *Cultural Trends* 17 (2008): 3–21; Raphael Samuel, *Theatres of Memory: Past and Present in Contemporary Culture* (London: Verso, 1994); John Urry, "How Societies Remember," in *Theorizing Museums: Representing Identity and Diversity in a Changing World*, ed. Sharon Macdonald and Gordon Fyfe (Oxford: Blackwell, 1996), 45–68. My thanks to Nick Merriman for sharing his ideas on museums and memory.

67. See, for example, Thomas Henry Huxley, "Suggestions for a Proposed Natural History Museum in Manchester," *Report of the Proceedings of the Museums Association* 7 (1896): 126–31; Flower, *Essays on Museums*.

68. MacGregor, *Curiosity and Enlightenment*; Albert Everard Gunther, *A Century of Zoology at the British Museum through the Lives of Two Keepers, 1815–1914* (London: Dawsons, 1975); Emma C. Spary, *Utopia's Garden: French Natural History from Old Regime to Revolution* (Chicago: University of Chicago Press, 2000).

69. Alicia Jackson, interview by the author, minidisc recording, 19 December 2006, Buxton, Derbyshire, *Re-collecting at the Manchester Museum Oral History Project*, disc V5, track 7, Manchester Museum Central Archive, University of Manchester.

70. Manchester Museum Committee Minutes, vol. 3 (25 March 1918), 365, Manchester Museum Central Archive, University of Manchester.

71. J. F. M. Clark, *Bugs and the Victorians* (New Haven, CT: Yale University Press, 2009); Lewis, *For Instruction and Recreation*; E. Ernest Lowe, *A Report on American Museum Work* (Edinburgh: Carnegie United Kingdom Trust, 1928); S. Frank Markham, *A Report on the Museums and Art Galleries of the British Isles (Other Than the National Museums)* (Edinburgh: Constable, 1938); Henry Miers, *A Report on the Public Museums of the British Isles (Other Than the National Museums)* (Edinburgh: Constable, 1928); Teather, "The Museum Keepers"; Alma Stephanie Wittlin, *Museums: In Search of a Usable Future* (Cambridge, MA: MIT Press, 1970).

72. Franz Boas, "Some Principles of Museum Administration," *Science* 25 (1907): 931; Conn, *Museums*; George W. Stocking, *The Ethnographer's Magic and Other Essays in the History of Anthropology* (Madison: University of Wisconsin Press, 1992).

73. Karen A. Rader and Victoria E. M. Cain, "From Natural History to Science: Display and the Transformation of American Museums of Science and Nature," *Museum and Society* 6 (2008): 152–71.

74. W. N. P. Barbellion [Bruce F. Cummings], *The Journal of a Disappointed Man* (London: Chatto and Windus, 1919), 61, 106, 132, 295. See also Clark, *Bugs and the Victorians*; Fortey, *Dry Store Room No. 1*.

75. "Review of *The Journal of a Disappointed Man*, by W. N. P. Barbellion," *Nature*, 10 July 1919, 363.

CHAPTER FOUR

Cultivating Genetics in the Country: Whittingehame Lodge, Cambridge

DONALD L. OPITZ

A persistent narrative in the history of the life sciences chronicles a progressive shift from older forms of field work to newer, laboratory-based disciplines at the close of the nineteenth century. Late Victorian naturalists seeking professional status promoted this shift, their attempts characterized by James Secord as a colonization of the field "by the laboratory approach."[1] Also, according to the dominant narrative, new institutional arrangements in academia reified a methodological progression from study of organisms' "life histories" within their habitats to microscopic study and experiments intended to penetrate underlying causes.[2] Spatially, the study of nature thus moved from itinerant observations in country or seaside places into the synthesizing, objective "truth-spots" of built academic institutions distanced from the natural environments of observation.[3] Despite the stubbornness of this progressive narrative, historians highly suspicious of it have urged a remapping of the conceptual and spatial boundaries between natural history and newly emerging biological disciplines like morphology, physiology, and bacteriology.[4]

To contribute to this remapping, this chapter considers a case in the early history of genetics, that of the foundation of the research station, Whittingehame Lodge, at the University of Cambridge. Unlike other departments' biological research laboratories, which were centrally clustered in the New Museums Site near the town center, the land allotted to genetics was located in fields outside of town, next to the University Farms and Observatory. The space for experiments was a two-acre plot consisting of gardens, greenhouses, and a laboratory in the professor's "lodge"—a marrying of field and laboratory that situated genetics in close alignment with the new School of Agriculture as opposed to the other biological departments.[5]

This chapter explains how genetics came to be cultivated in this distinctive spatial arrangement, which co-opted both field and laboratory within a purposefully built research site. Despite the Cambridge geneticists' departure from evolutionary morphology in favor of Mendelian studies, their development of a research program, led by William Bateson, claimed a continuity with gentlemanly natural history of the type practiced by Britain's champions of evolution and morphology—respectively, Charles Darwin and Francis Balfour. This claim bolstered the authority of the Mendelian approach in a socioeconomic milieu shaped by a widespread concern over the British agricultural depression and a perceived imperative to improve agricultural stocks by applying scientific knowledge.[6] Shepherded by a collaboration of Cambridge's aristocratic machine with private industry, the realization of Bateson's model and its promise to yield useful knowledge blurred the field/lab distinction sharpened by advocates of a laboratory approach. The case of Whittingehame Lodge thus presents us with another early twentieth-century research site "standing between the standardized space of the lab and the messiness of the field."[7]

FROM FIELD TO LAB:
THE MORPHOLOGY OF NATURAL HISTORY

Establishing genetics at the University of Cambridge followed an intensive period of laboratory-building and curriculum reform that bolstered the status of scientific teaching and research at Cambridge. Aristocratic leadership in lobbies, philanthropy, and administrative work underpinned many of the key developments. The establishment of the Cavendish Laboratory for experimental physics is a case in point, but the pattern also characterized the foundation and development of new academic departments and laboratory spaces in the biological sciences.[8] The Balfours and their relations were one of those "notable Cambridge dynasties" that turned their aristocratic influence to the many-faceted reform of the university's scientific culture in the 1870s and 1880s.[9] The clan of five brothers and three sisters were nieces, nephews, and cousins of the Hotel Cecil, from which Queen Victoria derived her long-term Conservative Prime Minister, Robert Gascoyne-Cecil, third Marquis of Salisbury. As such, the Balfours were members of a social elite known for its conservatism, evangelical Anglicanism, and intellectualism.[10] Their alliances, through marriage, to the Trinity College fellows, Henry Sidgwick and John Strutt (afterward Baron Rayleigh), further expanded their "intellectual aristocracy."[11] During this period, Charles Darwin's sons also came to Cambridge, and George Darwin befriended the Balfour brothers and

John Strutt on the tennis courts. Both Arthur Balfour and Strutt recalled memorable visits to Down House, the country site of Darwin's scientific work, but Francis Maitland Balfour was the more frequent guest. Charles Darwin wrote in 1882, "[Balfour] has a fair fortune of his own, so that he can give up his whole time to Biology. He is very modest, and very pleasant, and often visits here and we like him very much."[12]

Home-schooled in science, Francis Balfour literally transported his amateur practice of geology and natural history from his family's country house to his college rooms. The Balfour's country seat, Whittingehame, consisted of ten thousand acres of wood, glen, river, pasture, and prominent laccolith in East Lothian, Scotland. Adorning the estate was the neo-Grecian palace designed by Robert Smirke (also the British Museum's architect), more than five hundred acres of designed parks and arboretums replete with colonial flora, and kitchen gardens and glasshouses sporting experiments in fruit and vegetable cultivation.[13] When the brothers, Arthur, Francis, Gerald, Eustace, and Cecil, dispersed to Cambridge, London, and overseas, Whittingehame House served as a holiday retreat for knowledge production and, as such, joined the "society of . . . country houses" within which the Balfours, their relations, and their friends traveled, rested, socialized, and worked. Equipped with active laboratories, both Salisbury's Hatfield House in Hertfordshire and Rayleigh's Terling Place in Essex were among the Balfours' most frequented sites.[14]

At Whittingehame, bird-shooting, dredging, fossil-excavating, and bug-hunting provided rich natural history materials for a massive, collaborative cataloguing effort led by Francis Balfour, whose pursuit of geology and natural history staged his career in evolutionary morphology.[15] His natural history practice was enmeshed with gentlemanly traditions of sport and hunting. As Hayden Lorimer has suggested, nineteenth-century natural history sanctioned the "study of deer through a rifle's telescopic lens," which in turn contributed to "taxonomies detailing the statistical ephemera of the deer forest."[16] Consistently, in the 1870s the Balfours stalked deer at their Highland estate, Strathconan, and their annual kills, recorded by the head stalker, Mr. Campbell, disproved the myth that more female deer were typically calved than males.[17] The noted "keen sportsman" in Francis Balfour became part and parcel of his naturalist identity. Even when physically removed from the field, Balfour's laboratory-based dissections and microscopy were founded on field study: specimens obtained outdoors were prepared and preserved for analysis indoors. After giving up deerstalking in response to the outcry of anti-vivisectionists, he turned to mountaineering, an unfortunate choice that led to his famous, fatal Alpine accident in 1882.[18]

At Cambridge, Balfour's practice correlated well with that of an older generation of zoologists, particularly Alfred Newton. Newton, a "first-rate ornithologist," "naturalist," and "gentleman of the old school," defined morphology as a subdivision of zoology, which comprehensively attended to matters of "outward form, internal organisation, mode of development and habits of life."[19] Introduced to the formidable professor in zoology and comparative anatomy by his brother, Arthur Balfour, Francis taught his first "scanty band" of ten students in practical morphology in Newton's private room in 1875.[20] This number increased sevenfold by 1882. Propelled by his enthusiasm for the comprehensive study of nature, Balfour's interests participated in a broader trend toward the specialized laboratory study championed by Thomas Huxley. Michael Foster acted as the conduit between Huxley's South Kensington laboratories (where Foster was formerly a demonstrator) and the nascent experimental spaces in various Trinity College rooms (where he lectured in physiology). Foster steered Balfour's interests in the direction of embryology, beginning with the question of the primitive streak in chicks' eggs. According to Balfour's classmate, Walter Holbrook Gaskell, "It was a memorable day in the history of biology when Foster, talking in the little room of the philosophical library about his future career with Balfour, who wanted to devote himself to science, but was uncertain what line of research to follow, took up an egg, cracked it, showed him the embryo inside, and said 'What do you think of working at that?' "[21] Under Foster's direction, Balfour investigated the embryological development of a wide range of species, which, in turn, contributed to Foster's systematization of the teaching of embryology; the teacher-pupil team (aided by Alice Balfour's sketching skills) produced the standard text with which "an ideal course of embryology will begin," as recommended by Edward Bagnall Poulton.[22]

Like physiology, animal morphology was, to quote Helen Blackman, "a rather messy science" that met resistance as it sought institutional resources at Cambridge.[23] Its apparent departure from traditional field study drew upon two major influences: the marriage of Darwinian and von Baerian theory, and the introduction of Continental laboratory techniques.[24] Balfour described his own method most eloquently in his celebrated monograph on the Elasmobranch fishes ("one of the most primitive groups among Vertebrates"); through morphology, he intended to add "a few stones to the edifice, the foundations of which were laid by Mr Darwin in his work on the *Origin of Species*."[25] He explained the role of the "principle of natural selection" in guiding the study of embryos, but noted that the correspondence must necessarily be imperfect. Natural selection provided a "statement of what would occur without interfering conditions," but that "development as it actually occurs is the resultant of a series of influences of which that of heredity is only one." Thus, "the embryological record . . . is both imperfect and misleading":

> Like the scholar with his manuscript, the embryologist has a process of careful and critical examination to determine where the gaps are present, to detect the later insertions, and to place in order what has been misplaced.
>
> The scientific method employed . . . is that of comparison. . . . By this method it becomes possible with greater or less certainty to distinguish the secondary from the primary or ancestral embryonic characters, to determine the relative value to be attached to the results of isolated observations, and generally to construct a science out of the rough mass of collected facts.[26]

This appeal to fact-collecting underlined a methodology fundamental to the field-based natural history upon which Balfour built his naturalist career; it served as an ideal in the approach advocated by his successor, William Bateson.[27]

MORPHOLOGICAL LABORATORY VERSUS AGRICULTURAL STATION

In becoming "one of the greatest of the early proponents of Mendelian genetics," William Bateson, though agnostic toward natural selection, considered Darwin's model of scientific research as a foundation for his own: "To learn the laws of Heredity and Variation there is no other way than that which Darwin himself followed, the direct examination of the phenomena. . . . This can only be provided by actual experiments in breeding."[28] Two decades earlier, when concluding his *On the Origin of Species*, Darwin predicted that "the study of domestic productions will rise immensely in value."[29] It was, after all, through his observations of bred animals and hybridized plants that Darwin analogously argued for natural selection: "It seemed to me probable that a careful study of domesticated animals and of cultivated plants would offer the best chance of making out this obscure problem."[30] As a contemporary noted, Darwin's method was practiced in "the barn yard and the kitchen-garden," not within a designated laboratory space.[31] It was a method that earned him trust in some British scientific circles and, at the same time, distrust among foreign rivals like Julius Sachs, who privileged laboratory-based experiments.[32]

In Bateson's adherence to submitting "theory and hypothesis to the touchstone of fact," he may well have aspired to Darwin's method, yet he also had more immediate sources of influence.[33] Despite his apparent collusion in what Garland Allen has termed the "revolt against morphology," Bateson owed much to the Cambridge School of Morphology for absorbing its naturalist spirit.[34] It was at Cambridge, not the gardens of Down House, where Bateson, as undergraduate student and fellow, came into contact with gentlemen naturalists who, "born with an innate love of living things, quietly gather facts and work out truths."[35] And it was at Cambridge where Bateson studied under Britain's scientific heir-apparent to Darwin, Francis Balfour.

Upon Bateson's matriculation at St. John's College in 1879, he followed a course of study in zoology in preparation for the Natural Sciences Tripos, achieving firsts in Parts 1 (in 1882) and 2 (in 1883). He was a progeny of the Cambridge school of morphology headed by Balfour and, after Balfour's untimely death in 1882, Adam Sedgwick; as Robert Olby described it, "Bateson imbibed the heady wine of comparative embryology."[36] The reputations of Balfour and the Cambridge Morphological Laboratory attracted him to the discipline; as he wrote, students could "see methods applied and investigations carried on which they could hardly see anywhere else in the world."[37] He demonstrated his skill in and initial allegiance to the evolutionary morphology program in his careful study of the *Balanoglossus* worm. At the same time, he lost patience with the ambiguity of morphological interpretation and its fundamental inability to explain variation, a problem in which he became increasingly interested during his postgraduate field studies on the Chesapeake Bay and the Russian Steppes. Ultimately he rejected further attempts to observe the recapitulation of phylogeny through careful study of ontogeny. Despite Bateson's marked departure, morphology nevertheless retained its enduring grip on him in two critical ways. First, it impressed upon him a desire to find "a solution to heredity and variation which no less explained embryonic development."[38] Second, it predisposed him to the "investigation of the structure and properties of the concrete and visible world"—in other words, organs, tissues, and cells, not the cell nucleus and its components.[39] Owing to morphology's strong hold on Bateson, William Coleman argued for the importance of "conservative thought" in Bateson's distrust of chromosomal theory, noting that "he was before all else a morphologist" and that "in morphological conceptions he saw a solution to the problem—the nature of the hereditary mechanism."[40]

As Brian Hall has emphasized, it was precisely the hypothetical elements of evolutionary embryology that agitated Bateson and, at the same time, its emphasis on facts that appealed to him. In an oft-quoted passage, Bateson eloquently explained his growing impatience with the speculative nature of evolutionary embryology in his 1894 monograph on discontinuous variation: "The Embryological Method then has failed not for want of knowledge of the visible facts of development but through ignorance of the principles of Evolution."[41] He rejected starting from the central premise of Darwin—natural selection operating on small variations—and he rejected following Balfour's strategy of deducing hypothetical intermediary forms to be confirmed through observations of recapitulation. Bateson advocated, instead, for establishing patterns of variation and inheritance through careful compilation of "Specific Forms" of individual plants and animals which, he argued, "on the whole form a Discontinuous Series."[42] In essence he advocated returning to the task of collecting raw evidence of variation, "an empirical means of getting at the outward and visible phenomena which constitute Evolution."[43]

In this, although he suspended Darwin's view of natural selection, he nevertheless located in Darwin, "above all men a field-naturalist," the prototype for this new enterprise.[44] But Bateson warned that amateur naturalists' "indiscriminate accumulation of facts" would be insufficient: "For comparison we require the parent and the varying offspring together. To find out the nature of the progression we require, simultaneously, at least two consecutive terms of progression."[45] He thus promoted experiments in breeding to achieve this end: "The only way in which we may hope to get at the truth is by the organisation of systematic experiments in breeding, a class of research that calls perhaps for more patience and more resources than any other form of biological inquiry."[46] With this, he defined the future of his scientific endeavor, one that required expansive space in the country as opposed to the cramped laboratories in town. His concept of experimental breeding research drew upon the ideals of traditional natural history, in stark contrast to an approach he perceived to be narrow, misleading, and irrelevant.

THE DISTRIBUTED GEOGRAPHY OF BATESONIAN RESEARCH

Bateson had begun establishing the patterns for such a research enterprise within a distributed geography of sites that included the kitchen garden of St. John's College, a rented allotment at the Cambridge Botanic Garden, and breeding cages in the Department of Animal Morphology. After his marriage in 1896, his relocation to a small house near the Botanic Garden offered additional space.[47] Gradually he recruited other naturalists and amateur breeders to his cause. Printed circulars and memoranda of the Evolution Committee of the Royal Society (ECRS) requested readers to note observations and contribute records in support of Bateson's research agenda.[48] This expanded the geography for his science beyond the confines of Cambridge and linked together a network of domestic sites that paralleled Balfour's society of country houses. He derived his most significant assistance, however, from the women students of Newnham College who attended his lectures in studies of heredity and variation, and he obtained further recruits from his circle of family, relations, and friends. Among his most important colleagues were Edith Saunders and his sister Anna Bateson, respectively demonstrator and assistant in botany at Cambridge's Balfour Biological Laboratory for women students, named in Francis Balfour's honor.[49] His wife, Caroline Beatrice (née Durham), previously a student at Bedford College in London, became a constant partner in his research at home. As Marsha Richmond has noted, a "domestication of heredity" characterized Bateson's enterprise, consisting of a full private network of Cambridge "disciples," not unlike the informal collaborations characteristic of private research at country-houses.[50]

The inconvenience of scattered sites and a family outgrowing Bateson's small house compelled Bateson to centralize his efforts at a larger house on five acres of land outside Cambridge, in nearby Grantchester. As he reported in early 1900, "At Mich[aelmas] 1899 in order to work under better conditions I took a house with some grounds in the country and have moved all the zoological subjects thither, leaving the garden ground in Cambridge free for Miss Saunders' botanical experiments."[51] Here, at Merton House, he came closer to enacting his vision for a centralized breeding station and replicating, on a small scale, the type of country-house science that was typical of Balfour's massive Whittingehame estate and Darwin's modest twenty-eight acres at Down. Bateson's little enterprise, however, proved much more intensive than these precedents. Reginald Crundall Punnett joined the research staff in 1903 and, in the next year, won a Balfour Studentship, supported by the fund intended to further animal morphology along the lines Balfour established. Punnett later recalled, "My initiation into these studies was a strenuous one."[52]

In addition to the differences between Bateson's and Balfour's approaches to evolutionary problems, a fundamental social rift distinguished their capacities to marshal finances and garner institutional support. As a scion of the English aristocracy, Balfour's "circumstances delivered him from the hard necessity of working for his daily bread."[53] He turned his fortunes to building Cambridge's capacity for supporting natural science students and laboratory research. By contrast, Bateson, whose parents belonged to the merchant and professional classes, struggled to piece together funds that could subsidize his ever-growing—and controversial—experimental agenda. A Fellow of St. John's College (since 1885), in 1887—upon his second application—he won the Balfour Studentship based on the strength of his morphological study of the Chordates.[54] But Bateson turned his use of the legacy Balfour Fund away from its established purpose. He correctly surmised that "my work is of a kind with which the Cambridge people have little sympathy," and in July 1890, he reported to his sister Anna, "Newton tells me that they don't want to renew the Balfour."[55] Adam Sedgwick, who headed Cambridge's Department of Animal Morphology, found Bateson's critique of morphology "stupid & narrow" (though perhaps not "unprofitable") and made it clear that he would not consider him for a lectureship in advanced subjects of animal morphology should it become vacant.[56] Despite these challenges, Bateson remained connected to Cambridge zoology via Newton who, finding morphology "a subject of which he . . . was quite without knowledge," supported Bateson's alternative program of study.[57] For Bateson this was fortunate because, amid the developing friction with Sedgwick, Newton proved to be "very pleasant" and appointed Bateson as his deputy in 1899.[58] In addition to these sources, Bateson had since 1892 steadily earned a salary as St. John's College Steward. Supplemented by small Royal Society grants and private donations, this patchwork of funding only partially met his financial needs, particularly as his research agenda and family size continued to expand.

The distributed, decentralized geography of Bateson's research sites and his limited resources compelled him to seek support for the establishment of a large-scale research station equipped with land and buildings for breeding experiments. Country parks such as the Royal Horticultural Society's Wisley Garden, the Duke of Bedford's Woburn Abbey, and the Darwin estate at Down in Kent posed tantalizing possibilities, but none came to fruition.[59] A promising opportunity arose in the negotiations conducted between 1903 and 1905 over the late Frederick James Quick's £50,000 bequest for "study and research in animal and vegetable biology" through creation of a professorship at Cambridge.[60] When Bateson made his hopeful plea that the benefaction would be applied to the "Study of Heredity," he famously suggested to Adam Sedgwick that "Genetics" might be a preferred term for the professorship's title.[61] But his petition, supported by many Cambridge friends, including Sedgwick, now reconciled to the cause, proved unsuccessful. The chair and its research fund went instead for the study of protozoology, and George Henry Falkiner Nuttall was elected in 1906.[62]

Following this failure, the Cambridge University Association drew attention to Bateson's needs in its "Plea for Cambridge" circulated widely in the April 1906 issue of *The Quarterly Review*, emphasizing the scale of resources required:

> A branch of experimental science dealing with the study of variation and heredity in plants and animals has recently arisen, and has already attained very considerable proportions in Cambridge. It seems indeed that we are entering on a period when such studies will absorb the energies of most of the younger biological students. Under Mr Bateson some twelve researchers are already at work following out Mendel's law in many varieties of plant and animal. The extreme importance of these studies, which, if they prove a key to heredity, will place in man's hands an instrument as powerful as Watt's application of steam, is shown by the fact that Mr Biffen has already discovered that susceptibility to rust in wheat is Mendelian, and is thus a property which may be eliminated by breeding. For all these studies land is required, as well as a greenhouse, outbuildings, and a trained gardener. None of these is as yet attainable.[63]

The reference to Rowland Henry Biffen's discovery would be cited repeatedly in further public appeals, as the Cambridge agricultural botanist provided the most tangible evidence of the practical benefits that a Mendelian understanding of heredity promised to yield.[64]

The approaching centenary of Darwin's birth provided Bateson and his supporters with a fortuitous context for this campaign, which they argued promoted "subjects which were the chief-concern of Darwin's life-work."[65] By February 1908, Cambridge's Council of the Senate reported that they were entertaining "a generous offer made to the University by a Member of the University who wishes to remain anonymous."[66] The donor—believed by historians to be Arthur Balfour—provided start-up funds for a five-year professorship, in which the occupant would "teach and make researches in that branch of Biology now entitled Genetics (Heredity and Variation)."[67] Bateson was elected in June. Because the position had only temporary funding, limited salary, and no research endowment, Bateson returned to campaigning. As Beatrice Bateson put it, "his position had more of dignity than security."[68]

A new spate of "begging letters" ensued.[69] Bateson again turned to the ECRS, which wasted no time and issued a circular that same June. Under the guise of promoting "investigation of evolutionary problems," it pleaded for funds "urgently needed for the purpose of erecting and maintaining a building of moderate size in Cambridge, where breeding experiments with small animals, such as rabbits, canaries and insects could be carried on."[70] The committee estimated that the cost of the accommodations and labor needed to carry out the experiments "for a period of five years" would amount to £20,000. Shortly afterwards, Albert Charles Seward, Cambridge professor of botany, wrote to Bateson about the Royal Society Council's concern that such a petition represented a conflict of interest; whereas the Royal Society strove to promote studies in heredity and variation generally, Bateson sought through it a solution to a particular institution's predicament. As one of the secretaries of the Darwin celebration committee, Seward was sympathetic to Bateson's campaign but suggested a different tactic. He recommended utilizing anew "the begging Association" (the Cambridge University Association), and he enclosed a circular that he drafted for the purpose. The circular borrowed much of the language from the ECRS circular but sharpened the goal: "to build and equip a *School of Genetics*" (emphasis in original).[71]

As Richmond has shown, Darwin's commemoration in late June 1909 left a "favorable impression" of Mendelism on the audience.[72] Balfour's brother-in-law Rayleigh, as university chancellor and president of the Cambridge University Association, took the opportunity to make another explicit appeal in favor of funding a permanent chair. During a stay at Balfour's London townhouse, Rayleigh incorporated the plea into his welcome speech, an act that brought the campaign directly into aristocratic hands:

> During the last generation Cambridge, especially since the time of Michael Foster, has been active in biological work. We have the men and the ideas but the difficulty has always been lack of funds. At the present time it is desired among other things to establish a chair of genetics—a subject closely associated with the name of Darwin & of his relative Francis Galton, and of the greatest possible importance, whether it be regarded from the purely scientific or from the practical side. I should like to think that the interest aroused by this celebration would have a practical outcome in better provision in the further cultivation, in Darwin's own university & that of his sons, of the field wherein Darwin laboured.[73]

Bateson had a mixed reaction to the occasion. He wrote to Major Charles Chamberlain Hurst, one of the hybridizers in his research network, that "from the Mendelian's standpoint the gathering was rather tantalising," though he lamented the lack of time in the program for more explicit discussion of Mendelism.[74]

By that time, however, Bateson was already considering an alternative solution. Through a bequest by John Innes, the late London realtor, the means arose to found a research station outside Cambridge at Merton Park in Surrey. As Olby reported, the council of scientists and bureaucrats formed to launch the institution invited Bateson in October 1909 to serve as its director. After reworking the terms of the position to his satisfaction, he accepted the appointment and moved to Merton Park in early 1910. Cambridge thereby lost its chief figure for the Mendelian cause. Merton Park offered Bateson precisely what he strove to achieve for so many years at Cambridge: a lodge, indoor laboratory space, outdoor breeding grounds, staff, and the financial resources and flexibility to make his experimental enterprise viable for many years.[75] He wrote to Charles Davenport, director of the Biological Laboratory of Cold Spring Harbor in New York, "The opportunities afforded at the Innes Institution are so great that I did not feel I could refuse. My professorship here [at Cambridge] would, I imagine, have been continued, but I could see no prospect of any real increase in facilities."[76] After Bateson resigned the temporary professorship created for him, his most devoted student, Punnett, filled the vacancy and took over the work of advocating for further support of the Mendelian research program.[77] Despite losing Bateson, the Cambridge lobbyists persisted in the hope that Bateson might be lured back to his home institution once the professorship received a permanent endowment. But achieving such an end required the further summoning of Cambridge's aristocratic machine.

SITUATING GENETICS IN THE COUNTRY

Successive iterations of this endowment movement strengthened a developing, central motif: an apprehension of the Mendelian principles at work in species promised powerful, practical applications in the improvement of agricultural productivity. As noted, this motif was often articulated through citations of Biffen's successful wheat studies. According to the 1908 ECRS circular, "As a result of the investigations carried out by many workers, . . . it cannot be doubted that by their practical application discoveries of far-reaching consequence, both to science and to the practical business of animal-breeding can be made."[78] Along a similar vein, Seward's letter elab-

orated, "It is now generally admitted that the most promising method of attacking evolutionary problems is by experimental observations conducted
over a succession of years. By the application of these methods results of the
highest importance, both from the point of view of pure science and from a
practical standpoint have already been obtained."[79]

In a nation continuing to suffer from a dire agricultural depression, landowners and politicians were particularly concerned to discover ways to give
British agriculture a competitive edge over foreign imports. The government
therefore promoted scientific and technical education in rural areas to
stimulate better farming practices. Despite the slowness with which the
government responded with endowments for research, by the early 1900s the
creation of the Development Commission provided a significant means by
which research in the agricultural sciences—including the scientific study
of animal breeding and plant hybridization—could advance. These issues
and events did not go unnoticed by the Cambridge scientific community or
the Cambridge aristocrats who struggled to keep the farms on their estates
afloat.[80] Centrally situated within both groups, Arthur Balfour kept abreast
of the advances in Mendelian genetics, Bateson's deficient situation at
Cambridge, and the national agricultural issues. Whereas he previously
attempted to lend his help through private and anonymous means, now
Balfour publicly stepped forward and took over leadership of the campaign
to endow a school of genetics at Cambridge.

On his part, Balfour showed an enduring interest in promoting hybridization and breeding experiments, and he found Bateson's program particularly deserving of support. An enthusiast of arboriculture and horticulture,
Balfour turned his own woodlands to experiments in the growth rates of
hardwoods.[81] In 1899, he privately (and anonymously) donated £500 in support of the horse-breeding experiments conducted by University of Edinburgh naturalist James Cossar Ewart, who famously demolished the idea of
telegony; Balfour was then chancellor.[82] Such acts of philanthropy, including his anonymous endowment of the temporary professorship of biology at
Cambridge, fitted within the broader pattern of the Balfour family's support
of biological research in the memory of their late brother Francis.[83] More
generally, Balfour advocated for the British government's increased endowment of science and medicine, which he viewed as a prerequisite to Britain's
endeavor to keep apace intellectually and industrially with competitor nations.[84] As an agriculturalist whose estate suffered from the depression of
the 1880s and 1890s, Balfour sought a tariff solution to the damaging effects
of cheap imports, an issue that fueled much political unrest during his tenure
as prime minister. An alternative solution manifested itself in agricultural

eugenics, an area in which Balfour became increasingly interested.[85] A known skeptic of natural selection and adversary of biometrician Karl Pearson, Balfour sympathized with the Mendelian project of the like-minded Bateson.[86]

Motivated along these lines, and drawing directly from Bateson's and Punnett's advice, in July 1910 Balfour drafted and circulated a petition for contributions to an endowment of a permanent professorship and research institute, estimating the required funds at £20,000.[87] Balfour's circular inaugurated the final, successful lobby that placed the Cambridge chair on a permanent footing and, at last, established a research station of the scope Bateson had, for years, envisioned. In making his plea, Balfour utilized a familiar discourse that cast the importance of Mendelism in terms of its theoretical and practical potential, but in more certain terms:

> No more important group of scientific problems await solution than those connected with Heredity. These problems touch closely on every great question connected with life—scientific, philosophical, sociological, and political; on every practical art connected with the cultivation of crops or the raising of stock;—nay on every sport connected with the breeding of animals. . . .
>
> Let it be remembered that the funds required to endow the Professorship and the Experimental Station are to be used in no doubtful or uncertain enterprise. The results already attained afford a sure prospect of future success; the workers are competent and willing; all that is required are sure endowments as may be necessary to enable them to continue what they have so admirably begun.[88]

He then hosted a meeting at his London townhouse on 21 November 1910, inviting "a few representative members of the University of Cambridge and others interested in this subject."[89] The attendees proposed the "Arthur Balfour Chair of Genetics" as a "testimonial to him as a man and Minister."[90]

One attendee (and lifelong friend), Reginald Brett, Viscount Esher, sent Balfour's circular to associates in his political and defense circles, including Prime Minister Herbert Asquith and Sir Francis Trippel, an army major connected to the City Guilds, to recruit "others who might feel disposed to do honour to Mr Balfour, and to further a great work in which he is so deeply interested."[91] This effort yielded the desired result. Esher excitedly wrote to Balfour in February, 1912, "I think I have the money for our Chair of Genetics. Asquith has been very good and has helped me! The name of the benefactor is not to be indulged to anyone."[92] The donor was later identified to be William George Watson, founder and chairman of the highly successful

Maypole Dairy Company, Ltd.[93] He communicated his pledge to Cambridge genetics on 9 February 1912: "Confirming our conversation this morning I am willing to give £50.000/-/- to be divided as agreed and used to prove an endowment of the Study of Genetics in the University of Cambridge & for the benefit of the enlarged British School [of Art] at Rome."[94] Prime Minister Asquith wrote to Esher, "Regard is had to the public-spirited employment of wealth!"[95] Watson's endowment—as did Quick's and Innes's bequests—underlined the reality that in this period the *nouveau riche* among business-men increasingly surpassed the landed aristocracy's capacity for *noblesse oblige.*[96] Amid the proprieties imposed by anonymity, Balfour and Watson managed a private meeting in Cannes.[97] A *Times* letter to the editor pub-licly announced the anonymous benefaction on 13 March 1912.[98] In early April Esher made the formal proposal to the vice-chancellor of the univer-sity, Stuart A. Donaldson: "It would meet the wishes of our benefactor, and probably those of most Cambridge people, and I think would be generally appreciated, if the new Chair could be called the *Arthur Balfour* Professor-ship of Genetics in the University of Cambridge" (emphasis in original).[99]

Also by this time, Balfour and Esher began negotiations for establishing a research station to be connected with the chair. They received advice from the Cambridge scientists, who differed in their views. The physicist Joseph Larmor recommended closely allying the chair and its experimental space with the other biological laboratories:

A Professor of Genetics at Cambridge would be more effective if it were in close touch with the existing biological department as now, than if it were isolated in a separate department and *permanent* building of its own. I think there is a tendency to be guarded against for each depart-ment to establish itself complete and self-contained, which means some duplicating as well as isolation of effort, whereas the notion of *Univer-sity* should mean combined activity [emphasis in original].[100]

The Mendelians, however, opposed this view and advised in favor of ex-perimental conditions not unlike Bateson's former arrangement at Merton House. Based on the opinions of Punnett, Biffen, and their colleagues, Ray-leigh, under the auspices of the Cambridge University Association, negoti-ated the particulars with Esher:

You will see that they [the biologists] don't take Larmor's view that the professor should be tied to the laboratories. How could he do his real work there? The point about centralising is very important. At present, time is apt to be wasted by, say, going to attend to the proper cross-fertilisation of a plant in the Botanic Gardens, and then rushing off to kill a chicken on the University Farm, beyond Girton!

The efficient pursuit of Genetic research entails close and constant supervision of the experiments, whether plant or animal. The investigator ought, as far as possible, to live with his material. For this reason a *house* has been added to the capital estimate. The suggested *laboratory* would consist of a large and several smaller rooms for the breeding of small animals and insects, as well as a general room with conveniences for the use of investigators working in the garden. As there must be one man (gardener) living on the spot, a cottage has been estimated for [emphasis in original].[101]

The recommendation clearly privileged a domestic arrangement for experimental genetics, in which the lobbyists deemed an out-of-doors agricultural station, supplemented with an indoor laboratory, to be the more appropriate choice than the crowded New Museums Site on Downing Street.[102] Esher confirmed this decision with Balfour: "I came to the conclusion that the right thing to do is to put the Professor of Genetics into the closest possible touch physically with his work. We selected one acre of ground close to the University Farm as a site."[103] A contract of purchase was signed on 9 April 1912, placing nearly two acres of land on Storey's Way into Esher's and Balfour's hands.[104]

Even before the official purchase, Esher contracted the architectural firm, Dunnage and Hartmann, to draw up designs for the built structures. From the start, he felt that the house "ought to be of course commodious enough to hold a married man," though Punnett later complained that he felt the living space being constructed was inadequate and needed expansion.[105] Professor of Agriculture Thomas Barlow Wood recommended that laboratory space be built such that it could be later converted into domestic space: "Would it be possible to build the laboratory and animal house as a sort of annexe to the dwelling-house? It would look like a billiard room, and might possibly be used for that purpose if it ever became necessary for the Balfour Professor to move to another site."[106] Again, disagreement over spatial dimensions fueled more debate. Punnett shared the views held by his Mendelian cohort:

I think he [Herbert Hartmann] has designed an excellent house, but I and several of my colleagues are afraid that the study and laboratory have been so much reduced in size that they will hardly serve the purpose for which they are intended. We feel however that we cannot ask the anonymous benefactor to increase the very generous sum he is already expending. . . .

We are very anxious that the building should meet the wants of the Professor and one of my colleagues is prepared to provide the £150 necessary to pay for an extension of 10 feet. This would make an addition of 5 feet to the length of each of the two rooms.[107]

Although Esher adopted the laboratory recommendations, in the case of the living quarters he kept the original design.

As building progressed, Vice Chancellor Donaldson created, "by grace, 3 June 1912," the Arthur James Balfour Professorship of Genetics—Britain's first academic chair bearing the name of the new science.[108] With this confirmed, Balfour wrote Bateson to gauge the possibility of luring him back to Cambridge. Under the conditions of the post, Balfour and the prime minister (in this case, Asquith) were named the hiring authorities. Balfour explained his reasons for favoring Bateson's resumption of the chair:

> To you, more than to anybody else, is owed the impulse which started the Mendelian School at Cambridge, and, through that school, the foundation of the new Chair. Would you be prepared to resume your investigations as its occupant? If you would, I think it would be a great gain, both to Cambridge and to science; and I think it very probable that the Prime Minister would take the same view, though on this, of course, I cannot speak with any assurance.
>
> There is, so far as I know, nothing in the Trust Deed founding the Professorship which *would* confine it, or indeed *ought* to confine it, to Genetic investigations on Mendelian lines; but there is certainly a great deal to be said, on *every* ground, for making the first Professor one who would carry on with originality and power the line of investigation which has already proved so fruitful [emphasis in original].[109]

Bateson replied promptly, declining the offer and recommending Punnett for the position. He explained,

> I value very greatly the suggestion conveyed in your letter, and the appreciative expressions you have used in reference to my work, but I feel that I must abide by the decision then made.

The opportunities here [at John Innes] are already great and they will increase. Though we cannot have a constant succession of students yet the influence which this Institution will ultimately exert on biological science must be very considerable.

If there were difficulty in finding a man thoroughly qualified, the case would be different; but I am sure that Mr Punnett, who worked with me for several years and succeeded me in the Professorship of Biology is in every way worthy to be appointed to the new post. If he is thus chosen, as I trust he may be, I should have perfect confidence that the school of Genetics in Cambridge will develop rapidly and on right lines. If more information as to Punnett's qualifications were desired I would, of course, gladly supply it.[110]

Punnett was elected the first Arthur James Balfour Professor of Genetics on 22 November 1912.[111]

By late 1913, the building approached completion, stalled only by an unfortunate act of vandalism by suffragettes.[112] In the summer Punnett proposed Gravel Hill Lodge as a name for the house, given its proximity to the university's farm, Gravel Hill. But Esher and Balfour had another idea in mind and settled on Whittingehame Lodge, after Balfour's own country house. Punnett gave his approval: "Many thanks for your note with the name of the house on Storey's Way. It sounds very happy and is certainly much prettier than the only suggestion we felt able to make."[113] Academic genetics was thus modeled on an older form of country-house natural history in rationale, design, and now name. Punnett moved into the house on 20 October 1913, and Balfour and Esher transferred the properties to the university on 19 January 1914 (fig. 4.1).[114] It continued in use as a genetics research station until 1962 when the genetics department sold it to Churchill College, although by the 1930s its design came to be viewed as a real hindrance to the progress of genetics research, which by then shifted back to the laboratory bench.[115]

The imperatives for Bateson's Mendelian program did not simply derive from his admiration of an older style of natural history modeled on the experiences of the tradition of gentlemen-amateurs conducting experiments at their country seats, as epitomized by Charles Darwin's studies at Down House. His actual practice showed clear parallels—particularly when compared to the scientific experiments on domesticated products performed on estates like the Duke of Bedford's Woburn Abbey or Ewart's private farms.

Fig. 4.1. Whittingehame Lodge, Cambridge, early 1920s, artist unknown. The standing men are possibly R. C. Punnett (*left*), first Arthur James Balfour Professor of Genetics at Cambridge, and J. B. S. Haldane (*right*), an early population geneticist who was Reader in Biochemistry at Trinity College, Cambridge. Courtesy of Churchill College Archives, CCPH 5/15.

Bateson's cry for large-scale experiments in animal breeding and plant hybridization—believed to be promising for the improvement of agriculture—was, at the foundation, a utopian vision for overcoming the limitations of a distributed geography of distant and confining experimental spaces. Again, at the level of his practice, Bateson's method carried forward elements of country-house natural history, which he encountered in Francis Balfour's studies of animal morphology—fact-driven, field-based, and collaborative methods. Bateson's strategic appeal to this form of study, particularly by invoking Darwin's name, aligned Bateson with Cambridge's highly effective aristocratic machine. As his wife quipped, if not by birth an aristocrat, Bateson was nonetheless "intellectually an aristocrat."[116]

The cultivation of genetics in the Cambridge countryside involved the marrying of Victorian field-based methods with the developing Mendelian experimental program that, by necessity of the research questions, required centralization of field and laboratory within a single site where the researcher could "live with his materials."[117] The Mendelians' study of domestic productions thus situated genetics in a domesticated country setting, a feat successfully accomplished through a partnership between the *nouveau riche* and *noblesse oblige* motivated by the promise of science, in its Mendelian form, for bettering the state of British agriculture and society.

NOTES

This chapter builds upon chapter 8 of my doctoral dissertation, "Aristocrats and Professionals: Country-House Science in Late-Victorian Britain" (University of Minnesota,
2004), 221–250. Archival research was supported in part by a grant from the National
Science Foundation, SES-0094442 (2001) and a DePaul University Competitive Research
Grant (2007). For permission to quote from unpublished manuscript material, the author
gratefully acknowledges the American Philosophical Society, Mr. Andrew Michael
Brander, Churchill Archives Centre, the Syndics of the Cambridge University Library,
and Viscount Esher.

1. James A. Secord, "The Crisis of Nature," in *Cultures of Natural History*, ed.
N. Jardine, James A. Secord, and Emma C. Spary (Cambridge: University of Cambridge
Press, 1996), 449.

2. Lynn K. Nyhart, "Natural History and the 'New' Biology," in *Cultures of Natural
History*, ed. N. Jardine, James A. Secord and Emma C. Spary (Cambridge: University of
Cambridge Press, 1996), 426.

3. Dorinda Outram, "New Spaces in Natural History," in *Cultures of Natural
History*, ed. N. Jardine, James A. Secord and Emma C. Spary (Cambridge: University of
Cambridge Press, 1996),249–65; Thomas F. Gieryn, "Three Truth-Spots," *Journal of the
History of the Behavioral Sciences* 38, no. 2 (2002): 113–32.

4. See Nyhart, "Natural History," and David E. Allen, "On Parallel Lines: Natural
History and Biology from the Late Victorian Period," *Archives of Natural History* 25
(1998): 361–71.

5. Helen J. Blackman, "The Natural Sciences and the Development of Animal
Morphology in Late-Victorian Cambridge," *Journal of the History of Biology* 40 (2007):
71–108.

6. Paolo Palladino, "Between Craft and Science: Plant Breeding, Mendelian Genetics,
and British Universities, 1900–1920," *Technology and Culture* 34 (1993): 300–23.

7. Christopher R. Henke, "Making a Place for Science: The Field Trial," *Social
Studies of Science* 30 (2000): 484; Robert E. Kohler, *Landscapes and Labscapes: Exploring
the Lab-Field Border in Biology* (Chicago: University of Chicago Press, 2002).

8. Gerald L. Geison, *Michael Foster and the Cambridge School of Physiology: The
Scientific Enterprise in Late-Victorian Society* (Princeton, NJ: Princeton University Press,
1978), 79–190.

9. Christopher N. L. Brooke, *A History of the University of Cambridge*, 4 vols.
(Cambridge: Cambridge University Press, 1993), 4:180.

10. Andrew Roberts, *Salisbury: Victorian Titan* (London: Phoenix, 2000); Kenneth
Young, *Arthur James Balfour: The Happy Life of the Politician, Prime Minister, Statesman, and Philosopher, 1848–1930* (London: G. Bell and Sons, 1963).

11. N. G. Annan, "The Intellectual Aristocracy," in *Studies in Social History: A
Tribute to G. M. Trevelyan*, ed. J. H. Plumb (London: Longmans, Green, 1955), 243–87.

12. Arthur James Balfour (first Earl of Balfour), *Chapters of Autobiography*, ed. Mrs.
Edgar Dugdale (London: Cassell, 1930), 37; Robert John Strutt (fourth Baron Rayleigh),

Life of John William Strutt, Third Baron Rayleigh, ed. John N. Howard (Madison: University of Wisconsin Press, 1968), 45; Francis Darwin, ed., *The Life and Letters of Charles Darwin, Including an Autobiographical Chapter,* 3 vols. (London: John Murray, 1887), 3:250; R. B. Freeman, *Charles Darwin: A Companion* (Folkestone: Dawson, 1978), 127.

13. D. Croal, *Sketches of East Lothian,* 4th ed. (Haddington: Haddingtonshire Courier, 1904); Anonymous, "Whittingehame," *Transactions of the East Lothian Antiquarian and Field Naturalists' Society* 1 (1929): 132–33; Paul Harris, *Life in a Scottish Country House: The Story of A. J. Balfour and Whittingehame House* (Haddington: Whittingehame House Publishing, 1989); Land Use Consultants, *An Inventory of Gardens and Designed Landscapes in Scotland,* vol. 5, *Lothian and Borders* (Perth: Countryside Commission for Scotland, 1987), 242–46.

14. Balfour, *Chapters of Autobiography,* 230. For more on the Balfours and country-house science, see Donald L. Opitz, "'This House Is a Temple of Research': Country-House Centres for Late-Victorian Science," in *Repositioning Victorian Sciences: Shifting Centres in Nineteenth-Century Scientific Thinking,* ed. David Clifford, et al. (London: Anthem Press, 2006), 143–53, and Simon Schaffer, "Physics Laboratories and the Victorian Country House," in *Making Space for Science: Territorial Themes in the Shaping of Knowledge,* ed. Crosbie Smith and Jon Agar (Basingstoke: Macmillan, 1998), 149–80.

15. G. W. Balfour and F. M. Balfour, "On Some Points in the Geology of the East Lothian Coast," *Geological Magazine* 9 (1872): 161–64; George Leslie, "The Invertebrate Fauna of the Firth of Forth," *Proceedings of the Royal Physical Society of Edinburgh* 4 (1881): 68–97; Alice Blanche Balfour, "Butterflies and Moths Found in East Lothian," *Transactions of the East Lothian Antiquarian and Field Naturalists' Society* 1 (1930): 169–84.

16. Hayden Lorimer, "Guns, Game and the Grandee: The Cultural Politics of Deer-stalking in the Scottish Highlands," *Ecumene* 7 (2000): 417–21; William Scrope, *Days of Deer Stalking in the Scottish Highlands* (London: Hamilton, Adams and Co., 1883).

17. Thomas Speedy, *Sport in the Highlands and Lowlands of Scotland with Rod and Gun* (Edinburgh: William Blackwood and Sons, 1884), 255.

18. *Balfour Memorial: Undergraduate Meeting at the Union, Cambridge, 30th October, 1882,* (Cambridge: Fabb & Tyler, n.d.), 9; H. N. Moseley, "Francis Maitland Balfour," *Fortnightly Review* 32 (1882): 573; H. Blackman, "A Spiritual Leader? Cambridge Zoology, Mountaineering and the Death of F. M. Balfour," *Studies in the History and Philosophy of Biological and Biomedical Sciences* 35 (2004): 93–117.

19. J. Stanley Gardiner, *The Zoological Department, Cambridge* (Cambridge: W. Heffer and Sons, 1934), 2; Alfred Newton, *Zoology* (London: Society for Promoting Christian Knowledge, 1878), 2; Blackman, "Natural Sciences," 78.

20. *Balfour Memorial,* 5; A. F. R. Wollaston, *Life of Alfred Newton* (London: John Murray, 1921), 252.

21. W. H. G[askell], "Sir Michael Foster, 1836–1907," *Proceedings of the Royal Society* 80 (1908): lxxix; F. M. Balfour, "On the Disappearance of the Primitive Groove in the Embryo Chick," *Quarterly Journal of Microscopical Science* 13 (1873): 276–80.

22. E. B. P[oulton], "The Study of Embryology," *Nature* 36 (1887): 601. On Alice

Balfour's sketching, see Donald L. Opitz, "'Behind Folding Shutters in Whittingehame House': Alice Blanche Balfour (1850–1936) and Amateur Natural History," *Archives of Natural History* 31 (2004): 334.

23. Blackman, "Natural Sciences," 81; Mark N. Ozer, "The British Vivisection Controversy," *Bulletin of the History of Medicine* 40 (1966): 158–67; Mary Ann Elston, "Women and Anti-Vivisection in Victorian England, 1870–1900," in *Vivisection in Historical Perspective*, ed. Nicolaas A. Rupke (London: Croom Helm, 1987), 259–94.

24. Theodore Edmund Alexander, "Francis Maitland Balfour's Contributions to Embryology" (PhD diss., University of California, 1969), 52–3; A. H. Sykes, "Foster and Sharpey's Tour of Europe," *Notes and Records of the Royal Society of London* 54 (2000): 47–52. Lynn K. Nyhart, *Biology Takes Form: Animal Morphology and the German Universities, 1800–1900* (Chicago: University of Chicago Press, 1995).

25. Francis M. Balfour, *A Monograph on the Development of Elasmobranch Fishes* (London: Macmillan, 1878), v.

26. Ibid., 3–4.

27. For more on Balfour's animal morphology, see Alexander, "Francis Maitland Balfour's Contributions to Embryology," and Brian K. Hall, "Francis Maitland Balfour (1851–1882): A Founder of Evolutionary Embryology," *Journal of Experimental Zoology* 299B (2003): 3–8.

28. William Coleman, "Bateson and Chromosomes: Conservative Thought in Science," *Centaurus* 15 (1970): 228; William Bateson, "Heredity and Variation in Modern Lights," in *Darwin and Modern Science: Essays in Commemoration of the Centenary of the Birth of Charles Darwin and of the Fiftieth Anniversary of the Publication of the Origin of Species*, ed. A. C. Seward (Cambridge: Cambridge University Press, 1909), 92.

29. Charles R. Darwin, *On the Origin of Species by Means of Natural Selection, or the Preservation of Favoured Races in the Struggle for Life* (London: John Murray, 1859), 486.

30. Ibid., 4; James A. Secord, "Nature's Fancy: Charles Darwin and the Breeding of Pigeons," *Isis* 72 (1981): 163–86; James A. Secord, "Darwin and the Breeders: A Social History," in *The Darwinian Heritage*, ed. David Kohn (Princeton, NJ: Princeton University Press, 1985), 519–42.

31. Asa Gray, "Darwin on the Origin of Species," *Atlantic Monthly* 6 (1860): 110.

32. Soraya de Chadarevian, "Laboratory Science Versus Country-House Experiments: The Controversy between Julius Sachs and Charles Darwin," *British Journal for the History of Science* 29 (1996): 17–41.

33. J. B. F[armer], "William Bateson, 1861–1926," *Proceedings of the Royal Society of London* B101 (1927), ii.

34. Garland E. Allen, *Life Science in the Twentieth Century* (New York: Wiley, 1975).

35. M. F[oster], "Francis Maitland Balfour," *Proceedings of the Royal Society of London* 35 (1883): xxvi.

36. Robert Olby, "Bateson, William (1861–1926)," in *Oxford Dictionary of National Biography*, ed. Colin Matthew and Brian Harrison (Oxford: Oxford University Press, 2004).

37. [William Bateson], "Discussion of Reports," *Cambridge University Reporter*, Cambridge (February 20, 1894), 494.

38. Coleman, "Bateson and Chromosomes," 242.

39. Beatrice Bateson, *William Bateson, F.R.S., Naturalist: His Essays and Addresses* (Cambridge: Cambridge University Press, 1928), 392; Brian K. Hall, "Betrayed by Balanoglossus: William Bateson's Rejection of Evolutionary Embryology as the Basis for Understanding Evolution," *Journal of Experimental Zoology* 304B (2005): 10.

40. Coleman, "Bateson and Chromosomes," 261 and 253.

41. William Bateson, *Materials for the Study of Variation: Treated with Especial Regard to Discontinuity* (London: Macmillan, 1894), 9; Hall, "Betrayed by Balanoglossus," 7–10; and Alan G. Cock and Donald R. Forsdyke, *Treasure Your Exceptions: The Science and Life of William Bateson* (New York: Springer, 2008), 3–110.

42. Bateson, *Materials for the Study of Variation,* 5.

43. Ibid., 7. See G. Evelyn Hutchinson and Stan Rachootin, "Introduction," in William Bateson, *Problems of Genetics,* ed. Hutchinson and Rachootin (New Haven, CT: Yale University Press, 1979), viii–xi.

44. Bateson, *William Bateson,* 35–6.

45. Bateson, *Materials for the Study of Variation,* 7.

46. Ibid., 574.

47. Marsha L. Richmond, "The 'Domestication' of Heredity: The Familial Organization of Geneticists at Cambridge, 1895–1910," *Journal of the History of Biology* 39 (2006): 570–71; David Lipset, *Gregory Bateson: The Legacy of a Scientist* (Englewood Cliffs, NJ: Prentice-Hall, 1980), 31.

48. Printed circular, Evolution Committee of the Royal Society, March 1897, Royal Society Archives, MM/15/73; W. Bateson, "Memorandum from the Evolution Committee of the Royal Society," *Entomologist's Monthly Magazine,* 2nd series, 11 (1900): 139–40.

49. Marsha L. Richmond: "'A Lab of One's Own': The Balfour Biological Laboratory for Women at Cambridge University, 1884–1914," *Isis* 88 (1997): 422–55.

50. Richmond, "The 'Domestication' of Heredity." For "disciples," see R. C. Punnett, "Early Days of Genetics," *Heredity* 4 (1950): 4. See also Marsha L. Richmond, "Women in the Early History of Genetics: William Bateson and the Newnham College Mendelians, 1900–1910," *Isis* 92 (2001): 55–90.

51. Quoted in Richmond, "The 'Domestication' of Heredity," 573.

52. Punnett, "Early Days of Genetics," 7; John Willis Clark, *Endowments of the University of Cambridge* (Cambridge: Cambridge University Press, 1904), 348.

53. *Balfour Memorial,* 4.

54. Hall, "Betrayed by Balanoglossus"; Roy MacLeod, "Embryology and Empire: The Balfour Students and the Quest for Intermediate Forms in the Laboratory of the Pacific," in *Darwin's Laboratory: Evolutionary Theory and Natural History in the Pacific,* ed. Roy MacLeod and Philip F. Rehbock (Honolulu: University of Hawaii Press, 1994), 151–53.

55. Bateson, *William Bateson,* 20–21, 42.

56. Walter Frank Raphael Weldon held the lectureship from 1884 to 1890. Sedgwick's own position was as lecturer until 1890 when a readership in animal morphology was created for him; see Geison, *Michael Foster,* 378–79; Karl Pearson, "Walter Frank Raphael Weldon, 1860–1906," *Biometrika* 5 (1906): 15n.

57. Wollaston, *Life of Alfred Newton,* 252.

58. Bateson, *William Bateson,* 42.

59. Anonymous, "British-Grown Fruits and Vegetables: A Great Exhibition at Chiswick," *Garden*, supp. (1903): 235; William Bateson to the Duke of Bedford, 5 December 1898, Cambridge University Library, Department of Manuscripts and University Archives (hereafter CUA), Bateson Papers, MS Add. 8634; Bateson, *William Bateson*, 71.

60. T. Baines, ed., *Annual Register: A Review of Public Events at Home and Abroad for the Year 1903* (London: Longmans, Green, 1904), pt. 2, 2.

61. William Coleman, *Biology in the Nineteenth Century: Problems of Form, Function, and Transformation* (New York: Wiley, 1971), 163–64.

62. Bateson, *William Bateson*, 93; G. S. Graham-Smith and D. Keilin, "George Henry Falkiner Nuttall, 1862–1937," *Obituary Notices of Fellows of the Royal Society* 2 (1939): 494.

63. Cambridge University Association, "A Plea for Cambridge," *Quarterly Review* 204 (1906): 521–22.

64. Palladino, "Between Craft and Science." For further information on Biffen's work in developing strong and productive wheat varieties within the context of intellectual property debates, see Berris Charnley and Gregory Radick, "Plant Breeding and Intellectual Property before and after the Rise of Mendelism: The Case of Britain," in *Living Properties: Making Knowledge and Controlling Ownership in the History of Biology*, ed. Jean-Paul Gaudillière, Daniel J. Kevles, and Hans-Jörg Rheinberger, Preprint 382 (Berlin: Max-Planck-Institut für Wissenschaftsgeschichte, 2009), 51–56.

65. Bateson, *William Bateson*, 112.

66. Ibid.

67. Ibid. Robert Olby suggested Balfour's identity in "Scientists and Bureaucrats in the Establishment of the John Innes Horticultural Institution under William Bateson," *Annals of Science* 46 (1989): 499; this view is carried in Cock and Forsdyke, *Treasure Your Exceptions*, 306.

68. Bateson, *William Bateson*, 121.

69. Bateson self-styled his pleas "begging letters." W. Coleman, "Bateson Papers," *Mendel Newsletter* 2 (1968): 3; Bateson, *William Bateson*, 74–75.

70. "Experiments in Genetics," June 1908, American Philosophical Society, William Bateson Collection (hereafter APS WBC), BALTO#15.

71. A. C. Seward to W. Bateson, Cambridge, 14 July 1908, APS WBC, BALTO#15.

72. Marsha L. Richmond, "The 1909 Darwin Celebration: Reexamining Evolution in the Light of Mendel, Mutation, and Meiosis," *Isis* 97 (2006): 470.

73. Rayleigh to Arthur James Mason (Vice-Chancellor), London, 17 May 1909, CUA Conf. I.1.

74. W. Bateson to C. C. Hurst, 25 June 1909, CUA Add. 8634/D21.A; Richmond, "The 1909 Darwin Celebration," 470.

75. Olby, "Scientists and Bureaucrats."

76. R. D. Harvey, "The William Bateson Letters at the John Innes Institution," *Mendel Newsletter* (1985): 5–6.

77. Cock and Forsdyke, *Treasure Your Exceptions*, 381–86.

78. "Experiments in Genetics," June 1908, APS WBC, BALTO#15.

79. A. C. Seward to W. Bateson, Cambridge, 14 July 1908, APS WBC, BALTO#15.

80. T. W. Fletcher, "The Great Depression of English Agriculture, 1873–1896,"

Economic History Review 13 (1961): 417–32; Robert Olby, "Social Imperialism and State Support for Agricultural Research in Edwardian Britain," *Annals of Science* 48 (1991): 509–26.

81. James Winter, *Secure from Rash Assault: Sustaining the Victorian Environment* (Berkeley and Los Angeles: University of California Press, 1999), 98–99.

82. F. S. Parry to J. C. Ewart, 29 June 1899, University of Edinburgh Library (hereafter UEL), Ewart Papers, Gen.137/4/114. On telegony, see Richard W. Burkhardt Jr., "Closing the Door on Lord Morton's Mare: The Rise and Fall of Telegony," *Studies in the History of Biology* 3 (1979): 1–21, and Harriet Ritvo, "Understanding Audiences and Misunderstanding Audiences: Some Publics for Science," in *Science Serialized: Representations of Science in Nineteenth-Century Periodicals*, ed. Geoffrey Cantor and Sally Shuttleworth (Cambridge, MA: MIT Press, 2004), 331–49.

83. Opitz, "Aristocrats and Professionals," 185, 237–38.

84. Arthur James Balfour, *Arthur James Balfour as Philosopher and Thinker: A Collection of the More Important and Interesting Passages in His Non-Political Writings, Speeches, and Addresses, 1879–1912*, ed. Wilfrid M. Short (London: Longmans, Green, 1912), 316. See also Robert John Strutt (fourth Baron Rayleigh), *Lord Balfour in Relation to His Science* (Cambridge: Cambridge University Press, 1930).

85. Jason Tomes, *Balfour and Foreign Policy: The International Thought of a Conservative Statesman* (Cambridge: Cambridge University Press, 1997), 90–97; Balfour, *Arthur James Balfour as Philosopher and Thinker*, 211–17.

86. L. S. Jacyna, "Science and Social Order in the Thought of A. J. Balfour," *Isis* 71 (1980): 11–34; Robert Olby, "The Dimensions of Scientific Controversy: The Biometric-Mendelian Debate," *British Journal for the History of Science* 22 (1989): 299–320; Theodore M. Porter, *Karl Pearson: The Scientific Life in a Statistical Age* (Princeton, NJ: Princeton University Press, 2004), 268–69.

87. Arthur James Balfour, "Endowment in the Study of Genetics in the University of Cambridge," Typescript dated June 1910, Churchill Archives Centre (hereafter CAC), The Papers of Viscount Esher (Reginald Brett), ESHR 19/2; W. Bateson to B. Bateson, London, 30 June 1910, John Innes Foundation Historical Collections, John Innes Centre Library, Bateson Papers, G3E-8J.

88. Arthur James Balfour, "Endowment in the Study of Genetics in the University of Cambridge," Typescript dated June 1910, CAC ESHR 19/2.

89. Esher, typescript, 22 November 1911, CAC ESHR 19/2.

90. Ibid.

91. Ibid.; Esher to Sir Francis Trippel, London, 22 November 1911, CAC ESHR 19/2.

92. Reginald Baliol Brett, *Journals and Letters of Reginald, Viscount Esher*, ed. Oliver, Viscount Esher (London: Ivor Nicholson & Watson, 1938), 3:80.

93. "William George Watson," Typescript MS, n.d., CAC ESHR 19/2.

94. W. G. Watson to Esher, Reading, 9 February 1912, CAC ESHR 19/2.

95. H. H. Asquith to Esher, London, 14 February 1912, CAC ESHR 19/2.

96. W. D. Rubinstein, *Men of Property: The Very Wealthy in Britain since the Industrial Revolution* (New Brunswick, NJ: Rutgers University Press, 1981), 213–18; David Cannadine, *The Decline and Fall of the British Aristocracy*, rev. ed. (New York: Vintage Books, 1999), 90–92, 308–10, 357–59.

97. Esher to W. G. Watson, London, 12 February 1912, CAC ESHR 19/2; W. G. Watson to Esher, Reading, 3 March 1912, CAC ESHR 19/2.

98. Reginald Baliol Brett (second Viscount Esher), "Balfour Professorship of Genetics: Munificent Gift to Cambridge," *Times* (London), 13 March 1912.

99. Esher to Vice-Chancellor, Callander, 3 April 1912 (copy), CAC ESHR 19/2.

100. J. Larmor to Esher, London, 8 March 1912, CAC ESHR 19/2.

101. H. A. Roberts to Esher, Cambridge, 9 March 1913, with enclosure, "Estimate for a Small Institute for the Study of Genetics," CAC ESHR 19/2.

102. Blackman, "Natural Sciences."

103. Esher to A. J. Balfour, London, 14 March 1912, CAC ESHR 19/2.

104. "Contract of Purchase," 9 April 1912, CAC ESHR 19/2.

105. Esher to T. B. Wood, 14 March 1912, CAC ESHR 19/2; R. C. Punnett to Esher, Cambridge, 17 February 1913, CAC ESHR 19/2.

106. T. B. Wood to Esher, Cambridge, 23 March 1912, CAC ESHR 19/2.

107. R. C. Punnett to Esher, Cambridge, 7 August 1912, CAC ESHR 19/2.

108. "Balfour Professorship of Genetics," CUA CUR 39.47.

109. A. J. Balfour to W. Bateson, London, 25 June 1912, CUA Add. 8634/G.6.1 /12.

110. W. Bateson to A. J. Balfour, Merton, 26 June 1912 (copy of draft), CUA Add. 8634/G.6.1 /14.

111. "Balfour Professorship of Genetics," CUA CUR 39.47.

112. G. E. Dunnage to Esher, London, 17 May 1913 (telegraph), CAC ESHR 19/2.

113. R. C. Punnett to Esher, Cambridge, n.d., CAC ESHR 19/2; R. C. Punnett to Esher, Cambridge, 21 August 1913, CAC ESHR 19/2.

114. G. E. Dunnage to Esher, London, 28 October 1913, CAC ESHR 19/2; "Report of the Council of the Senate on an Offer Relating to the Arthur Balfour Professorship of Genetics," 19 January 1914, CUA CUR 39.47.

115. Plaque erected on Whittingehame Lodge, 17 February 1990 (still extant); Luigi Luca Cavalli-Sforza, "Recollections of Whittingehame Lodge," *Theoretical Population Biology* 38 (1990): 301–05; Joan Fisher Box, *R. A. Fisher: The Life of a Scientist* (New York: John Wiley & Sons, 1978), 398. See also Jennifer Marie, "The Importance of Place: A History of Genetics in 1930s Britain" (PhD diss., University College, London, 2004).

116. Bateson, *William Bateson*, 50.

117. H. A. Roberts to Esher, Cambridge, 9 March 1913, CAC ESHR 19/2.

CHAPTER FIVE

Scale and the Geographies of Civic Science: Practice and Experience in the Meetings of the British Association for the Advancement of Science in Britain and in Ireland, c. 1845–1900

CHARLES W. J. WITHERS

SITUATING THE BRITISH ASSOCIATION FOR THE ADVANCEMENT OF SCIENCE

At its foundation in York in 1831, the objectives of the British Association for the Advancement of Science (BAAS) were made clear: "To give a stronger impulse and a more systematic direction to scientific enquiry; to promote the intercourse of those who cultivate Science in different parts of the British Empire with one another and with foreign philosophers; to obtain more general attention for the objects of Science and the removal of any disadvantages of a public kind which impede its progress."[1] It was also decided that the association's meetings would always be held in different towns throughout Britain and Ireland.

Historians of the BAAS have pointed to the importance of this policy of geographical mobility in helping establish the wider role of the association in promoting science in Britain as a civic commodity and cultural resource. For Morrell and Thackray, the choice of locations was important to the vision the BAAS had for itself: "The adoption of a provincial stance by the British Association did foster a vision of social integration, for it assuaged the pride of those peripheral groups represented in the rank and file and thus aided the gradual and complete take-over by the Gentlemen of Science of the actual decision-making apparatus of the organization."[2] Charting the geography of BAAS meetings between 1831 and 1844, they distinguished between what they termed "the circuit of academic and metropolitan centres, 1832–1835" (Oxford in 1832, Cambridge 1833, Edinburgh 1834, Dublin 1835 and Bristol 1836), and the "circuit of provincial towns, 1836–1844"

(Liverpool 1837, Newcastle 1838, Birmingham 1839, Glasgow 1840, Plymouth 1841, Manchester 1842, Cork 1843, and York again in 1844).

These findings were advanced more than twenty-five years ago as part of work on the BAAS that focused upon the intellectual politics and biographical makeup of the BAAS as the "Parliament of Science" in the early years of the association, before 1845.[3] More recent work has revealed how the BAAS and its annual meetings were attracted to different towns after 1845. Civic deputations, university officials, and figures from leading scientific institutions came together to promote their town or city as an appropriate venue. Local scientific and other bodies were important in prompting an invitation to the BAAS, as was the presence of leading scientific men of national or international standing. Local sites of scientific interest were highlighted, and sites of industry or manufacturing were used to attract BAAS visitors. Once the meeting had been secured and during its week-long presence, handbooks (commonly written by local men of science) advertised these scientific sites and thematically addressed the scientific importance of the local setting to visitors and residents alike.[4]

The effect of the BAAS was not everywhere the same. For Edinburgh in 1871, for example, a local reporter commented thus on the association's civic presence:

> It fills our streets with its finger-posts; it takes our Courts of Law to lounge in; it seizes our University, and fills it with hurry and high debate; it soliloquises in evening dress in our public halls; it prescribes the preachers in our pulpits; it makes itself easy in our drawing-rooms; it raises commotion in our kitchens; it descends to the depths of our cellars; and exercises itself in a great variety of liberties which we are not in the habit of permitting to anybody, and all as if it thought that we ought to consider ourselves particularly well off in being utilised by so potent and august a visitor.[5]

He was circumspect, however, about the wider national effect:

> No one can question the great scientific services rendered by the British Association. It has undoubtedly raised the country from a state of apathy into one of appreciation for scientific pursuits; but that form of action which has enabled it to produce a powerful temporary impression ought to be followed up by another better fitted for continuing and promoting the impression already produced. The Association may, in fact, be compared to [a] gigantic boa-constrictor, which takes one hungry meal a-year,

and lies in a semi-dormant state during the rest of the period. . . . The energy is magnificent, but, at the same time, discontinuous and spasmodic, and though the inhabitants of cities such as Dundee and Bradford may for once in a generation receive a visit from the Association, for twenty or twenty-five years they are left to grow up—and they do grow up—in ignorance of the very existence of this great peripatetic body.[6]

If the national remit of the BAAS as a peripatetic and provincial body had such variable local expression, it is pertinent to ask questions about its modus operandi and to consider the connections between its national mobility and local civic settings. Since, from 1884, the BAAS held seven meetings overseas, such scrutiny might also be applied to its international and imperial dimensions.[7] If the BAAS succeeded by 1871 in shifting the attitude of the British public toward science from apathy to appreciation, was this the case in every town? Perhaps not, since in the eyes of this reporter at least and for smaller towns less frequently visited, the BAAS was a scientific "boa constrictor," its peripatetic nature meaning that science was only occasionally active. How did the BAAS experience vary locally, by and perhaps even within, different towns? Since the BAAS was made up of different scientific sections, each of which had programs of lectures and displays, was the experience of attending a BAAS meeting dependent not just on which town but on which section or activity one attended?[8] Is it possible to capture something of the experience of being at a BAAS meeting, either for the scientist or for those attending?

These questions have significance beyond the better understanding of the workings of a national association for the promotion of science in Britain in the nineteenth century. Others have considered the city as a venue for science and begun to address the use made of particular urban spaces and localities as scientific venues and to examine the cultures of expertise in urban science.[9] Such attention to the urban setting for science and its different expression is part of that wide and established field of research in the spatial nature of science as a whole, understood either in terms of the local sites of its production, in relation to science's mobility and the different forms of its transmission (as published books, private correspondence, public speeches and so on), or the different contexts and forms of science's reception.[10] Questions concerning the replicability of science and its mobility and standardization have likewise been brought into view.[11]

Precisely because the terms *production* and *reception* are problematic in relation to the making, circulation, and what we might term the civic "purchase" of science, questions have been raised about the issues inherent in what Secord terms "knowledge in transit."[12] For some people, for example, scientific knowledge was being "produced" as they read a journal or scientific book, or received and mediated as ideas as they attended others' lectures on the topic. In criticizing the local emphases of work within the geographies of science, Steven Harris has further noted that this has "not only predisposed researchers to choose research sites that are spatially and temporally circumscribed, it has also encouraged the selection of scientific practices that were themselves spatially and temporally circumscribed."[13]

In the nineteenth century, perhaps especially, science had to compete as a form of cultural good in a competitive marketplace as perhaps never before. The result was, often, that "popular" science was more popular than scientific. Illuminated lectures and managed displays would dazzle audiences, but they might not instill much comprehension of scientific principles. Science, either in the specialist laboratory or on a theater stage in front of public audiences, was a matter of site, but it was also a matter of practice, of particular epistemic cultures, each of which either individually or in combination could illustrate the rhetorical, textual, and embodied nature of claims to knowledge through persuasive rhetoric, display, repetition, and experimentation.[14] Certain subjects were seen as difficult or inappropriate for certain sections of society: botany, for example, was regarded as particularly suitable for women, for instance, provided the content was not overly and overtly sexualized.[15] While public display and experimentation could mask comprehension of cognitive content, it is also the case that science was not one thing and so, in turn, local sites for science could afford variable experiences of science: "Not only may a site be experienced differently by different groups of visitors at any one point in time, but visitors at different points in time are almost certain to have different experiences."[16]

Important questions thus remain within work in the geographies of science concerning the nature of "the local," the mobility and communication of science, and the public experience and understanding of what was taken to be science at any one moment or in any one place. Neither is it necessarily clear what the geographical scale should be at which to conduct studies of the geographies (or the urban history) of science: as Livingstone notes, "Precisely what the correct scale of analysis is at which to conduct any particular enquiry into the historical geography of science—site, region, nation, globe—has to be faced."[17] Scale, like place and space, is a complex geographical term.[18] Scale has been the subject of considerable theoretical reappraisal within human geography.[19] It is not my purpose to review this work in detail, but, unlike some who have called for the rejection of a hierarchy of scale from the "local," even the embodied, to the "global," I want to retain its analytic usage, not least for its potential in continuing to develop strong links between scholarship on the geography and the history of science. My concern here in examining the historical geographies of the BAAS and what they reveal of nineteenth-century civic science in Britain is not to suggest *the* correct scale of analysis. It is, rather, to address the importance of *different* scales of analysis and to illustrate scale as a relational matter by addressing the possibilities of working *across* different scales in exploring the workings of a national body that met in thirty different towns between the 1845 Cambridge meeting and the 1900 meeting in Bradford, and in whose fifty-four annual meetings in this period (two of which were overseas), different activities were incorporated in order to advance science differently to different audiences.

PROMOTING PROVINCIAL CIVIC SCIENCE

What Morrell and Thackray described for the hosting of BAAS meetings between 1831 and 1844—everywhere "Local pride, civic rivalry, competitive emulation, and the desire for spectacle united to ensure participation, achieve harmony, and make manifest the resources of science"[20]—was more complicated than this phrase suggests, especially in the second half of the century.

Decisions about which places would host a visit were based on the views of BAAS officers that situating a meeting there should bring scientific and civic benefits and on assessment of civic invitations that stressed the scientific capacity of the location, the educational advantages for the local inhabitants, and the financial support that local civic bodies would give the association. Preference was given to places not visited before if there was any conflict over the possible location. When Leeds and Manchester were discussed as potential sites for the 1858 meeting, for example, both towns were held to have more or less equal advantages. What helped to swing the decision toward Leeds was the fact that it was at the center of "a great district never yet visited by the British Association viz . . . the largest district of the Kingdom uncultivated by the Association."[21]

In his 1853 *On the Prospective Advantages of a Visit to The Town of Hull by the British Association for the Advancement of Science*, Charles Frost, president of the Hull Literary and Philosophical Society, noted that the principal benefit of having the BAAS come was the stimulus it afforded local science: "Herein we have a proof afforded of the utility of the Association in calling into action native talent, and exciting such of the inhabitants as possess a taste for science, to qualify themselves in advance, for taking an active part in the preparation of the intellectual treat to be placed before their philosophical guests."[22] Tourism would be developed; local facilities (such as, in Hull, new meeting rooms for the Literary and Philosophical Society) would often follow the acceptance of an invitation; and locals would benefit from the presence of scientific visitors: "It is scarcely possible to appreciate too highly the mind-purifying and soul-ennobling effects of coming into familiar contact with a vast assemblage of the master spirits of the age."[23] Yet it was also the case, he observed, that the coming of the association did not necessarily mean that it engaged with that town's daily life or, indeed, that residents could engage with the BAAS: "In short, the Association, when in the height of its activity, may be compared to a little commonwealth, which has parasitically located itself in the midst of the visited town, and there acquired, for a brief space, a local habitation—with a population of its own—engaged only in its own pursuits—and governed only by its own peculiar laws and customs."[24]

By the 1860s and 1870s, plans in certain towns to consider hosting a BAAS annual meeting for a second or subsequent occasion drew upon the experience of earlier visits. The BAAS looked favorably upon possible return visits if the earlier meeting had been successful and if science there had developed in the interim. Newcastle's invitation to the association in 1862 stressed how facilities had changed since the 1838 meeting and highlighted the emergent industrial development of the region: "The Scientific Interest of the neighbourhood of Newcastle has increased since 1838 in equal proportion."[25] This fact, together with the geographical location of recent meetings, was helpful in securing Newcastle's case: "There will be of course other claimants, but the ground is all fair for canny Newcastle; & as several later meetings have been far south of her, she has a clear locus standi, & the prestige of a never-to-be forgotten success. One thing: bring a delegation and documents and proof of space in rooms for a large meeting."[26]

In Leeds in respect of the 1858 meeting, the visit was reflected upon in the following year by executive members of the Leeds Philosophical and Literary Society, which body had been influential in bringing the meeting to the city.

> The Local Committee believe that the benefits which it [the meeting] conferred on the town of Leeds were neither few nor trifling. It called forth a large amount of public spirit and of individual energy, and was the means of eliciting from several of our townsmen very valuable contributions to our stock of scientific knowledge. It awakened a new and

lively interest in science and scientific men amongst considerable numbers of our population, and can hardly have failed to create in many cases a desire for more extensive and accurate information. It brought together the theorist and the practical man, who commonly move in separate and remote spheres. It established friendly and personal relationships with many of the distinguished leaders in science, from which the town has already reaped valuable results; and it made known to large numbers of the most educated class the true position of Leeds, as a seat of manufacturing industry and enterprize. The Meeting of the Association, however, they feel, should not be regarded primarily with reference to the benefit which it may have conferred on ourselves. Science is the foundation of the wealth and prosperity of Leeds and it was fitting that, when the opportunity offered, its citizens should welcome and honor the Masters of Science.[27]

Reflecting thirty-two years later upon the 1890 BAAS Leeds meeting, members of the Leeds Geological Association considered the particular benefits afforded to local geological science by the activities of section C:

The special feature of the present year has been the visit of the British Association to Leeds. Much was hoped for from the stimulus which, it was expected, would be given to scientific pursuits by the presence in our town of many eminent scientists. Though there has been no great accession of activity in our own Association as the result of these meetings, the Council feel that the prominence given to Yorkshire Geological work, especially in the Boulder and Photographic departments, ought to encourage the members to a determined endeavour for the attainment of a still higher standard of work in the future. They record with satisfaction that Prof. A. H. Green, F.R.S., F.G.S., one of our Honorary Members, presided over the Geological Section and that our ex-President Mr J. E. Bedford, F.G.S., acted as one of the Secretaries. Papers were read before this Section by two of our members Mr B. Holgate, F.G.S., and Rev. E. Jones, F.G.S. To Mr Bedford was entrusted the leadership of the British Association excursion to Ingleborough.[28]

In sum, the effective working-out of the BAAS's foundational policy of being nationally peripatetic depended always on local circumstances. Hosting an annual meeting involved at least a three-year cycle of negotiations (often more) between BAAS General Committee officers and representatives of local civic and scientific bodies. Places or regions not before visited were favored over other venues, some towns or cities being returned to because of the presence there of local scientists and of potential audiences was greater than elsewhere and because that town or city had earlier success-fully hosted a meeting. Where some local authorities used the BAAS meet-ing as an occasion for urban civic display unrelated to science, it is also the case that places benefited from the prospective presence of a BAAS annual meeting—Dundee's Albert Buildings were built specifically to house the 1867 BAAS meeting, for example—and that having existing civic space was a factor in choosing host towns. In Leeds's case, discussions with a view to an invitation and civic deputation had begun as early as 1850, and invi-tations were renewed annually, without success, between 1851 and 1857. What held Leeds back was the lack of a building in which to host a BAAS meeting. In the view of members of the Leeds Philosophical and Literary Society, the completion of the town hall rather than the fact that the whole region had not been visited by the BAAS was crucial in influencing the BAAS General Committee in favor of Leeds in 1857: "It was pointed out that Leeds had long desired a visit, and now that the only difficulty in the way was clearly removed by the erection of the Hall, it was contended that our town was clearly entitled to take precedence of Manchester, which had already received the Association."[29]

Such evidence about particular buildings, rather than BAAS national policy or the circumstances of individual towns, suggests that it is impor-tant to consider the conduct of science at the scale of individual meetings. To do so and to look at the workings of BAAS meetings in their urban set-tings and at the undertaking of BAAS science in its different sections afford different perspectives into what science was, and how it was performed.

PRACTICING CIVIC SCIENCE: SECTION E PRESIDENTIAL ADDRESSES AND THE PERFORMANCE OF GEOGRAPHY

BAAS annual meetings evolved a more-or-less standard format: a public ad-dress by the BAAS president on a scientific topic, sectional programs of papers (themselves usually begun by that section's presidential address), a *conversazione* or other formal social occasion, and day trips or longer field excursions. BAAS meetings were not so much one event as several, each with different specific settings—museums or town halls for official ad-dresses and section papers, field sites for excursions, industrial plants for themed visits—in which different practices such as lecturing, displaying, collecting, and so on took place.

In illustration of these different activities, I here examine the presidential addresses to section E, Geography, between that of Sir Roderick Murchison to the 1858 Leeds meeting and Sir George Robertson's address to the Bradford meeting in 1900, and look at elements of the spoken papers within section E in this period. My concern is with the addresses' cognitive content, with shifts in the nature of geography as a science within the BAAS, and with the related reception of these presidential talks and the sectional papers as reported in contemporary subject-based professional journals.[30] Reports on section E proceedings were regularly carried in geography journals, notably in the *Geographical Journal* (the outlet of the Royal Geographical Society) and in the *Scottish Geographical Magazine* (the outlet of the Royal Scottish Geographical Society). These journals also carried the published versions of section E papers, as did the journals of other subjects and the transactions of numerous regional societies.[31]

Most published reports on BAAS meetings were short, summarizing the president's speech and the content of sectional papers delivered. Reporting upon the 1887 Manchester meeting, for example, Hugh Robert Mill offered no view upon Sir Charles Warren's presidential address other than to complain that the concurrent timing of most sections' presidential addresses meant that one could not learn of other sections' work, but he did praise Sir Francis de Winton's public evening lecture on African exploration as "a model of what a popular lecture should be." Mill also noted how the meeting "was of more than usual interest to geographers, for, besides a number of valuable papers read to section E, there were many bearing directly upon the subject submitted to other sections." Noting too that "very important work was got through on Monday"—which included discussions on the scope and teaching of geography and on the work of the Ordnance Survey—Mill reports upon scenes of "extremely animated" debate, hurried papers because of lax timetabling, and, in closing, criticizes the poor quality of the BAAS handbook: "a sorry contrast both in bulk and quality to the Birmingham hand-book of last year."[32]

We can use this evidence to reveal something not just of what was said and by whom, but how it was said, and to what purpose, and, even, to learn something of the reaction from the different audiences in question. Table 5.1 documents the subject matter and the speakers of the forty section E presidential addresses given between 1858 and 1900. The principal subjects covered were: geography and exploration (with different emphases at one time or another upon Africa, the Americas, Asia, or the polar regions); geography and ethnology (in terms of the subject matter of human cultures and their environmental bases, and the strained intellectual and social relations

TABLE 5.1. BAAS section E presidential addresses, 1858–1900: Speaker and subject

Year	Location	Speaker	Principal subject matter
1858	Leeds	Sir R. I. Murchison	Relationships between geography and ethnology
1860	Oxford	Murchison	African exploration; definition of geography
1861	Manchester	Sir John Crawfurd	Ethnology and physical geography
1863	Newcastle-upon-Tyne	Murchison	Geography and exploration; role of section E
1864	Bath	Murchison	Geography and exploration; role of section E
1865	Birmingham	Sir Charles Nicholson	Geography and ethnology; exploration
1867	Dundee	Sir Samuel Baker	Geography, exploration, and commerce
1868	Norwich	Capt. Richards RN	Geography and exploration
1869	Exeter	Sir Bartle Frere	The progress of geographical discovery
1870	Liverpool	Murchison	Geography and the role of section E
1871	Edinburgh	Sir Henry Yule	Physical geography of the Far East
1872	Brighton	Francis Galton	Nature of section E; Ordnance Survey mapping
1873	Bradford	Sir Rutherford Alcock	Scope and application of geography
1874	Belfast	Maj. Charles Wilson	Physical geography and military operations
1875	Bristol	Lt. Gen. R. Strachey	Geography, exploration, and scientific specialization
1876	Glasgow	F. J. Evans	African exploration; *Challenger* expedition
1877	Plymouth	Sir Erasmus Ommanney	Advances in geographical science since 1841 [last Plymouth meeting]
1878	Dublin	Sir C. Wyville Thomson	*Challenger* expedition
1879	Sheffield	Clements Markham	"The objects and aims of geographers"
1880	Swansea	Lt. Gen. Sir J. H. Lefroy	Exploration in North America
1881	York	Sir Joseph Hooker	Geographical distribution of animals and plants

TABLE 5.1. *(continued)*

Year	Location	Speaker	Principal subject matter
1882	Southampton	Sir Richard Temple	Central Plateau of Asia
1883	Southport	Lt. Col. H. Godwin-Austen	Physical geography of the Himalayas
1884	Montreal	Lefroy	Recent progress in geographical science
1885	Aberdeen	Gen. J. T. Walker	Scientific geography and mapping
1886	Birmingham	Maj.-Gen. F. J. Goldsmid	Popularizing geography through schools
1887	Manchester	Col. Sir C. Warren	Geography education in schools
1888	Bath	Col. Sir C. W. Wilson	Geography and trade; exploration
1889	Newcastle-upon-Tyne	Sir Francis de Winton	"Science of applied geography, commerce"
1890	Leeds	Lt. Col. Sir Lambert Playfair	The Mediterranean
1891	Cardiff	E. G. Ravenstein	"The field of geography"
1892	Edinburgh	James Geikie	Geography of coastlines
1893	Nottingham	Henry Seebohm	Nature of geography; polar exploration
1894	Oxford	Capt. W. Wharton	Hydrography
1895	Ipswich	H. J. Mackinder	Geography, exploration, and university education
1896	Liverpool	Maj. L. Darwin	Polar exploration; geography and trade
1897	Toronto	J. Scott Keltie	Exploration; geographical education
1898	Bristol	Col. G. Earl Church	South America
1899	Dover	Sir John Murray	*Challenger* expedition; oceanography
1900	Bradford	Sir G. S. Robertson	"Political geography and the empire"

SOURCE: *Annual Report* . . . [Meeting of the BAAS]. Most section E presidential addresses were untitled: the subject matter has been here identified by reading the address in question. Although section E did have some presidential addresses before 1858 (i.e., Sir Roderick Murchison in 1848), complicated sectional relationships between geography and geology (section C) meant that section E addresses were not consistently given until 1858.

NOTE: Where a year is missing, no section E presidential address was given.

between proponents of the two human sciences in the BAAS); mapping as imperial and national survey; geography and education; and, least often, the effectiveness of section E as a forum for promoting geography.

Several addresses are notable in terms of the tones taken by the speakers in emphasizing their selected topics. At the 1864 Bath meeting, for instance, where crowds of thousands gathered to hear David Livingstone speak on his return from the 1858–63 Zambesi Expedition, Murchison used his presidential address to counter views that, since so few places remained to be known, geography was "exhausted." Exploration was key to geography; reports upon exploration key to the popularity of Section E: understood thus, geography was a "noble and unlimited science."[33] Eight years later, in Brighton, where Livingstone again attracted large crowds to section E, Francis Galton argued forcefully against this view, noting that "the career of the explorer, though still brilliant, is inevitably coming to an end" and that if the geography section was "to secure the attention of representatives of all branches of science," it was essential that geography turn to "principles and relations" rather than "primary facts."[34] This point about geography's credentials as a science (which for advocates ought to be apparent in its methods) and its connections with other sciences was emphasized by Strachey in his 1875 presidential address. In stressing that a "more scientific geography" should incorporate "the doctrine of evolution" in the study of human-environment relationships, Strachey demonstrated an awareness of and commitment to Darwin's work, and sought to situate geography in relation to wider debates with a view to that science's intellectual betterment: "The field of topographical exploration is already greatly limited. . . . The necessary consequence is an increased tendency to give geographical investigations a more strictly scientific direction."[35]

Given by men with geographical interests and credentials but not, in the modern sense of the term *professional* geographers, section E presidential addresses thus made important spoken statements, using varying specific content, about the nature and focus of geography's scientific basis and its connections with other scientific work and did so before the section's programs of papers and largely "professional" audience of paper presenters. The performance of such "disciplinary statements" and of the sectional papers and the reception of them had different audiences, however. Presentation and reception could be affected by the simple facts of timing. Reporting in 1887 upon what he saw as the "universal evil" of the lack of planning in conducting all sectional proceedings, H. R. Mill made the problem clear, as we have seen:

The long sittings, from eleven to three, and the impossibility of judging at what time a given paper will be taken, are productive of much dissatisfaction and annoyance. Only people of very great importance can command an audience in the section rooms between one and two o'clock; and there was hardly a day in any section on which the papers coming late on the list could be read properly, or discussed at any length. This was felt to be unjust, particularly by foreign members, who regarded the discussion of papers as the most important purpose of the Association. Some simple arrangements would greatly improve matters . . . [and] would also prevent the undignified devouring of sandwiches by the chairman while papers are being read. . . . The arrangements for garden parties, conversaziones, and excursions were, as they have been for many years, much more perfect than those for the scientific meetings.[36]

Two years later, reporting upon the Newcastle meeting, Mill struck a different note. Although the 1889 meeting of section E was designed to help promote commercial themes and the role of provincial geographical societies in this and in similar towns,[37] neither the papers nor the subsequent discussion met those objectives. "The address of Sir Francis de Winton as President was intended to strike a key-note of Commercial Geography, which it was hoped would give tone to the following papers and discussions. The result was disappointing. Except in one or two instances the papers which made any pretensions to the title of Commercial Geography contained very little Geography, and the discussion on several occasions became purely political, the addresses of some speakers being thoroughly inappropriate to a scientific meeting."

In other respects, however, the section's business was successful and warmly received: "Explorers always attract the public, and this year they were well represented. Dr. Nansen as he recounted his adventures on the Greenland ice must have felt that his heroic journey had been fully appreciated. The simple manly style in which he told the now familiar story, and his quaint and unexpected gleams of humour, commanded the closest attention of a crowded and enthusiastic audience."[38]

This last remark illustrates a feature distinctive to BAAS section E science. Popular lectures on exploration—just the topic that leading men of science saw as detracting from geography's scientific status—were well attended, especially if given by leading figures such as David Livingstone. More specialist papers and presidential addresses contributed to scientific knowledge but did so to much-reduced audiences. If explorers in person,

humor and a "manly style" attracted audiences to geography, so did the
visual cultures associated with the practices of telling and displaying—the
performance—of exploration narratives and tales of travel. A report upon
the 1894 Oxford meeting highlights this (and endorses the views of Galton
and Strachey decades before):

> The attendance was from first to last exceptionally good, and the audi-
> ences held together with quite remarkable constancy. That, however,
> was evidently due to the popular rather than to the strictly scientific
> character of the papers and discussions. It must always be remembered
> that in the British Association two audiences have to be catered for—the
> scientists and the amateurs—and that the financial success of the meet-
> ing depends much more on the latter than on the former. Complaints are
> often made—they have been made this year—that the treatment of the
> subjects in some sections is so technical that it repels the lay members,
> who form the bulk of the membership of the association. The objection
> may or may not be deserving of notice; but it certainly does not apply to
> the Geographical Section. That section is the happy hunting-ground of
> the unattached and amateur Associate. Thanks to the profuse and pro-
> miscuous use of the magic lantern, it has become the attractive show-
> room of the Association. But geography as a science, or the scientific
> aspect of geography, does not gain much by this ephemeral popularity.
> The audience is panting for sensations; the ubiquitous and irrepressible
> globe-trotter is the ideal of the hour, and the sensation is all the greater
> if the globe-trotter happens to be a woman. The paper which attracts a
> crowded audience may be a tedious narrative of a holiday spent in Ar-
> menia, or in Mexico, or in the desert of Libya, or in Montenegro, or in
> Arabia; but all its sins are forgiven if it is illustrated by what the official
> programme calls "optical projections," which means, in common par-
> lance, "lantern slides."[39]

Such evidence concerning the spoken papers and performances of BAAS
sessions is important for what it suggests more generally of the making and
reception of science, but there are interpretative dangers. Published reports
in professional journals do not declare the personal interests and status of
the reporters. By convention, reports on science did not always see it as
necessary to comment upon the rhetorical style of the scientist and commented only occasionally upon the reaction of the audience. The science
of geography attracted its largest audiences to its least scientific papers. In
contrast, in geography's presidential addresses and in reports upon them,

we can by the later nineteenth century note an emphasis upon geography as a science by leading geographers and other scientists. This schism in section E between specialist knowledge and disciplinary identity as a science and popular public science was heightened by the presence of leading explorers and by the use of technology ("magic lanterns," maps, and glass-plate slides), which lent an air of theatricality to the section's proceedings. This is not to state that specialist and technical papers always repelled audiences and that explorers' narratives and illustrated talks on popular subjects always attracted them. It is to suggest that since different locations could and did emphasize different topics, being at a BAAS meeting depended not just upon urban setting but upon what one attended in particular sites, upon whom you heard and what you saw.

EXPERIENCING CIVIC SCIENCE

Where printed, professional, and public reaction to BAAS science was by men writing most often in a disciplinary context, more private responses often survive only in letter or diary form. Many are by women. Women formed a large part of BAAS audiences, especially from midcentury. Like men, they paid for admission and so contributed to association finances, but they did not, in the main, contribute to the science itself, either as presenters or in public discussion. Women were nonetheless crucial to the operation of the BAAS as its audience: they were participants in the social circuitry surrounding the civic engagement with science, but not makers of science.[40]

The diary of Agnes Hudson, who attended the 1875 Bristol and 1879 Sheffield meetings, is illuminative in several respects. On the evening of 25 August 1875, she and her companion set out "to attend the first general meeting," but found it difficult to get to, all the carriages in Bristol having been requisitioned to convey people there. Even after arrival, things were little better: "When we reached the Hall which is a handsome, but insufficiently ventilated building we found the heat intolerable and we could not hear a word that Sir John Hawkshaw, the President [Hawkshaw was an engineer], said we soon withdrew, determing [sic] to read his address next day" [official addresses being printed in the following day's newspapers]. The next day,

> We went to . . . Section D. on Anthropology. We sat there a long while doing nothing, for the papers were not yet begun to be read till an hour after the right time, it was rather amusing when the proceedings did

begin for a gentleman in the audience had to rise and defend himself against some accusations made by the speaker, Mr. Pengelly, and they got quite excited over it. . . . When we had had enough of the Section we went to the Victoria Rooms and had lunch, . . . In the afternoon we walked up Brandon Hill which is quite close and whence you have a very good view of the whole of Bristol, on our way home we ordered some flowers for our hair and then dressed for dinner for the first time since our arrival here. At 8 p.m. we started in a cab for the Colston Hall where was held the microscopical Soirée, we encountered Mr Hart there who was much surprised to find that we had no one to look after us, and who informed us that some of the objects under the microscopes were well worth looking at, we found however that it required more patience than we possessed, the hall was crowded and very hot, which necessitated our imbibing a large quantity of ices, which were remarkably good for such an occasion. We arrived home safely at 10.30.

Next morning, Hudson recorded, the Anthropological Section sessions were so crowded that "several persons sat on the mantelpiece." Later in the week [31 August], "we went to the geographical Section where a paper was read on the Arctic expedition by Mr Markham [Clements Markham: president of section E in Sheffield in 1879] who had just returned to England on the Valorous, Colonel Montgomerie read a paper on the Himalayas, he pronounced it Himâlia, which was not at all interesting and which I do not think anyone would have listened to but that the Arctic expedition came next."[41]

Four years later, in Sheffield, she was again disappointed by section E proceedings: "Major Serpa Pinto [a leading Portuguese explorer] was to read a paper on his travels in Africa, but he deputed an English man to read it for him, and we did not find it so interesting as we had expected." An afternoon excursion to steel mills followed: "We returned to Sheffield at 8.20, and went home by ourselves, many of our fellow travellers hurried off to an evening lecture on radiant matter, but we went straight home."[42]

While Hudson's personal recollections are not a secure basis for wider generalization, they are nonetheless valuable in revealing one individual's experiences. Attendance at a BAAS meeting could be tiring, require a change of clothes (for women perhaps more than for men), and last well into the evening. Being present at paper sessions involved spectatorship of related discussions more than comprehension of the science. Motivation in going at all might stem from a combination of the speaker and the topic, one's engagement with the subject diminishing if the expected speaker did not appear.

There is supporting evidence about the audibility of speakers and about what we might think of as the celebrity status of some speakers. Caroline Fox attended the association meetings in 1836, 1837, 1852, and in 1857. Recounting her experiences at the 1836 meeting in Bristol, she recalled the crowds "at the large British Babylon" on 22 August that year. "By most extraordinary muscular exertions," she recalled, "we succeeded in gaining admittance. We got fairish seats, but all the time the people made such a provoking noise, talking, coming in, and going out, opening and shutting boxes, that very little could we hear."[43] In 1852, she spoke of the "great treat to be present" at discussions on the fate of Franklin and in 1857 was likewise thrilled to be present to hear African explorers, including David Livingstone: "tall, thin, earnest-looking, and business-like: far more given, I should say, to do his work than to talk about it."[44] At least Fox heard Livingstone: Lady Caroline Howard and her companions went to the geography session at the 1857 Dublin meeting only because they were "rather tired of geology" and found zoology "uninteresting," "but the room was greatly crowded and so we did not hear much." Her companions were even less fortunate in attending Livingstone's lecture on African discoveries: "Julie enjoyed herself so much and brought me back such an account of it that I felt quite in despair at being laid up. They however did not hear one word of the lecture as they got bad seats, and Dr Livingstone speaks in a whisper."[45] The reporter to the *Dublin Evening Post* explained Livingstone's inaudibility: "His voice had suffered severely from constant speaking under trees, which had no covering but the vault of heaven, and he regretted that he was not able to make himself better heard."[46]

What was significant, of course, was not Livingstone's audibility but his celebrity. Thousands went to the 1864 Bath meeting where Livingstone spoke after his Zambesi expedition and where the whole meeting of the geographical section was colored by the personal invective between Alfred Russel Wallace and John Crawfurd over their contrasting views on race, slavery, and the treatment of the Maoris, and by the death on the nearby Bathampton Downs, most likely by suicide, of the African explorer John Hanning Speke. On 16 August 1872, when Henry Morton Stanley spoke to more than three thousand people at Brighton, public excitement at seeing him was heightened when Stanley departed from his prepared speech and inveighed against his detractors. The public audience applauded Stanley. His chairman, Francis Galton, rebuked the public cheering: "I must beg to remind you," Galton declared, "that this is a serious society constituted for the purpose of dealing with geographical facts and not sensational stories."[47] In these words, Galton echoed his own presidential address wherein, as we have observed, he had seen section E to be not a geographical society [that function belonging to the Royal Geographical Society, of which he was a fellow] but "a constituent of a great scientific organization" and a forum that should enable geographers to "secure the attention of representatives of all branches of science."[48] But what Galton saw as an institutional space for the development of geography's scientific credentials, Stanley used—to the public's delight—as an opportunity for sensational narratives, personal invective, and public acclaim.

Experiencing geographical science depended, then, not upon differences between "professionals" and "amateurs," but between leading proponents of geography's still-uncertain disciplinary status and celebrity practitioners whose renditions of exploration were rejected by the discipline's proponents yet eagerly anticipated and raucously received by public audiences. The effects of celebrity were not confined to geography: in Dublin in 1878, the huge crowds that attended Thomas Huxley's defence of Darwinism left straight after: as the *Irish Times* recounted "after him the deluge, or rather the evaporation."[49]

This study of the nature of BAAS science in its later British meetings reflects, if unequally, those themes of the rise of urban expertise, science, and the representation of the city, places of knowledge in urban context, and knowledge from the street proposed by Dierig, Lachmund, and Mendelsohn in reviewing the urban history of science.[50] In realizing its overarching policy of ensuring the mobility of science as a national cultural good, the BAAS drew upon local expertise in different towns, upon scientific institutions whose members formed civic delegations to invite the association, or who led BAAS excursions, gave papers at its annual meetings, and conducted its week-long business in a variety of civic sites devoted, in the main, to different social and scientific practices. The BAAS was concerned always that science should have civic benefit. How that was interpreted and carried out varied by town, subject, and sectional activity, as well as by audience and within the ranks of delegates—a term which does not easily equate or translate into *professionals* and *public, experts* and *lay*—for whom analytic and conceptual distinctions between the production and the reception of science would not have been meaningful.

Several things follow from this evidence. One important point to note is that questions of geographical scale matter in interpretation of how the BAAS worked, since questions of *how* civic science worked are inseparable from *where.* One consequence of examining the different places and sites in which science was made in BAAS meetings in Britain between 1845 and 1900 is that the city or the town is not the necessary unit of analysis in any urban historical geography of science. Another related consequence is that the BAAS itself ought not to be seen as a single body, since different sections, practitioners, and epistemic cultures could and did give shape to what the science was—and, in relation to some audiences, shaped whether or not what was heard or seen was reckoned science at all.

Understanding the geographies of civic science demands attention, then, to the relationships between and within given levels of geographical scale and the nature of the activities undertaken there: from national policy, to local city-based realization, to the nature of particular buildings and sectional activities, and even to individual rooms (whether they were big enough for science's audiences, or too hot, or if people had to sit on the mantelpiece, or could hear at all), and so to the embodied experiences of different individuals.

Different practices in different settings—waiting for a lecture whose timetabling and audience behavior meant that hearing particular topics was a matter of luck, conversing with one's fellows, viewing specimens without comprehension, going to lectures to seek sensation or instrumental mediation through lantern slides not understanding of scientific principles—were all elements in the making and reception of association science. The experience of BAAS science could vary between participants at the level of "organizers" and "audience." In Glasgow in 1876, for instance, leading university figures and men of science thought the meeting a great success. By contrast, many townsfolk reckoned it "exceptionally dull" and, therefore, a failure. But, noted one reporter, it ought to be remembered "that the object of the Association is not to amuse the public, but to advance science, and that it may happen that the most exciting meetings may be the least productive of good results." As he continued in illustration of his point, "There was no morning, for example, so popular as in that in which we were told about the dancing tables, and the spirits that animated gin-and-water, and in which Wallace and Carpenter [a reference to the debates between Alfred Russel Wallace and the physician Alfred Carpenter over ethnology and Darwinian thinking] battled with words and still more with eyes, yet perhaps there was no morning more conspicuous for its absence of science."[51] To the editor of the Edinburgh-based *Scotsman*, there was little point in having the BAAS come to one's city if there was to be "nothing in the nature of a *quid pro quo* meant or involved." Reflecting that "the inhabitants of Glasgow, as might have been expected, had fed them well, and in every way treated them as handsomely as they could," it was nevertheless the case that "to the great majority of the outside public, the whole proceedings, with the exception of the festivities, must have seemed a sort of intellectual jungle, into which to plunge is to be lost."[52] For such reasons, we must be cautious about seeing BAAS meetings as split only along lines of socioscientific status, between "experts" and "public," "practitioners" and "delegates." These categories were shaped by different practices in different sites.

The fact that different contemporaries saw the BAAS as "a gigantic boa constrictor," a "parasitic little commonwealth" and an "intellectual jungle" is informative to a certain degree of contemporaries' reactions toward the annual meeting of this scientific body. Rather more significant, perhaps, in modern analysis of the work of the BAAS is the fact that our understanding of how institutional pretensions to civic science were realized depends to an extent upon the geographical scales at which we look and upon the practices and differences within its meetings. Historical understanding of science's making over time must always recognize its variant making in place, but not upon "correct" scales of reference. In considering the ways in which the BAAS meetings worked and for whom, this chapter has both established the importance of local work and tried to place local work as, always, part of bigger narratives. What civic science was, as evident in the workings of the BAAS meetings, could and did have different meanings for different people. That there was such variety serves to remind us that while it may be difficult to capture fully the exchanges, glances, and daily goings on in the public's engagement with BAAS science, we must be mindful as we seek to do so that such experiences and practices were always influenced by what the science in question was and how and where it was encountered.

NOTES

I am grateful to participants at the Geographies of Nineteenth-Century Science meeting, to seminar audiences in Edinburgh and in Toronto for comments on earlier spoken versions, to the readers appointed by the University of Chicago, and to the librarians and archivists of those institutions cited here for access to the material in question. The research on which this chapter is based was undertaken through ESRC grant RES-000-23-0927, and I gratefully acknowledge this support.

1. Roy MacLeod and Peter Collins, eds., *The Parliament of Science: the British Association for the Advancement of Science* (Northwood: Science Reviews, 1981), i.

2. Jack Morrell and Arnold Thackray, *Gentlemen of Science: Early Years of the British Association for the Advancement of Science* (Oxford: Clarendon, 1981), 98.

3. See, notably, Macleod and Collins, *Parliament of Science*, and Morrell and Thackray, *Gentlemen of Science*; but see also A. D. Orange, "The British Association for the Advancement of Science: The Provincial Background," *Science Studies* 1 (1971): 315–29; J. M. Edmonds and R. A. Beardmore, "John Phillips and the Early Meetings of the British Association," *Advancement of Science* 12 (1955): 97–104.

4. Charles W. J. Withers, Rebekah Higgitt, and Diarmid Finnegan, "Historical Geographies of Provincial Science: Themes in the Setting and Reception of the British Association for the Advancement of Science in Britain and Ireland, 1831–c.1939," *British Journal for the History of Science* 43 (2008): 1–31. For a fuller account of the BAAS's civic geographies in the century from 1831, including its international dimensions, see Charles W. J. Withers, *Geography and Science in Britain, 1831–1939: A Study of the British Association for the Advancement of Science* (Manchester: Manchester University Press, 2010).

5. "The British Association in Edinburgh," *Scotsman*, 3 August 1871.

6. "The British Association in Edinburgh," *Scotsman*, 9 August 1871.

7. The BAAS held seven overseas meetings (not discussed here): Montreal 1884; South Africa 1905, 1929; Winnipeg 1897; Toronto 1909, 1924; and Australia in 1914. For the BAAS overseas, see also Michael Worboys, "The British Association and Empire: Science and Social Imperialism," in MacLeod and Collins, *Parliament of Science*, 170–87; Saul Dubow, "A Commonwealth of Science: The British Association in South Africa, 1905 and 1929," in *Science and Society in Southern Africa*, ed. Saul Dubow (Manchester: Manchester University Press, 2000), 66–99.

8. Section A was Mathematics and Physics; B Chemistry; C Geology; D Zoology; E Geography; F Economics; G Engineering; H Anthropology; I Physiology; J Psychology; K Botany; L Education; and M Agriculture.

9. S. Dierig, J. Lachmund, and A. J. Mendelsohn, "Introduction: Toward an Urban History of Science," *Osiris* 18 (2003): 1–19.

10. See, for example, Simon Naylor, "Introduction: Historical Geographies of Science," *British Journal for the History of Science* 38 (2005): 1–12; David Livingstone, *Putting Science in Its Place: Geographies of Scientific Knowledge* (Chicago: University of Chicago Press, 2003); Charles W. J. Withers, *Geography, Science and National Identity: Scotland since 1520* (Cambridge: Cambridge University Press, 2001), 1–28; A. Simoes, A. Carneiro, and M. P. Diogo, eds., *Travels of Learning: A Geography of Science in Europe* (Dordrecht: Kluwer Academic, 2003); Crosbie Smith and Jon Agar, eds., *Making Space for Science: Territorial Themes in the Shaping of Knowledge* (Basingstoke: Macmillan, 1998); Jan Golinski, *Making Natural Knowledge: Constructivism and the History of Science* (Cambridge: Cambridge University Press, 1998); Adir Ophir and Steven Shapin, "The Place of Knowledge: A Methodological Survey," *Science in Context* 4 (1991): 3–21; Steven Shapin, "Placing the View from Nowhere: Historical and Sociological Problems in the Location of Science," *Transactions of the Institute of British Geographers* 23 (1998): 5–12; James Secord, "Knowledge in Transit," *Isis* 95 (2004): 654–72; Diarmid A. Finnegan, "The Spatial Turn: Geographical Approaches in the History of Science," *Journal of the*

History of Biology 41 (2008): 369–88; Richard Powell, "Geographies of Science: Histories, Localities, Practices, Futures," *Progress in Human Geography* 31 (2007): 309–29.

11. See the essays in Marie-Noëlle Bourguet, Christian Licoppe, and H. Otto Sibum, eds., *Instruments, Travel and Science: Itineraries of Precision from the Seventeenth to the Twentieth Century* (London: Routledge, 2003); Kapil Raj, *Relocating Modern Science: Circulation and the Construction of Knowledge in South Asia and Europe, 1650–1900* (Basingstoke: Macmillan, 2006).

12. Secord, "Knowledge in Transit." Secord has done much to establish the importance of different geographies of interpretation and reception of science in his *Victorian Sensation: the Extraordinary Publication, Reception, and Secret Authorship of "Vestiges of the Natural History of Creation"* (Chicago: University of Chicago Press, 2000).

13. Steven J. Harris, "Long-Distance Corporations, Big Sciences and the Geography of Knowledge," in "The Scientific Revolution as Narrative," ed. Mario Biagioli and Steven J. Harris, special issue, *Configurations* 6 (1998): 297.

14. For example, Bruno Latour, *Science in Action: How to Follow Scientists and Engineers Through Society* (Cambridge, MA: Harvard University Press, 1987); Karin Knorr-Cetina and M. Mulkay, eds., *Science Observed: Perspectives on the Social Study of Science* (London: Sage, 1983); Karin Knorr-Cetina, *Epistemic Cultures: How the Sciences Make Knowledge* (Cambridge, MA: Harvard University Press, 1999); Andrew Pickering, ed., *Science as Practice and Culture* (Chicago: University of Chicago Press, 1992).

15. On these points, see the essays in Aileen Fyfe and Bernard Lightman, eds., *Science in the Marketplace: Nineteenth-Century Sites and Experiences* (Chicago: University of Chicago Press, 2007); and Bernard Lightman, *Victorian Popularizers of Science: Designing Nature for New Audiences* (Chicago: University of Chicago Press, 2007).

16. Aileen Fyfe and Bernard Lightman, "Science in the Marketplace: An Introduction," in Fyfe and Lightman, *Science in the Marketplace*, 5. For a preliminary discussion of the reception of BAAS science, see Charles W. J. Withers, "Geographies of Science and Public Understanding? Exploring the Reception of the British Association for the Advancement of Science in Britain and in Ireland, *c.*1845–1939," in *Geographies of Science*, ed. Peter Meusburger, David Livingstone, and Heike Jons, (Heidelberg: Springer Science, 2010), 185–97.

17. David Livingstone, "Text, Talk and Testimony: Geographical Reflections on Scientific Habits: An Afterword," *British Journal for the History of Science* 38 (2005): 99.

18. On "place" and "space," see Charles W. J. Withers, "Place and the 'Spatial Turn' in Geography and in History," *Journal for the History of Ideas* 70 (2009): 637–58.

19. On this debate, see Sallie Marston, "The Social Construction of Scale," *Progress in Human Geography* 24 (2000): 219–42; Sallie Marston, John Paul Jones III, and Keith Woodward, "Human Geography without Scale," *Transactions of the Institute of British Geographers* 30 (2005): 416–32; Eric Sheppard and R. B. McMaster, eds., *Scale and Geographic Enquiry: Nature, Society and Method* (Oxford: Blackwell, 2004).

20. Morrell and Thackray, *Gentlemen of Science*, 129.

21. Bodleian Library, Oxford, British Association for the Advancement of Science Archive, MS Dep. BAAS 18, f.119, Dublin, 31 August 1857 (hereafter MS Dep. BAAS).

22. C. Frost, *On the Prospective Advantages of a Visit to the Town of Hull by the*

British Association for the Advancement of Science (Hull: Privately printed, 1853), 26. On this pamphlet, see also Phillip Lowe, "The British Association and the Provincial Public," in MacLeod and Collins, *Parliament of Science*, 118–44, especially 123–6: the quotation is from p. 124.

23. Frost, *On the Prospective Advantages of a Visit*, 32.

24. Ibid., 26.

25. Committee Minutes of the Literary and Philosophical Society of Newcastle upon Tyne, 18 September 1862, n.p. (University of Newcastle Library).

26. Correspondence Books of the Literary and Philosophical Society of Newcastle, 28 August 1862, n.p. (University of Newcastle Library). As members of the society pointed out in their deputation, "Since that time [1838], accommodation in Newcastle appropriate to holding a further BAAS meeting had greatly increased: its own Lecture hall could now hold 800 persons, the town's Library 1000."

27. Leeds University Library, Leeds Philosophical and Literary Society, MS 28 A, pp. 16–17.

28. Leeds University Library, Leeds Geological Association Minutes, Dep/052, Box 5, Minutes Book, Report of Council for the Session 1890–1891.

29. Leeds University Library, MS 28A, Reports and Minutes of the Leeds Philosophical and Literary Society, 1859, p. 3.

30. On the cognitive content of section E from 1831 and its relationships with other BAAS sections, see Charles W. J. Withers, Diarmid A. Finnegan, and Rebekah Higgitt, "Geography's Other Histories? Geography and Science in the British Association for the Advancement of Science, 1831–c.1933," *Transactions of the Institute of British Geographers* 31 (2006): 433–51.

31. In the two-year period 1889–1891, for example, 98 papers given at the BAAS section E (geography) sessions were published in other societies' journals: 49 by the Royal Scottish Geographical Society; 35 by the Manchester Geographical Society; 3 by the Tyneside Geographical Society; 3 by the Glasgow Philosophical Society; 2 each by the Belfast Natural History and Philosophical Society and the Birmingham Natural History and Microscopical Society; and single papers by the Essex Field Club, the Leeds Geological Association, the Somersetshire Archaeological and Natural History Society, and the Warwickshire Naturalists' and Archaeologists' Field Club. The Corresponding Societies of the BAAS (formally begun in 1884) played an important role in this: on the Corresponding Societies, see Roy M. MacLeod, J. R. Friday, and C. Gregor, *The Corresponding Societies of the British Association for the Advancement of Science, 1883–1929: A Survey of Historical Records, Archives and Publications* (London: Mansell, 1975).

32. Hugh R. Mill, "Report to Council," *Scottish Geographical Magazine* 3 (1887): 525, 527, 529.

33. Roderick I. Murchison, "Transactions of the Sections: Geography and Ethnology," *Report of the Thirty-Fourth Annual Meeting of the British Association for the Advancement of Science* (London: John Murray, 1865), 130–35.

34. Francis Galton, "Transactions of the Sections: Geography," *Report of the Forty-Second Annual Meeting of the British Association for the Advancement of Science* (London: John Murray, 1873), 198–202.

35. Lt. Gen. Richard Strachey, "Transactions of the Sections: Geography," *Report of the Forty-Fifth Annual Meeting of the British Association for the Advancement of Science* (London: John Murray, 1876), 180–88.

36. Mill, "Report to Council," 524.

37. John M. Mackenzie, "The Provincial Geographical Societies in Britain, 1884–1914," in *Geography and Imperialism, 1820–1940*, ed. Morag Bell, Robin Butlin, and Michael Heffernan (Manchester: Manchester University Press, 1995), 93–124.

38. Hugh R. Mill, "Report to Council on the British Association Meeting at Newcastle, 1889," *Scottish Geographical Magazine* 5 (1889): 606.

39. Walter S. Dalgleish, "Geography at the British Association, Oxford, August 1894," *Scottish Geographical Magazine* 10 (1894): 463.

40. On this point, see Rebekah Higgitt and Charles W. J. Withers, "Science and Sociability: Women as Audience at the British Association for the Advancement of Science, 1831–1901," *Isis* 99 (2008): 1–27.

41. Bodleian Library, MS. Eng.e.386, Diary by Agnes M. Hudson of her attendance at British Association meetings in Bristol and Sheffield and her journeys to France and Belgium, 1875–1879, ff.5v, 6–7, 11v–12v.

42. Ibid., ff.60v, 61v.

43. H. Pym, ed. *Memories of Old Friends, Being Extracts from the Journals and Letters of Caroline Fox* (London: Smith Elder and Co., 1882), 4.

44. Pym, *Memories of Old Friends*, 313.

45. National Library of Ireland, MS 4792, 31 August 1857.

46. *Dublin Evening Post*, 1 September 1857.

47. Timothy Jeal, *Stanley: The Impossible Life of Africa's Greatest Explorer* (London: Faber, 2007), 140.

48. Galton, "Transactions of the Sections: Geography," 198.

49. *The Irish Times*, 17 August 1878.

50. Dierig, Lachmund, and Mendelsohn, "Introduction."

51. *Glasgow News*, 14 September 1876.

52. *Scotsman*, 14 September 1876.

CHAPTER SIX

Islanded:
Natural History in the British Colonization
of Ceylon

SUJIT SIVASUNDARAM

Arcadias, utopias, Edens, and prisons: the cultural history of the island metaphor has included all of these representations. The island's distinctive boundaries between sea and land make it particularly prone to imagined escapisms and entrapments. In the entanglement of science and empire, islands were crucial sites for experimentation and for the study of natural diversity and evolution. The study of the environment of islands often saw the demarcation of zones: coastal and interior, with separate sets of associations. The idea of "the little England" was a pervasive one that became linked with islands across the British empire, from the Caribbean to the Pacific. This essay is an exploration of one island colony in the Indian Ocean, namely Sri Lanka, and the impact that its physical, political, and cultural geography had on the contours of British natural history.[1]

In thinking geographically about science, this chapter is an attempt to be island-centered and to move this little space into the center of the map of historical interpretation for once. Scholars often look at territories such as Sri Lanka from afar, be it from Europe or even from India.[2] Yet when the site of Sri Lanka is taken seriously, for itself, a multiplicity of layers of meanings attached to nature, as well as clashing political and topographical regimes of knowledge, come into view. Instead of science being solely the product of an Enlightened Europe, it emerges in processes away from Europe in the entanglements and tussles between scientific agencies on the island itself, much as in other colonies.

In order to take such a view, it is necessary to be expansive about what counted as science in the colonial context of the early nineteenth-century British empire. This age was characterized by rival traditions of science in the peripheries of the British empire, in turn religious and secular, professionalizing and amateur, elite and popular.[3] In Sri Lanka, there was friction between colonizing and colonized traditions of natural contemplation, respectively between British and Kandyan knowledge. Kandy was the last independent kingdom of the island.[4] The British had wrested the coasts of Ceylon, fearing that they might fall to the French, and by 1802 the "maritime provinces" had become a Crown colony. But the kingdom of Kandy continued to hold strong and only fell to the British in 1815.[5]

In siting Sri Lanka at the focus of the chapter, it is possible to address its internal topography rather than considering the island simply in terms of its relative placement on the world map. At the same time, it is possible to bring extant traditions of nature from Kandy into the history of science and to contest the continuing dominance of a narrative that sees such traditions as inferior folk knowledge when compared with proper European science. Even as this island-centered perspective is taken, the wholeness of this space fragments into competing terrains, which are coastal and highland.

Most of Sri Lanka's surface consists of plains between thirty and two hundred meters above mean sea level. But on either side of these plains are two distinct types of landscape. Hugging the ocean is a coastal belt about thirty meters above sea level. In the interior of the island is a series of valleys, plateaus, and mountains. The distinction between these two topographies is central to making sense of the placement of attitudes to nature in the period immediately before the British taking of Kandy and in the first two decades after the kingdom's fall. Before the kingdom of Kandy fell, the defense of its inaccessible highland terrain was connected to its attention to nature. The name Kandy is a Europeanized rendition of *kanda uda rata*, literally "the kingdom of the mountains," denoting the centrality of topography and nature to political identity. The inhabitants of the island referred to *uda rata* and *pata rata*, which translate as "up country" and "low country," and this nomenclature overlapped with the different territories occupied by colonizers and Kandyans. The political fallout between the British and the Kandyans might thus be cast in relation to the clash of two different networks of natural contemplation. One way of expressing or consolidating a hold on territory was by mobilizing natural knowledge out from the rival political centers of the highlands and lowlands. Seen in this way, the island space fragments into competing kinds of activities related to nature and the different spatial politics of knowledge: resistance came from the higher altitudes and colonialism from the coastal areas.

In turning traditional geographies of science inside out, new spaces in the history of science emerge. Yet some words of caution are also necessary. First, it is not the claim of this essay that Sri Lanka's past can be reduced to its physical geography; rather space mattered to the production of scientific knowledge about the island, and such spatiality had a physical basis in topography. But physical topography cannot be extracted from political structures or discursive imaginings of nature. Second, this is not an argument for a dialogic—or even circulatory—view of colonial science, to follow some recent claims in the literature.[6] A great deal of theft, contestation, and forgetting occurred at the interface of British and Kandyan attitudes to nature, and this would be lost if the word *dialogue* were used, without the underlying dynamics of power, in describing the encounter between British and Kandyan views. Similarly, because this is an island-centered approach, the wider imperial context is pushed aside. There can be no doubt that British accounts of the natural history of the island circulated far and wide, but this is not of concern here since such an argument proceeds from a different view of Sri Lanka's interest as a site in the geography of science. Third, this essay avoids any simplistic dichotomy over continuity or change from the precolonial to the colonial. In a place such as Sri Lanka, where there were many kingdoms and numerous waves of migration, it is difficult to locate a truly indigenous or even precolonial knowledge; indeed, the last kings of Kandy themselves were migrants from South India. Instead of stressing either continuity or change, the argument traces the meeting on the island of two different but equally hybrid traditions of science, represented by the Kandyan and British cases.

Inasmuch as space is thus critical to the analysis, it is also pertinent to point to the importance of time. This is an account of the early expansion of the British empire, when Britons still did not have supremacy over the tropical and densely forested interior of the island of Sri Lanka, and where their science of acclimatization was undeveloped.[7] Within two decades of the fall of Kandy, however, the terrain looked very different. The building of a road into the highlands of Kandy, the take-off of the plantation complex, and the reorganization of the provinces of the island, which made Kandyan territories lose their separateness: all these circumstances changed the geography of knowledge and linked the island into a global network of capitalism and standardized governance. This later history of the island has attracted sophisticated attention on the part of historical geographers, particularly by James Duncan, who places Sri Lanka in an account of fractured and negotiated modernity.[8] Yet in looking at the period from the late eighteenth century until 1830, when the spatial politics of knowledge were different, it is possible to see how colonialism and science both struggled to tame an island space and to see how, in doing so, these structures had to first trump and control local networks of knowledge.

The chapter begins with an account of the contest between Kandyan and British ideas of natural contemplation before turning to analyze two of the icons of the low country and up country, namely, the coconut tree and the hill station of Nuwara Eliya. The discussion begins several decades before Kandy falls to the British in 1815 and ends in the 1830s with the establishment of Nuwara Eliya as an ideal retreat for the British, an event which saw the conquest of the deepest reaches of the island.

NATURAL CONTEMPLATION AND KINGLY PATRONAGE IN KANDY

Toward the end of the *Culavamsa*, an eighteenth-century Buddhist chronicle, is an account of the life of King Kirti Sri Rajasiṃha (r. 1747–81) as the ideal ruler. According to more recent historians, the reign of Kirti Sri marked a period of cultural revitalization in the highland kingdom; his interests in gardening fit into this program of rejuvenation.[9] The king's efforts at building a temple at Gangarama, just outside the capital, occupy a prominent place in the *Culavamsa* narrative. A statue of the Buddha, a temple, courts, and paintings were undertaken at a great sum, "on a fair spot" near the island's principal river. For the benefit of the custodians, this temple was granted villages, fields, and gardens.[10] Kirti Sri's commitment to attaching gardens to temples was evident elsewhere too. Southeast of the capital of Kandy, he erected a temple and made certain to give it an attractive view of a garden. This garden was "adorned with bread-fruit trees, mango trees, cocopalms and other fruit trees."[11] The restoration of another temple, which the author of the *Culavamsa* calls *Meddepola vihara*, shows the king's interest in the symbolic value of nature.[12] The monarchical practice of gardening was thus closely tied to the site of the temple in the Kandyan kingdom. By restoring temples and their gardens across the kingdom, the king hoped to consolidate his state around his person.

Kirti Sri's name is linked with the magnificent religious paintings that he commissioned, which appeared on the rock faces inside the caves that housed the temples of this period (fig. 6.1).[13] Some of the topics that were staples of this religious art necessitated a command of trees: for instance, the Bo tree (*Ficus religiosa*) was constantly depicted because it was the tree under which Buddha was said to have attained enlightenment. Another scene was the holy mountain of Sripada, which by the British period was called Adam's Peak, where Buddha had set an imprint of his foot, and which was depicted in these paintings as a stylized natural scene, with trees growing on the sides of the peak.[14] This interest in naturalia is also evident in other

Fig. 6.1. An eighteenth-century temple mural showing the Buddha gazing at the Bo tree
(*Ficus religiosa*) in the weeks after his enlightenment, set upon a lotus pedestal,
from the Dambulla temple, Sri Lanka. Photo by Roshan Perret. Reproduced by
permission of Studio Times, Ltd., www.studiotimes.net.

objects of craft from this period, such as doorways, boxes, panels, and pillars
within these temples, which were made of wood and ivory.[15]

The natural emphases in Kandyan art may usefully be contextualized
within the rituals of the temple. The prominence of natural emblems
matched the fact that devotees brought flowers to the temple as offerings.
One temple, Danakirigala, was particularly renowned for a royal garden

called Daswatta, which supplied a thousand flowers for worship and contains a striking scene of landscape with trees in bloom.[16] The traditional colors of Kandyan art were red, yellow, black, and white, and yet natural historical subjects were sometimes rendered in green and grey, and would therefore have attracted special attention. Attitudes to nature on the island mirrored multiple registers of religious sensibility: Sinhalese Buddhism permitted a pantheon of gods. The guardian gods of the island were subject to Buddha, and under them were the regional gods and devils of the people. The symbolism attached to trees fitted this hybrid religious landscape.[17] In addition to Buddha's Bo tree, there were others such as erabadu (*Erythrina indica*) and diwul (*Feronia elephantum*), believed to be the abodes of certain gods and devils. Tales of plants that grow in the homes of the gods, such as *Parasatu* and *Kusa*, were well known.[18]

Knowledge of nature was crucial for the practice of indigenous medicine and for rituals of exorcism.[19] Some of the most well known natural historical lore was brought together by the Royal Asiatic Society of Ceylon at the end of the nineteenth century. One example refers to rice: "The plant has one name while living and another when dead; the fruit is known by one name, but when eaten by another; the coat has a name and the kernel has another; the person who propounds the meanings of these lines will indeed be wise."[20] Stressing how Kandyans, and the Sinhalese people more broadly, engaged with nature in temples, through religious rituals, and by utilizing different types of narrations must not allow us to place this tradition of natural knowledge in opposition to that of Europe. There was a vibrant tradition of natural contemplation in the island, aesthetic as well as religious, courtly as well as popular, which was practiced within and without Kandyan terrain. There were differences between this knowledge and British natural science, in terms of the religious and intellectual culture which gave rise to it. Yet it is vital not to create a simple dichotomy between colonial and colonized knowledge.

THE POLITICS OF BRITISH BOTANICAL GARDENS

Kandy was not the only power attempting to link political control to natural knowledge in the late eighteenth century. Taking an island-centered approach to knowledge-production brings British colonial figures into view, and suggests how the tussle to take Kandy was connected to the creation of a system of natural history that was superior and British. The colonial personnel on the island were not famous men of science, like their counterparts in London, yet they played a role in the birth of modern science.

In particular, they were among the agents who minted a new science by incorporating and overtaking local knowledge.

Before seizing the highland kingdom of Kandy, the British had begun their own program of natural history on the coasts, following in the footsteps of the Dutch.[21] Just as the kings of Kandy kept their own gardens, so the first Crown governor, Fredrick North, kept his. General MacDowall, the leading military figure on the island, also had his own garden, where he received boxes of exotic plants from the botanist William Roxburgh in Calcutta.[22] By 1812, the island had an official botanic garden in Colombo, undertaken with the blessing of Sir Joseph Banks. Banks made particular note of the importance of topography in writing of the principles on which a botanic garden ought to be founded in the island:

> In all hilly countries near the Line, there are a variety of climates in which the Plants of different countries will if properly attended to, succeed to perfection. The European Strawberry is abundant in Jamaica, towards the summit of the lowest ridge of Hills; at a somewhat higher elevation Apples, Pears and the Fruits of cold climates attain a considerable degree of perfection; in their arrangement of the Intertropical Plants we find that Coffee and Pimento thrive best in elevated stations, while sugar requires low land and the Cocoa Nut which bears abundantly near the sea, becomes sterile when removed to the first slope of the hills. . . . These observations point out the necessity of small subsidiary Gardens in various parts of the Island under the direction of Native Foremen, but under the superintendence of the Royal Gardener.[23]

The royal gardener was picked by Banks. This was William Kerr, a collector for Kew Gardens, who had served in Macau and Canton. But Kerr did not survive long in Ceylon and died within two years. By 1815, the garden was criticized for being ill-conceived. Banks was informed that its "situation was very flat and not sufficiently elevated above the surface of water for the purposes of a garden." Plants were said to "thrive" for a short time, but when their roots extended deeper, they encountered a soil that was too damp.[24] This being so, the search for an ideal space for the British botanical garden continued. Before his death, Kerr oversaw the inauguration of a new garden, south of the city in Kalutara, on a site of 560 acres, which the government bought from Messrs Layart and Mooyart and which included a plantation of coconut trees.[25] When Kerr's successor, Alexander Moon, a Scots-born working gardener from Kew recommended by Banks, took up his post on the island, he developed the Kalutara gardens.[26]

The British annexation of the Kandyan kingdom provided Moon with a new terrain for botanical exploration. In 1818, one year after his arrival, Moon proposed a journey "through the Kandyan provinces for the purposes of obtaining more correct knowledge of the vegetable productions of that interesting country."[27] The following year he wrote of the botanical promise of the Kandyan kingdom: "I entertain great hopes that if the Kandyan country (where European plants of every description thrive beyond every expectation) is supplied with garden seeds for a few years it may at no distant period be rendered independent of extraneous aid."[28] By 1822, Alexander Moon's interest in the interior had come to fruition in the prospect of a further botanical garden, situated outside the Kandyan capital in the highlands in Peradeniya. Political control and botanizing went hand-in-hand. In making the case for this garden, Moon differentiated the climate of the Kandyan kingdom from that of the "maritime provinces": "for never or seldom any sort of European vegetables can be brought to such perfection on the sea coast as to produce seeds, and in my humble opinion a garden in Kandy under the liberal views of Government offers the only source." The garden in Peradeniya arose then from the British notion of distinct topographical spaces; the highlands were better suited to European botany. After 1835, the search for the ideal climate for botany continued. One of Moon's successors, J. G. Watson, compared the climate of "Poosalave" with that of Peradeniya and Nuwara Eliya, and urged that the former was superior, because it was in "an intermediate station and climate."[29]

The site of the Peradeniya gardens also provides further evidence for the claim that there was a theft of Kandyan traditions of natural knowledge on the part of the British and that an island-centered approach may reveal better the local entanglements of natural knowledge. Contemporary and recent histories of the Peradeniya botanic garden often link its inauguration with Moon and the date of 1822. In fact, the new botanical garden was established on a historic site, part of which belonged to the Dalada Maligawa, the temple of Buddha's tooth relic in the capital of Kandy, and part of which belonged to another shrine and took the form of a garden. Moon wrote of his plans, "I am of the opinion that the site of the late Kandyan King's Garden at Peradenia is better adapted than any other place for the proposed Botanic Establishment." In the original proposal for the garden's establishment, it was envisaged that some stones from the ruins in the area could be reused to build Moon's quarters[30] (fig. 6.2). The establishment of a colonial botanical garden on an existing site allows the claim that there were symmetries in how nature was used by Britons and Kandyans: both sides sought to exert their control over territory by establishing a network of gardens.[31]

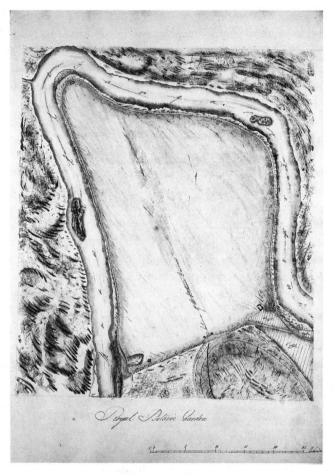

Fig. 6.2. Undated early image showing the layout of the Peradeniya botanical gardens,
with the Kandy road, coffee plantations, and entrance to the Gardens.
From a book of drawings entitled, *Sinhala Drawings*. Lindley Library
of the Royal Horticultural Society, London.

One reason why Alexander Moon is cast as the founder of the Peradeniya
gardens lies in the fact that he published a botanical work that attracted attention: his *A catalogue of the indigenous and exotic plants growing in
Ceylon, distinguishing the several esculent vegetables, fruits, roots and
grains together with a sketch of the divisions of genera and species in use
amongst the Sinhalese, also an outline of the Linnean system of botany in
the English and Singhalese Languages for the use of the Singhalese,* was published by the Wesleyan Mission Press of Colombo in 1824. The *Catalogue*

is curiously hybrid at a number of levels. It contains both English and Sinhala script, as well as plants organized according to Linnean and Sinhala modes of classification. It ends with two indexes to all the specimens catalogued in the book: one in Latin and one in Sinhala, both according to the Linnean and the Sinhalese classificatory systems. The Sinhala index appears both in Sinhala script and as a transliteration in Roman script. A total of 1,127 species were brought together in the *Catalogue.*

Governor Edward Barnes wrote approvingly of Moon's work and took some share of the praise upon himself in noting that his government had subscribed a sum of five hundred rix dollars in making its appearance possible. One of his dispatches to London included six copies of the work.[32] Moon's work was incorporated into European botanical knowledge. By 1817, Moon had prepared "hundreds of species of the choicest seeds" for the emperor of Austria and for the Kew Gardens. The next year he transmitted the "journal of [his] proceedings" to Banks, together with specimens kept safe in coconut shells, enclosed in baskets made from "Ceylon Rattan for the convenience of slinging them on board ship, as well as for the protection of the plants from injury."[33]

Tracing in these ways the flow of this information out from the island into wider spheres of natural knowledge need not mean that its location within the island is misplaced. In his preface, Moon acknowledged the help of an important early orientalist, Rev. B. Clough.[34] There was, however, no mention of his closest aide, Harmanis De Alwis Seneviratne, who had served as his "writer" from 1818, first in Slave Island, then in Kalutara, and finally in Peradeniya.[35] There is no information about De Alwis's background, but it is known that he was an Anglican at the time of his death.[36] De Alwis worked closely with Moon in putting the material together for the *Catalogue.* It is not clear why his work is omitted from Moon's acknowledgments. It may have been because Moon paid personally for De Alwis's expenses, including his lessons in art: Moon may have felt that he "owned" De Alwis's labors. [37] By 1823, the colonial state promoted De Alwis from the position of Moon's "writer" to that of "draftsman," a post he held for thirty-eight years.

In 1831, Governor Barnes took note of De Alwis's accomplished drawings in awarding him the title Mohandiram. It has been suggested that an extant image of De Alwis dates from this point (fig. 6.3). It is difficult to locate De Alwis in the early correspondence concerning botany, as Moon had several assistants of indigenous origin. Only Moon's helpers of European or Malay origin were ever named—facts again pointing to the hybridity of colonial practice. In 1820, his local assistants at the garden of Kalutara in-

Fig. 6.3. Portrait of de Alwis Seneviratne, from the frontispiece of *Sinhala Drawings*.
Lindley Library of the Royal Horticultural Society, London.

cluded a "Mohandiram," who was paid eighteen rix dollars; an "Overseer"
(paid the same); and a "Native Gardener" and "Clerk" who were paid fifteen
rix dollars. The clerk may be De Alwis.[38] De Alwis is easier to identify once
he bore the title of draftsman.[39] The draftsman was expected to attend work
"precisely from 10 till 5 o'clock," and his pay was to be reduced accordingly if he was five minutes late, absent for half a day, or all day.[40] In 1840,
Superintendent H. F. Normansell wrote to the colonial secretary of the island, noting that the draftsman had completed seventy drawings in three

months on average[41] (fig. 6.4). By 1845, Superintendent George Gardener noted that the draftsman was paid only three pounds a month. Yet his ability was unquestioned: "His equal as a native botanical artist, I am certain does not exist in India."[42] De Alwis's two sons, William and George, and three of his grandsons became botanical illustrators, taking up this tradition until after the First World War, and suggesting that the work may have become a caste occupation for this family.[43] De Alwis was a successful intermediary who learned the practices of British botany, but his position and

Fig. 6.4. "Rhododendron Aboreum," a drawing in all probability completed by de Alwis Seneviratne, place marked Nuwara Eliya. From a book of drawings entitled, *Sinhala Drawings*. Lindley Library of the Royal Horticultural Society, London.

role speak to a whole host of other now nameless indigenous collaborators who made contributions not just to British botany in Ceylon but to science elsewhere.

The importance that attends to avoiding an argument about the mobility of British botany on the island into wider networks, by keeping the analysis island-centered, is also evident with respect to the arrangement of the botanical gardens at Peradeniya. The gardens were not organized according to the Linnean system. In 1830, James McRae noted how the garden had divisions for ornamental gardening; an orchard for the improvement of fruit trees and for the introduction of species; an area for medicinal plants peculiar to the island, "of which a great many are used by the Natives"; a nursery for raising coffee, cocoa, and other valuable plants; a kitchen garden "for the culture of European and Native Vegetables"; and a considerable portion of land devoted to raising teak "and other valuable timber."[44] In 1838, J. G. Lear wrote of how it was the "original intention" for the gardens to have "a systematical arrangement usual in the general formation of such establishments," but that the death of Moon had meant that this intention had not been realized. Lear also commented on the lack of experience of those keepers who followed Moon, noting that the gardens had "frequently" been left to the charge of "native people" and were now in "such a confusion," that they would "place a stranger in a situation to ask for what [they were] intended."[45] It is intriguing to reflect on the relationship between the layout of these gardens during the time of the kings of Kandy and during the early British period.[46] What is clear is that the British did not successfully impose the model of botanical arrangement with which they were familiar.

The British network of botany emerged from a contest with Kandyan ideas of nature. This was not a dialogue or the simple passage of the precolonial into the colonial: Kandyan peasants did not willingly provide the British with all their information about nature. There were key collaborators who allowed this appropriation of information to proceed and for the Kandyan institution of the garden and Sinhala systems of naming to be incorporated into British natural history at just the time when Kandy was conquered in political terms.

THE COCONUT TREE, TROPICALITY, AND THE LOW COUNTRY

As much as an island-centered perspective allows a different account of the collision of different traditions of knowledge to emerge, it is possible to fragment the story further by taking more seriously the distinction between the highlands and the lowlands. In Ceylon, the British differentiation of the temperate and the tropical mapped onto an extant dichotomy between *uda rata* and *pata rata*. Indeed the ideas of temperate and tropical were symbolized by particular natural icons.

The coconut tree was an important subject of British colonial natural historical research. This is evident from a bibliography of articles on the island's agriculture and botany published in 1915, which listed a total of 185 essays and articles on the subject.[47] An important early essay on the coconut tree was published by Henry Marshall, a military surgeon in Ceylon. Marshall's essay, entitled "Contribution to a Natural and Economical History of the Coco-nut Tree," was based primarily on his experience of the island, although he also incorporated knowledge about the tree from other colonial territories. Marshall commented that the tree was "nowhere so extensively cultivated as in Ceylon."[48] From the early period of British expansion, the colonizers were obsessed with this tree, and it became one of the prime natural icons of the coastal belt.

British commentators who discussed how the island looked from the sea invariably noted the coconut tree. Take, for instance, the civil servant J. W. Bennett's *Ceylon and Its Capabilities: An account of Its Natural Resources, Indigenous Productions and Commercial Facilities* (1843). Bennett described how the sight of the coconut tree refreshed the explorer fatigued after a long journey. He contrasted the sight of Ceylon's coast with the "barren sand-hills, parched trees, and sun-burnt fields" that the traveler might have expected from a stretch of territory so close to the equator.[49] In making this contrast, Bennett privileged a mode of writing about Ceylon as an island. The parched and arid landscape that he had in view belonged to India. His message was that as soon as you saw Ceylon from the sea you knew you had arrived somewhere different.[50] A "verdant island" was signified by coasts populated with "myriads" of coconut trees "to the very verge of the sea." Popular writers on Ceylon frequently made the link between the sight of coconut trees and ideas of the romantic and picturesque. James Cordiner, for instance, a clergymen who wrote a popular narrative of Ceylon in 1807, noted that the southeast coast was particularly "picturesque and romantic" because of the border of coconut trees framing the hills and mountains. Implicit then was the idea that coconut trees were the first layer of vegetation in a composite tropical scene, and that the coconut tree was the emblem of the coasts.[51] In keeping with the idea of the picturesque, Marshall's article compared a grove of coconut trees to "the long aisles and Gothic arches of a cathedral."[52] The aesthetic pleasure that observers saw in coconut trees paralleled a fascination with their utility. Popular accounts on Ceylon are full of details about how each element of the coconut tree can be put to use.[53]

By contrast, coconut trees were not common in the interior. The natural history of this tree species fitted then into a bio-geographical classification of tropicality: "The Coco tree seems to require for its perfection a mean temperature of not less than 72° FAHRENHEIT, the proper climate for it will therefore be from the Equator to the 25th parallel latitude, and in the Equinoctical Zone to an altitude of about 2900 feet."[54] The absence of coconut trees from the interior, where Europeans had only recently ventured, meant that the tree was often taken as a sign of civilization. For James Emerson Tennent, who wrote perhaps the most important nineteenth-century account of Ceylon, published in 1859, a coconut tree was always indicative of "the vicinity of man."[55]

The meaning that the British gave the coconut tree as an icon of the coastal belt can be linked to preexisting traditions attached to this tree on the part of islanders. British orientalist travelers were fascinated by a particular statue that was connected to the coconut tree. Major Jonathan Forbes, in his account of Ceylon, written after serving in the island for eleven years, wrote of this issue:

> Before entering Belligamma (which is seventeen miles from Galle,) I stopped to examine a figure called the Kustia Raja: it stands on the roadside, is twelve feet in height, and forms part of a great mass of rock in which it is sculptured. One tradition affirms that the statue represents the Prince from the continent of India, who introduced the cocoanut-tree, and taught the Cingalese its many uses: another and more probable account states, that a King afflicted with leprosy established himself at this place for the convenience of worshipping at the neighbourhood wihare of Agra Bodi.[56]

In another account, J. W. Bennett wrote of how he came to know the tale of Kusta Raja from the head priest of the Karangoda Vihare in Saffregam. Tellingly, Bennett had been taken ill with what he called "a severe and intermittent attack of fever," and he relied on the priest and his attendants to nurse him. In his book, Bennett cites the high priest's account of the Kusta Raja statue and notes how the coconut is a cure for leprosy. Kusta Raja is cast as a Sinhalese king from the interior, where coconut trees do not grow. In his leprous condition Kusta Raja has a trance that lasts for several days and sees what amounts to the ocean, describing its margin as "covered with groves of trees of a rare kind, such as he had never seen; for instead of branches in various directions, as trees had in his country, their tops appeared crowned with a tuft of feathery leaves." The father of Buddha appears to Kusta Raja and gives him directions to go and seek the fruit of the coconut tree: "The inside of transparent liquid, and of innocent pulp, must be thy sole diet, till thrice the Great Moon (Maha Handah) shall have given and refused her light."[57] The King tasted the coconut and found it "sweet and delicious, and pure as crystal itself . . . whilst the fleshy part of it was a cooling and grateful food."[58]

This account, albeit abbreviated, points in one sense to the problem of translation. It is clear that the colonial voice dominates that of the priest and that the visions of the coconut and the idealization of its fruit refer to colonial modes of representation. However, the account of Kusta Raja emerged out a process of discussion with the high priest, and a separate indigenous meaning cannot be extracted and simplified. The British attended to existing utilitarian ideas of the coconut tree; local legends interpreted it as a fabulous feature of the coastal belt, and these accounts framed colonial interest.[59]

According to a different argument, the coconut tree and the coconut plantation might be cast as at once romantic, scientific, and utilitarian aspects of British colonialism, linked to the emergence of the modern empire and its exportation of natural historical regimes out from the center.[60] The mobility of the emergent science of colonial natural history is clearly discernible in this narrative and supports such a view. Yet this account of colonial science does not tell the whole story: for the iconicity of the coconut tree was also shared by those who inhabited the island. In seeing the tree as a curiosity of the tropical coasts and casting it as a valuable commodity, the British attributed new meanings to the landscape of the coasts additional to those already ascribed.

ANGLICIZATION, VESTIGES OF THE PAST, AND THE HIGHLANDS

For the British in Sri Lanka, topography was classified according to temperature, altitude, and distance from the ocean. Where the coconut tree thus became an emblem of a coastal zone, icons were also found for the highlands. Since the interior of the island was integrated into British control later than the "maritime provinces" and since British political control was more assured by this time, the imaginative geography that Britons attached to this region was more inventive. Even in the higher altitudes, however, colonial science did not operate on a land empty of meaning. By looking separately at British practices of engaging nature in the interior, it is possible to further fragment the status of the island as a unit of analysis in the geographical history of science.

This claim is evident in consideration of the events leading up to the creation of the hill station at Nuwara Eliya. The Briton who is usually given credit for having "discovered" the plains of Nuwara Eliya is John Davy, military surgeon on the island and the younger brother of the chemist Humphry Davy.[61] While on a tour through the highlands in 1819, Davy came across what he recorded as "Neuraellyia-pattan," a tract of land that was said to be fifteen or twenty miles in circumference lying about 5,300 feet above the level of the sea.[62]

It took another decade for Britons to pay due attention to Nuwara Eliya. In August 1828, Governor Barnes wrote to London stating that he was building a new military post at Nuwara Eliya. He noted its strategic significance: in times of urgency, it could supply troops "on either side of the great range of mountains that separate the island."[63] Barnes also praised the utility of this spot as a site for convalescence: "I look to it as being the means of saving many lives as well as the inconvenience and expense to individuals of a Voyage to Europe for the recovery of Health."[64] Two years later, Barnes's vision of Nuwara Eliya was at least partway to realization. A Colonial Office report noted the positive results of the climate of Nuwara Eliya by contrasting the soldiers who had recently arrived in the plains from the "low country" with those who had resided in the uplands for a while: "The former appear sallow and debilitated having that characteristic unhealthy countenance of Europeans living in tropical climates, whilst the latter, seem to possess that robustness of frame which we commonly meet with among the natives of an English agricultural district."[65]

Nuwara Eliya also became an ideal retreat for tired soldiers because European vegetables could be grown there. According to one military captain, writing in 1831, "the finest vegetables grown in England" were said to "thrive" in Nuwara Eliya. [66] Governor Barnes established his own residence—"a large and substantial country-house"—in Nuwara Eliya and there cultivated vegetables, flowers, and fruits.[67] By 1836 the road to Nuwara Eliya had been completed, making access to the hill station easier.[68] The station quickly became a small town: one commentator noted how Nuwara Eliya consisted of "military and public buildings, many whitewashed cottages and smoking chimneys."[69] By 1891, the population of Nuwara Eliya numbered 2,726 persons.[70]

These early successes inspired an eccentric attempt at "settlement" in Nuwara Eliya in 1845, led by Samuel Baker, later well known as an African explorer. Baker bought land from the government and obtained machines, stock, seeds, and hounds from London on board the *Earl of Hardwicke*; he persuaded his brothers to join him. His own account of the setting-up of his settlement is idiosyncratic and idealistic, although his family managed to build a relatively successful estate in Nuwara Eliya that remained through the course of the century.[71] More generally, the British saw the town of Nuwara Eliya as a new and welcome addition to the island. For Lieutenant Augustus de Butts, "Nuwera Ellia is, in truth, a new creation, and still in a state of transition from the majesty of 'nature unadorned' to the less sublime, yet equally pleasing, charms that belong to cultivation."[72] The British conception of the settlement of the highlands was linked to the vocabulary and practice of how nature could be improved through human intervention. For Samuel Baker,

> How changed are some features of the landscape within the few past years, and how wonderful the alteration made by man on the face of Nature! Comparatively but a few years ago, Newera Ellia was undiscovered—a secluded plain among the mountaintops, tenanted by the elk and boar . . . Here where wild forest stood, are gardens teeming with English flowers; rosy-faced children and ruddy countrymen are about the cottage doors; equestrians of both sexes are galloping round the plain, and the cry of the hounds is ringing on the mountain-side.[73]

Civilizing the highland wilderness was a romantic project that drew its peculiar charm from the possibility of making in Ceylon a place which was a little Britain, a home from home. Nuwara Eliya was expected to provide solace from the tropics. A correspondent in the *Asiatic Journal* in 1834 on the "health-station" of Nuwara Eliya is striking in this regard: "I am now clothed from head to foot in broad-cloth, with flannel next to my skin, the room closely shut up, and I declare to you that I can scarcely hold my pen my fingers are so cold."[74] He carried on, detailing the temperature, the need for fires and blankets, and the availability of familiar vegetables. This representation of Nuwara Eliya was consciously set against its opposite, namely, the coastal belt of the island, which embodied tropicality. It is noteworthy how often popular writers on Ceylon used the word *tropics* in opposition to what Nuwara Eliya represented.[75] One example comes from Charles Pridham, who wrote how the station was "so entirely dissimilar from any view or sensation within the tropics, that the novelty is at first delightful and exhilarating."[76]

It is important not to cast Nuwara Eliya simply as an icon of British power. It was known among the inhabitants as a site before British occupation. John Davy wrote, "Our guides call it Neuraellyia-pattan."[77] He also commented that while the plain was uninhabited, it was visited by blacksmiths from Kotmale looking for iron, and by gem-hunters looking for precious stones. These views were reiterated in the manual to the district published at the end of the century.[78] Samuel Baker was also adamant that it was known to the Sinhalese. Nuwara Eliya was said to denote "Royal Plains," because the kings of Kandy had retreated here in the past. Baker wrote, "There are native paths from village to village, across the mountains, which, although in appearance no more than deer-runs, have existed for many centuries, and are used by the natives even to this day." Baker confessed that he had learned the lie of the land by using these paths and by observing "notches" on trees that had been made by travelers who had in the past traversed the plain. He went on to present his ideas as to why Nuwara Eliya had been important to the Kandyan kings. In ancient times, according to Baker, the waters of Nuwara Eliya were conveyed by complicated waterways down the hills. From examining the remains of such waterways, Baker postulated that "more than fifty times the volume of water was then required than in use at present, and in the same ratio must have been the amount of population."[79]

Baker's claims, while they may be exaggerated and certainly are in keeping with the ways in which Britons historicized lands in the subcontinent in this period, are nonetheless noteworthy in documenting prior land use and culture in sites of British settlement. A further account of the pre-British history of Nuwara Eliya was presented by a historian in the mid-twentieth century. He argued that the area around present-day Nuwara Eliya is linked with the epic historical poem the *Ramayana*, which sees Rama rescuing Sita from the clutches of the island king, Ravana. Sita is said to have been kept captive in Nuwara Eliya. In addition to this, Kotmale, within reach of Nuwara Eliya, may be the spot where one of the island's hero-kings, Dutugemunu, temporarily resided. The discovery of an inscription on stone suggests there was a Buddhist temple in the plains of Nuwara Eliya.[80] Despite the difficulties of verification—whether from the guides who accompanied Davy or of Baker's exploration of vestiges from the past—there is evidence to support the claim that colonial Nuwara Eliya had antecedents. At the same time, Nuwara Eliya provides rich evidence for how the British utilized a confident set of scientific and rhetorical tropes in remaking the land in the colonial period.

Too often the spaces of the colonial world are treated as if they are empty of meaning and habitation, and so devoid of impact on European rationalities.

This essay is an attempt to make an island "speak" into the geographical history of science. In particular, it is an intervention in characterizations of colonial science. By looking at the island as a space of scientific and natural production, it is possible to uncover layers of information. These entanglements contain an overarching system of power, which leads to colonial triumph and the overtaking of local natural knowledge. The tensions between different traditions of natural contemplation, between colonizing and colonized, deserve a place in our histories and geographies of science; for the island colony was an important site for the making of modern science. By considering the island of Sri Lanka in terms of its different topographies, it is possible to see how the land itself was imagined and navigated in different ways in the lowlands and the highlands. Colonial attitudes to nature in these spaces were conditioned by a distinction between the tropical and temperate, even as the local segmentation of space into *uda rata* and *pata rata* was also employed: neither interpretation should be lost from view.

Historical geographers have considered landscape as a site of symbolic production. Yet even when a landscape's iconography is studied, what often emerges are the layers of embedded signification, rather than a real thing called "nature" that lies underneath. This insight is important in the scientific study of the remakings of colonial landscapes. In Sri Lanka, even as the British brought a range of meanings to bear—which were tied to romanticism, the picturesque, and, as colonialism unfolded, an increasing Anglicization of landscape—those readings of and terms for nature found their place on top of existing ones. They were "further glosses upon an already deeply layered text."[81] Iconicity was critical to the colonial divide between the temperate and the tropical; the coconut tree could become a point of intense association for the coastal belt, just as the hill station was for the highlands.[82]

It is important to recognize and explore the limits of colonial scientific knowledge. In the early phase of colonial expansion in Ceylon, British knowledge was "hemmed in" by already existing meanings in and of land, and these local traditions had to be overtaken and incorporated into colonial structures. Only when that had happened did the island become fully colonized. This view is important because the character of colonial science at the end of the nineteenth century is sometimes stretched backward and made to describe the early British empire. This is the case, for example, in the historiography of hill stations in India, where that late nineteenth-century "moment"—when hill stations became places of inaccessible government, where the British retreated in season and which were linked as sites to strict regimes of race, climate, and class—is taken to stand for the character of these urban spaces

more generally. As Judith Kenny notes, "The hill stations thus served a particular role within the imaginative geographies of imperial discourse, a role
that enabled the imperialist mind to intensify its own sense of itself by dra-
matizing distance and difference."[83] This essay is not an attempt to deny this
narrative, but to show that the hill station did not emerge straightforwardly
out of a blank slate, an unnamed or unknown landscape.[84]

The application of science in imperialism has been a central theme of
this discussion. By coming to the history of botany from Ceylon and the
island, it is possible to view Kew Gardens in a different way. Richard Drayton has shown how Kew Gardens was a center for the collection of natural
historical information from across the world, an entrepot through which
plants moved south to north and east to west. Economic botany was an
engine of agrarian patriotism and natural improvement, driven as it was
by the natural theological language of Christian paternalism.[85] Yet when
that subject is turned inside out, it looks different if we have Peradeniya in
central focus rather than Kew. The fragility of Kew's reach is more apparent.
The nodes of this network of information were not connected seamlessly,
for Peradeniya's origin and early history show the persistence of established
forms and modes of classification, alongside insistent attempts to assert
Linnean classifications. By looking at the island and from the island, it is
possible to take locality seriously in the history of science on the global
stage. Such a view of locality, tied to the many hybrid ways of knowing in
a particular place, does not require us always to read the making of natural
knowledge in relation to processes of circulation. For if Sri Lanka's history
is only found in relation to global patterns, something of its specificity as a
site is lost. Yet at the same time, such specificity in one context should not
allow either the Kandyan or the British understanding of scientific knowledge to be proscribed as rigid, compartmentalized systems, the one denoting
indigeneity and the other, homogeneity. Colonialism occasioned contact,
collision, and violence between different orders and types of scientific information within specific places, and the geographies and histories of this
story require sensitive interrogation.

NOTES

1. On the role of islands in colonialism, see Richard Grove, *Green Imperialism:
Tropical Island Edens and the Origins of Environmentalism* (Cambridge: Cambridge
University Press, 1995), and Rod Edmond and Vanessa Smith, eds., *Islands in History and*

Representation (London: Routledge, 2003). On tropicality and empire, see Felix Driver and Luciana Martins, eds., *Tropical Visions in an Age of Empire* (Chicago: University of Chicago Press, 2005), and David Arnold, *Tropics and the Traveling Gaze: India, Landscape and Science, 1800–1856* (Seattle: University of Washington Press, 2005).

2. For islandedness in the Sri Lankan case, see Sujit Sivasundaram, "Ethnicity, Indigeneity and Migration in the Advent of British Rule to Sri Lanka," *American Historical Review* 115 (2010): 428–52.

3. Sujit Sivasundaram, *Nature and the Godly Empire: Science and Evangelical Mission in the Pacific, 1795–1850* (Cambridge: Cambridge University Press, 2005); Sujit Sivasundaram, ed., "Global Histories of Science," Focus section, *Isis* 101 (2010): 95–162.

4. For the standard history of Sri Lanka, see K. M. De Silva, *A History of Sri Lanka* (Delhi: Oxford University Press, 1981); see also Patrick Peebles, *The History of Sri Lanka* (Westport, CT: Greenwood Press, 2006); Nira Wickramasinghe, *Sri Lanka in the Modern Age: A History of Contested Identities* (London: Hurst, 2006).

5. For the history of the Kandyan kingdom, see Lorna Dewaraja, *A Study of the Political, Administrative and Social Structure of the Kandyan Kingdom of Ceylon, 1707–1760* (Colombo, Sri Lanka: Lake House Press, 1972).

6. Eugene Irschick, *Dialogue and History: Constructing South India, 1795–1895* (Berkeley and Los Angeles: University of California Press, 1994); Christopher A. Bayly, *Empire and Information: Intelligence Gathering and Social Communication in India, 1770–1870* (Cambridge: Cambridge University Press, 1996); Kapil Raj, *Relocating Modern Science: Circulation and the Construction of Knowledge in South Asia and Europe, 1650–1900* (Basingstoke: Palgrave Macmillan, 2006); Bernard Cohn, *Colonialism and Its Forms of Knowledge: The British in India* (Princeton, NJ: Princeton University Press, 1996); Gyan Prakash, *Another Reason: Science and the Imagination of Modern India* (Princeton, NJ: Princeton University Press, 1999).

7. For the science of acclimatization, see David N. Livingstone, "Tropical Climate and Moral Hygiene: Anatomy of a Victorian Debate," *British Journal for the History of Science* 32 (1999): 93–110.

8. James Duncan, *In the Shadow of the Tropics: Climate, Race and Biopower in Nineteenth-Century Ceylon* (Aldershot: Ashgate, 2007).

9. John Clifford Holt, *The Religious World of Kirti Sri: Buddhism, Art and Politics in Late Medieval Sri Lanka* (Oxford: Oxford University Press, 1996).

10. William Geiger, ed. *Culavamsa: Being the More Recent Part of the Mahavamsa.* [Vol. 50 of the Pali chronicles series.] (London: Pali Text Society, 1929–30), 201.

11. Ibid., 218.

12. Ibid., 250–51.

13. For commentary on this, see Senake Bandaranayake and Gamini Jayasinghe, *The Rock and Wall Paintings of Sri Lanka* (Colombo, Sri Lanka: Lake House Press, 1986); Sinharaja Tammita-Delgoda, *Ridi Vihare: The Flowering of Kandyan Art* (Pannipitiya, Sri Lanka: Stamford Lake, 2006).

14. Tammita-Delgoda, *Ridi Vihare*, 58, 92–93.

15. Ibid., 133.

16. Bandaranayake and Jayasinghe, *Rock and Wall Paintings*, plate 68.

17. This hierarchy is well explained in Gananath Obeyesekere, "The Great Tradition

and the Little Tradition in the Perspective of Sinhalese Buddhism," *Journal of Asian Studies* 22 (1963): 139–53.

18. This information is taken from W. A. De Silva, "A Contribution to Sinhalese Plant Lore," *Journal of the Royal Asiatic Society, Ceylon Branch* 12 (1891): 113–44. See also Andreas Nell, "Some Trees and Plants Mentioned in the Mahavamsa," *Ceylon Historical Journal* 2 (1953): 258–64.

19. This is a subject in my forthcoming book. See also C. G. Uragoda, *A History of Medicine in Sri Lanka* (Colombo, Sri Lanka: Sri Lanka Medical Association, 1987), 38–39.

20. De Silva, "Contribution to Sinhalese Plant Lore," 140.

21. For more on Dutch natural history, see K. D. Paranavitana and C. G. Uragoda, "Medicinalia Ceylonica: Specifications of Indigenous Medicines of Ceylon Sent by the Dutch to Batavia in 1746," *Journal of the Royal Asiatic Society, Ceylon Branch* 53 (2007): 1–55; Rohan Pethiyagoda, *Pearls, Spices and Green Gold: An Illustrated History of Biodiversity Exploration in Sri Lanka* (Colombo, Sri Lanka: W. H. T. Publications, 2007), 40–56.

22. James L. A. Webb, *Tropical Pioneers: Human Agency and Ecological Change in the Highlands of Sri Lanka, 1800–1900* (Delhi: Oxford University Press, 2002), 55; J. C. Willis, "The Royal Botanic Gardens of Ceylon and Their History," *Annals of the Royal Botanical Gardens, Peradeniya* 1 (1901): 2–3; Pethiyagoda, *Pearls*, 59; R. Desmond, *The European Discovery of the Indian Flora* (Oxford: Oxford University Press, 1992), 160. For discussion of the possible sites of the Ceylon garden, see T. Petch, "The Early History of Botanic Gardens in Ceylon with Notes on the Topography of Ceylon," *Ceylon Antiquary and Literary Register* 5 (1920): 119–24; T. Petch, "The Early History of Botanic Gardens in Ceylon," *Ceylon Antiquary and Literary Register* 7 (1921): 63–73.

23. Webb, *Tropical Pioneers*, appendix 1.

24. Desmond, *European Discovery*, 161–62; Pethiyagoda, *Pearls*, 60–62.

25. Lot 6/281, Sri Lanka National Archives, Colombo, Sri Lanka (hereafter SLNA). This evidence goes against claims that this was the site of a sugar estate: see Petch, "Early History of Botanic Gardens," 68.

26. "Mr. Moon of the Botanic Garden," *Asiatic Journal* 21 (1826): 91.

27. Ibid.

28. Alexander Moon to the Deputy Secretary of Government, 7 January 1819, Lot 6/282, SLNA.

29. J. G. Watson to Colonial Secretary, 19 November 1835, Lot 6/1322, SLNA.

30. T. Petch, "Early History of Botanic Gardens," 68–69. The secretary for the Kandyan Provinces noted that the plot of land utilized for the Peradeniya gardens comprised four gardens, one of which belonged to the temple, and another of which paid tribute to the temple.

31. A recent hand guide to the Peradeniya botanic garden tells this story in a way that no European history has ever done, although the details cannot be verified. The historical value of the gardens is dated to 1371, and Kirti Sri's making of a "Royal Garden" on this site is noted. Rajadi Rajasinha is said to have resided at the gardens in a temporary residence. The temple was destroyed by the British, and a priest was said to have been on site until Moon's arrival. See *Visitors' Guide to the Royal Botanical Gardens, Peradeniya* (flyer).

32. Dispatch from Governor Edward Barnes, 26 October 1824, CO 54/86, The National Archives, Kew, United Kingdom (hereafter TNA).

33. Alexander Moon to Sir Joseph Banks, 16 February 1818, Lot 6/282, SLNA.

34. Moon, *A Catalogue*, 3.

35. All uncited biographical information is from Henry Trimen's obituary of Seneviratne, *Journal of Botany* 32 (1894): 255–56.

36. Rohan Pethiyagoda, "The Family de Alwis Seneviratne of Sri Lanka: Pioneers in Biological Illustration," *Journal of South Asian Natural History* 4 (1998): 100.

37. Ibid., 100.

38. Alexander Moon to Chief Secretary of Government, 5 August 1820, Lot 6/283, SLNA.

39. James McRae to Deputy Secretary of Government, 9 December 1828, Lot 6/1027, SLNA.

40. G. Bird to Deputy Secretary of Government, 28 July 1830, Lot 6/1028; J. G. Watson to Chief Secretary of Government, 6 April 1832, Lot 6/1028, SLNA.

41. H. F. Normansell to Acting Colonial Secretary, 5 October 1840, Lot 6/1487, SLNA.

42. Pethiyagoda, "The Family," 102.

43. Ibid., 101. The idea of caste is my own. See W. H. Harvey, "Ceylon Botanic Gardens," in *Ceylon Almanac* (1854): 44, for mention of two native draftsmen at work at this point in the gardens.

44. "A Report on the Present State of the Botanic Garden at Ceylon," 26 April 1838, CO 54/107, TNA.

45. "Report on the Present State of the Royal Botanic Garden," Lot 6/1405, SLNA.

46. In originally establishing the garden, Moon wrote, "There are already a number of fruit and Forest Trees, common to the island dispersed all over the grounds, which will afford immediate shade and shelter to the more exotic tender Exotic and Indigenous Plants on their introduction." See Petch, "Early History of Botanic Gardens," 68.

47. T. Petch, *Bibliography of Books and Papers Relating to Agriculture and Botany to the End of the Year 1915* (Colombo, Sri Lanka: H. Ross Cottle, 1925), 23–30.

48. Henry Marshall, "Contribution to a Natural and Economical History of the Coconut-Tree," in *Memoirs of the Wernerian Natural History Society* 5 (1823): 113.

49. J. W. Bennett, *Ceylon and Its Capabilities: An Account of Its Natural Resources, Indigenous Productions and Commercial Facilities* (London: W. H. Allen, 1843), 14.

50. See also Robert Percival, *An Account of the Island of Ceylon Containing Its History, Geography, Natural History* (London: C. and R. Baldwin, 2006), 37, for contrast with India.

51. James Cordiner, *A Description of Ceylon: Containing an Account of the Country, Inhabitants and Natural Productions* (London: Longman, Hurst, Rees, and Orme, 1807), 7. For a similar idea of coconut trees serving as the foreground of a tropical scene, see Percival, *An Account*, 36.

52. Marshall, *Contribution*, 114.

53. Percival, *An Account*, 215–27.

54. Marshall, *Contribution*, 112.

55. James Emerson Tennent, *Ceylon: An Account of the Island*, 2 vols. (London: Longman, Green, Longman, and Roberts, 1860), 1, 119.

56. Jonathan Forbes, *Eleven Years in Ceylon, Comprising Sketches of the Field Sport and Natural History and an Account of Its History and Antiquities*, 2 vols. (London: R. Bentley, 1840), 2, 170.

57. Bennett, *Ceylon and Its Capabilities*, 334.

58. Ibid., 333.

59. For an interpretation of the sculptural tradition in which this statue is placed, see J. E. Van Lohuizen-de Leeuw, "The Kustaraja Image: An Identification," in N. A. Jayawickrama, ed., *Paranavitana Felicitation Volume on Art and Architecture and Oriental Studies* (Colombo, Sri Lanka: M. D. Gunasena, 1965), 253–61. See also W. A. De Silva, "A Probable Origin of the Name Kushtharajagala," *Royal Asiatic Society, Ceylon Branch* 28 (1919): 86. He suggests that the name originates from the deity of the coconut in Travancore, south India.

60. On the colonial economy of the coconut, see Colvin R. De Silva, *Ceylon under British Occupation*, 2 vols. (Colombo, Sri Lanka: Colombo Apothecaries, 1953–62), 2, 471–73.

61. Robert Knox, a sailor in the captivity of the king of Kandy, noted Nuwara Eliya in the map that accompanied his work. See Robert Knox, *An Historical Relation of the Island of Ceylon* (London: R. Chiswell, 1681).

62. John Davy, *An Account of the Interior of Ceylon and of Its Inhabitants* (London: Longman, Hurst, Rees, Orme & Brown, 1821; repr., Dehiwela, Sri Lanka: Tisara Prakasakayo, 1983), 340–41 (1983 ed.).

63. Dispatch from Governor Edward Barnes, 11 August 1828, CO 54/101, TNA.

64. Ibid.

65. M. McDermott, report, "Some Remarks on the Soil, Climate and Temperature of Nuwera Ellia," 29 November 1831, CO 54/114, TNA.

66. Letter from F. C. Barlow, 9 December 1831, CO 54/114, TNA.

67. Forbes, *Eleven Years*, 133; Webb, *Tropical Pioneers*, 184.

68. For an estimate, see dispatch from Governor Edward Barnes, 19 December 1833, CO/54/130, TNA; for completion of the road, see Forbes, *Eleven Years*, 133.

69. Forbes, *Eleven Years*, 133.

70. C. J. R. Le Mesurier, *Manual of the Nuwara Eliya District of the Central Province of Ceylon* (Colombo, Sri Lanka: G. J. A. Skeen, 1893), 75.

71. Samuel Baker, *Eight Years' Wanderings in Ceylon* (London: John Murray, 1855; repr., Dehiwela, Sri Lanka: Tisara Prakasakayo, 1983).

72. Augustus De Butts, *Rambles in Ceylon* (London: W. H. Allen, 1841), 212.

73. Baker, *Eight Years*, 20.

74. "Asiatic Intelligence—Ceylon," *Asiatic Journal*, n.s., 13 (1834): 172–73.

75. See Forbes, *Eleven Years*, 132: "Perhaps there is in no country a climate more congenial to the natives of Great Britain, both as regards salubrity and comfort, than Nuwara-ellia; the temperature of the air never approaching to what is called tropical heat."

76. Charles Pridham, *An Historical and Statistical Account of Ceylon and Its*

Dependencies, 2 vols. (London: Boone, 1849), 2, 678. For another example of how Nuwara Eliya was placed in opposition to the tropics, see "Remarks upon the Comparative Healthfulness and Other Local Advantages of Nuwera Eliya in the Island of Ceylon and the Neilgherry Hills in Hindoostan," *Colombo Journal* 1 (1832): 472–73.

77. Davy, *An Account*, 340.

78. Mesurier, *Manual*, 64.

79. Baker, *Eight Years*, 25.

80. James T. Rutnam, "Ancient Nuwara Eliya, Part 1: The Ramayana Legend," *Ceylon Fortnightly Review*, 21 June 1957, 25, 27, 36; "Ancient Nuwara Eliya, Part 2: A Link with King Dutugemunu," *Ceylon Fortnightly Review*, 19 July 1957, 19, 21, 33; "Ancient Nuwara Eliya: Part 3: A Lithic Record," *Ceylon Fortnightly Review*, 23 August 1957, 17, 19.

81. Denis Cosgrove and Stephen Daniels, "Introduction," in Denis Cosgrove and Stephen Daniels, eds., *The Iconography of Landscape* (Cambridge: Cambridge University Press, 1988), 8.

82. Driver and Martins, *Tropical Visions*.

83. Judith K. Kenny, "Climate, Race and Imperial Authority: The Symbolic Landscape of the British Hill Stations in India," *Annals of the Association of American Geographers* 85 (1995): 696.

84. See also Dane Kennedy, *Magic Mountains: Hill Stations and the British Raj* (Berkeley and Los Angeles: University of California Press, 1996).

85. Richard Drayton, *Nature's Government: Science, Imperial Britain and the "Improvement" of the World* (New Haven, CT: Yale University Press, 2000).

PART TWO

Practices and Performances

Part 2 focuses on the ways in which science was undertaken, promulgated, defended, and made known. However it may have been seen—as a means to disclose Nature's workings, the expression of God's creativity, as a profession, a method, a source of amusement—science was in the nineteenth century a commodity as perhaps never before. Science, or, perhaps better, the sciences and scientists—and we mean by this use of plurals different subjects as they began to be defined and variant methods, as well as different people—had to compete for public attention as well as for the notice of fellow practitioners.

What better way than public oratory to practice a science or, at least, to declare in public competence with respect to its content? Speaking could at once help secure status (for the speaker and the science), impart knowledge, and lend personal dynamism through the performance. But while access to science's speech sites, such as the Music Hall in Edinburgh's George Street favored by the Edinburgh Philosophical Institution, was restricted to members of that body and their guests, the conduct of science's speech acts themselves had to be regulated, often subtly so, in order that science be delivered appropriately. As Diarmid Finnegan shows, not every speaker among those who lectured before the audiences of this Institution illuminated science's mysteries through verbal artistry. And even among speakers whose delivery, rhetoric, and platform manner was admired, such as Thomas Henry Huxley, the content could be received and interpreted in different ways. Lecture programs, speech styles, and the delivery of scientific discourse—not just the discourse itself thus merit attention for what is revealed about the conduct of science's practitioners and for insights into audience expectations.

Science's interpretive communities are always sited somewhere, even if distributed over space rather than for a moment concentrated in place in (say) the confines of a lecture hall. They also worked to constitute themselves as communities, albeit with divisions and differences of emphasis or procedure. Consideration of the means by which science's interpretive communities give meaning to knowledge can reveal what David Livingstone calls "reputational geographies." Using nineteenth-century readings of the seventeenth-century "proto-anthropological" thinker Isaac La Peyrère and documenting the contours of Charles Darwin's critical reception by scientific commentators of shared religious affiliation within late nineteenth-century Ireland, Livingstone argues that the relationships between how and where ideas were read, and by whom, must be examined closely if we are to comprehend the ways in which science moved over space.

Electricity had an almost elemental reputational geography: simply, was it safe or not? Answering that question depended greatly upon certain sites and social spaces. To an important degree, electricity's associations with danger and with utility had their obvious geographical expression in city streets. But for Graeme Gooday, moral panics and technical concerns about this new technology were most evidently played out indoors—in the home and on the theatrical stage. Determining electricity's safeness required its demonstration in those venues. Gooday illuminates both domestic and public exhibition spaces and shows how, in theaters particularly, the performance of electricity as a safe form of power, light, and civic display—even of bodily decoration—was put to the test. This was, at least initially, a form of social and technical spectacle. In the longer run, it also provided a public warrant for a certain form of technical knowledge and demonstrated that electricity's power, for many an almost mysterious force whose creative origins remained obscure even after its consequences were released at the flick of a switch, could be harnessed by domestic servants as well as by theater managers.

Electrical safety and electrical demonstrations involved performative trials in space. So too did the development of engine technology and advances in the use of high-pressure steam for power. But drawing boards and dry docks were inadequate sites for the large-scale experimentalism upon which shipping companies depended for their reputational geographies to be made known—as safe, reliable, and cost-effective. The performance of marine technology additionally required testing over space with questions of speed, efficiency, and reliability as key determinants of success. That is the message of Crosbie Smith's discussion of the marine steam engine between 1850 and 1885.

Questions of tolerance were always to the fore in expeditionary science: could one get on with one's fellows? Would the climate be tolerable, the locals welcoming, results meaningful? Knowing why one was on an expedition was usually the easiest part. Objectives were set in advance and usually elsewhere, just as the success of expeditions rested as much in the published results and spoken papers delivered as in the safe return of the explorers. Inasmuch, then, as expeditions had a pre-fieldwork planning phase and a public afterlife, published and orated, they were performed not just before, during, and after the "field stage" but in fields altogether different from those distant geographical places and spaces through which explorers trekked. Such movement between field and metropolis, as Lawrence Dritsas has it in examining the uncertain place of the Central African lakes, also allowed that different routes to scientific knowledge could be employed, and that different reputations for authority could be made or broken in relation to which methods had been practiced in securing what knowledge. This performance of reputational status about geography—and, in the longer-run, about the very shape and content of Africa itself—took place in the debating chambers of the Royal Geographical Society and on the shores of the lakes themselves. Despite later protestations about the virtues of actual observation, claims from the field were critically dependent upon trust in native testimony, just as the stay-at-home geographers relied upon textual exegesis. To convince London's audiences about the nature of Africa's geography, "actual observers" and "critical geographers" alike had to declare who had acted as their guides, and, more exactly, how they did so.

CHAPTER SEVEN

Placing Science in an Age of Oratory: Spaces of Scientific Speech in Mid-Victorian Edinburgh

DIARMID A. FINNEGAN

In October 1888 the *Pall Mall Gazette* offered to its readers "hints by the best public speakers of the day" on "how to become an orator." The venerable Thomas Henry Huxley, lauded as the "first of scientific orators," was among those who contributed advice. For Huxley, successful oratory depended upon the hard work of writing out an argument and resisting the "strange intoxication which is begotten by the breathless stillness of a host of absorbed listeners." Even if the prepared script was not consulted during delivery, it was only the laying out of "dry bare propositions" in writing that would prevent an orator from losing control of his argument.[1] This solution to the perils of public speaking presented reasoned speech as a disembodied communicative act that successfully resisted external pressures and overcame physiological hindrances. Huxley's ideal, however, did not capture the messier reality of efforts to communicate science during the preceding decades. During that period, logic and location along with propositions and performances were tightly bound together in the delivery of science lectures.

This historical and geographical reality is fleshed out here by a close examination of lectures on scientific topics given in mid-Victorian Edinburgh. Organized by the Edinburgh Philosophical Institution (EPI), a little-studied society active throughout the Victorian period, the lectures made up a portion of an annual series of talks addressing a wide range of subjects. The series served the overall aim of meeting "every kind of intellectual craving" of the city's educated classes. Reflecting a liberal ethos evident in similar societies elsewhere, the EPI selected speakers and topics carefully to avoid controversy and maintain neutrality. Even so, as enunciated in the hall of the institution, science no less than any other subject was knotted together with local conditions, politics, and protocols.

153

This chapter seeks to explore the places, practices, and personalities associated with the Institution's science lectures in the period 1850–1870. In so doing, it follows a more general line of argument that where science is constructed or communicated can have a significant bearing on its content and wider connotations, and that the making or marketing of science can significantly reconfigure the particular places in which those activities occur. Before developing this argument with reference to the science lectures delivered at the EPI, a number of related historiographical issues deserve further comment.

PLATFORM SCIENCE AND THE GEOGRAPHY OF SPEECH

By the time Huxley's advice was published, it was a cliché that "in these democratic days the future will lie with the articulate."[2] A growing body of historical research suggests that it is difficult to underestimate the cultural significance of public speech throughout the Victorian period.[3] This observation applies as much to science lecturers as to politicians and pulpiteers. Yet much of the pioneering work on science lectures in the nineteenth century, as Martin Hewitt has recently observed, concentrated on the period around the First Reform Act and cast science as a vehicle for social mobility attractive to marginal men.[4] According to these accounts, by the "age of equipoise" platform science had lost much of its radical edge and social utility and had become a form of mainstream and middle-class entertainment. More recent work, however, has demonstrated how much remains to be known about the character and significance of science lectures throughout the Victorian period.[5]

Taking seriously what Hewitt terms "platform culture" allows scientific speech to be analyzed alongside other verbal performances and placed within local manifestations of a ubiquitous culture of public speaking. One of the key issues raised by a focus on public speech is the positioning of science lectures within a wider, competitive cultural field. Much recent work on nineteenth-century science lectures has focused on the technologies of display employed in order to compete with other forms of visual entertainment.[6] The expectations and technologies of a wider visual culture were undoubtedly an important factor in shaping science lectures in a variety of social contexts. From the comportment and general appearance of the lecturer to experimental demonstration and scientific spectacle, making science visible was crucial in efforts to create, maintain, and enlarge science's place in civic culture. Yet visual appeal was only part of such efforts. In Edinburgh and elsewhere the ear as well as the eye had to be appropriately trained in order to participate in cultural life. Arguably, a concern with hearing aright was given particular emphasis in a self-consciously middle-class intellectual culture. The auditors attracted to the lectures of the Edinburgh Philosophical Institution were expected to be attuned to the nuances of sound, whether discerned in a musical performance, a public reading, or an intellectual address.[7] A mark of cultural refinement was the ability to hear whether the grace notes in an oratorio were properly executed or to follow by concentrated listening an intricate argument. Simply put, science had to sound right as well as look right to retain its place as part of intellectual culture in mid-nineteenth-century urban Britain.

In making this case I also want to argue that attention to the "voice of science" in the nineteenth century requires sensitivity to the geography of scientific speech. Suggesting that science did not have a single voice in the nineteenth century is a well-worn thesis, and the plurality of actors who spoke on behalf of science suggests that attention to the geography of scientific speech is of critical importance.[8] My aim here is to uncover the diverse ways in which science was articulated in a speech space whose limits and character can be studied in some detail. In approaching this particular task I have found helpful Charles Taylor's argument that any study of communicative action needs to be alert to the local customs and norms and to the background knowledge that make meaningful and innovative speech acts possible.[9] For Taylor, speech protocols are at once a cause and consequence of speech acts, acts which constantly renew but can also revise the linguistic codes on which they necessarily depend. For this reason, it is important to pay close attention to what Taylor terms "contexts of action," or the spaces of creative interaction between the structures and performances of communication. David Livingstone has given this argument a more plainly geographical inflection by calling attention to the relations between location and locution.[10] These more general insights can be sharpened and expanded by exploring the ways in which science lecturers at the Edinburgh Philosophical Institution simultaneously conformed to local expectations and customs, and reconfigured them according to agendas forged elsewhere. It follows that though my study is site-specific, it cannot be reduced to a set of local cultural or political coordinates. Rather, the speech space associated with the EPI was a dynamic entity that emerged from the interaction between local protocols and expectations and creative and diverse performances.

In what follows I explore the speech spaces created and regulated by the Edinburgh Philosophical Institution and examine the efforts of lecturers during the 1850s and 1860s to articulate scientific concerns without offending or fully replicating the assumptions and sensibilities of their Edinburgh audiences.

THE EDINBURGH PHILOSOPHICAL INSTITUTION

The Edinburgh Philosophical Institution was founded in 1846 to "enlarge the sphere" of the Philosophical Association, which had organized scientific and literary lectures for Edinburgh citizens since 1832.[11] Like other similar societies active elsewhere in Britain, the EPI aimed to provide a free platform for lecturers to supply a broad diet of subjects designed to meet the demands of local consumers of mental culture. Drawing on local talent, the directors of the Institution also prided themselves on attracting speakers with international reputations. Ralph Waldo Emerson included Edinburgh in his tour of Britain in 1847–48, giving four lectures under the auspices of the EPI. Six years later a youthful John Ruskin announced his dislike of Edinburgh's neoclassical architecture to an EPI audience. While avowedly more humanistic than its predecessor, the EPI included science as an integral part of its seasonal program of lectures. Among the better-known names appearing on the lecture programs in the1850s and 1860s were Edinburgh residents John Hutton Balfour, David Brewster, Samuel Brown, and Hugh Miller, as well as visiting speakers Edwin Lankester, Thomas Henry Huxley, John Tyndall, John Pringle Nichol, John Henry Pepper, John Lubbock, and Benjamin Waterhouse Hawkins.

The society's membership exceeded two thousand by 1860 and was drawn from Edinburgh's large middle class.[12] The institution, unlike its predecessor, was no longer the preserve of Edinburgh's petite bourgeoisie and included among its members representatives of a more socially elevated constituency.[13] The founding in 1864 of a Working Men's Club and Institute with the express purpose of providing for Edinburgh's lower classes a society equivalent to the EPI provided evidence, if it was needed, of the latter's role in defining a middle-class identity. Thus, when Thomas Henry Huxley delivered his notorious "men and apes" lecture to the EPI in 1862, his audience was not the "tough tartan workforce" evoked by Adrian Desmond.[14] Rather, Huxley was frequently applauded by ladies and gentlemen drawn from the city's genteel, well-to-do classes.[15] To attend the lecture, it was necessary to be a member of the institution, and while the majority of these were men in the year Huxley gave his address, there were also more than three hundred female subscribers.

The "presiding guardian" of the institution, the actuary and Fichte scholar William Smith, exercised considerable influence over the institution's affairs, being vice president from 1848 until 1886 and the long-standing chair of the Lecture Committee.[16] Smith, who was a Unitarian, a stalwart Liberal, and a key supporter of the Edinburgh Whig MP Thomas Babington Macaulay, saw the institution as a central player in Edinburgh's civil society. It was, Smith noted, "A recognized centre for the middle class intelligence of Edinburgh, a nucleus for the adult self-education of the city."[17] As such it educated Edinburgh's enfranchised population (and their wives and daughters) in the art and science of good government. Smith had no doubt that in bringing together Edinburgh's politically active citizens and "the very leaders of the thought of our age" to produce a more liberal "tone of thought," the institution played a vital role in creating and sustaining a healthy political culture. While Smith's liberalism did not necessarily represent the views of the institution's membership—which, if anything, had, a "legal-Tory tinge"—the message from the institution's platform was broadly consonant with the Whig-liberal politics represented by the institution's two figurehead presidents of the 1850s and 1860s, Thomas Babington Macaulay and Henry Brougham.[18] A formal nod was made toward the Tory constituency among members in the first few years of the institution's existence with the election of Christopher North (John Wilson) as president in 1847. Radical or independent liberalism, on the ascendancy in late 1840s Edinburgh, was also represented by the Edinburgh MPs Charles Cowan and Duncan McLaren, who served as extraordinary directors. Yet, for all this apparent parity, there appears to have been an overall lean toward a Whig-liberal position, at least among the institution's directors.

The EPI lectures generally took place in Queen Street Hall, a cube-shaped auditorium adjoining the institution's rooms and owned by the United Presbyterian Church. The hall was used by a number of Edinburgh's myriad voluntary associations, but every Tuesday and Friday evening in the winter it was leased to the EPI for its lecture series. It was never regarded as ideal and often was not big enough to accommodate the audiences attracted to the lectures. As well as being cramped, the hall at the back of 5 Queen Street was considered substandard in other ways. For example, George Wilson, an Edinburgh chemist and extramural lecturer, made a public appeal in 1853 for a more suitable venue.[19] A Greek amphitheatre, Wilson urged, was a more appropriate design for lectures on physical and literary subjects. Additional drawbacks were noted by others. The lack of ventilation and heating were "universally complained of," and the hall's single entrance meant that "a crush" was a regular feature of attending a lecture organized by the EPI.

The popularity of individual lecturers was registered by, among other things, the use of the nearby Music Hall. Part of the Assembly Rooms complex on George Street, the hall was able to accommodate an audience of fourteen hundred or more and was used by the directors of the institution anxious to avoid overcrowding. Despite the greater size of the properly ventilated hall, when George Wilson got to lecture there "by default" on the polarization of light in 1857, it is unlikely that he was fully satisfied.[20] The hall was custom-designed for musical evenings—the EPI organized an annual concert there for members—but it was still less than suitable for lectures of a literary or scientific kind.

Inadequate facilities and overcrowding were not the only practical problems with the two halls used by the institution. Poor acoustics meant that some speakers could not be heard. A case in point was the opening address by David Brewster delivered in 1851. Despite being an acknowledged expert on acoustics, Brewster gave his lecture rapidly and in "low tones," and his delivery, while eloquent, was difficult to hear.[21] Even when the speaker's voice carried, street noises often leaked into the auditoriums, much to the consternation of the institution's directors.[22] In 1865 the "thwack, thwack" of carpet sticks—the suspicion was that the timing was deliberate—disturbed a lecture delivered in the Music Hall by James Anthony Froude, a fact bitterly complained of by Henry Bowie, the long-standing secretary of the institution.[23]

From the institution's inception, regulating the auditory space of Queen Street Hall and the Music Hall was important to the directors, and stray noises from outside or inside the halls were taken seriously. Applause, cheers, and other controlled acclamations were acceptable forms of audience-generated noise, but beyond those conventional modes of participation, auditors were expected to refrain from the kinds of rowdy interruptions associated with less genteel settings. Most famously, William Thackeray when lecturing to the EPI in 1856 on the four Georges faced a sibilant chorus generated by members of the Scottish Rights' Association.[24] That this was a serious concern of the institution's directors is evident in the repeated efforts during the 1850s and 1860s to impose restraint on recalcitrant auditors. Measures to prevent the noisy "irregularities" of "certain parties" at lectures were discussed privately by the directors in 1856 and, in 1868, a report in the *Scotsman* made public the general disapproval of "puerile disturbances" among the audience throughout the lecture season.[25] The apparently trivial matters of reducing ambient sounds and policing audience behavior were symptomatic of the culture of listening fostered by the EPI.[26] The regulated space of the institution's lecture hall was managed in a manner akin to the National Gallery and other "ritual sites," which, as Nick Prior has argued, reinforced a distinctive middle-class disposition associated with moral seriousness, refined pleasures, and an emphasis on the aesthetic.[27] Yet for all that, audiences were never fully compliant with the tacit codes of conduct thought appropriate for encouraging the deep attention necessary for appreciating the finer points of an intellectual address. It was not until 1870 that the directors drew up a rule banning dogs from the hall during lectures and enlisted a policeman to enforce it. In the same year George Dawson stopped to draw attention to "some of his lady auditors plying knitting needles while he lectured."[28] Such behavior was deemed discordant with the serious moral and intellectual purposes of the institution. A quiet hall and rapt attention were necessary to hear what one member described as the "exquisite music of . . . great speech."[29]

Tacit codes of behavior also applied to lecturers. The range of opinions that could be expressed was as wide as the ability of the lecturer to couch them within the context of the good manners and social mores upheld and policed by the institution's directors and members. What could and could not be heard in the lecture hall was conditioned by the regulative ideals associated with the notion of a free platform. This notion did not translate into freedom to discuss any subject without fear of interference or censure. Certain topics—and the list varied depending on its originator—were classified as "sectarian" and were not permitted. Other topics, judged to undermine moral culture, could also be censured. John Tulloch, principal of St. Andrews University, expressed the basis of this restriction in his opening address to the institution in 1857, suggesting that "the highest freedom, depend upon it, is ever bound fast in moral law."[30] For Tulloch, moral earnestness and a spiritualizing effect were the marks of free intellectual inquiry, a sentiment frequently affirmed by lecturers, scientific or otherwise, but never precisely defined.

What all this makes clear is that speaking in Queen Street Hall required finesse. For some speakers, the prospect of delivering a lecture without giving offense or diluting "manly" resolve and conviction caused acute distress. Before his first lecture to the institution in 1854, Charles Kingsley "cried with fear up in his own room" and later declared that he had "felt more nervous than [he] had ever done in his life."[31] No doubt such nervousness was related to his struggles to overcome a stammer but it did not help that his lectures touched on a number of "delicate points" that he believed would "infuriate the Free Kirk."[32] In contrast, at least one of the lecturers appearing on the institution's platform was a master of the art of walking the line between offending and boring his audience. Ralph Waldo Emerson was well practiced in the art of "transpiercing" or punctuating his otherwise decorous lectures with heterodox aphorisms.[33] Emerson's radicalism, as one commentator has noted, was well adapted for the lecture circuit, and he self-consciously fostered a tension between convention and heterodoxy in his spoken discourse without overstepping the bounds of propriety.[34] Along with this technique, Emerson paid careful attention to tone and deportment. According to one commentator, Emerson read his lectures in Scotland "with a calm classical power and dignity" so that only two things remained "the silence and the voice that was passing through it."[35] In Edinburgh Emerson was warmly applauded by his audience and later denounced in print for his radical pantheistic views.[36]

The difficulties associated with uttering "free speech" within the lecture hall—appearing to avoid sectarian topics while upholding moral propriety—were compounded by pressures from without. The space of the lecture hall was not always privileged over other spaces of intellectual culture, despite the high profile given by the directors of the EPI to verbal instruction. Connop Thirlwall, Bishop of St. David's, in his inaugural address to the institution in November 1861 on self-culture, echoed a widely shared view when he opined that lectures were supplemental to reading, the latter providing a more effective means for supporting independent learning. However intended, Thirlwall's comment could easily be used to support the criticism, frequently enough made, that EPI lectures hampered true learning. A consequence of this aspersion was an increased pressure on speakers to provide lectures of substance that countered charges of superficiality. Other critics noted not the small but rather the enervative effects of fashionable mental culture. The EPI had a reputation for encouraging intellectual talk among educated females, leading a satirical account of Edinburgh society published in 1861 to single out "blue stockings" as a distinct Edinburgh "caste."[37] Around the same time, the institution was put forward as evidence of the decay of intellectual industry and the feminization of intellectual culture evidenced by the multiplication of passive consumers. The institution, along with other societies and the increase in public statuary, led to the conclusion that the city had "ceased to produce."[38]

In describing the delicately balanced speech space of the EPI, it becomes clear that speaking there presented a number of acute challenges. The speech space was carefully tuned, and it was easy to sound a discordant note. Building on this sketch of the general character of the speech space, I want to consider in particular the efforts by speakers to find an appropriate subject, pitch, and tone for science lectures.

SKEPTICISM, SECTARIANISM, AND SCIENTIFIC SPEECH

Science lectures were a feature of the winter program from the institution's inception. Rather than offer the extended courses of lectures on a single subject previously supported by the Philosophical Association, the directors of the EPI encouraged one-hour lectures on subjects with wide appeal. The rule of thumb for any lecture was that it should aim to develop moral sympathies and support self-culture while avoiding religious or political controversy. In one way or another, lecturers had to position their scientific discourse in the interval between sectarianism and skepticism, a task also colored by a concern with etiquette, aesthetics, and moral probity. My account of science lectures begins in the 1850s with two of the institution's most prominent and popular lecturers: Hugh Miller and George Wilson. Both Miller and Wilson tied science to Christian orthodoxy and skillfully avoided charges of sectarianism while upholding an image of science that was in harmony with moral meaning and aesthetic acumen. By the 1860s, the second period examined here, science lectures were more likely to be criticized for undermining the foundations of moral culture than for propagating sectarian ideas, and the perceived risk to moral order, rather than any poetic rendering of science, was an important reason for their popularity or notoriety.

SCIENCE AND THEOLOGICAL POETICS: THE 1850S

During the 1850s, effective science lectures at the EPI successfully integrated literary charm and moral sobriety, and delicately touched on controversial theological concerns without adopting a polemical tone. Among the most popular science lecturers in this period were Hugh Miller and George Wilson, and both were celebrated for their literary and scientific abilities. They also held in common a commitment to creedal Christianity, even if separated by denominational affiliation—Miller being a champion of the Free Church of Scotland, and Wilson flitting during his early years between Baptist and Episcopalian affiliation before settling in a Congregational church. For Miller and Wilson, science was, among other things, a useful vehicle for conveying the romance of theological orthodoxy, a romance that enlivened, and was enlivened by, their presentation of scientific discoveries.

Miller in particular used his EPI lectures to refute the charge that science lacked poetic power. In his third lecture of a series of seven given under the auspices of the institution in 1851, Miller prefaced his ongoing exposition of the geological history of Scotland with remarks on the connections between geology and poetry. In his preamble Miller spoke in somber tones of the recently deceased Scottish Poet David Macbeth Moir, praising his under-appreciated poetical abilities. Miller also used the occasion to take a different line on the relations between science and poetry to the one propounded by Moir a few months earlier in the same venue in which Miller now spoke. On that occasion Moir had directed members of the EPI to the injurious effects of science on the "career of poetry." In this, Miller urged, he was surely mistaken.[39] The poet had in fact nothing to fear from "the stony science" that supplied "dim and shadowy fields in which troops of fancy already walked."[40] As Miller made clear in a later lecture, it was his belief that nature's hieroglyphics, properly deciphered, would bring to light God's own artistry and that the basis for the substantial harmony between geology and poetry was the identity between the aesthetic and musical sense in the mind of God and the mind of man.[41] Pressing the parallelism further, Miller presented to his audience a history of the earth that was prophetic in character prefiguring "the advent of man," God's image-bearer. The ordered appearance of life, Miller urged, was but one aspect of a single "sublime scheme" that had at its apex the appearance of the "Divine Man."[42]

The literary and theological charm of Hugh Miller's writing was already well known, but his lectures to the EPI in 1851 displayed, as Ralph O'Connor has commented, the consummation of his dramaturgical skills.[43] Miller's transcendental geology communicated a Christian metaphysics in a literary mode that appealed to the audience of the institution, despite the much-discussed dissonance between the noise that emanated from the Free Kirk and the liberal tone of moderate Edinburgh.[44] There is no reason to doubt Lydia Miller's reminiscences, recorded in 1859 in a preface to a published version of Miller's 1851 lectures. There Lydia recalls that the lectures "excited unusual interest, and awakened unusual attention, in a city where interest in scientific matters, and attendance upon lectures of a very superior order, are affairs of everyday occurrence. Rarely have I seen an audience so profoundly absorbed."[45] Miller's subsequent appearances before the EPI in 1853 and in 1856 confirmed his popularity as a lecturer. In 1853 Miller gave two lectures on the geology of Edinburgh and its neighborhood to members of the institution in the Music Hall, a venue chosen to circumvent overcrowding.[46] The second lecture was "numerously attended" despite, as the report in the *Scotsman* noted, coinciding with a concert of Italian Opera performed in Edinburgh's Theatre-Royal on Shakespeare Square.

For all this, Miller's reputation as a lecturer was not universally acclaimed. According to one biographer, Miller's "manner was singularly ungainly at the reading desk on the platform [and] his pronunciation was harsh and intensely provincial."[47] Miller himself expressed doubt about the effectiveness of his lectures and judged that they were "better adapted for perusal in the closet than delivery in the public hall."[48] Miller's idiomatic speech was nevertheless in high demand. According to another acquaintance, Miller's public lectures were "melodious," and his Cromarty accent added zest to his literary style. For this admirer, Miller's verbal artistry was made more memorable by being heard, his northern accent marking out rather than detracting from the poetic style of his scientific lectures.[49] David Masson's celebrated pen portrait of Miller, published in *Macmillan's Magazine* in 1865, noted that "ladies would be dying to hear Hugh Miller talk," though they were often disappointed in private gatherings because Miller was "almost blockishly silent" in company.[50] Miller, it seemed, concentrated his energies on more solitary pursuits, including the careful crafting of popular lectures, and took great pains to ensure that his evocations of the sublimity of ancient landscapes were heard by the inner ear and fully absorbed by the imaginative faculty of his admiring hearers. Adapting his literary voice to the lecture hall, Miller carefully honed an approach that had been developed in solitude and successfully marketed in print. Unlike Thomas Carlyle, who tried his hand early at public lecturing only to become disillusioned, lecturing for Miller "was an experiment made late in life."[51] Yet for both men, writing with a rapt audience in mind helped them to enhance their widely acclaimed literary productions.[52]

Miller's concern with developing an appropriate style was matched by a concern about appropriate substance. This did not mean he truckled to what he regarded as opinions more liberal than his own. For all Miller's efforts to present geological science in a mode likely to win warm applause in the speech space of the EPI, his address contained notes of dissent from the metaphysical drift that set the general tone of the institution's platform culture. Miller, like his friend and biographer Peter Bayne, was Carlylean in literary style only and not in philosophical substance.[53] The direction of Carlyle's thought was, Miller felt, toward pantheism, and while he could assimilate and develop Carlyle's natural supernaturalism through his dramatic reconstructions of the geological past, he did so on the basis of a more traditional understanding of divine providence. This becomes clear, for example, in his 1856 lecture to the EPI on the paleontological history of plants. The classification of fossil plants, rather than being a convenient tool invented by the botanist, tracked more or less successfully an "external principle." This principle, Miller maintained, could not be explained

with reference to impersonal forces. Instead, it was a person who stood behind the ordered arrangement of fossil vegetation.[54] Yet for all Miller's attempts to steer his audience away from deistic or pantheistic schemes, he
proceeded cautiously and with due deference to those who might find such
appeals distasteful or out of place.[55]

Miller was not alone in successfully combining a public commitment to
beliefs that risked being condemned as sectarian with a verbal performance
warmly welcomed by the institution's directors and members. George Wilson, known in Edinburgh and beyond for, among other things, his defense of
Christian orthodoxy against the "infidel tendency" of the anonymous *Vestiges of the Natural History of Creation,* lectured frequently to the EPI from
its inception until his death in 1859.[56] His published lectures on the "five
gateways of knowledge" were described by one reviewer as "a hymn of the
finest utterance and fancy—the white light of science diffracted through the
crystalline prism of his mind."[57] John Cairns, Wilson's erstwhile mentor and
noted United Presbyterian clergyman, confirmed this opinion in an obituary
of Wilson in *Macmillan's Magazine.* Wilson's lectures had exhibited, not
unlike an evangelical sermon, "a high strain of moral eloquence that linked
every topic to man's joys, and sorrows, and deep enduring interests."[58] A
similar if briefer encomium was issued by William Smith to members of
the EPI, which remembered the "clear scientific exposition" and "graceful
play of fancy" that marked the discourse of the noble and true-hearted lecturer.[59] Wilson himself had declared in 1846 his long-held conviction that
science and poetry were like two binary stars exchanging rays and together
moving around and shining toward their source, "the Throne of Him who
is the cause of all beauty."[60] Ultimately, Wilson avoided a "frigid natural
theology" and cultivated a sublimity of thought that served his religious
mission.[61]

In contrast to Miller it is likely that Wilson accepted the honorarium
offered to lecturers, reliant as he was on income from his lecturing work.
As one commentator put it in a review of Wilson's memoirs, "The necessity
of living by science as well as for it involved him in the stern struggle for
place and distinction."[62] Even so, it is clear that he did not lecture to the EPI
for pecuniary reward alone. In 1856, a year after being appointed director of
the new Industrial Museum and Regius Professor of Technology at the University of Edinburgh, Wilson lectured to an overcrowded Queen Street Hall
on "the objects of technology and industrial museums."[63] Wilson used the
occasion to argue that the purpose of technology was wider than commercial utility. It also freed up time for the cultivation of the imagination and

provided occasions for exhibiting workmanship analogous to the creative acts of the deity. The lecture itself, described as "of the highest eloquence and poetical beauty," challenged objections that technology led to moral decline by "killing the conscience."[64] On the contrary, Wilson urged, technology was compatible with the highest intellectual and moral culture and thus could legitimately take its place among the other subjects tackled during the institution's program of lectures.

Wilson's popularity at the EPI was reinforced in other ways. For one observer, it was his abilities in charming Edinburgh's "lady philosophers" that aided his success as a lecturer. Wilson endeared himself to his female listeners in part because of his "cork foot," a material reminder of his long struggle with poor health that added pathos to his scientific poesy. Such popularity was hard earned. His lectures to the institution, given the expectation of "intelligent, educated and critically appreciative listeners," were regarded as different in scope and profundity to those delivered to other "popular assemblies" and took their toll on Wilson's health. His lectures to the EPI, his sister later recorded, were "almost invariably . . . followed by sharp illnesses."[65]

The earnest moral tone, the personal intensity of delivery, and the Carlylean tenor that characterized the scientific speech of Wilson and Miller resonated with the general intellectual and aesthetic sensibilities of members of the EPI.[66] Both lecturers successfully met the conditions for liberal, moral, and tasteful speech. Miller and Wilson, despite being separated by different denominational affiliations, shared a commitment to an artistry and theological poetics that cut across religious lines. Both could be heard with pleasure by EPI audiences, the warm applause and packed lecture halls testifying to the satisfaction taken in discourses that hinted at a theological commitment that was otherwise excluded from the institution's platform.

CREEDLESS SCIENCE AND MORAL CULTURE: THE 1860S

Wilson died at forty-one on Tuesday, 22 November 1859, nearly three years after Miller's suicide on Christmas Eve, 1856. Whether or not precipitated by their deaths, the early 1860s saw a change in the character of science lectures given to the EPI. More generally, the space for public lectures carved out by the institution arguably became less stable, and the tensions between different understandings of what constituted a "free platform" became more acute. Science lectures undoubtedly played a role in this, and an analysis of

their content, delivery, and wider impacts signals the dynamic and changing nature of the institution's speech space.

A striking indication of this shift is found in the lectures delivered by
David Page, former employee of W. and R. Chambers and author of several primers on geology. Page took Miller's place as the institution's local
lecturer on geological and ethnological topics, and his style and strategy
contrasted sharply with those perfected by Miller and Wilson. More purely
forensic and less epideictic, the aim of Page's rhetoric was to present in
plain terms the progress and results of modern geology. Sometimes mistak-
enly identified as the author of *Vestiges*, Page actively opposed attempts to
present science as a handmaiden to theology.

Page was not alone in striking a more secular note and offering talks
in a less effusive style. The scientific presentations of visiting speakers in
the 1860s avoided or explicitly distanced themselves from the theological
register and poetic vein in which the discourses given by Miller and Wilson
had been delivered. In the 1860s John Tyndall, Thomas Henry Huxley, John
Lubbock, and Benjamin Waterhouse Hawkins were the most noteworthy
science lecturers to appear on the institution's platform. Of these, Tyndall
and Huxley, the most resolutely secular and controversial, were repeatedly
invited back but did not return. Hawkins offered another series of talks to
the Lecture Committee in 1863 but was declined, and Lubbock did not return again until 1887.[67]

Hawkins merits brief and separate comment. He was arguably unusual
among EPI lecturers in appealing directly, as he himself put it, to "the eyes
rather than the ears of his audience." According to a report of his first lecture to the institution in 1862, his "rapidly executed drawings" illustrating the unity of plan of the higher vertebrates "afforded much amusement
and instruction."[68] But his privileging of what he termed "visual education"
over the listening ear contrasted with the verbal artistry beloved of his Edinburgh auditors, and Hawkins's emphasis on visual spectacle may have been
regarded by some as more suitable for a less-refined audience. Whether or
not this is correct, for a crowd used to dramatic readings, musical concerts,
and lengthy literary discourses, words as well as images were critically ap-
praised and politely or warmly applauded.[69]

A starker and more immediate contrast can be drawn between Hawkins's
explicitly anti-Darwinian stance and the much discussed pro-Darwinian
lectures delivered during the preceding season by Thomas Henry Huxley.
Of the institution's science lectures, Huxley's caused the greatest stir both
within and outside the institution. Controversial lectures that provoked
acerbic comment from the local and national press were not new, and Hux-

ley, like Emerson, received applause in the lecture hall while provoking the opprobrium of Edinburgh's evangelical press. Nevertheless, Huxley's 1862 lectures came to be remembered as among the most controversial ever delivered to the EPI.

It is important to note, however, that the immediate reception of Huxley's two lectures to the institution in January 1862, which took as their subject the evidence for the origin of man from the same stock as apes, cannot be easily recovered. Huxley, writing shortly after the lectures to Joseph Dalton Hooker, noted that his "announcement" had "met with nothing but approval," a fact that heartened Huxley, who took it as solid evidence of the "disintegration of old prejudices" and a sign of the advance of liberal thought in "saintly Edinburgh." Hooker, however, was not convinced. In a letter to Darwin, perhaps recalling his own trouble with Edinburgh's conservative churchmen nearly two decades previously, he poured cold water on Huxley's post-lecture elation: "Huxley writes to me in *Exuberat spirito* at the reception of his lecture in 'Saintly Edinburgh'—he quite forgets there are sinners enough in a population of 180,000 to make him an applauding audience. I do not think H. has the smallest idea in how small a circle he makes a noise."[70]

Huxley's intentions had been polemical. He had desired, as he put it, to disseminate his views "through regions which they might not otherwise reach" by provoking an outright attack by Edinburgh's evangelicals.[71] For Hooker, Huxley's efforts to stoke Edinburgh's already volatile religious atmosphere were pointless. Edinburgh, in Hooker's view, would remain split down the middle: "No one in Edinburgh who reads either side sees the other & no one out of Edinburgh reads either."[72]

Not surprisingly, perhaps, Huxley's view of the reception of his lecture by the audience was confirmed by the author of the attack published in the *Witness* newspaper, mouthpiece of the Free Church. Appalled at the approval Huxley received, the *Witness* suggested that the audience, rather than applaud, would have been better advised to shake the dust from their principled feet and exit the compromised auditorium. Yet the notion that Huxley's lectures were warmly welcomed was only one appraisal of the atmosphere in Queen's Street Hall during the two evenings in January. For William MacDonald, professor of natural history at St. Andrew's University, the audience had responded with a "respectful courtesy" due to Huxley's firm and settled tone but had not "expressed the usual forms of approval" that characterized those occasions when the relations between platform and hall were harmonious.[73] The audience had tolerated Huxley's views, admired his verbal performance, and respected his fervent tone, but had not approved of his argument.

In representing his lecture to Darwin and Hooker, Huxley made it clear that he had not held back in proclaiming an evolutionary view. To Darwin, he suggested that press reports of the lecture had played down his plain-spoken support for Darwin's "doctrines." Yet Huxley also confessed to avoiding points of difficulty in order to prevent the charge that they could not be answered. A lecture was not the place to counter serious objections. Huxley also found space to reassure attentive Edinburgh listeners that a "prodigious and vast gulf" separated his audience from "brute creation." The difference between "civilized man" and the gorilla or chimpanzee was not one of "substance or structure" but lay instead in the "divine endowment of intelligent speech."[74] His audience, representatives of civilization, could glimpse "here and there a ray from the sun of everlasting truth," a fact that allayed or postponed fears that Huxley's position threatened the very foundations of the moral culture the institution was designed to promote. While Huxley regarded his lectures as an opportunity to dilute if not demolish religious dogma, he was also anxious to affirm the moral and intellectual standing of the audience to which he spoke. Moreover, as in other lectures Huxley gave in the same period, his Edinburgh presentations did not eschew entirely the use of analogy or undermine the value of "legend, tradition and myth."[75] In upholding these forms of knowledge while accenting the importance of science, Huxley approached with some care the compound expectations of his Edinburgh audience, and this goes some way to explain the different accounts of his auditors' reaction to the lectures.

Although Huxley's polemical intent was muted in deference to the rules of polite and principled speech that operated with particular force in Queen Street Hall his "men and apes" lectures nevertheless prefigured a more combative and controversial tone for science lectures given at the institution for the rest of the decade. The first signs of this appeared a little over two months after Huxley's lectures when a decision was taken by the Lecture Committee to turn down an offer of a talk from James McCosh on Scottish Philosophy. Two years later the decision resurfaced and became a matter of public controversy. In July 1864 at meeting of the Evangelical Alliance (EA), McCosh spoke on the "religious tendencies of the present age." The address was delivered in Queen Street Hall but, for McCosh at least, it was a rather different speech space from the one policed by the directors of the EPI. Along with the publication of *Essays and Reviews*, McCosh cited the institution as a sign of the rise of a powerful enmity towards creedal Christianity. "I suppose," asserted McCosh, "that it would be reckoned a recommendation to a lecturer for the Philosophical Institution . . . that he was . . . dissatisfied with the old theology."[76] McCosh took the opportunity of saying to the EA what he apparently could not say to the EPI, and his remarks signaled a growing belief among Edinburgh's evangelicals that the institution was drifting toward the active promotion of theological hetero-doxy and unbelief.

The charge that the EPI contravened its own principles and applied an inverted religious test when deciding on names for the lecture program was repeated several times during the 1860s. Two lectures by David Page on the origin and antiquity of man in December 1866 that advocated the "develop-ment theory" and drew conclusions that questioned "the brotherhood of man" provoked a published attack by another local Edinburgh geologist, George Sleigh. In his response Sleigh accused the directors of the institution of knowingly inserting in the program lectures that had already appeared in print and whose conclusions threatened to undermine the basis of civil society.[77] Sleigh, "an old member of the Edinburgh Geological Society," found Page's dismissal of all that had been written with the purpose of rec-onciling revelation and scientific research galling, the more so because it undermined the belief that all men, whether "savage" or "civilized" could be "elevated" through exposure to divinely revealed truth. For Sleigh, the theory of development taken to its logical conclusion would overthrow the "order of creation" and by extension the order of society. The "great preor-dained laws" of development that Page offered as supplying a "nobler view of the Divine architect" implied for Sleigh the pursuit of success through the destruction of weaker members of societies. Moreover, Sleigh found in Page's arguments justification for the extirpation of apparently inferior races and for an invidious and "unmitigated selfishness." Sleigh was not the only person to find in Page's views grounds for scientific racism. The published version of Page's EPI lectures received warm support in the racist *Anthropological Review*.[78]

Concerns over the judgments being made by the EPI Lecture Commit-tee surfaced again during the institution's annual business meeting in 1867 when James Craufurd, an Edinburgh judge, called for a "sound and safe in-tellectual culture" that united intellect and heart through a Christian ethic that transcended sectarian differences. On the surface, Craufurd's purpose was to celebrate the institution's free platform, but his remarks were di-rected at the "past mistakes" made by the Lecture Committee in inviting speakers who "tended to shake the Christian faith of their hearers."[79] Crau-furd's efforts to steer the institution toward a position that acknowledged certain religious principles as nonsectarian and as guides for the "honest and reverent" pursuit of truth were resisted within and outside the walls of the institution. The EPI was now oscillating more obviously between a religiously motivated commitment to "safe intellectual culture" and a secular program aiming at "excluding theology altogether." In support of the latter, an editorial comment on Craufurd's statements appearing in the *Scotsman* argued that all "theological toes" deserved the kind of protection afforded by the "principle of exclusion."[80] Offering the example of a lecturer speaking on the subject of geology or ethnology (Page was surely in mind), the editorial stressed the impossibility of admitting any theological datum without offending at least two-thirds of the audience. The apparent earlier success of Miller and Wilson went unmentioned.

Further support for Craufurd's position was not slow in coming. The opening address for the winter season in 1868, delivered by William Thomson, archbishop of York, was firmly on the side of what the *Scotsman* snidely called Craufurd's "limited liability principle."[81] For Craufurd, participating in the marketplace of ideas required security against the high risks of "dark speculations" detached from shared sources of moral standards. Thomson, in arguing against the growing influence of positivism, similarly fenced off basic religious and moral intuitions from the imperialistic hubris of the self-assured positivist. By counting as knowledge only that which had been certified by the senses, the followers of the new philosophy were denying reality to aspects of human experience and desires that were vital for moral culture and a meaningful existence.

The concerted efforts of Craufurd and Thomson to provide a metaphysical-cum-theological basis for intellectual culture were met with yet further resistance. Two days after Thomson's address, Thomas Henry Huxley lectured to an Edinburgh audience on the physical basis of life in the Hopetoun Rooms located at the other end of Queen Street from the institution's lecture hall. Inaugurating a series of Sunday lectures organized by the renegade and heterodox Edinburgh churchman the Rev. James Cranbrook, Huxley used the occasion to attack Thomson's philosophy of science. Appealing to David Hume to distance himself from accusations of materialism and endear himself to an Edinburgh audience, Huxley shrouded "matter in itself" in darkness and argued that it was science, or a concern with the phenomena "that lay before us," and not the hopeless inquiries into ultimate end of matter or spirit that supplied the real meaning of man's existence. [82] From the Hopetoun Rooms, Huxley redrew the boundaries of what counted as scientific speech and mocked the "lunar politics" represented by Thomson's appeals to intuitive knowledge.[83]

If Huxley's tactics were to exert pressure from outside the EPI, his fellow X-Club member John Lubbock attempted a year later to shift the terms of the debate from the inside. In closing his second lecture to the institution on savages and civilization, Lubbock drew the audience's attention to his efforts to avoid "causing pain" or wounding any "reasonable susceptibility."[84] His purpose had been to present a truthful account uninhibited by theological views. This, Lubbock added, contrasted with the "dangerous" opposition to science lately expressed by James Moncrieff, an eminent Edinburgh lawyer, politician, and Free Churchman who had criticized Lubbock for using science "as a lever to upset the faith of the world."[85] In denying Moncrieff's charge, Lubbock presented science as unassailable and religion, properly understood, as something elevated and purified by scientific truth. It was religion and not science that required fundamental adjustment and science not religion that best represented the kind of free inquiry upheld by the institution.

During the 1860s, then, the primary issue being negotiated from the platform and in the press was not whether scientific speech could be accommodated within a program of nonsectarian but religiously inflected literary entertainment and instruction. Instead, it was whether or not science shorn of any theological resonance or commitments threatened the kind of moral culture upheld by the EPI. The religio-poetic approach of Miller and Wilson, which satisfied the aesthetic demands of an Edinburgh audience, was replaced in the 1860s by the creedless science of Huxley, Page, and Lubbock. If the scientific discourses of Miller and Wilson risked being criticized as sanctioning partisan theology, the resolutely secular or, in the ears of some, skeptical platform science of the 1860s was frequently suspected of instilling moral confusion and of severing the link between intellectual talk and moral culture.

An examination of the form and content of scientific speech heard in the lecture halls of the Edinburgh Philosophical Institution has taken us beyond the formal confines of a single society to other civic and private spaces where science and culture were fused and transformed in different ways and for different ends. Examining one location has ineluctably led to other kinds of connected sites and spaces. At the same time, the speech space fostered by the EPI has not been an incidental backdrop. Codes of verbal and bodily conduct and disquiet about sectarianism and skepticism were among the more or less tacit constraints on speech of any kind heard in Queen Street Hall. In the terms offered by Charles Taylor, the science lectures of the institution took place in a significant "context of action" set within a wider set of issues germane to nineteenth-century intellectual and platform culture. Dissecting this single though not monotone scientific speech space has brought to view the ceaseless interplay between custom and communicative action, and has moved us beyond the mapping of material spaces of science lectures to uncover the less visible geographies of speech marked out by contests over protocols, moral culture, and the nature and content of free expression. The institution's speech space was one in which science was differently rendered by speakers who in contrasting ways adjusted their verbal performances to meet and challenge the perceived expectations of the auditors. At the risk of papering over a diverse set of communications, Miller and Wilson adopted one strategy, while Huxley, Page, and Lubbock adopted another. All accommodated their speech to the space in which they wanted their message to be heard. At the same time, all were anxious to unsettle as much as conform.

Part of the reason for this dynamic was the volatile nature and position of the EPI itself. It represented in turns the continuing vibrancy of the "Athens of the North" and the saintly or reactionary city sarcastically and strategically called to mind by Huxley and Hooker. For William Smith, the institution embodied a rational liberalism fortified by a Fichtean emphasis on the scholar as a secular saint. As such it provided, as Robert Lowe put it to the institution in 1867, a harmonious society that "trenched on no man's feelings" and worked "fairly, purely."[86] Its directors looked to provide a platform that fostered a healthy civil society. Even so, within the institution disagreement about what constituted liberality and neutrality was endemic, and the feelings of men like James McCosh, whether justified or not, were assuredly trenched upon. The tensions were not only internally generated. From without, the directors faced increasing pressures to maintain the institution's position as the leading light in Edinburgh's intellectual culture. By 1872, the EPI had a rival in the Edinburgh Literary Institute, which offered a similar program of lectures and concerts for the growing middle-class population of South Edinburgh. The EPI felt itself increasingly marginalized to the point that one "life member" could describe in 1888 its present location in Queen Street as "out of the way, cold and cheerless—a sort of hole" that people naturally avoided.[87] A general narrative of decline had worried the institution from its beginning, and the directors faced a constant struggle to maintain its wide appeal for speakers and auditors alike. In pursuing that goal the EPI, always vulnerable to conflicting forces, set the conditions for successful public speech. Thus, without being able to fix the terms, it significantly shaped the ways in which science was articulated to a critical and divided public.

NOTES

I am grateful to David Knight, David Livingstone, Ralph O'Connor, Michael Taylor, and Charles Withers for generous commentary and constructive criticisms.

1. "How to Become an Orator," *Pall Mall Gazette*, 24 October 1888, 1–2.

2. Ibid.

3. Martin Hewitt, "Aspects of Platform Culture in Nineteenth-Century Britain," *Nineteenth Century Prose* 29 (2002): 1–32; Matthew Bevis, "Volumes of Noise," *Victorian Literature and Culture* 31 (2003): 578–79.

4. See, for example, Jo N. Hays, "The London Lecturing Empire," in *Metropolis and Province: Science in British Culture, 1780–1850*, ed. Ian Inkster and Jack Morrell (London: Hutchinson, 1983), 91–119; Ian Inkster, *Scientific Culture and Urbanisation in Industrialising Britain* (Aldershot: Ashgate, 1997).

5. For two examples of recent work that reopens research on nineteenth-century science lectures, see Jill Howard, "'Physics and Fashion': John Tyndall and His Audiences in Mid-Victorian Britain," *Studies in History and Philosophy of Science* 35 (2004): 729–58; and David Knight, "Scientific Lectures: A History of Performance," *Interdisciplinary Science Reviews* 27 (2002): 217–24.

6. On science lectures and visual culture, see Iwan R. Morus, "Seeing and Believing Science," *Isis* 97 (2006): 101–10.

7. On Charles Dickens's public readings, see Malcolm Andrews, *Charles Dickens and His Performing Selves* (Oxford: Oxford University Press, 2006). Dickens read several times to members of the EPI.

8. This point has been emphasized recently in Bernard Lightman, "Lecturing in the Spatial Economy of Science," in *Science in the Marketplace: Nineteenth-Century Sites and Experiences*, ed. Aileen Fyfe and Bernard Lightman (Chicago: University of Chicago Press, 2007), 97–132.

9. Charles Taylor, "Language and Society," in *Communicative Action: Essays on Jürgen Habermas's "The Theory of Communicative Action,"* ed. Axel Honneth and Hans Joas (Cambridge, MA: MIT Press, 1998), 23–35.

10. This is most fully worked out in David N. Livingstone, "Science, Site and Speech: Scientific Knowledge and the Spaces of Rhetoric," *History of the Human Sciences* 20 (2007): 71–98.

11. The latter was examined in Steven Shapin, "'Nibbling at the Teats of Science': Edinburgh and the Diffusion of Science in the 1830s," in Inkster and Morrell, *Metropolis and Province*, 151–78.

12. On one estimation, as early as 1830, 20.8 percent of Edinburgh's population were middle class compared with 5.3 percent for Scotland as a whole: see Richard Rodger, *The Transformation of Edinburgh* (Cambridge: Cambridge University Press, 2001), 18.

13. Shapin, "Edinburgh and the Diffusion of Science," 177, n. 85.

14. Adrian Desmond, *Huxley: From Devil's Disciple to Evolution's High Priest* (Reading, MA: Perseus Books, 1997), 300.

15. In 1857 only 7.8 percent of the EPI's subscribers belonged to the occupational

grouping "distribution and processing." See Graeme Morton, *Unionist Nationalism: Governing Urban Scotland, 1830–1860* (East Linton: Tuckwell Press, 1999), 117–18.

16. W. T. Gairdner, "The Late Dr William Smith," *Scotsman*, 30 May 1896.

17. "Presentation of a Bust of Mr William Smith to the Institution," *Scotsman*, 31 July 1863.

18. Morton, *Unionist Nationalism*, 122. On the Manchester Athenaeum, which the EPI set out to rival (Anon., "The Philosophical Institution," *Scotsman*, 3 October 1846), see Martin Hewitt, "Ralph Waldo Emerson, George Dawson, and the Control of the Lecture Platform in Mid-Nineteenth-Century Manchester," *Nineteenth-Century Prose* 25 (1998): 1–23.

19. "Philosophical Institution," *Scotsman*, 12 February 1853. The United Presbyterian Church was formed in 1847 and was an amalgamation of the United Secession Church and the Relief Church. It was the third largest Presbyterian denomination in Scotland.

20. See comment in Jessie Aitken Wilson, *Memoir of George Wilson* (Edinburgh: Edmonston and Douglas, 1860), 444.

21. "Philosophical Institution," *Scotsman*, 12 November 1851.

22. On the relationship between street noise and professional middle-class identity and behavior, see John M. Picker, "The Soundproof Study: Victorian Professionals, Work Space, and Urban Noise," *Victorian Studies* 42 (2000): 427–53.

23. Henry Bowie, "Music Hall: An Annoyance," *Scotsman*, 8 November 1865.

24. W. Addis Miller, *The "Philosophical": A Short History of the Edinburgh Philosophical Institution* (Edinburgh: C. J. Cousland and Sons, 1949), 16.

25. Edinburgh Central Library (hereafter ECL), Minutes of the Edinburgh Philosophical Institution, 19 March 1856, fol. 224; "Lecture on Sir George Harvey," *Scotsman*, 1 February 1868.

26. Sophie Forgan, "Context, Image and Function: A Preliminary Enquiry into the Architecture of Scientific Societies," *British Journal for the History of Science* 19 (1986): 102.

27. Nick Prior, "The Art of Space in the Space of Art: Edinburgh and Its Gallery, 1780–1860," *Museum and Society* 1 (2003): 63–74.

28. Miller, *Short History*, 26.

29. M. L. Vincent, "Professor Scott's Lectures at the Philosophical Institution," *Scotsman*, 5 January 1850.

30. "Professor Tulloch's Introductory Address," *Scotsman*, 7 November 1857.

31. Charles Kingsley, *Charles Kingsley: His Letters and Memories of His Life* (London: Kegan Paul, 1891), 321.

32. On Kingsley and public speaking, see Caroline Rose, "Charles Kingsley Speaking in Public: Empowered or at Risk?" *Nineteenth-Century Prose* 29 (2002): 133–50; and Louise Lee, "Voicing, Devoicing and Self-Silencing: Charles Kingsley's Stuttering Christian Manliness," *Journal of Victorian Culture* 13 (2008): 1–17.

33. On Emerson's lecture style and "transpiercing," see R. Jackson Wilson, "Emerson as Lecturer," in *Cambridge Companion to Ralph Waldo Emerson*, ed. Joel Porte and Saundra Morris (Cambridge: Cambridge University Press, 1999), 76–96.

34. Ibid., 87.

35. George Gilfillan, *A Second Gallery of Literary Portraits* (Edinburgh: James Hogg, 1852), 134.

36. Alexander Dunlop (pseud. "Civis"), *Emerson's Orations to the Modern Athenians; or, Pantheism* (Edinburgh: J. Elder, 1848).

37. John Heiton, *The Castes of Edinburgh*, 3rd ed. (Edinburgh: William P. Nimmo, 1861).

38. "Edinburgh Dissected," *Athenaeum*, 30 May 1857.

39. For a brief record of Moir's comments, see "Philosophical Institution," *Scotsman*, 8 February 1851.

40. Hugh Miller, *Sketch-Book of Popular Geology: Being a Series of Lectures Delivered before the Philosophical Institution of Edinburgh* (Edinburgh: Adam and Charles Black, 1863), 83.

41. Hugh Miller, *The Testimony of the Rocks* (Edinburgh: Thomas Constable, 1857), 242. It is not clear when this lecture (entitled "Geology and Its Bearings on the Two Theologies, Part 2") was given or if Miller remembered correctly when he stated that it was delivered to the EPI. In the preface to *Testimony*, Hugh Miller notes that lectures 1, 2, 5, and 6 were given to the EPI. From newspaper coverage, it is clear that lectures 1 and 2 were delivered in February 1856 (not 1855 as Hugh Miller asserts). Judging from the EPI's published program (*Scotsman*, 1 October 1851), we can surmise that lectures 5 and 6 were given late in 1851.

42. Miller, *Testimony*, 214.

43. Ralph O'Connor, *The Earth on Show: Staging Prehistoric Worlds for the British Public, 1802–1856* (Chicago: University of Chicago Press, 2007), 400–411. On Hugh Miller and aesthetics, see John H. Brooke, "Like Minds: The God of Hugh Miller," in *Hugh Miller and the Controversies of Victorian Science*, ed. Michael Shortland (Oxford: Clarendon Press, 1996), 171–86.

44. On the transcendental and Carlylean resonances of Miller's geological writings, see James Paradis, "The Natural Historian as Antiquary of the World," in *Hugh Miller and the Controversies of Victorian Science*, ed. Michael Shortland (Oxford: Clarendon Press, 1996), 122–50.

45. Miller, *Sketch-Book*, xiv, first published in 1859 by Thomas Constable.

46. "Philosophical Institution: Mr H. Miller's Lectures," *Scotsman*, 14 December 1853.

47. *The Life of Hugh Miller: A Sketch for Working Men* (London: Samuel W. Partridge, 1862), 122.

48. Hugh Miller, *Testimony*, ix.

49. Peter Bayne, *The Life and Letters of Hugh Miller* (Boston: Gould and Lincoln, 1871), 483.

50. David Masson, "Dead Men Whom I Have Known; or, Recollections of Three Cities," *Macmillan's Magazine* 12 (1865), 84.

51. Lydia Miller makes this observation in Miller, *Sketch-book*, xiv.

52. On Carlyle's creation of a literary voice manufactured from his mixed experience as a public lecturer, see Ivan Kreilkamp, *Voice and the Victorian Storyteller* (Cambridge: Cambridge University Press, 2005), 19.

53. Peter Bayne, *The Christian Life: Social and Individual* (Edinburgh: James Hogg, 1855), v.

54. Miller, *Testimony*, 16.

55. See, for example, ibid., 242.

56. On Wilson's opposition to *Vestiges*, see James A. Secord, *Victorian Sensation: The Extraordinary Publication, Reception and Secret Authorship of "Vestiges of the Natural History of Creation"* (Chicago: University of Chicago Press, 2000), 284–86.

57. Wilson, *Memoir of George Wilson*, 436.

58. John Cairns, "The Late Dr George Wilson of Edinburgh," *Macmillan's Magazine* 1 (1860): 202.

59. Wilson, *Memoir of George Wilson*, 495.

60. Ibid., 522.

61. Isabella Lucy Bird, "Christian Individuality," *North British Review* 37 (1862): 275.

62. Ibid., 279.

63. For more on Wilson and the Industrial Museum, see R. G. W. Anderson, "'What Is Technology?': Education Through Museums in the Mid-Nineteenth-Century," *British Journal for the History of Science* 25 (1992): 169–84.

64. "The Philosophical Institution: Technology," *Scotsman*, 20 February 1856.

65. Wilson, *Memoir of George Wilson*, 326.

66. Thomas Carlyle was elected president of the institution in 1868 on the death of Henry Brougham.

67. EPI Lecture Committee Minutes, 17 July 1863, fol. 85, ECL.

68. "Philosophical Institution," *Scotsman*, 5 November 1862. On Hawkins's Pestalozzian emphasis on visual education, see James A. Secord, "Monsters at the Crystal Palace," in *Models: The Third Dimension of Science*, ed. Soraya de Chadarevian and Nick Hopwood (Stanford, CA: Stanford University Press, 2004), 138–69.

69. It is noteworthy, as Ralph O'Connor points out, that when Miller's EPI lectures on the geological history of Scotland were published, they appeared without illustrations; see O'Connor, *Earth on Show*, 404.

70. Frederick Burkhardt, Joy Harvey, Duncan M. Porter, and Jonathan R. Topham, eds., *The Correspondence of Charles Darwin, 1862* (Cambridge: Cambridge University Press, 1997), 10:28.

71. Thomas H. Huxley, "Professor Huxley and His Critics," *Scotsman*, 24 January 1862.

72. Burkhardt et al., *Correspondence of Charles Darwin, 1862*, 10:28. Hooker was in the habit of playing down the effectiveness of Huxley's platform performances. For his comments on Huxley's inability to "command the audience" or "throw his voice" during his Oxford encounter with Samuel Wilberforce on 30 June 1860, see Frederick Burkhardt, Janet Browne, Duncan M. Porter, and Marsha Richmond, eds., *The Correspondence of Charles Darwin, 1860* (Cambridge: Cambridge University Press, 1993), 8:270.

73. William Macdonald, "The Origin of Species," *Scotsman*, 29 January 1862.

74. "Philosophical Institution," *Scotsman*, 11 January 1862.

75. See Gillian Beer, *Darwin's Plots: Evolutionary Narratives in Darwin, George Eliot and Nineteenth-Century Fiction*, 2nd ed. (Cambridge: Cambridge University Press, 2000), 76, 129–30.

76. "The Evangelical Alliance Conference," *Scotsman*, 7 July 1864, 3.

77. George Sleigh, *Strictures on the Lectures Lately Delivered in Queen Street Hall* (Edinburgh: MacLachlan and Stewart, 1867).

78. "Dr. David Page on Man, in His Natural History Relations," *Anthropological Review* 6 (1868): 109–14.

79. "Philosophical Institution," *Scotsman*, 27 March 1867.

80. [Editorial]. *Scotsman*, 29 March, 1867, 2.

81. Ibid.

82. "Professor Huxley on the Bases of Physical Life," *Scotsman*, 9 November 1868.

83. Thomas H. Huxley, "On the Physical Basis of Life," *Fortnightly Review* 5 (1869): 142.

84. "Sir John Lubbock on Savages," *Scotsman*, 4 November 1869.

85. "Mr Moncrieff MP on Modern Scientific Speculations," *Scotsman*, 14 January 1868.

86. "Mr Lowe MP on Education," *Scotsman*, 2 November 1867.

87. "The Philosophical Institution," *Scotsman*, 2 April 1888.

CHAPTER EIGHT

Politics, Culture, and Human Origins: Geographies of Reading and Reputation in Nineteenth-Century Science

DAVID N. LIVINGSTONE

MOBILIZING SCIENCE

It is precisely because scientific knowledge is *mobile* that it can be *mobilized* in different settings for different purposes. The realization that knowledge moves around the world in material forms has fostered what might be called a research program on the history and geography of print culture in its many shapes and sizes. This concern has focused fairly centrally on the book, and a whole raft of studies has been devoted to elucidating its nature. Now numerous other forms of text are being swept into the story. In his work on the English East India Company, for example, Miles Ogborn has shown how heraldic manuscripts, political pamphlets, stock listings, official regulations, and many more were implicated in the construction of economic, governmental, and trading knowledges.[1] For scientific enterprises, what Nick Jardine calls "routinely authored works—instrument handbooks, instruction manuals, observatory and laboratory protocols," all of which are basic to the regulation of empirical practices, have also attracted attention.[2]

My interest here is in the different ways texts have been mobilized for different purposes in different settings and thus in the hermeneutic dimensions of the subject.[3] As Roger Chartier has reminded us, "the process of the construction of meaning" is bound up with the ways in "which readers diversely appropriated the object of their reading." As he goes on, "We necessarily hold reading to be an inventive and creative practice that seizes commonly shared objects in different ways and endows them with meanings that cannot be reduced to the authors' and the publishers' intention alone."[4] Adrian Johns concurs, observing that "an apparently authoritative text, however 'fixed,' could not compel uniformity in the cultures of its reception. . . . Local cultures created their own meanings with and for such objects."[5] All of this resonates with Stanley Fish's suggestion that a "meaning experience" is derived neither from the text nor the reader per se, but from their joint location in interpretative communities sharing hermeneutic strategies.[6]

This is, in part, the terrain I wish to traverse here. My concern is to show how scientific meanings are imagined and reimagined through encounters with scientific texts and treatises. Two discrete, but related, cases will occupy our attention. Both revolve around what might be called the geography of theories of human origins, one dwelling on non-Darwinian conceptions of the genesis of the human species, the second focusing on the ways in which Darwinism was read in different venues. Taken in tandem, they draw attention to the cultural politics of origins narratives, whether creationist or evolutionary, throughout the nineteenth century, and to the located character of the making and mobilizing of scientific theories and reputations. At the same time they reveal something of the historical depth of the controversy over human origins—long pre-dating Darwin's interventions—and just how variable were the stances that partisans adopted. Besides, mapping the spaces of these encounters allows for comparisons both on a continental scale and within a single country.

The case of the seventeenth-century controversialist Isaac La Peyrère, who championed a polygenetic account of human origins, reveals how nineteenth-century commentators in Britain and America stage-managed his reputation to serve sectional interests. The cultural investments in whether the human species was of monogenetic or polygenetic origin, and the political imperatives that readers in different locations divined in historical anthropology will thus command our attention. For my second case, I turn to the geographies of scientific meaning within a single country—Ireland. How the contingencies of domestic cultural politics were critically implicated in what Darwin's theory of origins was made to mean in different localities is the focal point of concern. In a range of different Irish settings, the debate over Darwinism was conducted in very different registers, even among those with remarkably similar religious beliefs. What unites these two cases is the way in which a transcendental question that loomed large on the intellectual and social horizon of the nineteenth century—where did we come from?—was routinely translated into the idioms of local political dialect.

The value of a localizing strategy in seeking to make sense of Victorian science, of course, does not imply any straightforward spatial reductionism or local determinism. I do not claim that the figures on whom I dwell are

necessarily "typical" or "representative" of the spaces from which they issued their judgments or that no dissenting voices from the same locations
may be discerned. My argument, rather, is that local encounters with scientific works were shaped by how readers *conceived* of their local settings, by
what they *took* their situation to be, as much as by the ontological realities
of those venues. This means that the geographies of reading that I elaborate
here are no less intimately bound up with the geographical imaginations of
readerly communities than with the "real-world" circumstances of local
settings.

PRESENTING LA PEYRÈRE:
FRAGMENTS FOR A REPUTATIONAL GEOGRAPHY

Thanks to the work of historians like Jim Secord and Nicolaas Rupke, we
are beginning to map the geography of textual encounters and the ways in
which scientific meanings of both individuals and ideas are reconstituted
from time to time and place to place. Whether it was the staging of Alexander
von Humboldt's personae, or the local construction of Robert Chambers's
Vestiges, the importance of the geographies of reading and reception has
asserted itself.[7] I want to pursue this line of inquiry by developing what I
propose calling a rudimentary "reputational geography" of one of the critical figures in the development of modern anthropology, Isaac La Peyrère.

In 1655 La Peyrère put forward his audacious thesis about human origins. Drawing on the latest reports of geographical explorers, the increasing
availability of what were known as "pagan chronicles," together with an
imaginative rereading of St. Paul's epistle to the Romans, La Peyrère advanced the simple but shocking thesis that Adam was not the father of the
entire human race, but only of the Jewish people. It was, fundamentally, a
polygenetic account of human origins. His book *Praeadamitae* (1655) caused
a storm.[8] Elsewhere I have charted the serpentine genealogy of this daring
theory and its subsequent manifestations right up to the present day.[9] Here I
want to lay out something of how La Peyrère was staged in different settings
to suit different agendas during the nineteenth century.

HERETIC, HERO, HARMONIZER

Consider the following presentations of La Peyrère and his theory. First,
La Peyrère—heretic. From the original publication of *Praeadamitae*, La
Peyrère was routinely pilloried as a heretic. Forced to "recant" before Pope

Alexander VII in 1657, he was regularly typecast as an archetypal infidel whose querying of the biblical text had fueled skepticism. This reputation persisted throughout the nineteenth century so that even those who advanced similar ideas about the possibility of a pre-Adamic race sought to distance themselves from La Peyrère's profanity. Isabelle Duncan, for example, who brought out *Pre-Adamite Man* in 1860, conceded that her scheme bore some resemblance to that of "Isaac de Peyrère," but the similarities were so superficial, she insisted, as to bear not "the remotest analogy" to his views.[10] Time and again, La Peyrère was cast as a heretic. Thus Thomas Smyth, in 1850, announced that his theory of the plurality of human origins was "necessarily infidel. . . . It overthrows not only Moses, but the prophets and apostles also. And thus undermines the Scripture as a divine record, both of doctrines and of duties. It was for this purpose the theory was introduced by Voltaire, Rousseau, and Peyrere."[11]

Contrast this disparaging assessment of Peyrère's project with the following heroic judgment issued under the pseudonym Philalethes in 1864:

> After two centuries of neglect and oblivion, the name of Isaac de la Peyrère is once more received and honoured, as that of the first scholar who broke through the meshes of a groundless traditional prejudice, and proved that even in Scripture there are no decisive evidences of man's descent from a single pair; nay more, that there are distinct indications of non-Adamite races. . . . Peyrère was two centuries before his time; and whether we accept or reject his special theories, it is impossible not to admire his acumen, his candour, and his courage. Like all people who are wiser, fairer, and more keen-sighted than their contemporaries, he was of course persecuted and rendered as miserable as his theological adversaries, with their three favourite weapons—persecution, imprisonment, and fire—had it in their power to make him. He had dared to step out of the magic exegetical circle which theology had drawn around all the sciences, and his presumption was punished with prompt violence.[12]

To Philalethes, La Peyrère was a far-sighted free thinker who had begun the task of liberating anthropological inquiry from the strictures of religious prejudice and was thus to be celebrated as a heroic visionary.

Alexander Winchell's construction of La Peyrère stands in further contrast. Writing in 1877, Winchell urged that La Peyrère's "propositions were far in advance of the age, and . . . were defended with knowledge and candor which were not appreciated by the adversaries."[13] A couple of years later, and at rather greater expanse, he described La Peyrère, as neither a heretic nor heroic freethinker, but as "a learned and sagacious priest of the orthodox faith." He went on: "His sagacity surpassed his age; and I have come, not to bury him, but to honor him. His thesis was argued with soberness, candor and logic." To Winchell, La Peyrère was to be seen as a serious believer intent on finding harmony between scripture and secular learning, and he thus aligned himself with La Peyrère's perspective: "Peyrerius seems to have reached sound conclusions by a species of intuition. . . . The positions which, in the time of Peyrerius, were regarded as unscriptural, but which I am now prepared to defend on scriptural as well as scientific grounds . . . are mostly accepted by the modern church."[14]

None of these judgments, of course, was issued in a vacuum; they were articulated in specific historical-geographical settings during the third quarter of the nineteenth century. Thomas Smyth was writing in the antebellum American South from the perspective of a slave reformer who defended the institution on biblical grounds but was deeply troubled by thoughts of polygenesis. To him that "theory would be very inexpedient and suicidal to the South in the maintenance of her true relations to her colored population. It would be so because it is *novel*." He thus marshaled both scientific and scriptural arguments bolstered with historical, philosophical, and linguistic findings, to undermine Agasssiz-style polygenism and its Peyrèrean predecessor. In fact his quarrel with the American School of Anthropology championed by polygenists like Samuel George Morton, Josiah Nott, and George Gliddon, was in part because he felt their undertaking would actually undermine the traditional basis of southern slavocracy:

> The introduction in the South . . . of this novel theory of the diversity of races, would be a declaration to the world that its institutions could no longer rest upon the basis which has always been hitherto assumed, and that this theory has been adopted for mere proud, selfish, and self-aggrandizing purposes. This theory is further impolitic to the South, because of its immoral, anti-social, and disorganizing tendencies. It would remove from both master and servant the strongest bonds by which they are united to each other in mutually beneficial relations.[15]

Put simply, polygenist anthropology bestialized slavery; Adamic theology sanctified it. For Smyth, slavery had the benediction of Scripture, and any undermining of its authority was socially subversive. For generations, polygenism's spiritual infidelity was well known; now its seditious politics were laid bare. Christian slavery, to Smyth, did not mean that slaves were chattels; it gave rights and privileges; it was providentially ordered. But it was slavery all the same. No wonder polygenism was "impolitic to the South."[16] It would wreck the building blocks on which southern civilization had been raised. In Smyth's case, the typecasting of La Peyrère as a heretic was at once a theological *and* a political judgment.

The praise that Philalethes lavished on La Peyrère was issued from a very different stable—and a very different hermeneutic community. It appeared in the *London Anthropological Review*, the journal of the newly formed Anthropological Society.[17] This body had emerged out of the older, Ethnological Society of London, which itself had come into being in the 1840s as an offspring of the Aborigines Protection Society. But the Ethnological Society had gone into decline during the late 1850s until its reinvigoration by a newer set of anthropologists and archaeologists. Soon, however, friction with the humanitarian-Quaker elements led to the formation of the Anthropological Society in 1863, which gave institutional expression to a much tougher polygenetic and racialist line of thinking. In this environment, La Peyrère could be read as an inspirational resource for liberating the "science of man" from traditional restrictions. Thomas Bendysche, one of the Anthropological Society's vice presidents, outlined La Peyrère's crucial contributions in his memoir, "History of Anthropology," dwelling on his role in the inauguration of polygenism.[18] Philalethes staged him as another Galileo-like victim of prejudice and obscurantism, and thereby used his case as a foil for attacking benighted critics of Morton and Gliddon—including Thomas Smyth. Incensed, he alleged that Smyth "knew nothing whatever about La Peyrère, and had probably never read a line of his work." Smyth's tactics, he asserted, were typical of those who used the "base weapons of calumny and abuse . . . from the time of Peyrerius down to that of Vogt." This pressed him to the conclusion that "few scientific truths have ever been discovered—few discoveries have been made for the last five centuries, against which the combined forces of prejudice and ignorance have not marshalled their array of mistaken Biblical inferences. We leave it to others to write this sad, this humiliating, but instructive history."[19]

Winchell's harmonizing tactics were advanced in a different time and space—late 1870s Tennessee. A distinguished geologist with expertise in botany and zoology, Winchell occupied a specially created chair at Vanderbilt University. Later, when teaching geology and paleontology at the University of Michigan, he would assume the position of state geologist. At Vanderbilt, on the surface at least, Winchell was engaged in a project to reconcile science and religion by securing the orthodoxy of La Peyrère who, he insisted, "was less impious and mad than the bond slaves of dogma who

silenced his tongue."[20] Various commentators certainly read his undertaking that way. An anonymous reviewer in the July issue of *Appleton's Journal*
for 1880, for example, announced that "another step forward in the effort
to effect a reconciliation between the conflicting claims of science and religion is taken by Dr Winchell."[21] A decade later, in a belated commentary
in the *Methodist Review*, Henry Colman conceded that "the theory of pre-
Adamites conflicts with no biblical doctrine," and that La Peyrère was "an
orthodox Dutch [sic] ecclesiastic."[22]

Nevertheless, at the time, Winchell's scientific intervention trans-
gressed the boundaries of the hermeneutic community institutionalized at
Vanderbilt University, and he was dismissed from his post. Writing thirty
years after Thomas Smyth, he was evidently much better attuned to the
scientific politics of the Old South than to the interpretive community of the
local Methodist fraternity. His adoption of a modified version of Darwinism
was the presenting cause of his removal from office. There is good evidence
to suppose, however, that his promulgation of non-Adamic humanity was
also critically implicated. The reason was that his resurrection of Peyrèrean
anthropology was domesticated to a thoroughly racist scientific politics.
For part and parcel of his reinvention of La Peyrère (whose anthropology,
incidentally, was egalitarian) was a lurid account of black inferiority and
intense disgust at racial intermarriage. A lengthy chapter devoted to the
subject dwelt in unrelenting detail on the race's presumed anatomical,
physiological, and psychic inadequacies. Cephalic index, forearm length,
brain weight, mathematical competence, aesthetic judgment, racial inertia—
all were marshaled in support of his declaration that "just as far as the
African diverges from the style of a white man, he approximates the lower
animals." In the light of this litany of denunciation, the political need to
preserve the Adamic bloodline was an absolute moral necessity:

> I allow myself to pause here briefly, for the purpose of protesting against
> the policy of North American miscigenesis [sic], which has been recom-
> mended by high authorities as an eligible expedient for obviating race
> collisions. It is proposed to consolidate the conflicting elements by a
> systematic promotion of the interfusion of the white and black races. . . .
> The policy is not more shocking to our higher sentiments, nor more op-
> posed to the native instincts of the human being, than it is destructive
> to the welfare of the nation and of humanity.[23]

Advocates of racial intermixture were thus the special objects of his
scorn. Wendell Phillips was "in danger of acquiring the title of 'most elo-

quent platform virago,'" that is, if his "sex did not protect him." Bishop Gilbert Haven's sentiments were dismissed as "painful." Canon George Rawlinson was berated for having "added his name to this cluster of self-appointed conspicuities" by advancing proposals that would lead to the careless destruction of "ethnic pearls." All came under the whiplash of his tongue for their open advocacy of ethnic amalgamation. Science, by contrast, was on his side, and he presented his own "table of states of hybridization" and his natural law of "the mutual repugnance of races," to sustain his vision for national policy.[24] Staging La Peyrère as theologically orthodox was integral to Winchell's project to marshal pre-Adamite anthropology in the cause of postbellum southern segregationism.

TOWARD A GEOPOLITICS OF HUMAN ORIGINS

This emblematic reputational geography of La Peyrère gestures toward a more general geopolitics of human origins. Cultural investment in stories about the genesis of the human species was enormous and differed significantly from place to place as commentators issued their views on the meaning and implications of polygenism. Two different moments in time and place illustrate something of the different ways the idea of primitive human variety could be read. Taken from either end of the nineteenth century, they are intended to disclose something of what a historical geography of polygeny might begin to look like.

The second edition of Samuel Stanhope Smith's *Essay on the Causes of the Variety of Complexion and Figure in the Human Species*, arguably the first American treatise on anthropology, appeared in 1810.[25] It was a sustained defense of Montesquieu's climatic account of human diversity in support of traditional monogenism over against its opponents, Lord Kames in particular. Smith went out of his way to explain that human difference was literally only skin deep, and that environmental circumstances were entirely sufficient to explain racial differentiation. As president of the College of New Jersey, Princeton, it is not surprising that Smith would find polygenetic speculation heterodox, and he did not hesitate to portray Kames as engaged in "laudable attempts to disprove the truth of the Mosaic story" and to stage him as an "excellent specimen of the easy faith of infidelity!" Smith determined to pulverize all such new-fangled skepticism with the sledge hammer of scientific fact, and so he painstakingly rehearsed all the evidence he could muster from travelers, anthropologists, and medical practitioners for the claim that human hair, skeletal stature, forehead shape, physiognomic features, even mental powers and moral inclinations, were all correlated with "climates . . . some states of society, and modes of living." And yet above and beyond matters of scientific data and theological decorum alike was politics. At base, Smith's fundamental concern was the management of the moral economy. What was at stake in researches into historical anthropology was nothing less than the maintenance of the social order. Were there different species of humans, they would "be subject to different laws both in the physical and moral constitution of their nature."[26] The empire of climate at once preserved human nature from polygenetic subversion, secured the intellectual integrity of moral philosophy, and delivered the possibility of political stability:

> The denial of the unity of the human species tends to impair, if not entirely destroy, the foundations of duty and morals, and, in a word, of the whole science of human nature. No general principles of conduct, or religion, or even of civil policy could be derived from natures originally and essentially different from one another. . . . But when the whole human race is known to compose only one species, this confusion and uncertainty is removed, and the science of human nature, in all its relations, becomes susceptible of system. The principles of morals rest on sure and immutable foundations.[27]

At bottom *this* was why Smith found polygenetic anthropology so obnoxious. In the early days of the new American Republic, a confidence in a common human constitution was precisely the philosophy that was needed if public virtue were to be retained in a society "that was busily repudiating the props upon which virtue had traditionally rested—tradition itself, divine revelation, history, social hierarchy, an inherited government, and the authority of religious denominations."[28]

At the other end of the century, a markedly different geopolitics bred a markedly different attitude toward anthropological polygenism. Charles Carroll, a vicious racial supremacist, intent on literally bestializing the black races and excommunicating them from the human race, published his scurrilous assault on African Americans in St. Louis in 1900. Chief among his scientific sources were figures like Quatrefages, Broca, Topinard, Haeckel, Blumenbach, and Winchell. Winchell, in particular was repeatedly presented as a "great American scientist," and was quoted on no less than twenty-nine occasions on such subjects as the configuration of African skulls, their facial shape, cranial capacity, brain weight, chest circumference, and hair texture, as well as the racial distribution of various diseases and sundry medical statistics of insanity. Extended gobbets were extracted from

Winchell's writings to underwrite Carroll's supremacist politics, the object of which was to confirm from *both* science and scripture that the black and white races were of different origins. Indeed, it was a fundamentalist enterprise, for Carroll, staunchly hewing to a creationist account of human genesis, turned his guns on the theory of evolution by natural selection. But it was not simply a biblicist rejection of Darwinism; it was a political refusal. In his eyes, it was only through this atheistic theory of descent that "the Negro obtained his present unnatural position in the family of man." Darwinism's monogenism thus troubled him deeply, and he garnered every scrap of evidence he could find to confound it. Indeed, he attributed its delusions to fashionable notions about racial amalgamation that had in turn fathered religious skepticism and philosophical naturalism. His own book was thus a ferocious attack on the black race as "not an offspring of the Adamic family" and thus "not of the human family." On page after page it displayed its author's deep-seated phobia about "mixed blooded tribes" who, like the children of Cain, were "soulless" because "they were of amalgamated flesh."[29] Such—by now standard—pre-Adamite fodder was served up to Carroll's fairly wide readership.[30]

Carroll, it turns out, was far from a lone voice. A host of comparable writings, constituting a virulently persistent hermeneutic community, appeared in the decades around the turn of the century—and indeed continues to the present day among neo-Nazi racial supremacists in several of states of the United States. Just two emblematic examples will suffice to demonstrate its allure at the time. William H. Campbell's 1891 *Anthropology for the People*, for example, presented as the work of "Caucasian," blended scriptural commentary and scientific aspiration to produce what Paul Harvey has choicely dubbed a "truly miscegenated offspring," namely, "a mytho-scientific racism that blended folklore, Darwinian science, and biblical exegesis."[31] In a sequence of articles printed in *The Eastern Lutheran* and the *Chambersburg Democratic News* and drawn together into a book in 1898, Gottlieb Christopher Henry Hasskarl (1855–1929), a Pennsylvania Lutheran minister, similarly culled numerous scientific works to confirm that the black race was not Adamic and was therefore "inevitably a beast."[32] Like Carroll's, Hasskarl's articles were so laced with lengthy extracts from writers like Winchell that it is difficult to extract his own thoughts from the quagmire of quotation marks. But this itself is indicative of just how plentiful were those scientific works of polygenist anthropology that could readily be mobilized in the cause of a white supremacist credo.

Anthropological theories of human origins throughout the nineteenth century were invariably swept up into political arenas and were mobilized in the service of geopolitical agendas. What readers in particular settings thought about whether La Peyrère should be vilified or valorized, about the proper governance of republics or the relations between the races, about whether polygenism saved or subverted the Hebrew Scriptures—all had a bearing on how reported findings of anthropological science were locally put to work. The coming of Darwin's theory of evolution did nothing to disrupt such local constructions of scientific meaning. If anything, cultural particularities manifested themselves with even greater force as Darwin and his disciples broadcast evolutionary theory with proselytizing zeal. What differently located interlocutors took Darwin to mean, what they themselves felt comfortable saying, and what they imagined might flow from the adoption of his theory will be our next port of call.

MAKING DARWINIAN MEANING

In order to disentangle some of the hermeneutic threads in the local manufacture of Darwinian meaning, I want now to triangulate a number of Irish readings of evolutionary theory using three survey points located respectively in Dublin, Belfast, and Londonderry. My concern here is to map something of the distinctive reading spaces that may exist within a single country and thus to operate at a scale considerably more detailed than that of the state as a whole. By focusing on specific urban locations, I hope to move the debate beyond the standard practice of charting "national responses" to Darwinism. In each case I am confining my concerns to those who participated in the same religious tradition so as to avoid explaining the local constitution of Darwinism by resorting to religious stereotypes.

Consider, as base stations, then, the following settings. In Dublin in 1871, the distinguished Trinity College anatomist and prominent Presbyterian layman, Alexander Macalister—later professor of comparative anatomy and zoology at Cambridge[33]—expressed himself at length in the *Dublin Quarterly Review of Medical Science* on Darwin's *Descent of Man*. While Macalister was unconvinced by the extension of evolution into the sphere of psychic development and the tracing of religious and moral impulses among lower animals, he remained enthusiastic about the power of natural selection to account for both animal and human physiological evolution. In judicious tones, he began by noting that Darwin's new book "had not been hurriedly put together." He was careful too to discriminate other theories of evolution, such as those by Lamarck or De Maillet, from Darwin's and to insist that "those that confound the older and more modern theories betray a gross ignorance of the subject." He himself considered it entirely appropriate that the author of the *Origin of Species* should pursue his inquiries on into the human realm: "Man is sufficiently an animal to form part of the zoologist's domain, and if we profess to account for the origin of the other species we have no ground to put man out of the question."[34] As an anatomist, Macalister was drawn to the empirical evidence supporting the evolution of the human physical form and fastened upon several lines of inquiry justifying the theory. He emphasized the power of evolution to explain rudimentary structures and declared the superiority of an evolutionary narrative to traditional teleology: "How to account for these rudiments on any other but an evolution theory it is very hard to see. No teleological reason for their existence can be given, as they are for no end, perform no function; we can otherwise give no intelligible reason for their presence." It was the same for embryology. In sum, so far as the anatomical evidence for human development was concerned, Macalister concluded, "We cannot deny that the defenders of evolution have the best of the evidence on their side."[35]

Three years later in Belfast, in 1874, Macalister's fellow Presbyterian Josiah L. Porter, scientific traveler, scriptural geographer, professor of biblical criticism, and later president of Queen's College, announced his views on the subject. The distance between his views and Macalister's could not be more marked: "Darwin's book on 'The Origin of Species' is . . . to the logician . . . an utter failure." However acute an observer Darwin was, Porter was sure that "his reasoning faculty seems to have been completely overwhelmed by the force of one preconceived idea." Fundamentally, Porter could find no empirical evidence in support of the "essence" of Darwin's theory "that all forms of life, from the humblest zoophyte up to man, have evolved from one primordial germ." However striking and original its message, he was sure that "any one can see that it is not scientific." His own strategy was to allocate science and religion each to their own spheres, rigidly demarcated, and thereby to prevent the invasion of "Revelation . . . by crude theories and wild speculations."[36]

John Robinson Leebody, professor of mathematics and natural philosophy at the Presbyterian Magee College in Londonderry since 1865, had delivered his judgments in an article entitled "The Theory of Evolution and Its Relations to Religious Thought" in 1872. They were rather different from those of Porter, his fellow churchman. In this lengthy review, Leebody outlined the various evolutionary proposals of Laplace, Comte, Tylor, Vogt, Spencer, Haeckel, and Darwin expressing disquiet over any materialist implications. Of them all, Darwin's theory was by "far the ablest and most complete attempt to prove the absolute continuity of the chain which connects man with the lower animals." It certainly disclosed its fair share of empirical inadequacies, and Leebody was of the opinion that "we must decline, in the interests of science itself, to accept the Darwinian view of the origin of man's body, *until it is proved.*" Nevertheless, Leebody adopted a very different rhetorical cadence from that of Porter. *The Descent of Man,* he announced, was "a most valuable storehouse of facts in natural history, shewing almost unrivalled acquirements in this branch of science, and is a monument of painstaking industry." It was pleasingly "free from the offensive irreverence with regard to sacred things which disfigures the writings of Vogt, and Büchner, and even Häckel." In "the interests of both Science and Religion," Leebody concluded, "we welcome the appearance of the 'Descent of Man.' It enriches science by a vast number of valuable facts, and it will stimulate inquiry with regard to the theory of Evolution which may be expected to yield important results."[37]

Here we see advertised three rather different readings of Darwinian evolution by prominent Irish Presbyterian spokesmen during the early years of the 1870s. No doubt personal predilection and professional preoccupation had their parts to play in the judgments they delivered. But the different spaces they occupied in Dublin, Belfast, and Londonderry were, I want to argue, critically implicated both in the stances they assumed and the rhetorical tones they adopted in their public declarations. One thing is clear: belonging as they all did to the same religious denomination and sharing a seriously similar theology, their variant readings of Darwin cannot simply be attributed to the influence of their religious heritage.

A TALE OF THREE CITIES

In Dublin, Alexander Macalister was part of a progressive set of scientists congregating around Trinity College and the short-lived Royal College of Science, which championed the independence of science in general and evolution in particular.[38] He achieved considerable eminence by writing numerous major works, including *An Introduction to Animal Morphology and Systematic Zoology* (1876); *Zoology of the Vertebrate Animals* (1878), which was revised for American audiences by A. S. Packard; and *Physiology* (1895), and by securing election to the Royal Society in 1881. Chief among his hermeneutic community in Dublin were Ireland's astronomer royal Robert Ball; the botanist W. T. Thiselton-Dyer, who taught there prior to moving to Kew Gardens, where he later became Hooker's son-in-law and successor; and the anthropologist A. C. Haddon, who took up the chair

of zoology at the college in 1880 and during his tenure led the celebrated
Torres Strait Expedition of 1898–99.[39] The Trinity College geologist John
Joly and the professor of natural philosophy Francis Fitzgerald were no less
enthusiastic for evolution. And under Macalister's direction, the Trinity
Anatomy Department established its reputation as a stronghold of evolutionary physiology.[40] Writing in 1870 Macalister had already laid down the
principle that Darwin's theory "must be decided by work in the dissecting-
room, the field, the zoological garden, and the laboratory" and that the "shred
of old prejudice against researches in the principles of biology" that continued to be entertained by some religious partisans was the preserve of the
"narrow-minded." As we have already seen, he expressed his belief that
evolutionary theory vastly superseded the older teleology in providing a coherent explanation for rudimentary organs. Indeed he considered it "contrary to the spirit of true philosophy . . . to refer natural events to the direct
interposition of the Divine Power."[41]

Nevertheless, Macalister was a prominent Presbyterian layman and affirmed his faith in design in a forty-eight-page pamphlet published by the
Religious Tract Society in 1886, in which he used the human body as the
vehicle for confirming teleology. After a detailed excursus on the mechanical glories of human anatomy—in which the doxological rhetoric of the
wonderful and marvelous obtruded—he concluded, "The prominent lesson
which the examination of the human body impresses on us is that of perfect
adaptation of means to ends, of structure to function. The unprejudiced
mind cannot fail to read in every organ . . . the inscriptions of purpose, and
to learn thereby that they are the products of supreme power directed by supreme wisdom."[42] If teleology had no place in scientific explanation, it still
occupied a prominent location in human meditation on the natural order.
Indeed this tract-for-the-times was, in many ways, an expansion of a sermon
he had delivered at the annual meeting of the Jervis Street Mission Church
fifteen years earlier in 1871.[43]

If Macalister's position in the pantheon of late nineteenth-century scientific Dublin fostered his espousal of evolution, his local Presbyterian community did nothing to discourage it. No doubt this reflected the values of
a Dublin Protestant class enthusiastic about British science, particularly in
the face of Catholic opposition to "secular" education. After all the British
scientific establishment had long been expressing concern over the direction
in which the curriculum might move in a state controlled by Catholic
educationalists should Ireland achieve Home Rule.[44] The implications for
science as Irish nationalism became more Catholic profoundly troubled
those working for scientific modernization. Irish nationalism, it was feared,

would precipitate an increasing cultural isolationism that would breach the long-standing ties between Irish and British science—a connection that the *Dublin Daily Express* pointedly advertised when editorializing on Tyndall's Belfast Address by noting that Cambridge had elected to fellowship more Irishmen than the University of Dublin. The same day it told its readers that the Darwinian hypothesis should be accorded "free and fair discussion," for nothing had done "more injury to religion than the attempt to down new scientific theories as atheistic."[45] Huxley's virulent hatred of popery was well known, and the fact that he happened to be the leading champion of scientific modernization cemented evolution, anti-Catholicism, and scientific progressivism in the minds of observers.

Significant too was the fact that Macalister entitled his inaugural address for the 1882 session of the Dublin Presbyterian Association—an organization of which he had been president during the mid 1870s— "Evolution in Church History." In essence the lecture employed evolutionary vocabulary to chart the historical transformations of ecclesiastical polity and practice from early times. Darwinian metaphors obtruded. Thus he juxtaposed what he called the "archetypal theory," which postulated that the church "was primitively constituted as a perfect and complete institution," with the "evolutional theory," which took seriously the "mutations" responsive to specific "conditions" it had undergone. His own preference was for an evolutionary reading, and he thus felt it necessary to spell out just what its essential principles were: "The fundamental postulates are a capacity of variation in the train of sequences, and external modifying influences, and the latter may be either the direct action of the environments on the phenomena, or may be due to a power from without . . . Can we predicate these of the Christian Church? Has there been therein the variability? Has that variability, if it existed, related to changing external conditions?"[46] Just as Darwin used the historical diversification of language to illustrate the principles of evolutionary change, so Macalister found evolution a captivating trope to elucidate ecclesiastical history.

Porter's commentary was initially delivered on the evening of 30 November 1874, as the first lecture in a series staged to combat the malign influence of the recent Belfast meeting of the British Association, at which its president, John Tyndall, had sought to liberate science from ecclesiastical control. It subsequently appeared as a pamphlet and then as a chapter in the collected series of lectures. The ensuing volume, which drew together all the presentations that winter, made it clear that the lectures were intended to rebut the president's attack. Its editor protested that "while courtesy and precedent forbade any protest at the time, it was felt by many, and more especially by those resident in Belfast, that such teaching should not be permitted to pass unchallenged."[47] For they surely had taken the brunt of Tyndall's offensive salvo. Porter was deeply troubled by the British Association's posture, and in his inaugural lecture to his own students that winter, he paused to refer to the events that had recently "transpired at the British Association" and to attack "those pernicious dogmas" that its champions had issued from their privileged position.[48] Other educators took the same stance. Porter's colleague, Henry Wallace, also spoke to his students on the subject of the teachings of the British Association, and informed them that "Professor Tyndal's [sic] reasonings and inference from the real facts of science are as false as they are shallow." Tyndall-style science, Wallace went on, "can only sap the foundations of moral order, abolish the distinctions of virtue and vice, and 'guide' again to the atrocities of the French Revolution."[49]

Elsewhere I have reviewed the local machinations that transpired in the wake of Tyndall's infamous Belfast Address.[50] My advertising of this public spectacle is to redraw attention to the ways in which a distinctive expressive space was created by the oppositional tactics of Tyndall and his associates. They had violated local decorum: as a Belfast *Almanack* for 1875 put it, "Professors Tyndall and Huxley . . . exhibited very bad taste and less sense in propounding absurd theories on the subject of creation."[51] How Darwin's theory was read was profoundly shaped by that particularity and by the rhetorical resources that were available to local interlocutors as they engaged the debate about the meaning of Darwin's science.

The specter of the Tyndall episode in Belfast does not seem to have disturbed the calm assessment of the new biology that the Derry-based Leebody continued to issue. Writing a couple of years later, he announced that "there is nothing necessarily atheistical or contrary to Scripture in the theory of development as applied by Darwin to account for the variation of the species." Indeed, his attitude to Darwin himself was enthusiastic. "Mr Darwin's name is one destined to stand in the first rank among the leaders of intellectual progress," Leebody told his readers. "The Newton of biological science has yet, we believe, to arise, but scientific men are pretty generally agreed that Mr Darwin must at least be regarded as the Kepler."[52]

Plainly Leebody occupied a quite different rhetorical space from that of his Belfast colleagues. At least two factors seem to have contributed. Given his training in mathematics and physics (with strong testimonial support from both P. G. Tait and Wyville Thompson),[53] he was only too aware of the importance of keeping scientific inquiry free from unwarranted theological policing. In language markedly different from the ominous warnings issued to the Belfast faithful, he assured his readers that the "history of the past tells us that we need not dread that the religious beliefs of the community will be enfeebled or destroyed by the advance of science. It may tend to displace some of the traditional beliefs which are the excrescences on Scripture truth rightly formulated, but that will be no loss."[54] Important too was Leebody's institutional setting. On behalf of Magee College, of which he was to become president—an educational institution combining "secular" and "religious" education, though without state support—Leebody had been on a campaign to have its students placed on an equal footing with those in other Irish universities by being admitted to the University of Ireland degree examinations. In this setting, the need to defend nonsectarian scientific instruction was of paramount importance, and Leebody purposefully distanced himself from the proposals made by the Catholic church, which insisted that *all* subjects must be taught, as Cardinal Cullen put it, "on purely Catholic principles." Leebody strenuously demurred, quipping, "There is no Protestant Mathematics or Chemistry as distinguished from that taught in a Catholic college."[55] Experience, he insisted, had taught him that it was entirely "possible to associate daily with class fellows of a different creed without obtruding his special religious views upon them, or having their peculiar dogmas thrust down his throat; and that no sect, Protestant or Catholic, has a monopoly of intellectual ability or moral worth."[56] In the educational culture wars of late nineteenth-century Ireland, Leebody had to conduct a campaign that was seen to preserve the independence of scientific inquiry from too much denominational supervision.[57] The reputation of Magee College as a serious seat of learning and his own reputation as a man of science were both at stake as Leebody delivered his judgments on the Darwinian theory of origins.

In Belfast it took at least a generation to exorcise the ghost of the Tyndall drama. Robert Watts was still railing against Darwin, and Spencer, and even more against the efforts of those seeking reconciliation with the new biology throughout the 1880s. Henry Drummond in particular got under his skin with his effort to disclose "natural law in the spiritual world." In Watts's eyes, the Drummond strategy was simply "not an Irenicum between science and religion or between the laws of the empires of matter and of spirit" at all.[58] It was, rather, the colonization of faith by scientific imperialism and exhibited for all to see the aggressively expansionist character of natural law. William Todd Martin, another professor at the Assembly's College, issued in 1887 a three-hundred-page philosophical refutation of Spencer's cosmic evolutionism.[59] For his part, Leebody was even more supportive of evolution by the 1890s. In his inaugural lecture for the new session at the Magee College in October 1890, he proclaimed, "The theory of evolution, when first distinctly enunciated, some thirty years ago, was regarded as utterly antagonistic to religion. The storm of angry discussion raised by the publication of Darwin's "Origin of Species" is within the memory of men not yet past middle age." But now theologians, among whom he numbered Drummond, had announced their acceptance of it. Indeed, an entirely new attitude had developed: "Thoughtful people are beginning to recognise that the doctrines of evolution, when properly formulated, are quite compatible with Christian faith." His conclusion? "The difficulty of reconciling the theory of evolution with the Christian narrative of Genesis is not then a very serious one."[60]

By contrast, the medical practitioner Samuel B. G. McKinney of Sentry Hill—the family home of eminent Presbyterian gentleman-farming stock located in Carnmoney, just north of Belfast—rehearsed in 1895 many of the objections to evolutionism that had wended their way through the winter series lectures more than twenty years earlier.[61] McKinney, who had been a student at the Queen's Colleges and then in Edinburgh during the late 1860s and early 1870s, and was later the author of several theological works, was crystal clear in his rejection of evolution: "The weakness of the theory lies in the absence of any foundation of fact."[62] "There is not a single positive fact in the foundation of the evolution theory" he repeated. Several years earlier in 1888, he had resorted to phrenology to launch a rather offensive attack on Darwin himself:

A glance at any photograph of Darwin is sufficient to convince any one that his brain was so imperfectly developed that he was not naturally capable of exhibiting any higher functions of mind, and could only be a keen observer of facts and a steady plodder in experiments. Even his experiments on the influence of worms were due to the suggestions of another, and he originated nothing. . . . Although the evolution theory was contrary to reason and to scientific principles, the imperfection of his brain and the deficiency of his education in the knowledge of perfect archetypes made Darwin incapable of feeling the full force of the absurdity of his notion that poverty and ignorance are scientific indications of natural inferiority.[63]

Of course concerns about teleology, morality, and the links between mind and matter also typically intruded, but McKinney went further, drawing

on his travel experience to accuse evolutionary anthropologists of racism
and imperialism. "The theory of evolution is comforting to the slave-owner
and the despot," he announced, "who argue that the control of inferior races
is the birthright of the more highly-evolved." Indeed his time abroad as
a ship's physician during a number of voyages during 1877–79, which included a winter visit to the Cape of Good Hope, had sensitized him to the
qualities of races too comfortably relegated to the margins of evolutionary
progress by Darwinian enthusiasts:

> Englishmen used to fancy that an inferior race dwelt in New Zealand;
> but a few wars with the Maories [sic] proved that it is safer to boast
> of superiority in the security of an English study than on the field of
> battle. The scientific world was convinced that the New Zealander could
> not be many millions of years behind in evolution. There remained the
> consolation that the Negro may be bullied and insulted with impunity;
> and failure to resist tyranny is the great evidence of lowness of type ac-
> cording to modern teaching . . .
>
> Ignorant residents of European cities speak with great contempt of
> Hottentots as filthy wretches who hardly ever wash; but those accus-
> tomed from infancy to be waited on by servants, and to use sewers which
> they never see, and to find a bath easily obtained by turning a tap, are
> little better than children in a nursery as compared with independent
> men.[64]

Although written while in medical practice in London, where he moved in
1882, McKinney's reading of evolutionary theory was a compound product
of his Belfast Presbyterian heritage and his experience of overseas travel;
both coalesced to reinforce a staunchly anti-Darwinian stance.

The circumstances under which Darwin's theory was encountered in
different urban spaces in nineteenth-century Ireland, and the ways in which
commentators understood their cultural settings, had a major role to play
in what evolution was made to mean. Operating at the scale of the nation
state, denominational membership, political affiliation, and such like fails
to provide a sufficiently fine-grained account of the different ways in which
Darwinian meaning was manufactured in Victorian Ireland. Yet more local
circumstances and events need to be woven into the tapestry. This is not to
say, of course, that wider national, regional, religious, or other forces are of
no relevance in figuring out local encounters with scientific proposals. But
it is to insist that as these broader waves of influence swept into the rocks

of local particularity, their energies were modulated in highly significant ways.

How scientific claims about human origins were made mobile during the nineteenth century was critically bound up with two intimately intertwined sets of historical geographies: the geographies of reputation and the geographies of reading. The circulation of Isaac La Peyrère's polygenetic account of human beginnings discloses this association. For how his reputation was staged had a critical bearing on how his theological anthropology was read. At the same time the value of his achievements was no less conditioned by how readers understood the cultural setting within which they themselves were located. Some perceived in his message a monumental threat to civic stability; others embraced what they imagined were its racial imperatives—even if they ran against the grain of La Peyrère's original intentions. In either case, the meaning of Peyrèrean anthropology was differently constituted depending on the vicissitudes of cultural location, political positioning, religious persuasion, and his received reputation within the local community of discourse.

The geography of reading Darwin in Ireland was no less variegated. The opposition of those whose formative encounter with Darwinism was courtesy of the Tyndall debacle in Belfast seems all the more resolute. Those, like Leebody and Macalister, who had reflected on Darwin's proposals *prior* to the British Association shock and occupied different urban locations, responded with rather more poise. Moreover both had reputations to defend. Leebody was jealous for the academic reputation of Magee College and did all in his power to avoid any charge of antiscientific sentiment; Macalister established the scientific reputation of the Trinity Anatomy Department and labored to ensure that its students were kept abreast of the latest intellectual currents in a setting where fears of a clerically dominated curriculum were a preoccupation. It is significant too, I think, that Darwinian evolution caused little anxiety in Belfast prior to Tyndall's onslaught; indeed, there is evidence of some accommodation in the years before 1870. The local impact of public spectacle on the encounter with the new science was thus of major proportions, as Tyndall's reputation lingered long in many memories. The geography of Darwinism in Ireland, I suggest, was the compound product of long-standing feuds over who should control the curriculum, the iconic impact of Tyndall's attack, the institutional spaces occupied by commentators, and the relative security local spokesmen felt in their own sense of cultural identity.[65]

NOTES

I am most grateful to Mr. Wesley Bonnar and the staff at Sentry Hill for their help with my research on Samuel B. G. McKinney. The comments of Diarmid Finnegan, Nuala Johnson, Bernard Lightman, and Charles Withers on an earlier draft of this chapter are also much appreciated.

1. Miles Ogborn, *Indian Ink: Script and Print in the Making of the English East India Company* (Chicago: University of Chicago Press, 2007).

2. Nicholas Jardine, "Books, Texts, and the Making of Knowledge," in *Books and the Sciences in History*, ed. M. Frasca-Spada and N. Jardine (Cambridge: Cambridge University Press, 2000), 401. See also Jonathan R. Topham, "Scientific Publishing and the Reading of Science in Nineteenth-Century Britain: An Historiographical Survey and Guide to Sources," *Studies in History and Philosophy of Science* 31A (2000): 559–612.

3. David N. Livingstone, "Science, Religion and the Geography of Reading: Sir William Whitla and the Editorial Staging of Isaac Newton's Writings on Biblical Prophecy," *British Journal for the History of Science* 36 (2003): 27–42; David N. Livingstone, "Science, Text and Space: Thoughts on the Geography of Reading," *Transactions of the Institute of British Geographers* 35 (2005): 391–401.

4. Roger Chartier, ed., *The Culture of Print: Power and the Uses of Print in Early Modern Europe* (Cambridge: Polity Press, 1989), 4, 8. See also Alberto Manguel, *A History of Reading* (London: Flamingo, 1997).

5. Adrian Johns, *The Nature of the Book: Print and Knowledge in the Making* (Chicago: University of Chicago Press, 1998), 29.

6. Stanley Fish, *Is There a Text in This Class? The Authority of Interpretative Communities* (Cambridge, MA: Harvard University Press, 2003).

7. James A. Secord, *Victorian Sensation: The Extraordinary Publication, Reception, and Secret Authorship of "Vestiges of the Natural History of Creation"* (Chicago: University of Chicago Press, 2000); Nicolaas A. Rupke, *Alexander von Humboldt: A Metabiography* (Frankfurt am Main: Peter Lang, 2005; repr. Chicago: University of Chicago Press, 2008).

8. See Richard Popkin, *Isaac la Peyrère (1596–1676): His Life, Work and Influence* (Leiden: Brill, 1987).

9. David N. Livingstone, *Adam's Ancestors: Race, Religion and the Politics of Human Origins* (Baltimore, MD: Johns Hopkins University Press, 2008).

10. [Isabelle Duncan], *Pre-Adamite Man; or, The Story of our Old Planet and Its Inhabitants, Told by Scripture and Science*, 3rd ed. (London: Saunders, Otley, 1860), preface. On Duncan, see Stephen David Snobelen, "Of Stones, Men and Angels: The Competing Myth of Isabelle Duncan's *Pre-Adamite Man*," *Studies in History and Philosophy of Biological and Biomedical Sciences* 32 (2001): 59–104; Stephen Jay Gould, "The Pre-Adamite in a Nutshell," *Natural History* 108 (1999): 24–27 and 72–77.

11. Thomas Smyth, *The Unity of the Human Races Proved to Be the Doctrine of Scripture, Reason and Science with a Review of the Present Position and Theory of Professor Agassiz* (New York: Putnam, 1850), 337.

12. Philalethes, "Peyrerius and Theological Criticism," *Anthropological Review* 2 (1864): 109–10.

13. A[lexander] W[inchell], "Preadamite," in John McClintock and James Strong, eds., *Cyclopaedia of Biblical, Theological, and Ecclesiastical Literature* (New York: Harper and Brothers, 1877), 8: 484.

14. Alexander Winchell, *Preadamites*, 2nd ed. (Chicago: S. C. Grigg, 1880), 454–55, 457–58, 460.

15. Smyth, *Unity of the Human Races*, 333–34.

16. Ibid., 337.

17. For its history, see John W. Burrow, "Evolution and Anthropology in the 1860s: The Anthropological Society of London, 1863–1871," *Victorian Studies* 7 (1963): 137–54; George W. Stocking Jr., "What's In a Name? The Origins of the Royal Anthropological Institute, 1837–1871," *Man* 6 (1971): 369–90; Ronald Rainger, "Race, Politics and Science: The Anthropological Society of London in the 1860s," *Victorian Studies* 22 (1978): 51–70.

18. Thomas Bendysche, "The History of Anthropology," *Memoirs Read before the Anthropological Society of London* 1 (1863–64): 335–420.

19. Philalethes, "Peyrerius," 112, 113.

20. Winchell, *Preadamites*, 457.

21. "Professor Winchell's 'Preadamites,'" *Appleton's Journal: A Magazine of General Literature* 9 (July 1880): 86.

22. Henry Colman, "Pre-Adamites," *Methodist Review* 7 (1891): 902, 891.

23. Winchell, *Preadamites*, 253, 255, 81.

24. Ibid., 81, 82, 83, 85.

25. See John C. Greene, "The American Debate on the Negro's Place in Nature, 1780–1815," *Journal of the History of Ideas* 15 (1954): 384–96; Thomas F. Gossett, *Race: The History of an Idea in America* (Dallas, TX: Southern Methodist University Press, 1963); William Stanton, *The Leopard's Spots: Scientific Attitudes toward Race in America, 1815–1859* (Chicago: University of Chicago Press, 1960).

26. Samuel Stanhope Smith, *An Essay on the Causes of the Variety of Complexion and Figure in the Human Species* (Cambridge, MA: Belknap Press of Harvard University Press, 1965 [1810 ed.]), 129, 178, 8.

27. Ibid., 149.

28. Mark A. Noll, "The Rise and Long Life of the Protestant Enlightenment in America," in *Knowledge and Belief in America: Enlightenment Traditions and Modern Religious Thought*, ed. William M. Shea and Peter A. Huff (New York: Cambridge University Press, 1995), 100. See also Mark A. Noll, *Princeton and the Republic, 1768–1822: The Search for a Christian Enlightenment in the Era of Samuel Stanhope Smith* (Princeton, NJ: Princeton University Press, 1989).

29. Charles Carroll, *"The Negro a Beast"; or, "In the Image of God"* (St. Louis, MO: American Book and Bible House, 1900), 148, 39, 22, 48, 63.

30. One horrified reviewer, Edward Atkinson, described it as "the most sacrilegious book ever issued from the press in this country" and lamented that it "was securing a very wide circulation among the poor white of the Cotton States." See Edward Atkinson, "The Negro a Beast," *North American Review* 181 (1905): 202.

31. Paul Harvey, *Freedom's Coming: Religious Culture and the Shaping of the South from the Civil War through the Civil Rights Era* (Chapel Hill: University of North Carolina Press, 2005), 43.

32. G. G. H. Hasskarl, *"The Missing Link"; or, The Negro's Ethnological Status. Is He a Descendant of Adam and Eve? Is He the Progeny of Ham? Has He a Soul? What Is His Relation to the White Race? Is He a Subject of the Church or the State, Which?* (Chambersburg, PA: Democratic News, 1898), 9.

33. Something of Macalister's time at Cambridge is charted in Elizabeth T. Hurren, "A Pauper Dead-House: The Expansion of the Cambridge Anatomical Teaching School under the Late-Victorian Poor Law, 1870–1914," *Medical History* 48 (January 2004): 69–94. See also W. L. H. Duckworth, "Professor Alexander Macalister," *Man* 19 (November 1919): 164–168; G. E. S[mith], "Alexander Macalister, 1844–1919," *Proceedings of the Royal Society of London, Series B* 94 (1922–23), xxxiv, xxxv.

34. Alexander Macalister, Review of *The Descent of Man, and Selection in Relation to Sex* by Charles Darwin, *Dublin Quarterly Journal of Medical Science* 52 (1871):133–35.

35. Ibid., 141, 142.

36. J. L. Porter, *Science and Revelation: Their Distinctive Provinces. With a Review of the Theories of Tyndall, Huxley, Darwin, and Herbert Spencer* (Belfast: William Mullan, 1874), 20, 21, 23, 5.

37. J. R. Leebody, "The Theory of Evolution and Its Relations to Religious Thought," *British and Foreign Evangelical Review* 21 (1872): 12, 17–18, 33, 34.

38. See Greta Jones, "Darwinism in Ireland," in *Science and Irish Culture*, ed. David Attis (Dublin: Royal Dublin Society, 2004), 1: 115–37.

39. I have discussed Macalister in David N. Livingstone, "Science, Site and Speech: Scientific Knowledge and the Spaces of Rhetoric," *History of the Human Sciences* 20 (2007): 71–98.

40. See Greta Jones, "Scientists against Home Rule," in *Defenders of the Union: A Survey of British and Irish Unionism Since 1801*, ed. D. George Boyce and Alan O'Day (London: Routledge, 2001), 188–208.

41. Alexander Macalister, "Review of Works on Life and Organisation," *Dublin Quarterly Journal of Medical Science* 50 (1870): 131–32, 113.

42. Alexander Macalister, *Man, Physiologically Considered.* Present-Day Tracts No. 38 (London: Religious Tract Society, 1886), 47.

43. Alexander Macalister, "The Body—The Temple of God," *Plain Words* 9 (May 1, 1871): 137–40.

44. See Greta Jones, "Catholicism, Nationalism and Science," *Irish Review* 20 (1997): 40–61.

45. See the discussion and citation in Jones, "Scientists against Home Rule," 190.

46. Alexander Macalister, *Evolution in Church History* (Dublin: Hodges, Figgis, 1882), 9–10.

47. William Johnston, "Preface," in *Science and Revelation: A Series of Lectures in Reply to the Theories of Tyndall, Huxley, Darwin, Spencer* (Belfast: William Mullan, 1975), iii.

48. J. L. Porter, *Theological Colleges: Their Place and Influence in the Church and in the World; with Special Reference to the Evil Tendencies of Recent Scientific Theories.*

Being the Opening Lecture of Assembly's College, Belfast, Session 1874–75 (Belfast: William Mullan, 1874), 3, 8.

49. Henry Wallace, "Teachings of the British Association," Lecture manuscript held by the Gamble Library, Union Theological College, Belfast.

50. See David N. Livingstone, "Darwinism and Calvinism: The Belfast-Princeton Connection," *Isis* 83 (1992): 408–28; David N. Livingstone, "Darwin in Belfast: The Evolution Debate," in *Nature in Ireland: A Scientific and Cultural History*, ed. John W. Foster (Dublin: Lilliput Press, 1997), 387–408.

51. *McComb's Presbyterian Almanack, and Christian Remembrancer for 1875* (Belfast: James Cleeland, 1875), 84

52. J. R Leebody, "The Scientific Doctrine of Continuity," *British and Foreign Evangelical Review* 25 (1876): 759, 769

53. *Testimonials in Favour of John R. Leebody, M.A., Senior Scholar in Mathematics, in Queen's College, Belfast; and First Honorman and Gold Medallist in Mathematics and Mathematical Physics, in the Queen's University of Ireland. As Candidate for the Chair of Mathematics and Natural Philosophy in the Magee College, Derry* (Belfast: Printed by W. & G. Baird, 1865).

54. Leebody, "Scientific Doctrine," 773.

55. John R. Leebody [An Irish Graduate], "The Irish University Question," *Fraser's Magazine* 10 (1872): 60, 63. The "Irish Graduate" given as author was Leebody, who also published a history of his own college: *A Short History of McCrea Magee College, Derry, during Its First Fifty Years* (Londonderry: Printed at "Derry Standard" Office, 1915).

56. Leebody, "Irish University Question," 62.

57. For background, see T. W. Moody, "The Irish University Question of the Nineteenth Century," *History* 43 (1958): 90–109.

58. Robert Watts, "Natural Law in the Spiritual World," *British and Foreign Evangelical Review* (1885); reprinted in *The Reign of Causality: A Vindication of the Scientific Principle of Telic Causal Efficiency* (Edinburgh: T. & T. Clark, 1888), 328. See also Robert Watts, *Professor Drummond's "Ascent of Man," and Principal Fairbairn's "Place of Christ in Modern Theology," Examined in the Light of Science and Revelation* (Edinburgh: R. W. Hunter, n.d., circa 1894).

59. William Todd Martin, *The Evolution Hypothesis: A Criticism of the New Cosmic Philosophy* (Edinburgh: James Gemmell, 1887).

60. J. R. Leebody, "Evolution," *The Witness*, 10 October 1890, 3.

61. A little biographical information is available in Brian Walker, *Sentry Hill: An Ulster Farm and Family* (Dundonald: Blackstaff Press, 1981); I. R. Crozier, *William Fee McKinney of Sentry Hill: His Family and Friends* (Coleraine: Impact Printing, 1985).

62. S. B. G. McKinney, *The Origin and Nature of Man* (London: Hutchinson, 1898), 181. McKinney also authored *The Abolition of Suffering* (London: Elliot Stock, 1890); *Disease and Sin: A New Text-Book for Medical and Divinity Students* (London: Wyman & Sons, 1886); *The Revelation of the Trinity* (Edinburgh and London: Oliphant, Anderson & Ferrier, 1891).

63. S. G. B. McKinney, *The Science and Art of Religion* (London: Kegan Paul, Trench, 1888), 35–36.

64. McKinney, *Origin and Nature of Man,* 209, 211, 213.

65. Of course these particular instances in the geography of reading do not exhaust the settings in which Irish readers read and wrote about Darwin's theory. Spaces of Catholic encounter, for example, received a massive hermeneutic steer from the letter that was issued by the bishops and archbishops in 1874, in which they repudiated the "blasphemy upon this Catholic nation" that had recently been uttered by the "professors of Materialism . . . under the name of Science." See Paul Card Cullen et al., "Pastoral Address of the Archbishops and Bishops of Ireland," *Irish Ecclesiastical Record* 11 (November 1874): 49.

CHAPTER NINE

Electricity and the Sociable Circulation of Fear and Fearlessness

GRAEME GOODAY

It cannot be too strongly urged that the whole danger is due, not to the electricity, but to the want of experience in the workmen. Electricity is most tractable; if it can be led across the Atlantic without danger or difficulty, surely it can be led about our houses without danger.
—William Preece, quoted in "The Fire Risks Incidental to Electric Lighting," 1882[1]

Since the 1880s, electricity has arguably become the most ubiquitous techno-scientific feature of the modern world. Attempts to authenticate experimental claims about electricity have thus not been restricted to laboratories,[2] nor indeed to any other "truth spot" discussed by geographers of science.[3] This chapter offers a more inclusive geographical analysis of one particularly troublesome late Victorian question rarely discussed by historians: in what circumstances was electricity dangerous or safe? Electricity could evidently bring pain, injury, destruction, and death in some contexts and yet also enhance domestic illumination, corporeal health, theatrical entertainment, and urban street security in others. This topic arose in conversations among all social classes, and in locations both far and near from new installations of electric light. I show here how exhibitions, theaters, and homes were key venues for presenting the alleged "safety" of electricity to a variety of audiences. It will become clear, however, that counternarratives of fear circulated more readily than did irenic characterizations by promoters of electricity such as William Preece, the chief Post Office electrician quoted in the chapter epigraph. In contrast to his attempt to essentialize electricity as an inherently safe agency, I shall show how the body of the female theatrical artiste proved to be a key site for technocratic performance demonstrating the harmlessness of properly tamed electricity.[4] I conclude nevertheless by indicating how anxiety about electricity lingered among skeptical household servants well into the twentieth century.

SAFETY AND DANGER—PERFORMATIVE FACTS OF ELECTRICITY

> [Among the] reasons for slow progress, . . . there is the curious fear
> of electricity [especially among] domestic servants, whose lack of
> knowledge and consequent fear of apparatus are often the reason[s]
> why some of our larger homes do not adopt electrical methods.
> —Caroline Haslett, "Electricity in the Household," 1931[5]

In documenting the advent of incandescent electric lighting in the 1880s, historians have typically focused only on its advantages for the domestic environment: it posed fewer fire-risks than gas light; produced no smoke; consumed no oxygen, and emanated no unpleasant fumes.[6] Yet such claims for electric light's superlative qualities only became orthodoxy by the mid-twentieth century—by which time electric light had anyway replaced most gas lighting in urban Britain. Few historians of electrical science and technology have treated fear of domestic death by the intangible mysterious force of electricity as a significant social phenomenon, let alone as a topic worthy of detailed geographical investigation.

William Essig has analyzed the numerous *public* deaths of wiremen from accidental contact with electricity in New York in the 1880s—especially in the "wire panic" of 1889.[7] Yet his account leaves us wondering why such demonstrations of electricity's lethal powers did not deter U.S. householders from taking electric light into their homes thereafter. Carolyn Marvin notes householders' fears on this point, but suggests that electrical experts in the United States and United Kingdom used the press and public lecture demonstrations to ridicule popular anxiety about electricity and assert that it really was under their control.[8] By contrast, I argue that fear of electricity was not a transient marginal phenomenon but a widespread sensibility that prompted promoters of electricity to use a variety of techniques in different sites to try to win over the public to their controversial claims for the relative safety of electricity vis-à-vis its arch rival, coal-gas.[9]

To appreciate how spatial considerations entered into the rhetorical presentation of such matters, we should note that Cooter identifies the advent of traffic safety campaigns in the late 1870s as the "moment of the accident." For him it was the urban *street* as the locale in which (or for which) new "facts" of danger were constructed by a combination of social anxiety and concentrated press reportage.[10] Following in this vein, I argue that the home and the theatrical stage were the primary settings in which panics about the putative dangers of lighting were constructed, largely by widespread reporting in the press. Thus, for example, no public excitement was attached to the demise of the euphonium player Mr. Bruno after he touched hanging wires connected to an arc light at the Holte theater in Aston on 20 January 1880. Although the incident was reported briefly in the *Times* the following day, there was nothing like the strong response that followed subsequent incidents of electrically related death.[11] By 1890 an excitable press often attributed any death involving electrical technology to the same cause. Such was the complaint made by the electrical engineer James Gordon concerning an African American youth oiling an electric tramcar in 1891: his grisly death from oil-fire burns was widely misreported as electrical in cause. In a chapter on fire risks written especially for his spouse's widely read handbook *Decorative Electricity* of 1891, Gordon further complained that the more common and still lethal gas explosions received rather less press coverage.[12]

Although the public victims of electrical accidents were exclusively male in the nineteenth century, the major locus of concern in contemporaneous debates about the safety of illumination—electrical and otherwise— was the female body. Whereas no women died in the nineteenth century as a result of accidental electrocution,[13] many did die, often rather gruesomely, from contact with oil or gas lamps, in the home and on stage. Within the home it was commonly the responsibility of women—whether servants or the lady of the house—to manage illumination. In affluent homes where female servants had traditionally trimmed candles and maintained paraffin-, oil-, or gas-lamps, such employees were typically redeployed to dusting and changing the electric lamps.[14] With that responsibility came a heritage of gendered concerns about putatively vulnerable female bodies: terrible stories of women being burned to death in their own homes by paraffin lamps that were dropped, tipped over, or had simply exploded were commonplace. This fed, for example, the propaganda of Charles Marvin, a freelance consultant to the oil industry, whose sensationalist booklet *The Moloch of Paraffin*—with the catch line "one hundred inquests a week"—cast the chief culprit as the faulty design of the lamps, not the paraffin itself (fig. 9.1).[15]

Gas explosions were similarly dismissed as mundane hazards by male observers who did not have to face them as an everyday threat. In 1884, for example, at an international conference on the health benefits of electricity, the sole advocate of gas lighting noted, rather chauvinistically, that while gas explosions might occasionally "blow out windows," this hazard

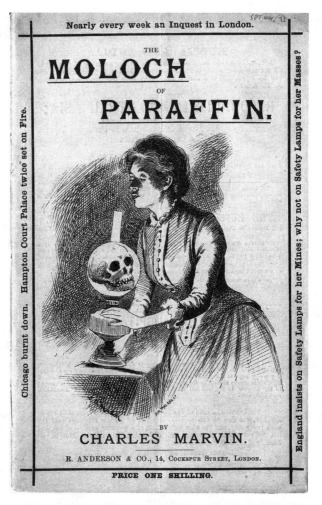

Fig. 9.1. Charles Marvin's sensationalist pamphlet *The Moloch of Paraffin*, 1886.
Institution of Engineering and Technology Archives, London.

generally affected "only the kitchen-maids."[16] The testimonies of female house servants on this point were typically either overridden or overtly stigmatized as the epitome of irrational objections to innovations in lighting.

It was in association with a domestic installation on Lord Salisbury's estate at Hatfield in mid-December 1881 that the first hysteria arose in Britain around electrically related death. But the death concerned not a female servant but a male laborer, William Dimmock.[17] While the *Times* published reports—almost on a weekly basis—of events pertaining to Lord or Lady

Salisbury or their residence, Hatfield House, some twenty miles north of London, it was unusual for the establishment newspaper to condescend to report deaths of unskilled laborers, let alone by name.[18] In the early 1880s, however, deaths linked to electricity were newsworthy largely by dint of their novelty and strangeness. Dimmock's death was not unprecedented, being the third electrical fatality recorded in the United Kingdom, and the second to be reported in the *Times* following coverage of Mr. Bruno's death.[19] As I have discussed elsewhere, the Hatfield death prompted extended press correspondence and remained thereafter a popular focal point of discussion regarding the dangers of electricity.[20] What made it so newsworthy was ongoing coverage of the recent Vienna Ringtheater explosion (see more below);[21] the unprecedented *domestic* location of Dimmock's death; the absence of any misadventure on his part; and the political significance of Hatfield House. At that celebrated venue for dinners and balls attended by senior politicians, diplomats, and royalty, death from accidentally touching an electrical wire threatened more than just the lives of humble laborers.[22]

A few days later, on 19 December 1881, the *Manchester Guardian* observed of the Hatfield incident and similar accidents at the recent Electrical Exhibition in Paris that the "dangers associated with the common use of electricity have been abundantly proved." The editorial argued that *educating* consumers about electricity and its associated technologies would eliminate the risks. For this purpose it heralded the opportunity of an imminent new exhibition at Crystal Palace to supply this need. Appealing to the authority of William Preece that such a display would be "admirably adapted" for this purpose, while implicitly diverging from his naturalistic narration of electricity's domestication, the *Guardian* noted that "a successful metropolitan exhibition may lead to provincial ones, at which the public may, whilst selecting the most efficient principle, learn so much of the practical working as will be necessary to render safe the introduction of electricity into the household."[23] Nevertheless, such faith in the power of exhibitions to disseminate knowledge of the putative "safety" of electricity was not fulfilled.

Exhibiting "Safety": The Paris and Crystal Palace Exhibitions

It is hardly too much to say that an apparatus for generating electricity, whether it be a galvanic battery or a rotating machine, is still looked upon by the majority of people as something more or less perilous and "uncanny," something which may take fire, or explode, or give off sparks, or in some way do unforeseen and direful mischief. Now, one great benefit of exhibitions will be to furnish evidence of the freedom from danger which attends upon the use of the force happily described by FARADAY as a "universal spirit in nature."[24]

Historians of electrification have noted the proliferation of electrical exhibitions from the 1880s to promote the new illuminant while understating the "experimental" nature of these displays.[25] The international Paris Electrical Exhibition of autumn 1881, followed next spring by a similar event at the Crystal Palace in south London, both continued a midcentury tradition of commercial techno-scientific display but now with a more didactic tone.[26] For the first time at Paris, visitors were offered detailed explanations of the equipment on display by companies seeking to persuade them that no risk would arise if they installed electrical light in their own homes. Thus, in October 1881, the *Times* could ingenuously remark that such exhibitions would furnish evidence of the "freedom from danger," a point echoed by the *Manchester Guardian* two months later. Favorably disposed toward electric lighting, *Times* journalists uncritically reported the propaganda of company personnel that their mode of electric lighting was safe, attractive, reliable, and economical.[27]

Yet it took only one or two outbreaks of electrical fires in the Paris displays to undermine these representations. Initially the *Times* correspondent at Paris dismissed these incidents on 13 September as minor diversions,[28] and an editorial two weeks later conventionally blamed the installing workmen for their "ignorance," emphasizing that in the surgery at least the intense electrical heating in wires could be tamed for the purposes of cauterization.[29] Yet it also noted more overtly that these fires had excited genuine "alarm" at the Paris exhibition. Another editorial admitted that the accidents had been both "ludicrous" and "alarming" owing partly to the "temporary character of the arrangements," hinting at joint-managerial culpability instead.[30]

While the *Times*' general strategy was to downplay the significance of no less than five fires that broke out at the Paris exhibition, electrical engineers and associated scientific practitioners were less stoical, for obvious professional reasons. Speaking to the Society of Arts concerning the recently closed Paris exhibition on 16 December 1881, Preece acknowledged that these fires from contact between live wires epitomized the "danger of electric lighting." Reiterating that electricity could be a "dangerous servant" in the wrong hands, he now conceded that the Paris exhibition demonstrated that electricity had to be actively contained in order to be safe. Indeed, in May 1882, Thomas Bolas, in a Society of Arts lecture entitled "The Fire Risks Incidental to Electric Lighting," noted that just by leaning against some or-dinary (bare) lighting leads at an exhibition stall, a certain M. Christophie had rather disconcertingly found that not only was his watch-chain melted, but his waistcoat was set on fire.[31]

Given the widespread knowledge of such electrical accidents by 1882, there was a strong imperative to make the Crystal Palace electrical exhibi-tion more obviously a display of how safe electricity could be. It opened unfinished and a month behind schedule on 14 January 1882, delays "in many instances" being attributed to the need for "improvements" to the ex-hibitors' equipment. Most tactically well-positioned was the British Edison Company, which demonstrated a miniature version of its entire distribution system within the palace, deploying this to greatest effect in the concert room—the sort of venue in which the comparative benefits of electricity over hot, fume-laden gas lighting were considered greatest by the electrical lobby.[32] Certainly the montage of the exhibition images presented in the *Illustrated London News* gave more attention to the Edison exhibit in its reportage than to others—two out of the eight images (fig. 9.2).

Although there were no reports of dangerous incidents among the exotic galleries, there was a major public relations disaster in mid-March. Half-way through the exhibition, insurers for the Crystal Palace company and the exhibitors collectively insisted that to cover the risks of an electrical exhibition, they would immediately double the premiums demanded for protection against fire. This was an "excessively high" rate, as judged by both the *Times* and *Engineering*, since it effectively added two thousand pounds to the bills of the exhibition. This was unequivocal evidence of con-cern at the dangers of so much electrical machinery gathered together in one confined space. A leading gas industry journal noted with characteristic *schadenfreude* the travails of those in the new, competing field that to com-pensate for this enormous increase in the costs of the exhibition, its Crystal Palace managers would need to attract forty thousand extra visitors—a tall order.[33]

Shrewdly picking up on the renewed anxiety about the fire risks of elec-tricity, Edison's officials presented their low-voltage system of incandes-cent lighting to the press and public as one bearing minimal hazards. It persuaded, for example, a visiting *Times* journalist in April 1882 that even "a child may hold the electrodes without danger."[34] Much was made of

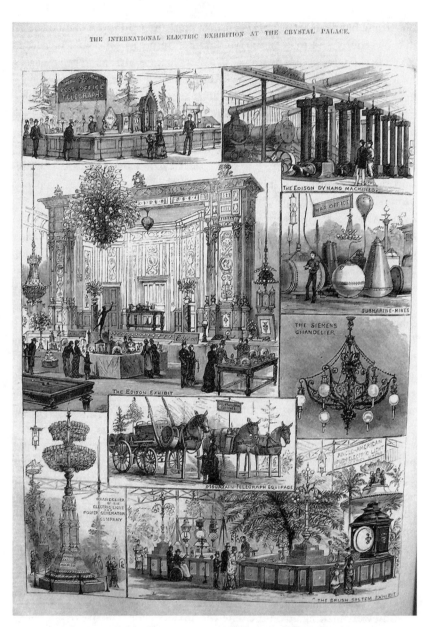

Fig. 9.2. The Crystal Palace Electric Exhibition, *Illustrated London News*, March 1882.

cooling fans, "small plugs" to act as safety fuses, as well as the use of wood and iron tubing to shield prospective householders from conducting copper wires. But the Edison company did not just rely on material culture to guarantee safety: the constant surveillance of an Edison engineer insured that circuits would be adjusted so that no excess current would overheat wires or destroy lamps whenever there was a sudden drop in demand, such as the end of a theater show or church service.[35] Employees of other companies were not always able to enjoy such utopian control. This was evidenced by the death of the unfortunate Mr. Henry Pink, who touched the thousand-volt Hochhausen alternating current machine that he had been hired to tend at London's International Health Exhibition in 1884.[36] Public displays purporting to show the safety of electricity could thus all too easily mutate into demonstrations of its deadly powers.

While public exhibitions did not unequivocally show electricity to be safe, the next section shows how the theater was a key domain in which such a demonstration could be made. This needs to be understood in terms of the previous grisly history of employee death from open-flame illumination in stage entertainment.

DEATH AND CONFLAGRATION IN THE THEATER

No one unacquainted with theatres would believe the amount of risk of life (to say nothing of property) which is incurred every night by unprotected gas. The wonder is that a ballet girl is not roasted every week, and a theatre burned down every year.
—Letter to the editor, "Burning to Death," 1863[37]

The hazards of working in, or even just attending, the nineteenth-century theater lit by candles, gas, or paraffin lamps have been much discussed. Hence the arrival of the electric light in the thespian world has been narrated as if it were a prerequisite of the modern "safe" theater. Until the arrival of incandescent electric lighting in the 1880s, the theater auditorium and stage—like the home—was generally lit with naked flames.[38] Female theatrical workers typically faced risks to life and limb on stage by wearing dresses that might just fleetingly brush past those flames. While skin burns and destroyed costumes were almost a mundane feature of theatrical life, occasionally the ignition of such materials led to an agonizing, fiery, and rather public death. These were typically viewed by the theater audience in ways that can only have added to the drama of the moment, and with some apprehension that they too might be caught up in the conflagration. As we shall see, the burning down of theaters was, if anything, slightly more common than the immolation of stage players.

The most widely cited case that epitomized the nonfictional tragedy of the Victorian stage was the dramatic blaze that killed the twenty-three-year-old Clara Webster in December 1844. She was the rising star of English ballet, and the story of her death has stayed in the folk memory of British theater ever since. While she performed in *The Revolt of the Harem* at Drury Lane Theatre in London, Miss Webster's muslin dress touched a stage oil lamp and rapidly caught ablaze. The fire buckets on stage turned out all to be empty, and her fellow dancers did not dare to help her lest they too suffered as she did. Eventually a heroic stagehand threw himself on her to smother the flames, the show continuing as she was carried backstage for medical care. Despite positive prognoses, she died three days later, to the great distress of her family and the theatrical community. At her inquest, the coroner lamented that no fire guards were placed around the oil lamps to prevent the accidental burning of dresses that touched them, and he concurred with several witnesses that wire lattices and starch could easily be added to dresses to minimize their flammability. While the jury decided on a verdict of accidental death rather than managerial negligence, it is telling that soon afterward theatrical costumiers began to produce flameproof dresses for stage artistes.[39]

This tragedy was soon romanticized to the point that Miss Webster's fate became the touchstone comparison for female death arising from clothes catching fire. Such was the case for the fiery deaths of women in London and Colchester in June and July of 1857 when their light muslin dresses accidentally touched flaming matches. As professionally elevated a journal as the *Medical Times* commented that such fiascos could be avoided by appropriate techno-sartorial fire prevention. In a brazen paragraph that was reproduced soon after in both the *Times* and *Theatrical Journal*, the *Medical Times* reported that

> it ought to be generally known that all ladies' light dresses may be made fire-proof at a mere nominal cost, by steeping them, or the linen or cotton used in making them, in a diluted solution of chloride of zinc. We have seen the very finest cambric [white linen], so prepared, held in the flame of a candle, and charred to dust, without the least flame; and we have been informed since, Clara Webster, a dancer, was burnt to death, from her clothes catching fire on the stage, the muslin dresses of all

the dancers at the best theatres are made fireproof. Our manufacturers should take the hint.[40]

Yet as Terence Rees has noted, even as the gaslight replaced the allegedly more treacherous oil lamp, deaths and disfigurations continued, since not all theaters availed their dancers of such protective costumes. On Friday 23 January 1863 in a pantomime at London's Princesses Theatre, the dress of dancer Ann Hunt caught fire as she was about to go on to the gaslit stage, as did that of her friend Sarah Smith when she sought to save her. Although Hunt subsequently survived with extensive burns, Miss Smith did not.[41] The following day, Janet Burchell burned to death in her own bedroom in Upper Harley Street as the result of hot coal falling from the fire grate onto her muslin dress.[42] The temporal juxtaposition of these events prompted considerable excitement in the press and its corresponding readers on both managerial indifference to the perilous situation of female performers wearing unprotected dresses, and (by way of moralized contrast) the apparent fecklessness of middle-class women who refused to avail themselves of chemically flame-proofed dresses.[43] Alluding to the death of Miss Burchell, one exasperated father argued that the only sure way to preempt tragedy was to install metal safety guards on fireplaces.[44]

Although legislating on such matters for the home was politically impossible, the Lord Chamberlain issued regulations in the following year requiring all British theaters to protect their staff members by technological enhancements. This required that gaslights should not only have guards on them but be located at least four feet from the stage level; moreover theaters were now obliged to have wet blankets and water buckets in the wings at all times. This placed a heavy burden of obligation on theater managers, and they responded by attempting to redistribute the responsibility to the actors and dancers who went out on the stage. In order to minimize their legal-financial liability, managers now advised performers that it was *their* misconduct that constituted the primary danger from fire. For example, a managerial safety notice in the Marylebone Theatre in 1864 listed six ways in which a female dancer could incur a fine or dismissal by breaching stage discipline in relation to lighting. These included instant sacking (on a second offence) for crossing the stage to take up a starting position after the gas had been lit; or for allowing her garlands or other accessories to catch fire from overhead gaslights.[45]

In documenting the social construction of "safety" versus fire hazard in Victorian theater illumination, we should not focus only on the performers, but look to the whole theater building and the audiences within it. Notwithstanding the implementation of the Lord Chancellor's requirements, from 1866 to 1885 there were usually at least two major fires a year in British theaters. While theater historians have argued that electrical incandescent lighting helped to bring this carnage to an end, given evidence of deaths at the Aston theater and at Hatfield House, electric lighting was not an immediately obvious solution for theater managers. Before public supply systems became the civic norm in the 1890s, the enormous cost of fitting theaters for electric light with stand-alone generators was daunting. Moreover, such was the unreliability of early stand-alone electric installations that gas had to be maintained in theaters as a back-up in case of electrical failure. For example, when Richard D'Oyly Carte's new Savoy Theatre opened in London in early October 1881, its much publicized fitting of an unprecedented eleven hundred Swan electric lights was not ready, since the dynamos were not capable of supplying so large a number effectively. So he had to endure the ignominy of having the world's first fully electrified theater lit by gas at its first public performance, writing to the press to apologize for this embarrassment. And when the first electric light was eventually shown on the Savoy stage in late November, audiences were disconcerted both by the whiteness of its rays and their occasional sudden darkening as basement dynamos stuttered and stalled. [46]

The electrification of the theater was thus not a simple or natural move to make: various kinds of spectacle were needed to create public interest in this new technological idiom as a theatrical display worth seeing in its own right. The chronology of the transition to theaters lit by electrical incandescence lights therefore needs careful consideration, noting several close consecutive incidents in December 1881. By far the worst ever European theater fire had occurred a few weeks earlier at Vienna's Ringtheater on 8 December. Although many bodies were burned beyond recoverability, later estimates were that at least 580 of the Ringtheater's cast and audience died that night following the explosion of a gas lighting installation behind the stage. Subsequent inspection of the building revealed that it had very few operational fire exits, so accusations of negligence were soon issued against the theater's management. [47] Thereafter the Austrian emperor invited the introduction of electric lighting in the nation's theaters, leading to a highly profitable enterprise for the English electrical engineering company Crompton. [48] Another after-effect of the Ringtheater fire was that a certain degree of panic broke out among theater managers across Europe, alarmed that this incident might have a negative effect on their audiences. This prompted some exaggerated claims by the manager of the Gaiety theater that were satirized in a scathing *Punch* article on "burning questions" two weeks later on 24 December 1881:

> The recent catastrophe at Vienna, and the comments thereon, will make us take our amusements timidly. . . . [Yet b]ecause gas has exploded, shall we abolish gas, and go back to oil and candles, with a double chance of fire . . . ? Shall we ask any more weak questions, or confess our helplessness, and sit down with the assurance that in England, at least, for nearly sixty years, although more than twenty theatres have been destroyed, there have only been, as the Manager of the Gaiety [theater] asserts, "two deaths from a fire—a Dresser and a Manager?"[49]

Clearly more deaths than this had arisen from accidental contact with oil or gas lighting in theaters during the previous two decades, primarily among dancers. Yet none of these inspired a national panic of the sort that arose from the electrical death at Hatfield House in December 1881.[50] Thus the safety of electrical lighting was still very much moot. And it was largely, I suggest, through the fortuitous way in which electrically lit theaters such as the Savoy suffered no injuries or deaths from electric causes that the reputation of electricity as safe was partly established. So confident was D'Oyly Carte about the safety of his new electrically lit theater that he invited a large deputation from London's Society of Arts on Saturday 12 November to witness the event.[51] Finally on Wednesday 28 December for the matinee performance of Gilbert and Sullivan's *Patience*, all 1,158 Swan lamps were in operation, and the entire stage was lit by electricity alone for the first time. As the electrophile *Times* reported, "the success of the new mode of illumination was complete." Not only was it absolutely steady, and able to give an almost "daylight" appearance to the color of costumes, but the control exercised over the levels of lighting by the use of adjustable circuit resistances entailed that there was no longer any advantage in gas—except for its greater reliability.[52]

It was not merely that electric light did not start any fires by accident. Even a deliberate attempt to destroy an electric light on stage brought no danger, and this soon became canonical as a display of the unparalleled safety of the incandescent light. As the *Daily News* reported, during the interval after the first act of *Patience*, D'Oyly Carte gave a brief speech on the advantages of electric light in "diminishing the risk of fire." He then presented a safety demonstration that at first looked so alarming that it appeared to "cause a visible movement towards the exits" among the more nervous-looking theater-goers. The theater electrician, Mr. Keppler, stood

Fig. 9.3. R. Hammond, *The Electric Light in our Homes*, 1884.

at the front of the stage holding a lit Swan lamp around which D'Oyly Carte placed some flammable cotton gauze. When the proprietor gave the signal, Keppler smashed the lamp with the stroke of a hammer, at which the light was extinguished without any hint of a fire—in implied contrast to the deadly devastation wrought by the shattering of gas equipment at the Vienna Ringtheater three weeks earlier. After loud applause, Carte invited all those with calling-card credentials to inspect the arrangements after the performance to be sure that no trickery had been perpetrated on the audience in this astounding display.[53]

Although applied initially to the Swan lamp, this histrionic experimental demonstration was subsequently appropriated by the Edison company for its own direct-current incandescent lamps. At its display at the Crystal Palace Electrical Exhibition in April 1882, the company's English representative, Edward Johnson, regularly demonstrated the harmless demolition of Edison lamps. As the *Times* reported with regard to the safety of the lamp "a handkerchief being placed over one of the globes and the glass being shattered, the instantaneous extinguishing of the light by the destruction of the vacuum is all that happens." The handkerchief itself was not even singed.[54]

Such was the viral popularity of this theatrical presentation that it soon appeared in other contexts too. At the Society of Arts' regular meeting on 3 May 1882 discussed above, the lecturer smashed a Swan filament lamp against a piece of fabric without conflagration.[55] This was repeated around the United Kingdom by the popularizer-entrepreneur Robert Hammond in his 1883 lectures (fig. 9.3).[56]

In addition to showing the danger-free consequences of smashing light-bulbs to invoke tacitly favorable comparisons with the Vienna Ringtheater catastrophe, theatrical advocates of electrification collaborated with electrical engineers in a new experimental demonstration that drew a further implicit contrast with the dreadful history of burned ballerinas in the oil- and gas-lit theater. Not only could electric lights be shown harmless to touch, but—apparently miraculously—they could even be worn directly on the body itself, specifically the body of the female theatrical dancer, without any burns or injury.

PROVING SAFETY: ELECTRICALLY LIGHTING THE FEMALE COSTUME

> Dress decorations can be carried out with tiny one-candle electric lamps, fed from a small secondary battery concealed in the dress, but perhaps they are now only suitable for fancy balls, as dress lights have become common [i.e., vulgar] since the theatres have adopted them; but years ago, when electric lights were quite new, we derived a great deal of amusement and also some tribulation from experiments on them.
> —Alice Gordon, *Decorative Electricity*, 1891[57]

The manufacture and ornamental display of miniature electric lights in the 1880s is a phenomenon noted in passing by several historians of electric lighting, but little significance has been attached to it beyond that of transient gimmickry and commercial advertising.[58] I suggest that its role was more important and subtle, helping to rehabilitate electricity as a safe natural agency in theater and home. In the first decade of electrical illumination, at least, the miniature ornamental light had a characteristically gendered significance as a form of "decoration" to be worn specifically by women—either upper-class ladies or theatrical performers.[59] The superficial purpose was to yield some kind of aesthetic benefit, presumptively, but not exclusively, for male observers. More interestingly and more specifically, it had the subtext of illustrating how little harm would accrue to women when such lights were worn close to the body. This was in complete contrast to the form of sartorial encounter with mitigated electrical hazard that working men would encounter as installation engineers or power-station operatives. They were typically expected to wear clothing of a very different sort—asbestos gloves and, latterly, metallic protection suits—to avoid all bodily contact with electrical equipment and thus avoid workplace death or injury. The asymmetry in how clothing and electrical technology were juxtaposed for the two sexes was reflected (somewhat contingently) in the comparative death rates for both.[60] Yet as we shall see, the mere absence of electrocution for women wearing decorative electric light bulbs did not constitute unequivocal factual evidence that it was a hazard-free encounter.

When the prince and princess of Wales visited the Crystal Palace exhibition in March 1882, the princess received from Edward Johnson, of the Edison company, a souvenir of a tiny electric chandelier fashioned like a bouquet of fern leaves and flowers, the buds of which were miniature incandescent lamps.[61] At around the same time, the young Alice Gordon received from Edison's rival, Joseph Swan, three or four dainty lamps each the size of a large pea which, she later recalled, looked "fairy-like" inside the petals of organic flowers.[62] This technique of naturalizing the electric light within quasi-organic displays of artificial flora was central both to upper class women's elaborate ballroom fashions of electrical jewelry in the 1880s, and to the aesthetic trend for domestic beautification later epitomized in Mrs. Gordon's Decorative Electricity. More strikingly, these tiny electric lights were deployed on the bodies of female theatrical performers, perhaps first used in French ballet in 1881 to adorn the breastpiece, necklace, and headwear of dancers.[63]

Given the potential harm to the women from electric shock, heat burns, and burns from battery acid—especially to the fast-moving dancers—these were not just arbitrary, whimsical uses of electric light. Since the theater had been the site of so many horrific deaths of actresses whose dresses touched naked gas footlights, and since the female body had become the most characteristic site of cultural anxiety about the danger of domestic illumination, the persistent noninjury of female performers wearing electric lights on their bodies was a powerful piece of visual rhetoric for promoters of both the theater and of electricity. In order to perform while wearing such lights, these dancers were typically obliged to wear either primary or secondary batteries hidden under their costumes. According to the editor of the French journal *L'Electricien,* Edouard Hospitalier, the production of miniature batteries for electric "jewellery" and theatrical body lights had been a specialty of the French makers Trouvé, Scrivanow, and Aboilard.[64] The risk of fire or electrocution was minimized by designing these batteries to operate with much lower currents than used for room lighting.[65]

The impresario and electrophile Richard D'Oyly Carte was the first British theater manager to employ female performers wearing electric ornamentation at the Savoy in 1882–83. Rumors of this and other displays prompted protests from one engineer—an Associate of the Institution of Civil Engineers [AICE]—who construed this plan as an inhumanely dangerous imposition on performers:

> It is no secret that more than one theatrical manager has in contemplation, or has already decided, to employ the electric light in the coming Christmas pantomimes or extravaganzas as a personal ornament for their hobgoblins and fairies. This can only be done by means of some form of accumulator carried by or attached to the individual, and whether the star light is to form an ornament for the head, or to glisten at the end of a wand, the constant danger is patent. There is no need to be an electrician to understand this. A grave responsibility will be incurred by those managers who permit it—a still graver one by those who force it on their employés [*sic*]. The life of even a super[numerary dancer] at 20s a week ought not to be placed in jeopardy by that autocratic production of our time, the theatrical manager, even if both are willing to run the risk.[66]

Having successfully managed the use of electric light in the main body of the Savoy theater without major incident for twelve months, D'Oyly Carte was not deterred by such jeremiads from *Times* correspondents. He included these lights in the final scene of *Iolanthe,* Gilbert and Sullivan's new "fairy" opera that heavily satirized the magical themes of Wagner's Ring Cycle, with the Fairy Queen as the electrified counterpart of Brunnhilde. Here, the association between the magical technology of electrical light and the supernatural power of fairies was forcefully illustrated. As the London *Morning Advertiser* reported of the opening night, Monday 27 November 1882, the dramatic *ennui* of the last scene was rescued by a startling innovation:

the Fairy Queen and her three chief attendants each wore an electric star in her hair, rendering a "very brilliant and original effect" in the final scene.[67]

The absence of any injuries to dancers or infelicitous failures of the lighting paved the way for further daring expansion of this apparently risky enterprise. By mid-February 1883, D'Oyly Carte increased the number of electrically lit fairies involved in the final scene to around thirty. As a reviewer for the *Times* noted, this innovation was entirely successful: the ladies' "flowing drapery" effectively concealing the accumulators carried on their backs to "maintain the incandescence of the tiny lamps on their foreheads."[68] Such apparent thespian frivolity even captured the attention of the general science journal *Nature* on 1 March 1883. It reported the well-controlled electric lighting and safety features in the Savoy Theatre's installation, noting that "all risk of fire" was apparently avoided by having all leading wires "thoroughly insulated" and the circuit fuses arranged to melt before heated wires could "cause any danger." In the context of explaining that there had been no electrical accidents at all for the first eighteen months of operations, the *Nature* reporter commented on the technocratic marvel of the electric fairies: "Each battery is provided with a switch, by means of which the light can be turned on or off by the wearer at their pleasure."[69]

Such was the success of this thespian practice that it was rapidly taken up by other dramatic enterprises. One commentator suggested that, by the latter half of the 1880s, few theaters in either London or the provinces would be without them, especially for the Christmas pantomime.[70] In New York, the Edison company supplied similar miniature forehead lights and electrically lit wands for female performers at the popular recreation ground Niblo's Garden.[71] A further cohort of "electric girls" was employed by Edison at the Philadelphia Exhibition in 1884, and the Electric Girl Lighting Company hired out their services for respectable social purposes.[72] By 1892, large-scale displays of women adorned—apparently voluntarily—in electric light were characteristic features of Edison publicity.[73] Whereas the Edison displays were stage-managed to display apparent safety, Wosk suggests that some female performers in the United States were employed in music-hall electrical displays that were deliberately designed to convey a thrilling sense of danger; the electrical press was of course sensitive to the ambiguity of such performances.[74]

During the 1880s, such electrical costuming was extended to women's fancy dress at high-society balls on both sides of the Atlantic, albeit in ways rather less fraught with hazard than bodily electric lighting. Wary of the singeing power of electrical wiring from an earlier domestic encounter,[75] Mrs. Cornelius Vanderbilt dressed as "the Electric Light" for a Vanderbilt ball on 26 March 1883 and posed to be photographed with a single glowing electric lamp held safely aloft in her gloved right hand.[76] At a less elevated social echelon, Mrs. E. E. Gaylord, spouse of the manager and electrician of the local electric light company, wore a costume covered in full-size electric bulbs to represent "electrical enterprise" at the "Greatest Event in the history of Brookings, South Dakota," held at the local opera house in 1890. A contemporary depiction shows her tense facial expression, perhaps reflecting the anxiety of having to stand on copper plates that connected wires in her shoes to an off-stage dynamo. One surmises she was not in a position to request the asbestos insulating clothing standardly available to male employees of her husband's company.[77]

The as-yet-untold story of women's experiences of displaying electric lighting on their bodies would probably reveal a much less cheerful picture than posterity has left us of D'Oyly Carte's electric fairies on the Savoy theater stage in 1882. A hint of this can be gleaned from Mrs. Gordon's explanation to readers of *Decorative Electricity* of why she had abandoned electric jewelry:

> In those days batteries were difficult to manage. Once the case was set up on the floor and the acids burnt a hole in the carpet. Sometimes the battery heated, and leaked, and once I well remember, the old lamps having worn out, I had some new ones given to me that were a wrong resistant [sic] for the battery. It heated, and we had barely time to cast the battery into the bath before the gutta-percha sides gave way, and the acids poured out, taking off all the paint. So having spoilt a dress, a carpet and a bath, I abandoned personal electric light decorations.[78]

Although accumulators (secondary batteries) had long made such risks less immediate, it is clear that in getting as far as Mrs. Gordon did, she had found that there were considerable hazards in trying to prove the safety of the electric light by bodily display—as "AICE" writing to the *Times* had hinted in 1882. While the use of small electric lights as bodily adornments for both theatrical dancers and society ladies went out of fashion in the later 1890s, the association with theatrical fairies stayed. Following the Savoy theater heritage of *Iolanthe*, these became known in Britain as "fairy lights"—safer in the family home than the Moloch of fiery destruction that was the candle-lit Christmas tree.

LINGERING FEARS OF ELECTRICITY IN
THE SERVANTS' QUARTERS

> In the apprentice days—or nights . . . of electricity in stage-land, its
> duties were more ornamental or fantastic than severely utilitarian.
> Corps de ballet were equipped with belts of now obsolete primary
> or secondary cells, lighting lamp-tipped magic wands. . . . Despite
> the many seemingly insurmountable difficulties, and the lamen-
> table failures which met the efforts of those who endeavoured to
> supersede with electricity older methods of lighting theatres, this
> form of illuminant has gradually come to be recognized by theatri-
> cal managers as the only safe and satisfactory one both for general
> lighting and for artistic stage effects.
> —"Electricity in Stageland," 1898[79]

The electrical industry might well have celebrated the success of the electric
light in the theater: at last it could plausibly claim that a domain of gaslight
formerly marred by accidents of fire and death was now free of fatalities
owing to the introduction of electricity. The extent to which the decreased
death rate of theater-going was attributable to other changes in flameproof
theater construction and management of safety exits was left somewhat
understated. But for the middle-class theater-going public, the association
between electricity and safety was much easier to make, and certainly since
the advent of electric lighting in British theaters in the 1880–90s there have
been no panics about the dangers of either undue proximity to stage illumination or the combustibility of theaters.

Yet we cannot end the story of the performativity of safety with the
middle-class theater-going public. It was not they who had to install, operate, and replace the household's electric lights—this was the prerogative of
(primarily) female domestic servants, who were long reluctant to risk use of
electric lighting. The acquiescence of servants in this scheme had long been
presumed. At the Society of Arts' Lecture in May 1882 discussed above,
the Edison company's senior company representative, Edward Johnson,
explained that the company's system was "to be used by uneducated and
unscientific persons, without the supervision of trained experts in the employment of the company. [The company] proposed to put the electric light
into houses in such a simplified form, and with such provisions, as to render
supervision entirely unnecessary; to bring the lamps within the care of ordinary house servants, no matter how ignorant they might be; and in such a
way that no damage or waste was possible."[80]

Some servants took a different view of this matter, however. Even if they had seen the electrically lit Gilbert and Sullivan operettas, they did not necessarily feel obliged to submit to their employers' whims in this matter—nor accept that they were "ignorant" of the dangers of electric lighting. They were the ones who had long had to deal with the prospect and likely unpleasant consequences of coal-gas explosions; and they were understandably concerned about the possible perils of any successor illuminant, however "scientific" it was claimed to be.[81] As Robert Hammond reported in his 1884 lecture tour, a new servant in his home was most alarmed to discover the house was equipped with electric light. Unable to sleep the first night, she showed a "haggard" countenance the following morning:

> This look of distress, indeed, did not wear off with a few days' experience, but got more marked. At last she begged to be allowed to leave a house where she was put into so much terror. It was in vain that the other servants explained to her that no danger could possibly arise, and showed her—for my servants are great authorities on the electric light question—that she had nothing to dread. She persisted in her intention of going, fearing evidently that some night she would find herself, bed and all, blown into the street.[82]

While Hammond caricatured this servant as the epitome of irrational fears about electric lighting, fears among servants about electricity lasted through another two generations, evidently communicated by conversation. That is why in September 1931 Caroline Haslett of the Electrical Association for Women reported to a celebratory meeting of the IEE that the fear of electricity among female staff was still often the reason why "some of our larger houses do not adopt electrical methods." Sharing the view of Johnson fifty years earlier that "ignorance" of servants was the key problem, she proposed a program of education to rid them of this fear—as if mere instruction in electricity could win their trust in the mysterious agency. Instead, in a mid-twentieth-century Britain in which electricity became increasingly a mundane household commodity after the arrival of the National Grid, it was only the death of these humble yet defiant servants that removed the fear of electricity from common discourse.

The primary locales for Victorian debates on the safety of electricity were neither the laboratory nor scientific society, but the domains of commerce, leisure, and domesticity. And it was by theatrical performance and sociable conversation rather than through abstruse technical publications that the

diverse views on this issue were circulated and conflicting views affirmed
and perpetuated. This chapter has focused particularly on the theater as a
site with a history of danger from gas and oil lighting, against which a well-
stage-managed series of demonstrations of smashed electric lightbulbs and
electric body lighting for women showed convincingly that death would not
arise from electricity in the major ways that were typical for comparable
kinds of exposure to older forms of illumination. Yet while these two major
techniques of demonstrating electrical safety circulated in the bourgeois
world of the theater, lecture hall, and ballroom, these forms of fearlessness
did not inhibit the resilient social circulation of stories among working-
class servants of the dangers of electric lighting. These two circulations impinged in two ways: when electrical promoters used derogatory accounts of
servant fearfulness to stigmatize the views from the servants' quarters, and
when servants in upper-class houses threatened to resign if electric lighting
were introduced.

Focusing on the spatial topography and dynamics of natural knowledge
reveals that differentiated processes of circulation occurred among different
social groupings quite independently of the workings of scientific institutions. Beyond just looking at the topic of electricity and fears of it, this
socially stratified approach to the construction and geographical mobility
of technical knowledge may reveal more than we had hitherto suspected of
counterhegemonic disbelief and skepticism among the less articulate and
least well-recorded in society. By these means we might develop a more
demographically and geographically inclusive appreciation of what Secord
has called "knowledge in transit."[83]

NOTES

1. William Preece, Discussion of "The Fire Risks of Electric Lighting," Society of
Arts, London, 3 May 1882, quoted in Thomas Bolas, "The Fire Risks Incidental to Electric
Lighting," *Journal of the Society of Arts* 30 (1882): 672. I have reconstructed Preece's
comments from the third-person narrative reported in the society's journal.

2. For discussion of laboratories as key sites of knowledge-making, see Robert E.
Kohler, "Lab History: Reflections," *Isis* 99 (2008): 761–68; for a contrasting view, see
Graeme Gooday, "Placing or Replacing the Laboratory in the History of Science?" *Isis* 99
(2008): 783–95.

3. See David Livingstone, *Putting Science in Its Place* (Chicago: University of Chicago Press, 2003); Gooday, "Placing or Replacing the Laboratory"; and Thomas F. Gieryn,

"City as Truth-Spot: Laboratories and Field-Sites in Urban Studies," *Social Studies of Science* 36 (2006): 5–38. For a portrayal of the Cavendish Laboratory in Cambridge as a privileged site for electrical science, see Simon Schaffer, "Late Victorian Metrology and Its Instrumentation: A Manufactory of Ohms," in *Invisible Connections: Instruments, Institutions, and Science,* ed. Robert Bud and S. E. Cozzens (Bellingham, WA: SPIE Optical Engineering Press, 1992), 23–56.

4. For a broader study of this topic, see Graeme Gooday, *Domesticating Electricity: Technology, Uncertainty and Gender, 1880–1914* (London: Pickering & Chatto, 2008); and Livingstone, *Putting Science in Its Place,* 156, 204. See also Simon Schaffer, "Self-Evidence," *Critical Inquiry* 18 (1992): 327–62; and Patricia Fara, *An Entertainment for Angels: Electricity in the Enlightenment* (Cambridge: Icon, 2002). For a more detailed survey of the techniques used to develop electrical safety, see Gooday, *Domesticating Electricity,* chapter 4.

5. Caroline Haslett, "Electricity in the Household," *Journal of the Institution of Electrical Engineers* 69 (1931): 1376–77.

6. David Nye, *Electrifying America: Social Meanings of a New Technology* (Cambridge, MA: MIT Press, 1990), 31.

7. Mark Essig, *Edison and the Electric Chair* (Stroud: Sutton, 2003), 212–23.

8. Carolyn Marvin, *When Old Technologies Were New: Thinking about Electric Communication in the Late Nineteenth Century* (New York: Oxford University Press, 1988), 119–22.

9. Mark Rose, *Cities of Light and Heat: Domesticating Gas and Electricity in Urban America* (University Park: Pennsylvania State University Press, 1995); Mary Ann Hellrigel, "The Quest to Be Modern: The Evolutionary Adoption of Electricity in the United States, 1880s to 1920s," in *Elektrizität in der Geistesgeschichte,* ed. Klaus Plitzner (Bassum: GNT-Verlag, 1998), 65–86; Anne Clendinning, *Demons of Domesticity: Women and the English Gas Industry, 1889–1939* (Aldershot: Ashgate, 2002).

10. Roger Cooter, "The Moment of the Accident: Culture, Militarism and Modernity in Late Victorian Britain," in *Accidents in History: Injuries, Fatalities and Social Relations,* ed. Roger Cooter and Bill Luckin (Amsterdam: Rodopi, 1997), 107–57.

11. "Death from an Electric Shock," *Times* (London), 21 January 1880. This death and that of a Russian sailor on the Tsar's yacht in the Thames were mentioned in William Siemens, "Electricity and Gas," letter to the editor, *Times,* 3 November 1880. See Gooday, *Domesticating Electricity,* chapter 3.

12. Mrs. J. E. H. [Alice M.] Gordon, *Decorative Electricity, with a Chapter on Fire Risks by J. E. H. Gordon* (London: Sampson and Low, 1891), 25–7.

13. Albert Gay and Charles H. Yeaman, *Central Station Electricity Supply,* 2nd ed. (London: Whittaker, 1906), 467–71.

14. Rookes E. B Crompton, "Artificial Lighting in Relation to Health," *Journal of the Society of Telegraph Engineers and Electricians* 13 (1884): 390–415. Lamp-trimming could also be a footman's duty; see Isabella Beeton, *The Book of Household Management; Comprising Information for the Mistress* (London: S. O. Beeton, 1861), 964, 994.

15. Charles Marvin, *The Moloch of Paraffin* (Privately published, 1886). Copy in Institution of Engineering and Technology Archives, London.

16. Crompton, "Artificial Lighting," 404–5.

17. "Fatal Accident with the Electric Light," *Times*, 15 December 1881.

18. Robert Arthur Talbot Gascoyne-Cecil, the third Marquis of Salisbury (1830–1903), was in late 1881 the Conservative leader in the House of Lords, and subsequently prime minister for three periods. See Lady Gwendolen Cecil, *Life of Robert, Marquis of Salisbury*, vol. 3 (1880–86), 5 vols. (London: Hodder & Stoughton, 1931); Andrew Roberts, *Salisbury: Victorian Titan* (London: Phoenix, 2000); Hannah Gay, "Science and Opportunity in London, 1871–85: The Diary of Herbert McLeod," *History of Science* 41 (2003): 427–58.

19. "Death from an Electric Shock," *Times*, 21 January 1880; "Notes," *Telegraphic Journal and Electrical Review*, 8 (1880): 54; "Fatalities from Electric Lighting," *Telegraphic Journal and Electrical Review* 26 (1890): 39.

20. See, for example, the *Manchester Guardian*, 12 December 1881, 8.

21. "The Electric Light and Its Friends," *Journal of Gas Lighting, Water Supply and Sanitary Improvement* 38 (1881): 1030–31.

22. On 7 January 1880 the Marquis arranged for six Jablochkoff arc lights to be hung around Hatfield House for the occasion of a grand ball. See "Notes," *Telegraphic Journal and Electrical Review* 8 (1880): 54; Cecil, *Life of Robert, Marquis of Salisbury*, 3:3–4.

23. "The Forthcoming Electrical Exhibition at the Crystal Palace," *Manchester Guardian*, 19 December 1881.

24. "The List of Awards to British Exhibitors at the Paris Exhibition," *Times*, 24 October 1881.

25. See Kenneth Beauchamp, *Exhibiting Electricity* (London: Institution of Electrical Engineers, 1997); Thomas Parkes Hughes, *Networks of Power: Electrification in Western Society, 1880–1930* (Baltimore, MD: Johns Hopkins University Press, 1983); Nye, *Electrifying America*; Charles Bazerman, *The Languages of Edison's Light* (Cambridge, MA: MIT Press, 1999), 199–217; Robert Fox, "Edison et la Presse Française à l'Exposition Internationale D'électricité de 1881," in *Science, Industry, and the Social Order in Post-Revolutionary France*, ed. Robert Fox (Aldershot: Variorum, 1995), 223–5; and François Caron and Christine Berthet, "Electrical Innovation: State Initiative or Private Initiative? Observations on the 1881 Paris Exhibition," *History and Technology* 1 (1984): 307–18.

26. Iwan Rhys Morus, *Frankenstein's Children: Electricity, Exhibition and Experiment in Early-Nineteenth-Century London* (Princeton, NJ: Princeton University Press, 1998); Beauchamp, *Exhibiting Electricity*, 165.

27. In September 1881 the *Times* reported almost daily on the exhibition.

28. "The Electrical Exhibition," *Times*, 13 September 1881.

29. "The Series of Highly Interesting Letters which We Have Published," *Times*, 27 September 1881.

30. "The List of Awards to British Exhibitors at the Paris Exhibition," *Times*, 24 October 1881.

31. Bolas, "Fire Risks Incidental to Electric Lighting," 665.

32. See "The International Electrical Exhibition," *Times*, 9 January 1882; "The Electrical Exhibition," *Times*, 27 January 1882; "The International Electrical Exhibition," *Times*, 27 February 1882.

33. "Electricity and Fire Insurance," *Times*, 14 March 1882; "The Fire Insurance Companies and the Electrical Exhibition," *Journal of Gas Lighting* 29 (1882): 509.

34. "The Edison Electric Light," *Times*, 13 April 1882.

35. Ibid.

36. "Inquest of Henry Pink," *Times*, 2 October 1884.

37. "Burning to Death," letter to the editor, *Times*, 28 January 1863.

38. Maureen Dillon, *Artificial Sunshine: A Social History of Lighting* (London: National Trust, 2001).

39. "The Late Miss Clara Webster: Coroner's Inquest," *Times*, 20 December 1844; see Ivor Guest, *Victorian Ballet Girl: The Tragic Story of Clara Webster* (London: Adam and Charles Black, 1957).

40. Quoting from *Medical Times* in *Times*, 7 July 1857, 10c; and "Our Little Chatter Box—'Fire-Proof Ladies Dresses,'" *Theatrical Journal* 18 (1857): 276.

41. Terence Rees, *Theatre Lighting in the Age of Gas* (London: Society for Theatre Research, 1978), 156–84.

42. "Death of a Young Lady by Fire," *Times*, 26 January 1863.

43. See correspondence in *Times*, 27 January 1863; 28 January 1863.

44. "Pater," "Fire Guards," letters to the editor, *Times*, 28 January 1863.

45. "Safety Regulation at the Marylebone Theatre, 1861," reproduced in Rees, *Theatre Lighting in the Age of Gas*, 162.

46. Rees, *Theatre Lighting in the Age of Gas*, 169–70.

47. "Burning of a Theatre in Vienna. Great Loss of Life," *Times*, 9 December 1881; "Theatres and the Electric Light," *Telegraphic Journal and Electrical Review* 10 (1882): 10.

48. Rookes E. B. Crompton, *Reminiscences* (London: Constable, 1928), 117–20. Crompton erroneously dates the Ringtheater episode to 1883.

49. "Burning Questions," *Punch* 81 (1881): 294.

50. Gooday, *Domesticating Electricity*, chapter 3.

51. Anonymous, "Electric Lighting at the Savoy Theatre," *Journal of the Society of Arts* 30 (1881–2): 6.

52. "Savoy Theatre," *Times*, 29 December 1881, 4.

53. "Electric Lights on the Savoy Stage," *Daily News*, 29 December 1881, 3, "The Savoy Theatre," *Electrical Review* 10 (1882): 10.

54. "The Edison Electric Light," *Times*, 13 April 1882.

55. This demonstration had been shown at the Savoy Theatre in late December 1881. See "The Savoy Theatre," *Electrical Review* 10 (1882): 10–11. It was subsequently repeated by Robert Hammond in his 1883 lectures; see R. Hammond, *The Electric Light in Our Homes* (London: Warne, 1884), chapter 3; and Bolas, "Fire Risks Incidental to Electric Lighting," 670–71.

56. Hammond, *Electric Light in Our Homes*, 59–61.

57. Gordon, *Decorative Electricity*, 121.

58. For the history of electric jewelry for men and women, see Maureen Dillon, "'Like a Glow-Worm That Has Lost Its Glow': The Invention of the Electric Incandescent Lamp and the Development of Artificial Silk and Jewellery," *Costume: the Journal of the Costume Society* 35 (2001): 76–81.

59. Edison's male workforce would occasionally be deployed marching in "electric torchlight parades," wearing helmets with full-size electric lamps glowing on top. By

1900 menswear such as ties, watch-chains, and scarf-pins could be purchased with built-in miniature electricity lights. See Nye, *Electrifying America*, 147–8.

60. See Gay and Yeaman, *Central Station Electricity Supply*.

61. Frank Lewis Dyer, Thomas Commerford Martin, and William Henry Meadow-croft, *Edison: His Life and Inventions* (New York: Harper & Bros., 1929).

62. Gordon, *Decorative Electricity*, 121.

63. For one well-known image, see Wolfgang Schivelbusch, *Disenchanted Night: The Industrialisation of Light in the Nineteenth Century* (Oxford: Berg, 1988), 71.

64. Since 1879, Trouvé had produced miniature electromechanical gadgets for party wear, such as a rabbits beating kettle drums, birds with moving wings, and skulls with rolling eyes: see Julie Wosk, *Women and the Machine: Representations from the Spinning Wheel to the Electronic Age* (Baltimore, MD: Johns Hopkins University Press, 2001), 73.

65. See the images in E. Hospitalier, *Domestic Electricity for Amateurs*, trans. C. J. Wharton (London: E. & F. N. Spon, 1889), 190–97.

66. "AICE," "Danger from Exposed Electric Light Wire," *Times*, 13 November 1882.

67. Review in *Morning Advertiser*, 27 November 1882.

68. "Savoy Theatre," *Times*, 17 February 1883.

69. "The Electric Light at the Savoy Theatre," *Nature* 27 (1882–3): 418–19.

70. Hospitalier, *Domestic Electricity for Amateurs*, 192.

71. Nye, *Electrifying America*, 30.

72. Beauchamp, *Exhibiting Electricity*, 173, 194; Marvin, *When Old Technologies Were New*, 137–38.

73. For the reproduction of "Electra" on the Edison float for the four hundredth anniversary Columbus Day Parade in 1892, see Marvin, *When Old Technologies Were New*, 108–9; Fred Nadis, *Wonder Shows: Performing Science, Magic, and Religion in America* (New Brunswick, NJ: Rutgers University Press, 2005), 55.

74. Wosk, *Women and the Machine*, 71–2, 254.

75. Gooday, *Domesticating Electricity*, 99–108.

76. This was arguably modeled on the classical Greek portrayal of the goddess of truth and justice; see Wosk, *Women and the Machine*.

77. Reproduced in Marvin, *When Old Technologies Were New*, 108–9, 138–39.

78. Gordon, *Decorative Electricity*, 122–23.

79. "Electricity in Stageland," *Electrician* 42 (1898): 336–37.

80. See Bolas, "Fire Risks Incidental to Electric Lighting," 104.

81. Crompton, "Artificial Lighting," 404–5.

82. Hammond, *Electric Light in Our Homes*, 71–2.

83. James A. Secord, "Knowledge in Transit," *Isis* 95 (2004): 654–72.

CHAPTER TEN

"The 'Crinoline' of Our Steam Engineers": Reinventing the Marine Compound Engine, 1850–1885

CROSBIE SMITH

Few subjects can be as ripe for spatial analysis as maritime history. Historians of nineteenth-century ocean steamships have tended to assume the inevitable rise of iron-screw (and later steel-screw) steamers of increasing size, power, and luxury. Even Pollard and Robertson's scholarly *British Shipbuilding Industry* claims that "the first major step towards reducing fuel consumption to an economical level was the perfection of the compound engine by John Elder and his partner Charles Randolph."[1] By focusing on case studies from the history of marine engineering, I show how and why different generations of compound engine projectors offered different readings of a new, often problematic (and therefore rather less-than-perfect) technology. I thus explore the ramifications of a spatialized alternative to "Whig" perspectives for an important sector of modern history of technology.

In their introduction to this volume the editors include within the defining features of "geographies of science" the making and meaning of science *in place*, and the movement of science *over space*. This chapter, concerned broadly with engineering science in practice, adopts a similar analytic framework adapted to the cultures within which marine engineering took place. On the one hand, "privileged sites" for the construction of ocean steamships took the form of shipbuilding yards and engine-building works situated in geographically specific locations on the banks of rivers such as the Clyde or Thames. These sites were connected with other privileged sites of scientific production, exhibition, and display: the London-based Institution of Naval Architects or the Institution of Engineers in Scotland, for example. Some institutions, especially the British Association for the Advancement of Science (BAAS), had a more mobile character, while journals, such as *The Engineer* and *Engineering*, circulated as mediating sites between technical contributors and more commercial readers.[2]

On the other hand, ocean steamships—embodiments of the knowledge, skills, and practices of shipbuilders and marine engineers—were highly "mobile" sites, carrying to far-flung places the authority of their builders, the prestige of their owners, and the ambitions of their nation. The credibility of their builders and owners rose or fell according to the degree to which these mobile sites, the largest moving artifacts on earth, fulfilled the promises ascribed to them by their projectors prior to their entry into service.[3] Internally, ocean steamers consisted of the planned juxtaposition of a variety of often-competing spaces, ranging from navigational, engine, boiler-room, freight, and mail spaces to passenger accommodation, itself inclusive of deck spaces, cabins, saloons, galleys, and (by the eve of the Great War) theaters, shops, swimming pools, and cafes.[4] Externally, ocean "liners," whether engaged primarily as cargo carriers or mail and passenger vessels, steamed on fixed routes according to regular, advertised schedules. Although long-distance sailing packets had previously laid claim to the title of "liners," steamship owners marketed their vessels as independent of the very head winds that so often rendered the passages of sailing ships long and irregular. The newcomers thus boasted a capacity to produce new forms of geographical space, modeled on railroad systems, across trackless oceans of the globe. Samuel Cunard, for example, committed himself in the 1830s to the optimistic belief that "steamers, over a route of thousands of miles in length, might start and arrive at their destination with a punctuality not differing greatly from that of railway trains. . . . The steam-ship, in fact, was to be the railway train *minus* the longitudinal pair of metal rails." Half in jest, however, he suggested that such rails were required only on "ugly, uneven land" but not on the "beautiful level sea."[5]

In *The Production of Space*, Lefebvre asserts that "nothing and no one can avoid *trial by space*—an ordeal which is the modern world's answer to the judgement of God or the classical conception of fate." Thus it is "in space, on a worldwide scale, that each idea of 'value' acquires or loses its distinctiveness through confrontation with the other values and ideas that it encounters there. . . . Ideas, representations or values which do not succeed in making their mark on space . . . will lose all pith and become mere signs, resolve themselves into abstract descriptions, or mutate into fantasies."[6] With respect to maritime history, Lefebvre's insights prompt the question of how nineteenth-century steamship projectors not only attempted to make their mark on space and therefore to win, at least for a time, the "trial-by-space," but also to what extent they reengineered geographical space for a new era of imperial commerce. While railways and telegraphs could literally embed themselves in rural and urban landscapes, steamship owners faced far greater challenges. Cunard might have whimsically reflected on ocean railways, but Isambard Kingdom Brunel, with the goal of extending his Great Western Railway westward from Bristol to New York,[7] eventually found the full joke was on himself. Ocean steamship projectors as often as not failed to leave their mark on an ocean neither so level nor so beautiful as the solid earth.

In the historical model adopted in this chapter, complementary "static" and "mobile" sites not only liberate maritime history from deterministic perspectives but also offer the possibility of reframing our understanding of the kinds of historical processes that engineered ocean steamships in the mid-Victorian period. Taking the "privileged site" (whether "static" or "mobile") as the fundamental unit of analysis, I address questions of the production, display, and communication of "power," simultaneously physical, economic, cultural, and political. Such sites embody the authority of human agents in networks of shipbuilders, engine-builders, shipowners, and merchants. The first section discusses the range of technological choices open to projectors of ocean steam navigation in the 1850s. The second section shows how Clyde shipbuilders and marine engineers prioritized economy in various forms of the marine compound engine and how they set about promoting their designs to wider communities. The third section argues against the view that in Scotland there was a simple steam lineage and a single reading of the marine compound engine.[8] In particular, John Elder's successor as Clydeside's premier shipbuilder and marine engineer, William Pearce, replaced Scottish values of economy with a new and spectacular priority on speed, power, and size in the design of his yard's principal products.

SPACES OF TRIAL

In the early 1850s little consensus existed among maritime communities as to the optimum design for ocean ships. Shipowners as well as discerning traveling publics knew, for example, that although the crack American wooden-hulled sailing packets to European ports had a safety record second to none, they had begun to lose their wealthier clientele to transatlantic mail steamers whose owners received substantial (and controversial) mail contracts from their respective governments.[9]

Over the previous decade, the founders of the Cunard line of wooden paddle steamers built confidence into the high-risk project of a year-round regular mail service between Liverpool and North America. Shared values of evangelical Presbyterianism served to consolidate a tight network based

232 CROSBIE SMITH

on mutual trust. Centered on the Clyde, that network consisted of engine builders (Robert Napier and his Glasgow works' manager and engine designer David Elder), shipbuilders (John and Charles Wood, Robert Steele, and Robert Duncan in down-river sites at Port Glasgow and Greenock), Glasgow steamship owners and merchants (especially George Burns and David and Charles MacIver), and visiting Halifax merchant Samuel Cunard himself. This network's agreed-upon and shared practices (engineering, management, and religious), both ashore and afloat, played a central role in the rise in the line's credibility to the point where one journal claimed that the Cunard ships, which prioritized safety and reliability over speed, luxury, and technological innovation, were as "safe as the Bank of England."[10]

Admiralty mail contracts also facilitated the establishment of steamship services between Britain and other parts of the empire. The Royal Mail Steam Packet Company linked Southampton to the West Indies and to the Isthmus of Panama. The line's reputation for safety and reliability, however, contrasted with that of Cunard, and it probably only survived thanks to the Admiralty's view of its strategic importance, backed by a high-value annual mail contract of some £240,000. The Peninsular and Oriental Steam Navigation Company (P&O) carried the mails to India with a transfer overland by way of Suez, gradually displacing the role of the venerable East India Company and receiving further lucrative contracts for extensions to the Far East. On these longer routes, sailing ships chartered through brokers such as William Lindsay earned rich returns carrying good-quality steam coal in huge quantities from South Wales or Northeast ports to far away coaling stations often located on exotic islands in the West Indies or in the Indian Ocean.[11]

A very different design of ship, the iron-screw (propeller-driven) steamer, was a comparative rarity in the early 1850s. Ever since the commissioning of Brunel's *Great Britain* in 1845, such steamers had been controversial. Insurers disliked iron hulls on account of the compass errors to which they were prone. The Admiralty had also concluded, on the basis of experimental research into the effects of gunfire on iron plates, that iron hulls should not be used on mail steamers liable to be requisitioned for war service. And although a small number of profitable iron-screw steamers—especially those of Liverpool's Inman Line—had appeared recently on the North Atlantic, they often depended on auxiliary sails to compensate for the low power of their engines. This economical combination of steam and sail made them suitable for the profitable emigrant trade but not for the regular delivery of mails.[12]

Pure sailing ships, however, promised to reign almost unchallenged on global routes. From the 1840s in America and from the 1850s in Britain, the tea clipper marked a new era for sail. Even here the choices included iron or wooden hulls—and later the "composite hull," constructed with wrought iron frames to enhance strength and teak planking sheathed in copper to protect it from fouling in tropical seas. Another major market for clipper ships commenced in the early 1850s with the discovery of gold in Australia. As a result, masters competed for record-breaking nonstop voyages with emigrants outward and a variety of valuable commodities, including gold, homeward.[13]

The auxiliary iron sailing ship emerged as one of the most economical compromises between sail and steam. A strong iron hull, fashioned for fast passages under sail, carried a small steam engine driving a screw used during calms or when entering or leaving port. The *Great Britain*, following her stranding in 1846, was eventually rebuilt as an auxiliary sailing vessel for the Australian emigrant trade. From the profits accumulated from his coal shipments, Lindsay too invested in auxiliary sailing ships with detachable propellers. One such vessel steamed several hundred miles up the Yangtze River for a cargo of tea. Delivery to London grossed more than £10,000 in freight, with comparatively little expenditure of coal.[14]

Brunel and Scott Russell's *Great Eastern* promised a very different future for steam navigation (fig. 10.1). Designed to steam to Australia and back without refueling, she was projected to "carry 12,000 tons of coal, and 8,000 tons of merchandise." As an advocate of such ocean giants explained, one "great object in carrying so large a quantity of coal is to avoid the enormous expense of foreign coaling stations." In 1851, for example, P&O "had in their employ 400 sailing vessels transporting English coals to their foreign depots between Southampton and Hong Kong . . . and making the average price of their coal 42s[hillings]. per ton, against 14s. the home price." The *Great Eastern*'s vast dimensions made her "double the length and breadth of Noah's Ark, as given in the book of Genesis, and four times the tonnage, . . . [with] room for a greater variety of character or specimens of natural history."[15] The five funnels, six masts, paddles, and screw constituted a veritable mobile factory for the production of motive power from the thousands of tons of coal filling the hull spaces. But mishaps and delays during construction, fitting out, and trials rendered the ship unfit for service until 1860, and even then financial and technical problems ensured that the ship never fulfilled her projectors' promises.[16]

The *Great Eastern* offered one engineering solution to the problem of coal supply and consumption. The Swedish-American inventor John Ericsson, backed by New York merchants, spent some $500,000 on a very different projected solution. He embodied his radical experiment in a 2,200-ton

Fig. 10.1. Isambard Kingdom Brunel and John Scott Russell's *Great Eastern*.
In the colored original, the "Leviathan's" funnels are painted, ironically,
red (with narrow black bands and black top), the proud scheme of Samuel Cunard's
British and North American Royal Mail Steam Packet Company—a line that
would never at the time have entertained owning such a monster of the ocean.
From Captain Felix Riesenberg, *Early Steamships. Currier & Ives Prints No. 4*
(London: Studio, 1933), plate 6.

ocean steamer, the *Ericsson*, driven by a hot-air ("caloric") engine. An early
demonstration trip prompted one of the guests, Professor J. J. Mapes, to
proclaim that "there were but two epochs of science—the one marked by
Newton, the other by Ericsson," while the *Tribune* declared Ericsson "the
great mechanical genius of the Present and the Future."[17] The projectors
promised that the ship's daily coal consumption would be little more than
10 percent of that of a typical Atlantic mail steamer.[18] The *Ericsson*, however, subsequently sank during a trial trip out of New York and, although
raised, was reengined as an ordinary steamer.[19]

Throughout the 1850s, Glasgow shipbuilders and engineers prioritized
economy of motive power and invoked physical theory and experimental
practice to try to realize it. Informed by the new science of thermodynamics,

the aspiring scientific engineer Macquorn Rankine and shipbuilder James Robert Napier (son of Robert) also projected an air engine for marine use, but difficulties with the experimental model precluded the construction of full-scale air-engines to their design.[20] In the same period, the first Clyde-built marine compound engines came into practical service and from the beginning promised significant reductions in coal consumption. Were those promises to be fulfilled, steamship builders could claim to have reinvented the steam engine, to have remade space for power on board ship, and to have opened the way for shipowners to command oceanic space itself.

"A Constant Succession of Unfathomable and Costly Experiments"

Son of Robert Napier's works' manager, John Elder entered into an engine-building partnership with Charles Randolph in 1852 and within a year had taken out a patent on a marine compound engine. The first steamship fitted with the new design, *Brandon,* received a favorable report on the trials in the pages of the *Nautical Magazine.* The new Glasgow firm then persuaded the Pacific Steam Navigation Company (PSNC), which had been carrying mail, passengers, and freight on the West Coast of South America since 1840, to adopt the engines for two new steamers. None of the early PSNC compound-engined vessels operated at spectacularly high steam pressures. In its 1854 report on the *Brandon,* indeed, the *Nautical Magazine* used her good performance to correct "a popular error which very generally exists, that there is a great advantage in using high-pressure steam, as being a saving of fuel." Arguing that the higher the steam pressure, the greater the quantities of heat required, the journal concluded that "it follows that there will be equal or even greater economy of coals raising steam at a low than at a high pressure."[21]

Less than four years later, one Clydeside shipbuilder, acquainted with Glasgow College natural philosophy, set out to demonstrate that high-pressure steam was very definitely a great advantage in the saving of fuel. The gentleman Greenock shipbuilder John Scott (known as John Scott, Yst [the youngest] on account of there being several John Scotts in the family) built the experimental steamer *Thetis* in 1857. Educated at Edinburgh Academy alongside Peter Guthrie Tait, James Clerk Maxwell, and Edward Harland, Scott attended Glasgow University during the early years of William Thomson's tenure of the natural philosophy chair, when Thomson was much occupied with the new science of thermodynamics. John's father,

Charles, meanwhile managed the firm's well-established engine-builders—Scott, Sinclair and Company of the Greenock Foundry—which had constructed railway locomotives for Brunel and supplied steam engines to the navy. In 1851 the family firm split: Charles and his son laid out a new shipbuilding site as Scott & Co. and built their first iron steamer soon after, while Charles's brother and nephew (both John) similarly launched into iron shipbuilding as John Scott & Sons until overtaken by financial troubles in 1861. Scott & Co. took over the Greenock Foundry in 1859 and prepared to build both hull and engines for a new generation of iron-screw steamers. A year later John Scott, Yst., became a founder member of the Institution of Naval Architects in London.[22]

Officially completed on 13 May 1857, Scott & Co.'s 523-ton iron-screw steamer *Thetis* immediately became a mobile experimental site and spent the next two years undergoing trials witnessed by a variety of experts. She embodied Scott's conviction, consistent with thermodynamics, that high-pressure steam was the key to economy in marine engines. The boilers operated at a terrifying 115 pounds per square inch (psi).[23] An early witness was Liverpool consulting engineer and shipowner Alfred Holt. He arrived the day before her completion to oversee the delivery of the first small steamer for his brother George's firm of Lamport and Holt, which, from its founding in the mid-1840s, had remained resolutely loyal to sail alone. Since overseeing work on auxiliary sailing vessels for Lindsay, Holt was a frequent visitor to the Greenock yard and became a close personal friend of John Scott and a strong admirer of his shipbuilding expertise.[24]

Rowan & Company's Atlas Works in Glasgow constructed the boilers, surface condensers, and compound engine for the *Thetis*. John Martin Rowan's son later explained that his father had been inspired by Thomas Craddock's 1848 designs for land-based compound engines, and that by 1856 Craddock had joined the Atlas Works as manager "in order that a definite trial of his plans, as applied to marine work, should be carried out." With the boiler tested to a pressure of 240 psi before the Board of Trade surveyor in April 1857, the subsequent trials of both engines and boilers showed "that there were grave defects inherent in their design, and the engines and boilers were subjected to substantial alterations." The changes were incorporated in an engine patent taken out in the names of Rowan and Horton in April 1858.[25]

Toward the end of 1858, Rankine, now professor of civil engineering and mechanics at Glasgow University, told his friend James Napier that "Rowan's engine burns . . . 1.018 lb. per I.H.P. [indicated horse power] per hour." Rankine was preparing for the engine-builders a formal report on performance measurements that he, Rowan, and Thomas R. Horton (Rowan's works manager) had recently carried out aboard the vessel "with the cordial co-operation of Mr John Scott of Greenock."[26] These measurements coincided with a surge of BAAS interest in a range of topics concerned with steamship performance, including tonnage measurement, strength of wrought iron and steel, and fuel consumption. As Marsden shows, in the autumn of 1858 Napier and Rankine promoted in *The Engineer* the Napier-built steamer *Admiral* as an exemplar of science-based naval architecture.[27]

"No other application of steam is yet so unsatisfactory as that to ocean steam navigation," ran the editorial in *The Engineer* on 5 November 1858, as though to prepare the ground for a new dawn in the world of steam-ship economy. "It is not so much the power required to drive the vessel as it is the fuel required to produce the power." Claiming that it was "very generally known that the consumption of coal in ocean steamers making long voyages is about six pounds per effective horse power per hour," the journal contrasted the recorded performance of a stationary Cornish pumping engine—the gold-standard for steam engineers in the nineteenth century—as the equivalent of a mere 1.57 pounds, while locomotives on the South Western Railway "burn scarcely more than 3 lb. of raw coal per horse power per hour." It therefore listed several desiderata for marine engine systems, including the use of high-pressure steam in the range 60–90 psi., surface condensation (allowing fresh, rather than salt, water to be used in the boilers), expansive working and superheating of the steam. That *The Engineer*'s editorial and Rankine's report to Rowan appeared in the same month (November) was hardly coincidental.[28]

In late spring 1859, the journal printed a full account of "Rowan's Marine Boiler Engines" already communicated to the Liverpool Polytechnic Society. At least three local members of the society, Thomas Arnott, Robert Lamont, and J. Gray, communicated their findings. Mr. Arnott first reported that the invention "was creditable to the ingenuity of the Clyde engineers, and at length promised to revive the progress and prospects of our mercantile steam marine, which was held in check by the cost of fuel, and consequent displacement of cargo at much sacrifice, especially on long voyages." With the boilers "economising fuel-heat," and the engine "working up the entire steam by the combined action of high and low pressure steam," the *Thetis* had delivered on these promises, with her boiler and engines occupying merely "a space 9 ft. square." His fellow members also testified that the new design met all the desiderata listed earlier by *The Engineer* (fig. 10.2).[29]

By early autumn of 1859, John Scott had printed "for private circulation only" two documents relating to the *Thetis* as a mobile site for performance

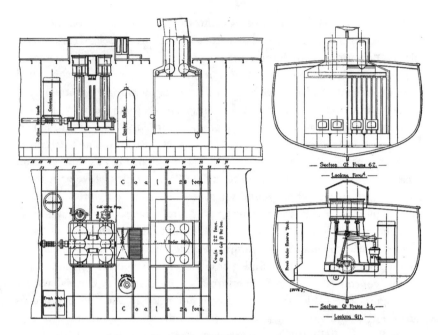

HIGH - PRESSURE MACHINERY IN THE "THETIS."

Fig. 10.2. John Scott, Greenock shipbuilder, constructed to his own account the
"experimental" steamer *Thetis* (1857). Rowan's Atlas Works in Glasgow supplied
the boilers, engines, and surface condensers to operate at unusually high steam pressures
for the time. Professor Rankine of Glasgow University undertook the performance
measurements on Clyde trials. His results showed a dramatic reduction in coal
consumption. From *Two Centuries of Shipbuilding by the Scotts at Greenock*
(London: Offices of Engineering, 1906), facing page 38.

measurements. The documents were clearly intended for prospective cus-
tomers, either directly or through agents, and for fellow shipbuilders and
marine engineers on the Clyde trusted to share knowledge and expertise.
The first was entitled "Notes of an Experimental Trial Made with Rowan's
Patent Expansion Steam Engines" on 12 August 1859 between Cloch and
Toward Lighthouses (a distance of 5.8 nautical miles) on the upper Firth of
Clyde in slack water. The document recorded every quantitative detail of
the vessel, her engines and boilers, particulars of the trial, power developed
(using four sets of indicator diagrams taken at half-hourly intervals during the trial), fuel consumed, and performance. At a boiler pressure of 90
psi., engines delivering 256 horse-power, and the vessel making 9.66 knots,
the coal consumption was calculated as 1.202 lb. per horse-power per hour.
Crucially, those on board to witness the trial included M. Forquement (a

French government naval engineer) and Thomas Stirling Begbie, who acted as London-based agent for a number of Clyde shipbuilders (among them Scott, Denny, and Caird).[30]

The second document presented the *Thetis*'s "commercial" performance over fourteen complete voyages, transporting in total 7,622 tons of cargo and consuming just over 121 tons of coal in total. In his qualitative introduction, Scott explained that because the vessel had too great a capacity for the trade, she had on occasions been steaming light, with the screw not fully covered. As a result, "the total quantity of cargo transported does not convey an accurate idea of her dynamical power." Indeed, "experience in her working amply proves that when the Screw propellor [sic] is properly covered the dynamic result is very superior." He therefore referred the reader to the tabulated results on each voyage. These showed, he stressed, that despite the less-than-full loading—together with other unfavorable circumstances such as waiting times for berthing, during which coal was still being consumed to maintain steam pressure—"the result still surpasses to a very marked extent the best performances hitherto obtained from ordinary marine engines of the most approved construction."[31]

The Engineer's editorial in January 1860, headed "Progress of Steam Navigation," offered a far more optimistic assessment compared to fourteen months earlier. "In regard to steam economy," it remarked, "some of the best results have been obtained within the year, and the direction of improvement in this respect has been precisely that which we have so often pointed out." Thus Randolph & Elder's engines were now "worked with superheated steam of 45 lb. pressure, expanded nine-fold, and the consumption of coal is about one-half of the usual expenditure in ocean steamships." An even more noteworthy performance was that of Scott's *Thetis*, with steam at 115 lb. "expanded fifteen-fold." These engines have thus "on a short trial, obtained the best result yet recorded, and have probably maintained a rate of about 2 lb. of coal per horse power per hour, in regular working, for a period of several months." The editorial therefore concluded that we could "look forward with confidence . . . for still further efforts and increased success."[32]

There was indeed much at stake. Very different readings of the "success" or otherwise of Rowan's new engine system ensued. Writing to *The Engineer* early in 1860, J. Frederick Spencer of London Bridge, for example, expressed the opinion that "combined cylinders or other complications . . . tend . . . to retard and not advance steam economy." He accused "two first-class engineering firms . . . [of] just now patenting arrangements for using four and six cylinders instead of two, to enable them to expand 'Comfortably.'"

Referring to the current Victorian fashion for hooped petticoats to make long skirts stand out, he asserted that such a patent "is the 'crinoline' of our steam engineers, and will have its day."[33]

Counterattacking, Scott insisted that he was "not aware that Mr Spencer has had any experience in the construction or working of combined cylinder engines, the introduction of which he so greatly deprecates." He also concluded with the weighty claim that there were now fourteen pairs of engines being constructed on Rowan's system, which "considering the short period the invention has been before the public, is no mean indication of the attention which it is attracting."[34] The "fourteen pairs" of engines consisted of seven sets of machinery constructed by Robert Stephenson of Newcastle-on-Tyne in accordance with Rowan and Horton's patent. Of these, four sets were designed for service on the River Ganges in India and were later credited with "remarkable success" in competing with the best of the older steamers. That "success," according to Stephenson's former engine designer, was in part due "to the circumstance that although they often ran in very muddy water on the Ganges, yet it was comparatively free from salt."[35]

Geography, however, could dramatically alter the reading of "success." Two other sets went to the Australian Steam Navigation Company to run coastwise between Sydney and Brisbane—with very different results. An engineer stationed at Sydney with the Royal Navy reported on the performance of the first steamer and recorded a coal consumption of only 1.865 lb. He nevertheless concluded that "mechanical difficulty may for a time prevent the practical working of this most economical steamer." The second ship apparently had to have the engines and boiler replaced due to the reported leakage of sea water, despite the adoption of surface condensers, into the system.[36] In contrast, the final vessel was a deep-sea ship. Built on the Tyne for the London and Mediterranean Steam Navigation Company, the *Sicilia* recorded on trials a consumption of 1.36 lb., but also suffered salt water leakage requiring the early replacement of boilers and engines. Her Scott-built sister, *Italia*, found her maiden voyage cut short with a return to Plymouth for repairs to a leaking boiler. Reported boiler corrosion was a refrain with respect to almost all the seagoing steamers fitted with Rowan and Horton's system, including a small coastal dispatch boat delivered by Scott to the French Navy.[37]

In a lengthy historical account to the Institution of Engineers and Shipbuilders in Scotland (IESS) in 1880, Rowan's son attempted to place the blame squarely on "the evils of faulty workmanship, insufficient arrangements, boiler corrosion, and the numerous minor difficulties of pistons and piston rod packing, and of lubrication in presence of steam of a high temperature." In the case of the *Italia* he even cited the "timidity of her captain and engineer" as one of the reasons for her aborted maiden voyage.[38] The ensuing discussion among IESS members, however, included a rival narrative.

James Howden, with a boiler-designing agenda of his own, professed to members that he had been something of a detached witness to the *Thetis* in 1859: "I was never in the engine-room, there being a considerable amount of exclusiveness in these [sic] days in regard to things of this sort; but I had a bird's eye view of the engines through the skylight, and the boiler was described to me." The *Thetis*, he emphasized, had a distinctive character: "Owing to the unusual pressure employed, the novelty of the machinery generally, and the high economy attained on the trials as certified by Dr Rankine, this steamer was the subject of much comment in her day." But it was to the *Thetis*'s immediate successors that Howden directed members' attention: "The result of the trials in these various steamers . . . was, on the whole, unsatisfactory—that is, none of them continued to work so that they could have been used for regular ocean service, such as ordinary steamers require to maintain." Complex boilers, surface condensers, and cylinder arrangements, he argued, together mitigated against the fulfillment of promises in each of these steamers (fig. 10.3).[39]

As testimony to the credibility of his analysis, he told members that the owners of the Napier-built *Athanasian* had, in late 1860, contracted him to fit new tubular boilers working at 100 psi and capable of using sea water if necessary. Having measured the horse-power of the engines by taking indicator diagrams while on passage from Dublin to Glasgow, he fitted the new boilers in May 1861. Such was the owners' satisfaction that they purchased the hapless *Sicilia* eighteen months later and refitted her with Howden's tubular boilers.[40]

Such lessons were evidently not lost on Alfred Holt as he began planning for the launch in the mid-1860s of his new Ocean Steam Ship Company between Britain and China to bring home tea in direct competition with the sailing clippers. Following Scott's example of using a full-scale ship for trials, he refitted his small steamer *Cleator* (1854) with an "experimental" compound engine of his own, designed, in contrast to Rowan's system, to be as simple as possible. It had but one high- and one low-pressure cylinder, the former vertically in line below the latter, thus minimizing engine room space. Opting for tried-and-tested locomotive technology, Holt employed a tubular boiler operating at 60 psi. The *Cleator*'s trials, including a series of longer commercial voyages, satisfied Holt and his closest associates.[41]

On behalf of Lamport and Holt, Alfred contracted with Andrew Leslie as shipbuilder and R. & W. Hawthorn as engine-builder for his brother's

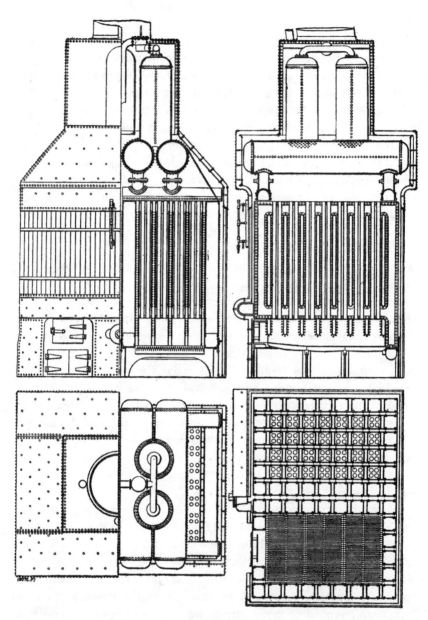

Fig. 10.3. Rowan and Horton's boiler designs (fitted to the *Thetis*) went beyond typical tubular (water-tube) boilers and received the epithet "tubulous" on account of the way in which four round tubes carrying hot gases from the furnace passed through each vertical water tube (shown in square cross section). While critics admitted that the arrangement offered the economy of high-pressure working, they claimed that the design proved extremely difficult to maintain, especially if the condenser water became contaminated by sea-water. From *Two Centuries of Shipbuilding by the Scotts at Greenock* (London: Offices of Engineering, 1906), facing page 35.

firm's first compound-engined vessels. One of these, *Humboldt*, sailed from the Tyne to Gibraltar and back to Liverpool in February 1866. "This was a test & trial trip with Alfred Holt's new engine," wrote the ship's master. "Mr Lamport [had initially] thought the engine was a failure. Alfred Holt thought otherwise and sent me to prove it was alright in the Humboldt."[42] Through this trial-by-space, Holt won the battle for high-pressure steam over the ever-skeptical Lamport. As Scott began to deliver the first three China steamers to Alfred Holt the same year, Holt published a circular highlighting their Clyde-built pedigree and making clear that they were very different in character from auxiliaries: "They are of full power; and will steam the whole passage, both out and home."[43] In his classic *History of Merchant Shipping* (1874–76), Holt's old friend Lindsay identified the fulfillment of this promise and rendered it a historical "fact" in his account of the rise of the steamship: "Starting from Liverpool they *never stopped till they reached Mauritius, a distance of 8500 miles*, being under steam the whole way, a feat hitherto considered impossible."[44]

Holt recorded the conclusion of one 1867 voyage with evident satisfaction: "I had great pleasure in hearing that the 'Ajax' had arrived in London the day before, just under 75 days from Shanghai, a marvellous passage leaving the [sailing] clippers altogether out of sight." Yet even Holt had occasional private doubts centered on the skills of the engineers. "I was much pleased with her performances," he wrote in 1868 at the end of the delivery voyage from the Tyne to Liverpool of his latest China steamer *Diomed*, "but only fear whether she is too scientific an instrument for those who will have to use her, [whereas] a careful watchful thoughtful man might take that ship to China and back at 9 knots an hour on 14 tons of coal a day . . . I look with great interest to her performance."[45]

Skeptical notes were also struck by close associates. Alfred and Charles Booth, for whom Alfred Holt was overseeing a new but growing fleet of steamers to the Amazon, had been experiencing the troublesome side of high-pressure steam, most notably when their first two steamers needed replacement boilers after only nine months service. "The fact is, AH [Alfred Holt] does everything by rule of thumb which comes right if he is always at it," Charles noted in 1869. But when Holt merely issued general directions, the problems tended to remain. "Steamship owning seems to be a constant succession of unfathomable and costly experiments," Charles ruefully concluded, "and can only be carried on when there are large earnings coming in."[46] It was an insight that would also sum up the new direction, away from prioritizing "economy" to an emphasis on power, wealth, and speeds well in excess of even the fastest sailing clippers.

"The Finest and Fastest Steamer Afloat"

Six hundred miles up the mighty Yangtze River, an extraordinary event occurred in the spring of 1882. Claimed to be the fastest ocean-going steamship afloat, and one of the largest in the world, the SS *Stirling Castle* had arrived at Hankow to load the season's first and finest China teas. Launched on the River Clyde only a few months earlier, the *Castle* had run repeated trials on the Firth of Clyde's measured mile off Largs with an average speed of well over eighteen knots, comfortably in excess of contract specifications. Her builders, John Elder & Co., had reason to feel pleased with their latest ship, which had been constructed in accordance with all that scientific engineering could offer. Having apparently failed to persuade the Glen Line, one of Glasgow's principal shipowners and leaders in the lucrative Far East trades, the shipyard's ambitious chairman, William Pearce, talked rival owner Thomas Skinner into challenging Glen for the high freights on offer to the ship that first delivered the new crop of teas to the drawing rooms of Victorian Britain. If successful, the yard could expect to sell many more such vessels—adapted to a variety of global routes—to envious owners, British and foreign, and it had even commenced negotiations to open a new shipbuilding facility on the Delaware River to supply the booming United States economy from within.[47]

In the mid-1860s PSNC failed to reach a favorable financial arrangement with the American-owned Panama Railroad (opened in 1855) for the transfer of mail, passengers, and freight across the isthmus. The line therefore projected a radical geographical alternative—the opening of a regular mail steamship service from Liverpool to Valparaiso by way of Rio de Janeiro, Montevideo, and the Straits of Magellan. In 1869 John Elder & Co. delivered four purpose-built, 2,800-ton ocean steamers, compound-engined for a service speed of twelve knots. But in September Elder died, aged forty-five. A new partner, Pearce, soon became the dominant figure in reshaping the company's reputation.[48]

Pearce hailed from a context far removed from Presbyterian Clydeside. Born in 1833 near Chatham in Kent, Pearce had learned the skills of shipwright and naval architect at the Royal Dockyard there, and in 1861 he oversaw the construction of the first ironclad warship built in the navy yards. Two years later he became a Lloyd's shipbuilding surveyor on the Clyde. By 1864 he had moved to Robert Napier's shipyard as general manager and five years later transferred to Elder's as a partner.[49] Pearce played a major role in increasing PSNC's ocean fleet. With weekly departures from 1873, the line required some seventeen vessels to maintain the 19,000-mile route, which the managing director hailed as "the greatest steam-line in the world." But the line did not pay, and by 1874 half the fleet of twenty-one ocean-going vessels had been laid up.[50]

Undaunted, Pearce encouraged the launch in 1877 of a new monthly service to Australia using the laid-up PSNC liners under the name Orient Steam Navigation Company. Two years later he delivered the 5,300-ton, 16-knot *Orient*, the largest vessel in the world (excluding the *Great Eastern*). Her first voyage of just under thirty-eight days from London to Adelaide broke all previous records for steam and sail.[51] Pearce regarded *speed* as one of the major qualities with which to persuade shipowners of the superior character of a Clyde-built ocean steamer. He therefore soon turned his attention to the Atlantic crossing and delivered the record-breaking *Arizona* to the struggling Guion Line in 1879. Within a few months of entering service, the *Arizona* ran at full speed head-on into an iceberg while homeward bound—and survived. Pearce's liners, it seemed, were not only fast and luxurious; they were also unsinkable.[52]

"Now he is not only a rich man, but a man whose importance it is impossible to rate at a higher value than he does himself," mocked the anonymous author of the satirical *Clydeside Cameos* in a chapter devoted to "Billy Drill," or William Pearce, in the mid-1880s. The chapter brilliantly portrayed Pearce's persuasive technique. One episode related to the building of the *Arizona*. During a slack period in the yard, Pearce "determined to utilise his men and machinery in producing on spec. [*i.e., speculation*] what had been his dream for years—an ideal of an ocean liner." The undertaking was "a big thing" but "Billy likes big things, and is not wanting in confidence in himself." As construction proceeded, he commenced negotiation with the managers of a line "who had rather fallen behind in the Atlantic race, and whose vessels were becoming out of date and unpopular." He thus "offered this coming triumph to them." And when they hesitated, he guaranteed that the ship "would prove the fastest and finest steamer afloat," while promising to retain a holding himself of £80,000 in the vessel. As it happened, "the new liner proved a perfect triumph" and paid "so splendidly that Billy's interest in her is said to be the best investment he ever made."[53]

The *Stirling Castle* (1882) embodied, perhaps more than any other of his mighty Victorian liners, Pearce's culture of speed and show. Costing some £140,000, the 4,400-ton ship was indeed expensive. A year earlier, Holt had paid only £37,750 for each of four 2,800-ton ocean steamers divided between Scott and Leslie.[54] "The new China clipper made her trial trip on March 18th, and proved herself to be the fastest steamer afloat," *The Marine Engineer* reported on 1 April 1882. Over a two-day period, the liner had been showing her paces. On the first day, she had averaged more than eighteen knots over distances exceeding forty miles. The second day was the occasion for displaying the new liner in action to a select group of witnesses, all of whom were Pearce's official guests. This "party of gentlemen" included Thomas Skinner (owner) and his two sons, A. B. Beldan (superintending engineer of the Skinner's London & China Line, better known as the Castle Line); Signori Mattei and Raggio (shipowners and merchants of Genoa); a Captain Steele of San Francisco; and W. H. Woodthorpe (Board of Trade surveyor).[55]

According to the report, the vessel "ran the measured mile [off Largs] six times, both with and against wind and tide, giving an Admiralty mean average of 18.418 knots, equal to 21.303 miles." Moreover, the trial speed was far in excess of the contract speed of 17.5 knots. Less scientific but more visible proof of the vessel's vibration-free performance was provided "by filling a wine-glass with water as full as it would hold and placing it on the saloon table. Even in the fastest run not one drop overflowed." After the conclusion of the trials, Pearce presided over a luncheon for his guests and delivered a speech that *The Marine Engineer* reported verbatim. His firm, he claimed, had designed not merely a vessel well adapted to the trade but one that could scarcely be surpassed in capacity and power for that trade. Denying that he was going to predict "what this ship will do" in the future, Pearce explained that that very morning she had fulfilled a promise that no other ocean steamer had ever done: with a cargo of three thousand tons on board, she had made an average speed of about 18.5 knots. He also drew attention to the versatility of the design, suited not simply to carrying teas but also for use as an Atlantic liner or, in time of war, as an armed cruiser. Moreover, she had "more than the usual number of bulkheads, rendering her unsinkable, and perfectly safe in case of collision" (fig. 10.4).[56]

Almost identical reports of the vessel's trial trip appeared in the English-language Far East Press. Thus, at the end of April, the *China Mail* reprinted an account originally published in the *London and China Express* containing Pearce's remarks and including a much fuller account of Thomas Skinner's reply, in which he cited Holt's "series of vessels built to go round the cape of Good Hope at the rate of nine or ten knots, with a very small consumption of coal." Since the opening of the Suez Canal in 1869, however, both Skinner's firm and the rival Glen Line had built ships "of a very superior class" to Holt's. These fast steamers were designed to engage in what shipowners called "the Derby of the Ocean—the Blue Riband of the sea," that is, the race to bring home the first teas of each season.[57]

Trials over space within the sheltered waters of the Clyde were one

Fig. 10.4. Delivered by William Pearce to the Cunard Line one year after the
Stirling Castle, the even larger and faster *Oregon* was a record-breaking Atlantic
"greyhound" that her original owner, the Guion Line, could not afford to complete.
Two years later, Edward Harland, builder for the rival White Star Line, predicted that
"ere long the 'greyhounds' of the Atlantic, from a commercial point of view, would be
proved an egregious mistake" (*Marine Engineer* 7(1885–86): 174). A few months later the
Oregon foundered after collision with an unidentified sailing vessel, thus undermining
Pearce's assertion that his liners were "unsinkable." From R. A. Fletcher,
Steam-Ships: The Story of Their Development to the Present Day
(London: Sidgwick & Jackson, 1910), facing page 250.

thing. Passages to and from China were another. Pearce and his associates
could no longer maximize their control over press accounts by offering well-
orchestrated coastal trips. After leaving the Clyde, the *China Mail* reported,
the ship was soon averaging around fifteen knots despite poor weather and
"rolling heavily." Conditions in the western Mediterranean worsened with
an easterly gale, and the ship eased down for some hours to "stow cargo
afresh," implying that the ship's violent movements had caused freight to
break loose with worrying consequences for the vessel's trim. On the following day, the "ship rolling and labouring very heavily," the engines were
stopped for 2.5 hours. After the Suez Canal, steam pressure was reduced to
80 psi., and a subsequent report from the *Shanghai Courier*'s Hankow correspondent revealed that "on her voyage out the new ship caught fire in her
bunkers, and it is to ascertain the amount of injury the Sterling Castle has received, that a surveyor has been sent to report on her present condition."[58]

The *Singapore Times* marked the new ship's first arrival in Singapore by
recording the formal verdict of those "whose position calls on them to watch

the vessel performance from day to day with scientific tests and caretaking interests." It was an authoritative if terse verdict "embodied in the trite and satisfactory report" of her (and the builders') engineers: "'More steaming can be got out of her than the builders' engineers ever anticipated.'" It also reported the master's expectation that his ship would take twenty-five days from Shanghai to London, eleven days less than the previous record passage in 1880.[59] To achieve this goal, the line appointed the most experienced and trustworthy men. Captain Marshall was "the oldest steamer Captain now in the Merchant fleet in these seas" and had been running to and through eastern waters since 1864. The second officer, the aptly named Fritz Showman, had been master of the celebrated Jardine Matheson sailing clipper *Fiery Cross.* Two other officers had transferred from transatlantic service. The second engineer came from Guion's *Arizona.* All told, there were more than one hundred crew aboard. Before departing from Shanghai, "the vessel will ship a double crew, the Captain's instructions being to run her 'full tilt all the way home.'"[60]

In mid-May the *Courier*'s Hankow correspondent predicted that the approaching *Stirling Castle* would soon load four thousand tons of tea at well over £7 (and possibly as high as £8) per ton, a figure which (as he reported later) "was too high to fill her up." Yet even a revised price of £6.10s was a good £2.10s above that offered to a rival Glen Line steamer and £3 above that for Holt's. The higher rate "no doubt finds many customers on account of the great speed she made during her outward voyage." Average betting was on a thirty-day passage, although one punter had opted for twenty-five days. Even the master "took a bet with a gentleman here [in Hankow] the other night, that he would be in Mr Skinner's London office on the 21st of June," equivalent to a thirty-day passage.[61]

The *China Mail,* however, received "private advices" from Hankow showing "that in deference to the unprecedented powers of the Sterling Castle [*sic*], all Tea steamers were quietly settling down to the prosaic and unsentimental but paying pace of 'say 10 knots an hour.'" And its correspondent confirmed that "the Glens are out of the race this year, for which I believe the captains are truly grateful." The *Mail* recalled remarks made by the Glen Line's managing partner that shipowners would "readily build ships with increased capacity for speed in proportion as merchants were desirous of paying for them." But Glen had declined Pearce's offer to build such a ship as the *Stirling Castle* "on the ground of the expensive coal-swallowing capacities of the vessel." Indeed, by the slowing down to ten knots of their own record-breaking tea steamer *Glenogle,* they might reduce coal consumption from more than 110 tons to as little as 35 tons per day compared to the 150 tons or more for Skinner's ship at full speed. The paper also relayed reports circulating in commercial maritime circles that the building and running of the *Castle* was "more like an experiment or trial trip," and concluded that it was "doubtful whether, under existing conditions of the China carrying trade, the powerful new steamer belonging to Messrs Skinner & Co. is the most suitable for the trade."[62]

In 1882 (with 50 stokers) and 1883 (with 111 stokers), the *Stirling Castle* steamed the 10,500 miles in just over twenty-nine days.[63] But Skinner's reputation was seriously weakened in 1882–83 by the loss of three steamers out of his eight-strong fleet, along with fifty-nine passengers and crew. The *Stirling Castle*, part-owned by her chief beneficiary Pearce, went to Italian owners soon after the conclusion of her second record-breaking voyage.[64] With Guion's transatlantic fleet, the pattern was very similar: Pearce supplied the expensive record-breakers, and the benefits accrued to him rather than to the managing owner. His real targets were the large lines, such as Nord-Deutscher Lloyd and Cunard as well as the Admiralty. Pearce was knighted in 1887, but his life was cut short at the age of fifty-five by a sudden heart attack in December 1888.

In the first three decades of Victoria's reign there was no single design of ocean ship that was a self-evident, long-term winner. Rival contenders, as mobile sites, would win or lose according to their respective performances, witnessed by select experts and duly reported in national or local newspapers and specialist journals. Fast-sailing clippers bringing home China tea or carrying gold-seekers to Australia had no rivals for speed and prestige. Brunel's radical "experimental" designs remained only "wonders of the age." Failing to win passenger confidence, they made their mark on space in very different ways from their projector's original vision—by the *Great Britain*'s role as auxiliary sailing ship to Australia and by the *Great Eastern*'s work as transatlantic cable-layer.

With the introduction of higher pressures in the compound steam engines, technological choices remained. Equating high pressure with minimizing waste, Holt's commitment to economy entailed a steadfast refusal to follow competitors into the race for greater speed and power. For Holt and his Liverpool Unitarian community, steamship lines meant regularity, economy, and reliability by direct analogy to the railway systems of the nineteenth century. Equating compound engines with that very speed and power, however, Elder's successor, Pearce, gave the Clyde a very different image as the birthplace of the ocean greyhound, fast and reliable but a vast consumer of the earth's material resources. With Pearce's ships, performance mattered: witnesses needed to see the facts for themselves in the form of power and speed. These mighty machines of the oceans had thus been shown to carry everything before them as they briefly made their mark on space—Atlantic records, tea races, even icebergs. But perhaps the final word should go to Sir Edward Harland, head of shipbuilders Harland & Wolff and contractor to the White Star Line, who predicted that "ere long the 'greyhounds' of the Atlantic, from a commercial point of view, would be proved to be an egregious mistake."[65]

NOTES

This chapter was made possible by funding from my Arts and Humanities Research Council Large Grant for the "Ocean Steamship Project" over the period 2001–7. I am indebted to all members of the Project team: Anne Scott, Ian Higginson, Phillip Wolstenholme, Trish Hatton, Don Leggett, Ben Marsden, and Will Ashworth. I am especially grateful to the archivists in the Merseyside Maritime Museum and Glasgow University Archives for their very positive support in every respect. Special thanks are also due to David Livingstone and Charles Withers for the invitation to the splendid Edinburgh conference, which saw a preliminary trial of this chapter, and to the many participants who offered suggestions for its development.

1. Sidney Pollard and Paul Robertson, *The British Shipbuilding Industry, 1870–1914* (Cambridge, MA: Harvard University Press, 1979), 15.

2. Ben Marsden, "The Administration of the 'Engineering Science' of Naval Architecture at the British Association for the Advancement of Science, 1831–1872," *Yearbook of European Administrative History* 20 (2008): 67–94. "Privileged sites" are analyzed in Crosbie Smith and Jon Agar, "Introduction: Making Space for Science," in *Making Space for Science: Territorial Themes in the Shaping of Knowledge*, ed. Crosbie Smith and Jon Agar (Basingstoke: Macmillan Press, 1998), 9–16.

3. See Steven Shapin, *A Social History of Truth: Civility and Science in Seventeenth-Century England* (Chicago: University of Chicago Press, 1994), 8, for "trust in affirmed actions."

4. Probably on this account, Michel Foucault, in "Of Other Spaces," *Diacritics* 16 (1986): 22–27, calls the ship the heterotopia *par excellence*. For discussion, see Smith and Agar, *Making Space*, 9.

5. Edwin Hodder, *Sir George Burns, Bart: His Times and Friends* (London: Hodder and Stoughton, 1890), 193–94; Stephen Fox, *The Ocean Railway: Isambard Kingdom Brunel, Samuel Cunard and the Revolutionary World of the Great Atlantic Steamships* (London: HarperCollins, 2003), never explores the ramifications of "railway" rhetoric.

6. Henri Lefebvre, *The Production of Space*, trans D. Nicholson-Smith (Oxford: Blackwell, 1991), 416–17.

7. Fox, *Ocean Railway*, 70.

8. Cf. Fox, *Ocean Railway*, 272–74 ("line of succession").

9. Two scholarly studies that have stood the test of time are Robert Greenhalgh Albion, *Square-Riggers on Schedule: The New York Sailing Packets to England, France, and the Cotton Ports* (Princeton, NJ: Princeton University Press, 1938), on sailing packets; and David Budlong Tyler, *Steam Conquers the Atlantic* (New York: Appleton-Century, 1939), esp. 154–68, on mail contract controversies.

10. "A Trip to America," *Engineering* 1 (1866): 337–38 (Bank of England reference); see also Crosbie Smith and Anne Scott, " 'Trust in Providence': Building Confidence into the Cunard Line of Steamers," *Technology and Culture* 48 (2007): 471–96.

11. See Stuart Nicol, *Macqueen's Legacy: A History of the Royal Mail Line*, 2 vols. (Brimscombe Port Stroud: Tempus, 2001); Freda Harcourt, *Flagships of Imperialism: The P&O Company and the Politics of Empire from Its Origins to 1860* (Manchester: Manchester University Press, 2006); William Schaw Lindsay, Journal, LND 35/2, pp. 9–10 (transcript), Lindsay Papers, Caird Library, National Maritime Museum, Greenwich. I thank Don Leggett for his work on mining the Lindsay Papers.

12. Ben Marsden and Crosbie Smith, *Engineering Empires: A Cultural History of Technology in Nineteenth-Century Britain* (Basingstoke: Palgrave Macmillan, 2005), 28–31; Alison Winter, " 'Compasses All Awry': The Iron Ship and the Ambiguity of Cultural Authority in Victorian Britain," *Victorian Studies* 38 (1994): 69–98 (compass deviation in iron ships); Tyler, *Steam Conquers the Atlantic*, 169–73 (Admiralty experiments on, and opposition to, iron plates); T. T. Vernon Smith, *The Past, Present and Future of Atlantic Ocean Steam Navigation* (Fredericton: Fredericton Athenaeum, 1857), 10–11, 17–18 (advocacy of screw steamers, especially their advantages when aided by sail over paddle steamers).

13. See, for example, Basil Lubbock, *The China Clippers* (Glasgow: Brown, Son & Ferguson, 1946), esp. 134–36 (composite construction), and *The Colonial Clippers* (Glasgow: Brown, Son & Ferguson, 1948), on Australian clippers.

14. Ewan Corlett, *The Iron Ship: The Story of Brunel's SS Great Britain* (London: Conway Maritime Press, 1990), 131–44; W. S . Lindsay, *History of Merchant Shipping and Ancient Commerce*, 4 vols. (London: Sampson Low, Marston, Low, and Searle, 1874–76), 1:428–29 (on auxiliary sailing vessels), 467–69 (on the Yangtze voyage). Powered by an 80 hp steam engine, Lindsay's *Robert Lowe* steamed the 608 miles from Shanghai to Hankow in ten days (including anchoring by night). The return, with the current, took only fifty-seven hours of steaming.

15. Smith, *Atlantic Ocean Steam Navigation*, 24. Smith also looked forward to the day when Atlantic liners would "give us an opportunity of dining one Sunday here [in British North America], and going to Church the next Sunday with our friends in England" (31). I thank Ben Marsden for this reference.

16. The promises and problems of the *Great Eastern* are summarized in Marsden and Smith, *Engineering Empires*, 101–7. See also George S. Emmerson, *John Scott Russell: A Great Victorian Engineer and Naval Architect* (London: John Murray, 1977), 65–157; Angus Buchanan, *Brunel: The Life and Times of Isambard Kingdom Brunel* (London: Hambledon, 2002), 113–33.

17. Quoted in Tyler, *Steam Conquers the Atlantic*, 189–90.

18. Ibid., 190.

19. Ibid. See also the discussions in Ben Marsden, "Blowing Hot and Cold: Reports and Retorts on the Status of the Air-Engine as Success or Failure," *History of Science* 36 (1998): 373–420, esp. 385–90; Marsden and Smith, *Engineering Empires*, 71.

20. Marsden, "Blowing Hot and Cold," 395–400; Marsden and Smith, *Engineering Empires*, 73–75, 113–16.

21. "The Screw Steamer 'Brandon,'" *Nautical Magazine and Naval Chronicle* 23 (1854): 507–9. See also W. J. M. Rankine, *A Memoir of John Elder: Engineer and Shipbuilder* (Edinburgh: Blackwood, 1871). The far-from-unproblematic introduction of Randolph and Elder's marine compound engine into PSNC service will form the subject of a separate study.

22. *The Edinburgh Academy Register: A Record of All Those Who Have Entered the School since Its Foundation in 1824* (Edinburgh: Constable, 1914), 106–12, 113–22; Johnston Fraser Robb, "Scotts of Greenock, Shipbuilders and Engineers, 1820–1920: A Family Enterprise," 2 vols. (PhD diss., University of Glasgow, 1993), 1:50–55 (on the family business in this period), 101 (Institution of Naval Architects).

23. Robb, "Scotts of Greenock," vol. 2, ship no. 40.

24. Scott & Co. to Lamport & Holt, 12 May 1857, Letterbook (1857–58), 24–25, GD 319/11/1/9, Scotts of Greenock Papers, Glasgow University Archives; Crosbie Smith, Ian Higginson, and Phillip Wolstenholme, "'Imitations of God's Own Works': Making Trustworthy the Ocean Steamship," *History of Science* 41 (2003): 379–426; see 399 for Holt's friendship with Scott.

25. Frederick J. Rowan, "On the Introduction of the Compound Engine, and the Economical Advantage of High Pressure Steam; with a Description of the System Introduced by the Late Mr J. M. Rowan," *Transactions of the Institution of Engineers and Shipbuilders in Scotland* 23 (1880): 51–97, on 52–53. The "defects" included the development of small holes in the tubes carrying the hot gases from the furnace through the square vertical water tubes that made up the boiler. See *Two Centuries of Shipbuilding by the Scotts at Greenock* (London: Offices of Engineering, 1906), 34–35; "Rowan and Horton's Steam Engines and Boilers," *Engineer* 10 (1860): 210 (Rowan and Horton's patent). Craddock's compound arrangements were somewhat dismissively treated in Rankine, *John Elder*, 25–27.

26. W. J. M. Rankine to James Robert Napier, 22 November 1858, DC 90/2/4/38, Napier Papers, Glasgow University Archives; Rowan, "Compound Engine," 52–59 (published versions of Rankine's 1858 reports); Marsden and Smith, *Engineering Empires*, 119 (*Thetis* summary).

27. Marsden, "Naval Architecture," 12–18.

28. "Editorial," *Engineer* 6 (1858): 357–58.

29. "Scottish Matters," *Engineer* 7 (1859): 375; "Liverpool Polytechnic Society," *Engineer* 7(1859): 381. See Smith, Higginson, and Wolstenholme, "Imitations of God's Own Works," 392 (Liverpool Polytechnic Society). Members of the society included the distinguished Henry Booth, former secretary of the Liverpool and Manchester Railway; iron shipbuilder John Grantham; and shipowners/merchants such as C. T. Bowring.

30. John Scott, "Notes of an Experimental Trial Made with Rowan's Patent Expansion Steam Engines," Privately printed sheets, DC 90/2/6/34, Napier Papers, Glasgow University Archives. See also Robb, "Scotts of Greenock," 1:508–10 (Begbie).

31. John Scott, "Abstract of the Performances of Screw Steamer 'Thetis,' Fitted with Rowan's Patent Expansion Engines, in the Glasgow and Liverpool Trade, between 14th May and 25th August, 1859," DC 90/2/6/34, Napier Papers, Glasgow University Archives. Scott's distinctive language of "dynamical power" and "dynamic results" echoes William Thomson's and Rankine's terminology, especially in "The Dynamical Theory of Heat" and "Thermo-dynamics." See for example Crosbie Smith, *The Science of Energy: A Cultural History of Energy Physics in Victorian Britain* (Chicago: University of Chicago Press, 1998), 107–25.

32. "Editorial," *Engineer* 9 (1860): 9.

33. J. Frederick Spencer, "Surface Condensation," (letter to the editor), *Engineer* 9 (1860): 22. Spencer was almost certainly referring to Rowan and Horton and to Randolph and Elder.

34. Ibid., 74. The exchange also focused on rival surface condensers.

35. Rowan, "Compound Engine," 64–76.

36. Ibid., 68–69.

37. Ibid., 69–70.

38. Ibid., 76, 69.

39. Ibid., 88–90 (Howden's contribution to the discussion). Howden's first contract for marine engines was for Hendersons' Anchor Line of Glasgow in 1859. These were apparently compound. In the mid-1880s he introduced commercially his (subsequently) famous system of hot-air forced draft in marine boilers. See "James Howden," *Marine Engineer and Naval Architect* 36 (1913–14): 165 (Howden's obituary).

40. Rowan, "Compound Engine," 90–93.

41. Smith, Higginson, and Wolstenholme, "Imitations of God's Own Works," 409–11.

42. Ibid., 401–11.

43. Alfred Holt, Circular Letter Dated 16 January 1866, Papers of Alfred Holt, 920 HOL/2, Liverpool Central Library; Smith, Higginson, and Wolstenholme, "Imitations of God's Own Works," 410–11.

44. Lindsay, *Merchant Shipping*, 4:434–35 (Lindsay's italics).

45. Alfred Holt, c. 3 September 1867 and c. 28 February 1868, Diary, Book A, Papers of Alfred Holt, 920 HOL/2/52, Liverpool Central Library.

46. Quoted in A. H. John, *A Liverpool Merchant House: Being the History of Alfred Booth and Company, 1863–1958* (London: Allen & Unwin, 1959), 57.

47. *China Mail* 27 April 1882 (trial trip); 4 May 1882 (the Delaware project); 30 May 1882 (offer to Glen Line).

48. N. R. P. Bonsor, *South Atlantic Seaway* (Jersey: Brookside, 1983), 146–47.

49. "Sir William Pearce, Bart, M.P.," *Marine Engineer* 10 (1888–89): 344 (obituary of Pearce); Anthony Slaven, "Pearce, Sir William, First Baronet (1833–1888), *Oxford Dictionary of National Biography*, Oxford University Press, 2004; online at http://www.oxforddnb.com/view/article/21692 (accessed 13 February 2007).

50. Rankine, *John Elder*, 67; Bonsor, *South Atlantic Seaway*, 147–49. PSNC's capital had doubled to £4 million in the same period. The latest PSNC liners were, at over 4,200 tons each, among the world's largest in the mid-1870s.

51. Peter Newall, *Orient Line: A Fleet History* (Preston: Ships in Focus, 2004), 13–15.

52. On the Guion Line and the *Arizona*, see R. A. Fletcher, *Steam-Ships: The Story of*

Their Development to the Present Day (London: Sidgwick & Jackson, 1910), 247–51; Fox, *Ocean Railway*, 282–89. Within two years of the dramatic collision, Pearce had begun to build similar-sized vessels for Nord-Deutscher Lloyd (NDL), one of the two premier German shipping lines on the Atlantic. NDL in fact took delivery of no fewer than seventeen similar liners from Fairfield between 1881 and 1891. The first nine vessels saw service speeds rise from fifteen to eighteen knots: see N. R. P. Bonsor, *North Atlantic Seaway*, 5 vols. (Jersey: Brookside, 1975–80), 2:551–56.

53. *Clydeside Cameos: A Series of Sketches of Prominent Clydeside Men* (Glasgow: Ranken, 1885), 31, 33–34. The same piece describes how Pearce extracted a high price from the Cunard Line for the contract for their *Umbria* and *Etruria*.

54. Malcolm Cooper, "Thomas Skinner's Castle Line," *Ships in Focus* 24 (2003): 250–59, esp. 252; *O. S. S. Co. General Book* (1865–1882), Ocean Archives, Merseyside Maritime Museum, Liverpool (statement of cost of each new Ocean Steam Ship Co., steamers in 1880).

55. "Trial Trips," *Marine Engineer* 4 (1882–83): 30.

56. Ibid. The report also carried Skinner's speech as managing owner.

57. *China Mail*, 27 April 1882. Extracts from the *China Mail* (and other Far East newspapers) relating to the steamer tea races are collected in "Tea Races, 1877–94" (unpublished typescript), OA/1B/68, Ocean Archives (Glen Line), Merseyside Maritime Museum. I am grateful to Phil Wolstenholme for researching the Glen Line archive and locating this typescript in March 2002. I also thank Ian Higginson for his contributions to our Project workshops on Far East shipping. Alfred Holt's critical perspective on fast steamers is summarized in Crosbie Smith, Ian Higginson, and Phillip Wolstenholme, "'Avoiding Equally Extravagance and Parsimony': The Moral Economy of the Ocean Steamship," *Technology and Culture* 44 (2003): 443–69, esp. 467–68.

58. *China Mail*, 28 April, and 4 and 22 May 1882, in "Tea Races, 1877–94." The reports of 4 and 22 May are reprinted from the *Singapore Times* and *Shanghai Courier*, respectively.

59. Ibid., 4 May 1882.

60. Ibid.

61. Ibid., 22, 23, and 30 May 1882 (replicating the *Shanghai Courier* of several days earlier).

62. Ibid., 30 May 1882.

63. See David R. MacGregor, *The China Bird: The History of Captain Killick and the Firm He Founded, Killick Martin and Company* (London: Chatto and Windus, 1961), 187–88. MacGregor quotes a later historian who in 1947 wrote of the 1883 voyage that the monsoon was "dead in her teeth" on the run to Suez. There were "stories to the effect that her funnels were red hot on more than one occasion. . . . At short intervals unconscious firemen would be hauled up in slings, revived with buckets of water and then [involuntarily] hustled below again to continue the never-ending task of feeding her thirty-six furnaces."

64. Cooper, "Castle Line," 252–53.

65. "The White Star Line," *Marine Engineer* 7 (1885–86): 173–74, at 174. Harland & Wolff designs promoted comfort and space rather than speed.

CHAPTER ELEVEN

Expeditionary Science: Conflicts of Method in Mid-Nineteenth-Century Geographical Discovery

LAWRENCE DRITSAS

I am thankful to old Nile for so hiding his head that all "theoretical discoverers" are left out in the cold.
—David Livingstone, August 1870[1]

During the middle decades of the nineteenth century, the British geographical community was absorbed by attempts to determine the source of the Nile. The rhetoric of the day framed the question as ancient, and in that sense British geographers were continuing the great tradition.[2] But over the course of the later 1840s and throughout the 1850s and 1860s, there was increasing excitement that the question would soon be answered. The solution mattered for reasons of commerce, geography, and national prestige. This chapter uses the context of the search for the source of the Nile to highlight an important moment in the history of geography as a science: the debate that placed "theoretical discoverers" (also known variously as "critical geographers," "armchair geographers," or "carpet geographers"), who worked with sources of text and testimony, in opposition to explorers who traveled to the field and used instruments and direct observation. My purpose in recounting this history is to describe the less-familiar methods of critical geography and to examine the debates held over the credibility of this form of geographical discovery in relation to exploration. The elaboration of this theme should also serve to indicate the located nature of the various practices of geographical discovery in the nineteenth century—linked as they were to the spaces of expedition, archive, discussion, and publication—and to demonstrate how geographical knowledge was constructed across and between these spaces, using social and methodological processes concerned with negotiating credibility—the "here and everywhere" nature of geographical discovery.[3] My concern, then, is with linking the geographies

of science in one particular context with the emergent science of geography
as it brought to bear its methods and procedures upon a pressing question in
the mid–nineteenth century.

Before beginning this study, I must note that the search for the source of
any river depends ultimately upon which definition of a source is used. At
each bifurcation of a river, the criteria employed to choose which tributary
is the main channel are not obvious. The width of a channel and its volume
of annual flow are two indicators that may have different results. Furthermore, the source could be defined as the point furthest away from the river's
mouth from which water flows to it. This arbitrary nature of determining a
river's source was not lost on the Victorians. In the discussion held in 1859
at the Royal Geographical Society (RGS) in London after a paper by James
MacQueen on the geography of central Africa, William Henry Sykes, a naturalist and fellow of the Royal Society argued that "every great river has
more than one source."[4] He then used the fingers on his hand as an analogy, all joining at the wrist to form the arm. Likewise a few years later, the
travel writer Laurence Oliphant acerbically compared the arguments about
the Nile's sources with simultaneous debates about that other great issue of
the early 1860s, the origin of species: "The 'species' can't discuss its 'own
origin,' without becoming so violently excited as to endanger its peace of
mind; and if it is any satisfaction to those who are still maintaining a bitter controversy as to 'the source of the Nile' to hear it, we can assure them
that they may fight about it for ever, for it is as impossible to discover in a
precise form the source of a mighty river as the origin of a race."[5] Samuel
Baker, who explored the regions north of Lake Victoria with his wife Florence in the early 1860s, followed such sentiments and wondered if the goal
were unobtainable. "I believe," he said, "that the mighty Nile may have a
thousand sources."[6]

Despite acknowledgments that the source of the Nile might not be
knowable, the question was yet too tantalizing to ignore. In the early 1850s,
the map of central Africa between 4° north to about 20° south latitude and
20° to 38° east longitude was largely blank. From within this uncharted
space flowed the Nile, the Zambezi, and the Congo. There was a strong
desire to map precisely these interwoven drainage basins. In 1855, a map
was displayed at the RGS that placed a vast inland sea in the centre of the
continent[7] (fig. 11.1). The three missionaries who had produced the map,
Ludwig Krapf, James Erhardt, and Johann Rebmann, had been living about
twenty-five miles inland from Mombasa since 1846. Their map collated
information from many travelers to suggest the existence of an inland sea.
This map, considered ridiculous by many who saw it at the time, nonethe-

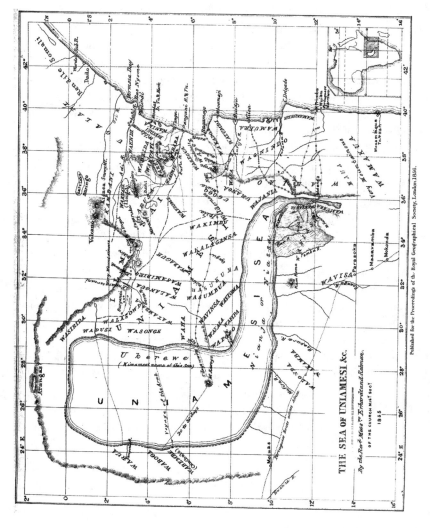

THE SEA OF UNIAMESI &c.

By the Rev.ᵈ F. Mᵍⁱⁱ Dʳ E. Harditz and Rebman.

OF THE CHURCH MISⁿ SOCⁱᵉ

1855

Published for the Proceedings of the Royal Geographical Society, London, 1856.

Fig. 11.1. The infamous "slug map," which accompanied James Erhardt's article "Reports Concerning Central Africa, as Collected in Mambara and on the East Coast, with a New Map of the Country." *Proceedings of the Royal Geographical Society of London* 1 (1855–57): 8–10.

less had the effect of increasing the interest in the region and pushed the search for the source of the Nile into its climax phase of the later 1850s and 1860s.

A further source of inspiration for those who would map the Nile's upper reaches was the existence of so many ancient Latin and Greek sources, and more recent ones written in Arabic and Portuguese that provided information about the region. By comparing and contrasting many such sources, the "theoretical discoverers" attempted to determine the source of the Nile from within their libraries. British geographers also interviewed anyone whom they felt might have some direct or indirect geographical information to convey, but these sources were, as we shall see, difficult to use. The interior of central Africa was in this way approached differently than the truly *terrae incognitae* of the New World, upon which the ancients had been silent. In the middle of the nineteenth century, the information provided by ancient geographers about the African interior was still very relevant and was discussed alongside modern and contemporary sources.[8]

The question of the Nile's sources began to be unraveled in the mid-1870s with a series of expeditions made by Verney Lovett Cameron and Henry Morton Stanley that carefully mapped Africa's great lakes.[9] But the search for the source of the Nile should not be seen as the slow triumph of direct observation over wild speculation. In this chapter the story is approached as a dialogue between methods: those used by explorers and those used by the "critical geographers"—as they were termed at the time. It was the latter methods of "theoretical discovery" that Livingstone resented, preferring the on-the-ground methods of exploration: dead reckoning, astronomical positions, careful recording of paths taken and distances traveled, and, above all, direct observation. I analyze these disagreements of method in detail. But first, in order to understand the conflicts, it will be necessary to take a brief look at the sources and less-familiar methods the critical geographers used to determine the source of the Nile: the analysis of ancient, Arab, Portuguese, and African written and oral sources.

THE SOURCES AND METHODS OF CRITICAL GEOGRAPHY

The most frequently mentioned ancient accounts of the source of the Nile in the debates of the mid–nineteenth century are those of Ptolemy.[10] The rediscovery of his work in the early fifteenth century through its translation into Latin by Jacobo d'Angelo was a revelation to European cartographers. There was, however, a paradox in the Renaissance cartography of Africa that persisted almost unchanged until the question of the source of the Nile was solved in the 1870s. As Francesc Relaño puts it,

> The Portuguese exploration along the west coast of Africa and the revival of Ptolemy were contemporaneous events. The Portuguese added to knowledge of the continent from perceptual experience—through what they saw or thought they saw—whereas the Renaissance mapmakers' reading of Ptolemy led them in the opposite direction, back to classical times, as they made the almost obligatory references to the ancient author. Thus fifteenth-century cartographers were faced with two different and contradictory sources: what the classical authors said and what modern mariners and pilots claimed to have seen.[11]

The result of this contradiction was that the outline of the African continent was drawn ever more accurately due to recent observations of explorers, while Ptolemaic information about the interior of the continent continued to be reproduced *inside* the modern outline. The three most commonly represented Ptolemaic attributes were the presence of two lakes at the same latitude near the equator, the so-called mountains of the moon associated with the source of the Nile, and the source of the Nile being located south of the equator. These attributes were found on maps from the fifteenth through the early eighteenth centuries until empiricist cartographers led by Jean Baptiste Bourguignon d'Anville cleared the African interior of unsubstantiated features, leaving it almost blank.[12] This "positive geography" left great gaps to be filled, but the sources available to fill in the blanks could not be substantiated and cited, thus leaving problems for mapmakers.[13]

A leading analyst of this problem in the early nineteenth century was William Desborough Cooley. Active in the RGS from its beginning, Cooley went on to found the Hakluyt Society in 1846. A lawyer by training, he had begun to write on travel and exploration for the *Foreign Quarterly Review* and the *Atheneum* in the 1820s.[14] During the 1830s, 1840s, and 1850s, Cooley endeavored to utilize all available sources to piece together the geography of the African interior through a meticulous process of cross-examination and corroboration. He never visited Africa.

Cooley's strongest analysis of ancient sources was published in 1854. In *Claudius Ptolemy and the Nile*, he directed his readers to use, but also to doubt, ancient authorities; he wrote, "Geographical statements deficient in recognizable truth soon become unintelligible and consequently worthless."[15] His target was the famous "mountains of the moon." The accepted interpretation at the time was that Ptolemy indicated a mountain range, the mountains of the moon, which were located near the equator in central east Africa. From these mountains flowed the first waters of the Nile.[16] Soon after leaving these mountains, the rivers entered two large lakes which were, in turn, reservoirs from which flowed the main tributaries to the Nile. Cooley attacked this claim through philology, textual criticism, and comparison to other sources. He argued two iconoclastic points in the book. First, Ptolemy was always conveying information about the region around the Blue Nile in modern-day Ethiopia, not the White Nile of central Africa. Second, Ptolemy himself never mentioned the "mountains of the moon"; the phrase was inserted by later Arab geographers in the tenth or eleventh century. The scholarship was solid and convincing, but there were equally strong dissenting opinions.[17] Despite Cooley's work, references to the mountains of the moon returned again and again in the 1860s, particularly in the writing of Richard Burton.[18] Even though there was disagreement about what Ptolemy knew about central Africa, Cooley did show that the ancient and classical sources for the geography of central Africa were less reliable the further south one went. The next places to turn for information were mediaeval Arab geographers and later accounts by Portuguese travelers.

Early accounts of sub-Saharan Africa in Arabic were largely influenced by Ptolemy. This changed after the tenth century because of the spread of Islam south into west Africa and the appearance of travel accounts with descriptions of the continent's river systems.[19] Cooley had used these sources for *The Negroland of the Arabs* (1841), which was in many ways a masterpiece of critical geography.[20] In the introduction Cooley advised his readers that in order to understand the Arab texts, it was necessary to compare them with present information and to use logic to remove the falsities; what remained would be the best truth that could be known. To follow this method, Cooley began by removing any doubtful information from his geography. He then read every source he could, checking each for internal consistency, and then read all the sources against one another, bearing in mind more recent discoveries. This method also allowed for intervention into the texts that were studied. That is to say, Cooley adjusted any distances reported for a day's travel that he thought were excessive in others' accounts by comparing these distances with more recent reports made by British travelers who experienced similar conditions. Such meticulous, painstaking work was only possible due to Cooley's skills as a linguist. *Negroland* was received as a "great achievement."[21] Other scholars pursued similar lines of research with Arab sources and used similar methods. Charles Tilstone Beke and Antoine Thomson d'Abbadie, for example, had both traveled in Abyssinia and collected much oral information about the Nile's southern reaches. They used this experience to inform further analyses of ancient and Arab sources about the source of the Nile. Despite their use of similar methods, they disagreed in the late 1840s about whether the Nile's sources were above or below the equator.[22]

Further sources of geographical information were the accounts written in Portuguese, but these were seldom definitive. Despite the long residence of Portuguese settlers on the coast of west and east Africa and inland along the Zambezi and Sofala rivers, the Portuguese sources available to British geographers were regarded as lacking the epistemic qualities from which to draw firm conclusions. Of particular interest to those engaged in the Nile question were many tales of great lakes that were found in the interior of the continent. In his study of these reports Cooley wrote, "The object proposed in this paper is to collect and compare the several statements extant respecting the great lake in the interior of Africa, to determine their true meaning and value, and thus, with the aid of new particulars derived from original sources, to endeavour to establish the geography of that region on a firm and consistent basis."[23] The available Portuguese accounts were difficult to use for two reasons. First, as they were often narrative descriptions of journeys taken for trading purposes, they lacked consistent evidence of the use of astronomical positioning by sextant that empiricist geographers desired. Second, the accounts often appeared decades after the journeys took place. Nevertheless, these sources were attractive; they were produced by Europeans and often presented a wealth of detail. Critical geographers sought ways around the shortcomings. Reading deeply into the account of two Portuguese army officers who traversed the continent, James MacQueen, for example, estimated daily distances traveled by comparing the description of the terrain given in the account with his own experience of traveling or with the reports of others who traveled in similar conditions.[24] Later, in 1859, MacQueen used the recent data supplied by Burton and Speke, along with the 1853–54 journey of Silva Porto, to construct a new map of central Africa. He was able to "fix" positions on Porto's route by corroborating them against those made with instruments by Burton and Speke.[25] In 1855 Cooley uncovered a travel narrative dating from 1843 of a journey made through central Africa by the Portuguese Joachim Graça, but in analyzing the details, he found that it was difficult to rectify this account with those of David Livingstone and Robert Moffat then appearing.[26]

In further support of their textual analyses, and when the chance arose, the critical geographers interviewed informants who might have further information about the Nile. Such oral testimony was difficult to work with and was not a substitute for direct observation, but nonetheless it could lead to successful conjectures. Earlier in the century, James MacQueen correctly suggested—his contention was by direct observation—that the river Niger flowed into the Atlantic, and did so on the basis of interviews with slaves in Grenada and textual analysis.[27] It is clear in reports about oral testimony that if it was to be useful, its collection and interpretation required the use of careful methods as a sanction of its credibility. Beke applied just such methods and safeguards when he made extensive use of local informants for his 1843 paper "On the Countries South of Abyssinia." When visiting the Abyssinian market town of Yejubbi, Beke interviewed numerous travelers. It was precisely because he had interviewed so many that he could claim to construct a reliable geography: "[In Yejubbi] I had frequent communication with individuals of all tribes, . . . who had visited all parts of the Galla country and the adjoining states; and from them I obtained a mass of information, which, although sometimes differing in minor details (a circumstance which was expected), is, in all the main points, perfectly consistent, and in various parts mutually corroborative."[28] Beke carefully named his informants and described their backgrounds. He noted their intelligence and their experience of distant travel. In these ways he allowed his readers to follow his method by indicating not only where something was said, but how it was said as a basis for its warranted credibility. In this way, in addition to the method of corroborating multiple oral sources, Beke placed emphasis upon analysis of the person as a truthful agent: "With respect to the information furnished to me by Omar, I am bound to say that I have every reason to give him credit for veracity. He answered all my questions with the greatest readiness, explained cheerfully any apparent discrepancies, and sometimes called on me to say that he had been speaking to Ali (who frequently took part in our conversations), and found that he had been mistaken in something he had told me, &c."[29] As others have shown, the collection of geographical information via oral testimony was a process structured by assumptions about class, race, education, and language. To be credible, such information required assessment of both the interviewer and the interviewee.[30]

In applying these methods, critical geographers were concerned to build the best maps that they could with the available information. They were also concerned to indicate where relevant geographical information already existed—making the determination of the Nile's sources an archival problem as much as it was a problem of how to visit remote places, observe them appropriately, and return safely.[31] These two approaches could be compatible, but, as we shall see, they could also result in conflict.

"CRITICAL GEOGRAPHY OR ACTUAL OBSERVATION"

The methods of the critical geographers were acknowledged to be imperfect. Summing up developments concerning the Nile before 1847 in the *Journal* of the RGS, Charles Beke wrote that "actual results" obtained by travelers were still insufficient to determine the location of the source of the river and cautioned therefore that the methods of the critical geographer must be used, guardedly:

> [That] speculation [on the source of the Nile] must still, at times, come in the aid of facts is unavoidable; but it will be our endeavour to confine this speculation within legitimate bounds, and to limit it, indeed, to the reconciling of seemingly contradictory statements and to the arranging and combining of isolated and unconnected facts, where actual information is still insufficient and unsatisfactory. To say that we shall, on all points, come to definite results, is more than is warranted by the imperfect nature of the premises.[32]

If the solution was that more explorers needed to visit the region, this too would result in problems. Critical geographers were aware that direct observation by trained observers was a reliable—if not the best—method. Yet the time and energy that critical geographers committed to their exegetical work encouraged the belief that their methods were approximating the truth. It was in part because of their personal interests in these methods that the results of critical geography could not be simply set aside by any field observation. There was, rather, an iterative process of analysis and debate that brought the two types of results together. This process is revealed by examining the reception of the results of expeditions through consideration of the processes by which "in-the-field" information was amalgamated with the corpus of knowledge produced by critical geography.

One such example concerns the reception of one seemingly insignificant piece of information, the hydrography of the northern end of Lake Nyassa (now Lake Malawi) as presented by David Livingstone's Zambesi Expedition. This expedition was funded by the government and sent out with a broad brief steeped in the ideology of the "civilising mission." The members of the expedition collected data and specimens relating to botany, zoology, geography, geology, geomagnetism, ethnography, and economics, and even political intelligence.[33] The fieldwork lasted from early in 1858 until the middle of 1864, when Livingstone returned to Britain, the last of the expedition's members to do so. Much of the scientific work appeared across a wide range of specialized journals. Dispatches and letters from the field were frequently read aloud at meetings of the Royal Geographical Society, and as the explorers arrived home, they attended the meetings themselves and presented their findings. What transpired at one such meeting, held on 13 June 1864, provides the setting for this first example of the differing reception of contrasting geographical information.[34]

The meeting was choreographed to stimulate interesting discussion. Two maps were on view that represented the same object: Lake Nyassa and its position in south-central Africa (fig. 11.2). It was obvious to everyone in the room that the two maps differed in their representation of the position, size, and orientation of the lake. The first map to be discussed was based on Cooley's research.[35] The second map was based upon the recent findings of the members of the Zambesi Expedition who had visited the lake and sailed on its waters. In the room representing those explorers was Dr. John Kirk, recently returned from central Africa and the expedition's surgeon and naturalist.

Cooley was the first to read his paper, presenting his views on the geography of central Africa. The chair of the meeting was the leading member of the Royal Geographical Society, Sir Roderick Impey Murchison. The obvious differences between the maps concerning the orientation of the lake were such that Murchison foresaw the potential difficulty for Cooley's conclusions in the light of the expedition's recent and direct observations. Stepping in before the audience heard the latest findings from the Zambesi Expedition, Murchison requested that "all deference should be shown to Mr. Cooley's powers as a critical geographer, for he was sure the Society desired to do justice to every man, whatever his labours might be, whether in critical geography or actual observation."[36] The RGS secretary, Clements Markham, then proceeded to read out Livingstone's most recent report, which was accompanied by a map drawn by John Kirk. Discussion followed the papers. Murchison began the debate by stating the obvious: "[There was] a great discrepancy between the observations of the Portuguese who visited the country many years ago, and the *de facto* recent observations of Dr. Livingstone and Dr. Kirk."[37] This discrepancy laid the foundation for the heated debate that then ensued about the geography of the region. The discussion continued with comments by John Kirk and the explorer John Speke, together with some of the great critical geographers of the RGS such as James MacQueen, Charles Beke, and, also present for the occasion, Francis Galton. Although the critical geographers argued that direct observation was inherently limited to what the explorers themselves saw directly, Kirk's *de facto* description of the lake's dimensions and situation were regarded by

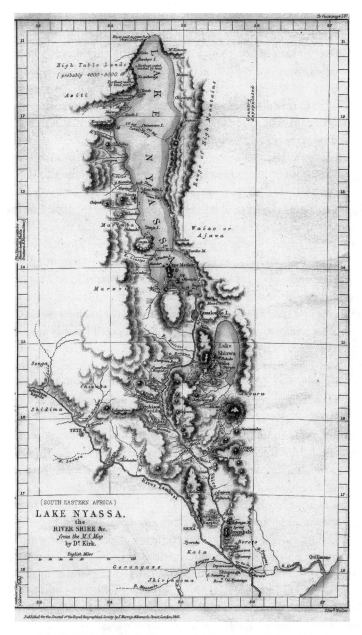

Fig. 11.2. John Kirk's map of Lake Nyassa, which accompanied David Livingstone's "Letters from the Zambesi to Sir R. I. Murchison, and (the Late) Admiral Washington." *Proceedings of the Royal Geographical Society of London* 8, no. 6 (1863–64): 256–63.

the others then present as being sufficiently thorough to be accepted over Cooley's analyses of Portuguese manuscripts. Confirming this view, Murchison drew the discussion to a close by restating the society's ideology of field-based exploration: "When gentlemen go into such countries, risking their lives to search out the truth and making astronomical observations which fix latitudes and longitudes it is obvious that all preceding accounts, derived from Portuguese and Arab travellers who did not make such observations, must give way to facts."[38]

But Kirk's victory was not total. Direct observation would only succeed as a geographical method where and when it occurred properly, the critical geographers argued. Kirk, Livingstone, and their fellow explorers had succeeded in mapping the general position and outline of the lake, but they did not themselves actually observe the northern end of Lake Nyassa (fig. 11.3). For this information they relied instead upon local testimony, provided in this case by Chief Mankambira, who lived about forty-five miles south of the northern end of the lake. This failure by the expedition's members to observe directly the northern limit of the lake meant that the question of whether or not a river flowed into or out of the lake at that end, and, therefore, whether or not any such unseen river connected Lake Nyassa to Lake Tanganyika remained open so far as the critical geographers were concerned. Moreover, reliance in the field upon such testimony had the effect of confirming rather than denying their critical forms of evidence and argument. Using local testimony was an accepted component of the methods and standards of critical geography, as we have seen, but Chief Mankambira was only one source, and multiple informants were preferred. Kirk and Livingstone believed that only a small river entered the north end of the lake and that this had no connection with Lake Tanganyika, further to the northwest. In the published account of the Zambesi Expedition, they provided the reasons by which they judged Mankambira to be a reliable informant: "Mankambira had never heard of any large river in the north, and even denied its existence altogether; giving us at the same time the names of the different halting-places round the head of the lake, and the number of days required to reach the coast opposite his village; which corresponded, as nearly as we could judge, with the distance at which we have placed the end [of the lake]."[39]

The question about the direction of the river and if it connected the two great lakes was of great interest because it was directly relevant to the much wider discussion of the search for the source of the Nile. By now, the Nile "controversy" had reached a high point, and any potentially relevant piece of hydrological or geographical information was scrutinized thoroughly and

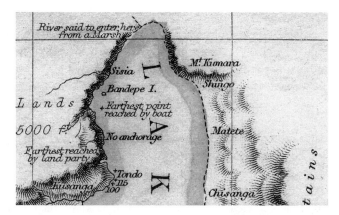

River said to enter here from a Marsh

Sisia

Bandepe I.

M. Kumara

Shungo

L a n d s

5000 ft.

+Farthest point reached by boat

No anchorage

Matete

Farthest reached by land party

kusanga

Tondo
+115
100

Chisanga

t a i n s

Fig. 11.3. Detail of Fig. 11.2, the northern end of Lake Nyassa, showing the limits of direct observation by members of the Zambesi Expedition.

debated energetically. The explorers of the Zambesi Expedition knew that failing to reach the end of the lake themselves would leave a gap in their data. This was the reason they tried so hard to get there so as to see things for themselves. They knew, too, that any information they did acquire would be compared, contrasted, and correlated with such knowledge as had already been accumulated and collected by the explorers Richard Burton, John Hanning Speke, James Grant, and Samuel Baker, who had been exploring the regions further north, *and* assessed in relation to the voluminous work of the critical geographers.

The still-remaining and key question in 1864 was whether Lake Tanganyika drained to the north, which would connect it to the Nile, or to the south, which would connect it to the Zambezi. This is why it was important to know if there was a large river at the northern end of Lake Nyassa. We now know that Lake Tanganyika has an outlet to the west, to the Congo, but in the 1860s this idea did not receive much attention and was excluded from debate.[40] After the public discussion between Cooley and Kirk, the question resurfaced elsewhere. At the 1864 Bath meeting of the British Association for the Advancement of Science (BAAS), Murchison, chair of the geographical section (section E), referred to the Nile question as one of the greatest geographical puzzles of the time and proceeded to promote the role of the critical geographers as those whose work posed the questions that most needed answering. Endorsing there the validity of Cooley's analyses, Murchison asked the audience at the section E session, "Are we not at this moment most anxious to determine, by positive observation, whether there exists a great series of lakes and rivers proceeding, as Cooley has suggested,

from Tanganyika on the north to Lake Nyassa on the South?"[41] Despite losing out to Kirk's *de facto* observations on the situation of Lake Nyassa,
Cooley's idea that Lake Tanganyika drained south into Lake Nyassa via
a river connecting the two remained a plausible theory that needed to be
tested. Chief Mankambira's testimony and Livingstone and Kirk's assessment of it was not enough to rule out a river entering the north end of Lake
Nyassa.

Modern readers may wonder at the ability of homebound geographers to
contradict field reports, especially, as was the case at the RGS meeting, since
the recently returned explorers were in the room, but two concerns made
such discussions necessary. First, the scientific credibility of the RGS relied
in part "upon the knowledge it produced being open to rational public scrutiny."[42] Such debates were intrinsic to the way the RGS made knowledge
and decided what and who were credible. Second, because the geography of
Africa was known to ancient authorities and, more recently, to Portuguese
settlers, discussions within the RGS of recent observations necessarily debated the balance of authority between what Murchison called "critical
geography" and actual "observation." Explorers were permitted extensive
authority only when they directly and *appropriately* observed geographical phenomena and when their observations were not totally at odds with
any previous, credible observations. Such "ocular demonstration," as others have shown, was powerful evidence, but it was not immediately and
unequivocally deemed complete in terms of geographical fact and regional
description, since by its very nature it was limited to the explorer's line-
of-sight.[43] In any analysis beyond reporting direct observations (direct observation in this case meaning observed firsthand by Europeans of credible
status), critical geographers of long-standing respect, such as MacQueen
and Cooley, possessed an equivalent authority in the corroboration of other
forms of geographical evidence.

Observed facts could only be regarded as such when they were warranted
elsewhere. It was at the stage of metropolitan analysis, when the reports of
expeditions were discussed and disseminated, that the practices of expeditionary science were subject to domestic inquiry. But in such cases and
places, critical geographers could also combine any such new data with past
observations in order to perform the kind of textual comparisons that underlined geographical methods. This combination of processes could have
the effect of relegating geographical explorers to the status of "mere" botanical collectors, who sent dried specimens to herbaria where the experts
analyzed them under appropriately controlled conditions.[44] Many geographical explorers resented such associations between "in-the-field" empiri-

cism, the time-limited nature of such study, and the truth claims that could be made of their work. Explorers understandably felt that they could infer much about a place or region from time spent in the field, even if they did not directly observe something. Richard Burton certainly bristled against the critical geographers of the RGS:

> Modern "hinters to travellers" direct the explorer and the missionary to eschew theory and opinion. We are told somewhat peremptorily that it is our duty [as explorers] to gather actualities, not inferences—to see and not to think; in fact to confine ourselves to transmitting the rough material collected by us, that it may be worked into shape by the professionally learned at home. But why may not the observer be allowed a voice concerning his own observations, if at least his mind be sane and his stock of collateral knowledge be acceptable?"[45]

Accepting the validity both of Burton's disinclination merely to collect data and the methodological concerns of the critical geographers, hindsight allows us now to review the situation of the science of geography being forged in debates over competing methodologies. The problem with the hydrography of central Africa around 1860 was that no one person or interest group had enough information to win any argument categorically. The disagreements here exposed reveal a deeper division among contemporary geographers over methods, standards of evidence, and institutional authority. Concerning debates over the northern end of Lake Nyassa—which nobody directly involved in the debate had ever seen—the "actual observers" and the "critical geographers" were not working in a common intellectual environment in terms of their understanding of the geographical problem in hand. Different practitioners used, and were committed to using, different types of indirect observations upon which to base their claims. Given their reliance either on text or on personal experience, different geographical authorities were not able to coordinate their responses to the available information about the region; they differed because they disagreed over the validity of one another's data and methods. Kirk and Livingstone were deeply aware that they had not directly observed the northern end of Lake Nyassa, but they nonetheless felt that they possessed enough direct experience of the lake as a whole to infer that a large river did not enter the northern end. Native testimony could be called upon as additional support to their views. Conversely, Cooley and other critical geographers placed strong faith in the texts of Portuguese and Arab travelers and questioned explorers' inferences beyond direct observation. Their detailed method of textual exegesis and cross-source corroboration was designed to produce credible geographical information. There were, in other words, strong social interests at work in shaping geographical knowledge; interests that concerned the credibility of methods and the credibility of the analyst. Critical geographers were concerned to protect their methods in part because they were concerned to protect the role of the "at home" metropolitan analyst in the construction of geographical knowledge. Read this way, Murchison's caveats and pleas for civility at the beginning of the 1864 RGS meeting appear as what they were: attempts to keep the discussion focused upon comparing and contrasting the different observations that were available concerning the geography of central Africa.

A similar debate occurred over a river at the northern end of Lake Tanganyika, which an RGS-funded expedition led by Richard Burton and John Hanning Speke had visited in early 1858.[46] Their purpose was to find out if the lake was a source of the River Nile. Because of logistical difficulties, Burton and Speke never made it to either the northern or southern end of the lake. As a result, the critical information concerning the presence of rivers entering and exiting the lake had to be acquired by interviewing local informants. One crucial informant in this case was Sheik Hamed bin Sulayyim, who lived on an island off the western shore of the lake across from Ujiji. Speke visited the sheik without Burton, to interview him about trade routes and geography. What Hamed bin Sulayyim told Speke, or, more precisely, what Speke reported he was told about the rivers entering and exiting Lake Tanganyika, was evaluated and reevaluated for years afterward because it was crucial to determining the hydrography of the region.[47] In essence the Tanganyika problem centered upon the consequences of the rivers' directions of flow. If a large river ran into the north end of the lake, then Lake Tanganyika was not part of the Nile system. Conversely, if a river exited the northern end, then it could contribute to the Nile. River systems mattered for their commercial potential as much as for their geographical importance. John Speke argued that the northern end of Lake Tanganyika was surrounded by mountains and only small rivers entered it. In this way it followed that Lake Victoria, which he claimed to have discovered, was the more likely source. Burton opposed Speke's opinions about Lake Victoria; he thought that there were no mountains surrounding the northern end of Lake Tanganyika and that the lake drained to the north, possibly contributing to the Nile's waters.

The details of this argument need not concern us here; they are tortuous. What is important is the status of local testimony in the debate. The issue at hand was whether or not local information was credible, or if it could be dis missed. For instance, James MacQueen argued that Africans often spoke of rivers running in the opposite direction to their true course and that it was valid to disregard their testimony when making maps in the light of conflicting reports and historical research because "we have so much experience of this mode of reversing the course of the river by native Africans."[48] And Speke decided in the years after his interview with the sheik that his testimony about the direction of particular rivers around Tanganyika could indeed be reversed if doing so would help corroborate other evidence in the circumstances.[49] Informants could be correct about one thing and wrong about another, and thus the testimony could be selectively edited when needed.

Who the informant was constituted yet another factor in the acceptance of testimony, together with what they said and how they said it. In the writings of the members of the Zambesi Expedition, many types of "locals" can be identified, and their origin or status contributed to the explorers' recognition of them as informants. In brief, the more "Portuguese" a local appeared to the Zambezi explorers, the greater their potential credibility; in other words, phenotype mattered. Knowledge of the Portuguese language was also an important signifier of status.[50] In east Africa, Speke and Burton differed with each other over which type of local (and of which race) best provided accurate testimony. Burton preferred those with whom he could speak in Arabic, which he spoke freely, or Swahili.[51] Speke did not posses Burton's formidable linguistic skills and, through interpreters, used any testimony that would fit with his own ideas of the region's geography as drawn from his direct experience of the terrain.[52]

The modulation of local testimony to fit particular hydrographical theories and thus, particular views about Africa's geography, should not be dismissed as bad practice. We need to set these practices in their contemporary epistemic and institutional context. African informants presented huge epistemological problems for British geographers. The societies visited by Burton, Speke, Kirk, and Livingstone were almost totally unknown in Britain. Little could be said about their representatives' status as truth-tellers: their languages were unknown, and their histories were unwritten. How was this "vulgar knowledge" to be included, if at all?[53] Sharing knowledge successfully involves sharing common concepts, values, and standards of truth.[54] There was little such common ground between British geographers and the local residents of central Africa, just as there was no shared ground within the geographical community. Meanwhile, the manuscripts, atlases, and travel narratives perused by the critical geographers were well accepted sources found in libraries and archives. There was no trouble in bringing these forms of evidence into play. Moreover, the textual sources of critical geographers could be quoted and made to enter discussion in spaces of debate in ways that interviews with African informants could not. Testimony was always viewed at a distance via an interlocutor whose own credibility was subject to scrutiny; even explorers could occasionally be caught in a lie.[55]

Recognition and reconciliation of these epistemological problems lay in the methods of corroboration advanced by critical geographers. Where informants were interrogated and shown to have beliefs similar to those held by the explorers as a result of direct observation, they might be trusted to "fill in" the spaces that the explorers themselves did not view directly. This language of mediation is evident throughout the narratives of exploration. Burton refers to his "stock of collateral knowledge." Kirk argued that because Chief Mankambira knew much of what they already had observed themselves, his testimony on the unseen northern end of the lake was "tolerably definite concerning rivers entering the lake on the north-west."[56] Speke went to great lengths to check that his informants agreed with his own findings and the findings of other informants.[57] He would dismiss any parts of testimony that he found spurious. Of Hamed's testimony Speke wrote, "All that he said with regard to the southern half of the lake is very near the truth, for it is an exact corroboration of many evidences."[58] Differences occurred when critical geographers and rival explorers used competing evidence to corroborate local testimony and so came to different conclusions about its credibility. Unbeknownst to them and thousands of miles away, African informants were the champions of some and the foils of others.

In his analysis of the discovery of the source of the Nile, Baker approached the problem as a "violent clash between theory and fact."[59] He contrasted "theoretical geographers" with "practical explorers." The present chapter demonstrates that, while this dichotomy might look appealing, the distinctions understood in the mid–nineteenth century were more subtle. Critical geographers and explorers had differences of method and opinion, but within the discussion spaces of the RGS and the BAAS, their evidence and methods were considered carefully, and each contributed to the process of geographical discovery. Some methods were shared by both. The corroboration of oral testimony was important to critical geographers and explorers alike. In cases where their methods produced different results, what led to conflict was the degree of personal investment in certain methods.

While each worked in different spaces and in differing ways, critical geographers and explorers worked to construct knowledge that would be credible and accepted among their fellow British geographers. They were working toward similar ends while inhabiting very different methodological spaces.

On the one hand, the process of discovery was distributed in and between places where specific rules applied to those settings—that is, expeditions, their discussion in geographical societies, and their representation in maps and in journals. On the other hand, the process occurred in that single contextual space of European geographical exploration.

From this evidence, we can begin to see how it is that activities and practices located either in the field, the metropolis, or somewhere in between were essential to the process of geographical discovery. All such activities have a geography, and we must consider the many scales at which science works and the purposes of the different methods employed. We must also keep in mind the unity of institutional purpose that linked dislocated activities together coherently. In the nineteenth century, men of science had to correspond and extend their concepts in print and via speech to local audiences, and they had to secure facts. Where one was and who one was strongly structured contributions to the knowledge of distant geographies and how they—persons and facts alike—were accepted. The Nile's sources looked very different from a debating chamber in London than they did from a small sailboat on the northern end of Lake Nyassa or to a local inhabitant on its shores.

NOTES

1. David Livingstone in August 1870, cited in Horace Waller, ed., *The Last Journals of David Livingstone, in Central Africa, from 1865 to His Death: Continued by a Narrative of His Last Moments and Sufferings*, 2 vols. (London: John Murray, 1874), 51.

2. An early survey of this history is found in Harry Hamilton Johnston, *The Nile Quest: A Record of the Exploration of the Nile and Its Basin* (London: Lawrence and Bullen, 1904).

3. I use this expression in the fuller sense of the sociology of scientific knowledge and so ignore boundaries between contexts and justifications for discoveries. On this, see Steven Shapin, "Here and Everywhere: Sociology of Scientific Knowledge," *Annual Review of Sociology* 21 (1995): 289–321.

4. Sykes's comment is recorded in the discussion that follows James MacQueen, "Observations on the Geography of Central Africa," *Proceedings of the Royal Geographical Society of London* 3 (1858–59): 208–14.

5. [Laurence Oliphant], "Nile Basins and Nile Explorers," *Blackwood's Edinburgh Magazine* 97 (1865): 100–17.

6. Samuel W. Baker, "Account of the Discovery of the Second Great Lake of the Nile, Albert Nyanza," *Journal of the Royal Geographical Society* 36 (1866): 15.

7. James Erhardt, "Reports Concerning Central Africa, as Collected in Mambara and on the East Coast, with a New Map of the Country," *Proceedings of the Royal Geographical Society of London* 1 (1855–57): 8–10.

8. For the debate between ancient authority and the early exploration of the African coasts, see Francesc Relaño, *The Shaping of Africa: Cosmographic Discourse and Cartographic Science in Late Medieval and Early Modern Europe* (Aldershot: Ashgate, 2002).

9. Cameron's discovery that Lake Tanganyika has a western outlet that joins the Congo was definitive: see Verney Lovett Cameron, "Exploration of Lake Tanganyika: Letter from Lieut. V. L. Cameron, Describing the Discovery of an Outlet," *Proceedings of the Royal Geographical Society of London* 19 (1874–75): 75–78.

10. For a summary of Ptolemy's legacy, see David N. Livingstone, *The Geographical Tradition: Episodes in the History of a Contested Enterprise* (Oxford: Blackwell, 1992).

11. Francesc Relaño, "Against Ptolemy: The Significance of the Lopes-Piggafetta Map of Africa," *Imago Mundi* 47 (1995): 49.

12. Margarita Bowen, *Empiricism and Geographical Thought: From Francis Bacon to Alexander Von Humboldt* (Cambridge: Cambridge University Press, 1981).

13. Matthew H. Edney, "Reconsidering Enlightenment Geography and Map Making: Reconnaissance, Mapping, Archive," in *Geography and Enlightenment*, ed. David Livingstone and Charles W. J. Withers (Chicago: University of Chicago Press, 1999), 186–67.

14. Cooley's life has been examined in detail in Roy C. Bridges, "W. D. Cooley, the RGS and African Geography in the Nineteenth Century," *Geographical Journal* 142 (1976): 27–47, 274–86; see also Roy C. Bridges, "William Desborough Cooley (1795–1883)," *Geographers Biobibliographical Studies* 27 (2008): 43–62.

15. William Desborough Cooley, *Claudius Ptolemy and the Nile; or, an Inquiry into That Geographer's Real Merits and Speculative Errors, His Knowledge of Eastern Africa and the Authenticity of the Mountains of the Moon* (London: John W. Parker and Son, 1854), 1.

16. Christopher Ondaatje equates the Ruwenzori Mountains of Uganda with Ptolemy's mountains of the moon: Christopher Ondaatje, *Journey to the Source of the Nile* (Toronto: HarperCollins, 1998).

17. Murchison agreed; Beke did not. Roderick Impey Murchison, "Presidential Address," *Journal of the Royal Geographical Society* 33 (1863): 184; Charles Tilstone Beke, *The Sources of the Nile: Being a General Survey of the Basin of That River, and of Its Head-Streams; with the History of Nilotic Discovery* (London: James Madden, 1860).

18. See, for example, Richard Francis Burton, "On Lake Tanganyika, Ptolemy's Western Lake-Reservoir of the Nile," *Journal of the Royal Geographical Society* 35 (1865): 1–15. A similar story can be told about the mountains of Kong in West Africa: see Thomas J. Bassett and Phillip W. Porter, "From the Best Authorities: The Mountains of Kong in the Cartography of West Africa," *Journal of African History* 32 (1991): 367–413.

19. Nehemia Levtzion, "Arab Geographers, the Nile, and the History of Bilad Al-Sudan," in *The Nile: Histories, Cultures, Myths*, ed. Haggai Erlich and Israel Gershoni (Boulder, CO: Lynne Reinner, 2000), 71–76.

20. William Desborough Cooley, *The Negroland of the Arabs Examined and*

Explained: Or, an Inquiry into the Early History and Geography of Central Africa (New York: J. Arrowsmith, 1841).

21. "The Negroland of the Arabs by W. D. Cooley [Review]," *Journal of the Royal Geographical Society* 12 (1842): 120–25.

22. Frederick Ayrton, "Observations upon M. D'abbadie's Account of His Discovery of the Sources of the White Nile, and upon Certain Observations and Certain Objections and Statements in Relation Thereto, by Dr. Beke," *Journal of the Royal Geographical Society* 18 (1848): 48–74.

23. William Desborough Cooley, "The Geography of N'yassi, or the Great Lakes of Southern Africa, Investigated; with an Account of the Overland Route from the Quanza in Angola to the Zambezi in the Government of Mozambique," *Journal of the Royal Geographical Society* 15 (1845): 185–235.

24. James MacQueen, "Notes on the Geography of Central Africa, from the Researches of Livingstone, Monteiro, Graça, and Others," *Journal of the Royal Geographical Society* 26 (1856): 109–30.

25. MacQueen, "Observations on the Geography of Central Africa."

26. William Desborough Cooley, "Journey of Joachim Rodriguez Graça to the Mwata Ya Nvo," *Proceedings of the Royal Geographical Society of London* 1 (1855–57): 92–93.

27. Charles W. J. Withers, "Mapping the Niger, 1798–1832: Trust, Testimony and 'Ocular Demonstration' in the Late Enlightenment," *Imago Mundi* 56 (2004): 178.

28. Charles Tilstone Beke, "On the Countries South of Abyssinia," *Journal of the Royal Geographical Society* 13 (1843): 254.

29. Ibid., 266.

30. Charles W. J. Withers, "Authorizing Landscape: 'Authority,' Naming and the Ordnance Survey's Mapping of the Scottish Highlands in the Nineteenth Century," *Journal of Historical Geography* 26 (2000): 532–54; Charles W. J. Withers, "Reporting, Mapping, Trusting: Making Geographical Knowledge in the Late Seventeenth Century," *Isis* 90 (1999): 497–521.

31. Driver has shown how field methods were themselves subject to scrutiny: see Felix Driver, *Geography Militant: Cultures of Exploration and Empire* (Oxford: Blackwell, 2001), chap. 3.

32. Charles Tilstone Beke, "On the Nile and Its Tributaries," *Journal of the Royal Geographical Society* 17 (1847): 2.

33. I have analyzed this expedition in detail elsewhere; see Lawrence Dritsas, *Zambesi: David Livingstone and Expeditionary Science in Africa* (London: I. B. Tauris, 2010).

34. The events of this meeting, including the discussion after the papers, are recorded in David Livingstone, "Letters from the Zambesi to Sir R. I. Murchison, and (the late) Admiral Washington," *Proceedings of the Royal Geographical Society of London* 8 (1863–64): 256–63.

35. Bridges, "W. D. Cooley"; William Desborough Cooley, "On the Travels of Portuguese and Others in Inner Africa," *Proceedings of the Royal Geographical Society of London* 8 (1863–64): 255–56.

36. Livingstone, "Letters from the Zambesi," 256.

37. Ibid., 258.

38. Ibid.

39. David Livingstone and Charles Livingstone, *Narrative of an Expedition to the Zambesi and Its Tributaries, and of the Discovery of the Lakes Shirwa and Nyassa, 1858–1864* (London: John Murray, 1865; repr., London: Dover, 2001), 291.

40. For an example of how a western exit for the Congo was ruled out see, Alexander George Findlay, "On Dr. Livingstone's Last Journey, and the Probable Ultimate Sources of the Nile," *Journal of the Royal Geographical Society* 37 (1867): 193–212.

41. Roderick Impey Murchison, "Geography and Ethnology," *Report of the Thirty-Third Meeting of the British Association for the Advancement of Science* (London: John Murray, 1864), 122–23.

42. Clive Barnett, "Impure and Worldly Geography: The Africanist Discourse of the Royal Geographical Society, 1831–73," *Transactions of the Institute of British Geographers* 23 (1998): 242.

43. Withers, "Mapping the Niger."

44. P. D. Lowe, "Amateurs and Professionals: The Institutional Emergence of British Plant Ecology," *Journal of the Society for the Bibliography of Natural History* 7 (1976): 517–35.

45. See the preface in Richard Francis Burton, *The Lake Regions of Central Africa, a Picture of Exploration* (New York: Harper & Brothers, 1860).

46. The accounts of Burton and Speke of this visit appear in Richard Francis Burton, "The Lake Regions of Central Africa, with Notices of the Lunar Mountains and the Sources of the White Nile," *Journal of the Royal Geographical Society* 29 (1859): 1–454; John Hanning Speke, "Journal of a Cruise on the Tanganyika Lake, Central Africa (Speke's Journal Part 1)," *Blackwood's Edinburgh Magazine* 86 (1859): 339–57; John Hanning Speke, *What Led to the Discovery of the Source of the Nile* (Edinburgh: Blackwood, 1864).

47. Examples of this reevaluation are to be found in Findlay, "On Dr. Livingstone's Last Journey."

48. James MacQueen, "Notes on the Present State of Geography of Some Parts of Africa," *Journal of the Royal Geographical Society* 20 (1850): 246.

49. John Hanning Speke, "The Upper Basin of the Nile, from Inspection and Information," *Journal of the Royal Geographical Society* 33 (1863): 322–46.

50. For a similar point about the language and social position of local informants (albeit in the Scottish Highlands), see Withers, "Reporting, Mapping, Trusting."

51. See Dane Kennedy, *The Highly Civilized Man: Richard Burton and the Victorian World* (London: Harvard University Press, 2005), chap. 4.

52. Speke's reflections on the discovery of Lake Victoria were published in two parts: John Hanning Speke, "Captain J. H. Speke's Discovery of the Victoria Nyanza Lake, the Supposed Source of the Nile: From His Journal—Part 2," *Blackwood's Edinburgh Magazine* 86 (1859): 391–419; John Hanning Speke, "Captain J. H. Speke's Discovery of the Victoria Nyanza Lake, the Supposed Source of the Nile: From His Journal—Part 3," *Blackwood's Edinburgh Magazine* 86 (1859): 285–312.

53. Withers, "Reporting, Mapping, Trusting."

54. David Bloor, "Toward a Sociology of Epistemic Things," *Perspectives on Science* 13, no. 3 (2005): 285–312.

55. Johannes Fabian, *Out of Our Minds: Reason and Madness in the Exploration of Central Africa* (Berkeley and Los Angeles: University of California Press, 2000).

56. Livingstone, "Letters from the Zambesi," 262.

57. John Hanning Speke, *Journal of the Discovery of the Source of the Nile* (Edinburgh: Blackwood, 1863).

58. Speke, "Journal of a Cruise on the Tanganyika Lake," 352.

59. John Norman Leonard Baker, "Sir Richard Burton and the Nile Sources," *English Historical Review* 59 (1944): 49.

PART THREE

Guides and Audiences

Part 3 is centrally concerned with the texts, printed and nonprinted, that constituted certain forms of scientific knowledge and certain audiences, "professional" or otherwise, during the nineteenth century. While the dissemination of scientific knowledge from field practitioner to public audience might require rhetorical skill and command of the conventions of platform science, the transfer of such knowledge from field to cabinet required the discipline of observational skills. One had to be taught not just what to look at, but how to see. Guidance in such matters was perhaps especially a feature of the nineteenth century, given changes then in the nature of objectivity in science and science's visual depiction, and, arguably, it was most evident in the botanical and field sciences.

Such issues are at the heart of Anne Secord's examination of the transference of botanical knowledge between the field and the cabinet in ways that centered upon texts as guides: guides to specimens, to districts, and to procedure in and away from the field; guides to the collecting, mounting, and displaying of actual individual plants. The production of satisfactorily scientific observation in the botanical world demanded that books should discipline visual habits. Here, then, the spaces of the book illustrate not just the products of observation, but also its processes.

Our understanding of the key role of different technicians of print and of the technologies used to overcome distance from the immediacy of the real world is enhanced by an appreciation of the role of translators and reviewers as guides, and of book merchants as key agents in the dispersion of printed knowledge to different audiences. Jonathan Topham's essay centers upon this idea of knowledge in transit and on the ways in which printed matter crossed borders—political, linguistic, epistemic. Looking at several case studies—mainly at Laplace's physics, Lavoisier's chemistry, Lamarck's natural history, and Cuvier's geology—Topham demonstrates (echoing Livingstone's examination of La Peyrère and Charles Darwin), how the circulation of scientific knowledge was critically dependent upon the mediation of others between author and audience: publishers, booksellers, translators, and reviewers.

Buying a science book or reading about one in the burgeoning periodical literature of the nineteenth century, even pressing botanical specimens into service, was altogether easier than making specimens of mineralogical geography or making the topographic expression of regional geology become mobile and thus easily accessible to dispersed audiences. As Lawrence Dritsas shows for the delineation of central Africa, so Simon Naylor emphasizes for subterranean Cornwall: maps had an effective visual economy, guiding the curious to sites/sights of interest and simultaneously rendering local space and deep time accessible as portable paper products for easy use in the field. For the men of the Royal Geological Society of Cornwall, surveying their county above and below ground depended on the reductive visual authority of the map as a guide to what could and could not be seen: to the links between the country rock, soil, and agricultural productivity; to subsurface ore deposits; to natural districts; and, they hoped, to the prominent place of Cornwall on the map of Britain's natural sciences in the nineteenth century.

Science's guidebooks provided instruction on procedure and on the location of specimens and noteworthy features. As Aileen Fyfe makes clear, they also drove and reflected a rapidly expanding popular science and scientific tourism market. Thus, as a genre, they began to be structured along similar lines: small enough to be portable; cheap enough to be affordable; synoptic, not inclusive; and, commonly, presenting the scientific merits of regions or places through a carefully ordered textual sequence of geology, natural history, antiquities, notable features, and the like. In this move to textual standardization and the provision of accessible scientific knowledge, there lay a danger: format might overrule content, and content in general might hinder individual encounter.

The spaces and objects of scientific encounter were often so large, however, that guidebooks and maps could not easily contain them, and neither public audience nor professional scientist could venture there. Knowledge of the American West after 1850—a shimmering "empty" interior more encountered in transit by settlers than understood by scientists—required direct encounter, observational training, and dependence upon indigenous informants. But even as expeditionary efforts opened up America's heartland, the knowledge gathered required analytic space for its accumulation and interrogation, for measured reflection and, if possible, for the education of audiences keen to know their country's shape and contents. Thus, argues Sally Gregory Kohlstedt, "The Smithsonian Institution contributed to the public instantiation of the idea of the American West not as a fixed place but as a reflection of the nation itself." For audiences who could not encounter the West, the Smithsonian would become the West, displaying in its metropolitan collecting spaces the products of expeditions, managed observations, and accumulated textual guides, all under the guidance of distinguished curatorial staff.

CHAPTER TWELVE

Pressed into Service:
Specimens, Space, and Seeing
in Botanical Practice

ANNE SECORD

Our understanding of scientific practice has been greatly enhanced by "geographies of reading and writing," which have shown that the production and circulation of texts and the cultural location of readers are "at least as significant as the global-scale cycles of circulation from periphery to metropolis, from field site to centre of calculation, from ecological niche to taxonomic cabinet."[1] So far, however, the fruitfulness of this approach for investigating past science, for all that it considers the book as a material object, has paid little attention to the book itself as a performative space produced by the practices and actions of both writer and reader. Books have long been analyzed as far more than repositories for an author's text, with marginalia and interleaving being the most obvious ways in which readers insert themselves into books.[2] There has, surprisingly, been less emphasis on the ways in which scientific authors—especially botanists—adopted and adapted such practices in order to promote specific practical skills. It is, however, this very element of nineteenth-century botanical books that alerts us to their role as guides (regardless of whether they are explicitly labeled as such) by drawing attention to an awareness of a large lay audience that the compilers of these texts wished to prompt into action in specific ways. Attending to the "geographies of reading and writing" in botany in relation to the microspaces of botanical books shows that such practices are not only as significant as the global-scale cycles of circulation, but also integral to them.

Recent investigation by Robert Mayhew of the "physical space of the book as a key part of its ability to express meaning" has revealed the importance of paratextual devices, such as prefatory material, notes, size of margins, and graphic configurations, all of which aim to engage a reader's attention in a specific way. Since the meanings of texts change when the paratextual elements are altered (in later editions, for example), these spatial features provide crucial evidence as to the purpose of a book.[3] Applying this form of analysis to botanical books is particularly fruitful, as it allows us to see that a range of genres was developed in order to encourage specific ways of seeing, not least in those areas of botany that pressed the limits of observation. In addition, because just about anything flat can be incorporated into a book, botanists, more than other naturalists, left traces of their own practices within the spaces of the botanical works that they owned. This essay examines the manipulation of the physical space of the book by authors and readers alike in order to extend our understanding of the spatial arrangements of the informal networks that characterized British nineteenth-century botany and to show how published botanical monographs and local floras can be regarded as spaces of active scientific engagement at a variety of levels.

Botany, like other natural historical practices, is a subject in which a wide diversity of spaces are brought into conjunction, both in terms of the objects dealt with (distribution and classification of plants) and with respect to practitioners' presence in different places (the field, taxonomic cabinet, and library). These various spaces of natural history have usefully been analyzed in terms of the different modes of observation they engender. The steady gaze of the sedentary naturalist in the cabinet, whose quest, through the close examination and comparison of specimens, was an overview of nature, has been contrasted with the fleeting and fragmented observations of the field naturalist, especially the explorer, whose encounter with unbounded space and the vast range of natural objects could lead to the point of sensory disorganization.[4] For all the importance of these observational differences between field and cabinet work, this characterization does not distinguish the visual experience of travelers' transitory encounters with new floras in unexplored regions from that of local collectors whose knowledge of growing plants was based on long and repeated exploration of confined areas, nor does it appreciate that while most botanists made collections of dried specimens not all were engaged in taxonomic pursuits. The observational experience of most local botanists consisted of a constant interplay between the visual skills required in the field and in the cabinet. This becomes clear if, instead of associating different forms of visual expertise with specific scientific aims (taxonomic work or the exploration of new floras), we consider Charles Withers and Diarmid Finnegan's argument that "fieldwork as practice cannot simply be separated from the justificatory cultures of its display" (be those publications, museums, or herbariums), and that it "was a practice rooted more in accuracy than in collecting *per se*."[5]

Accuracy was essential because the transmission of knowledge was not a one-way process but a circuit in which many of those consuming the works of experts were encouraged to produce information that was, in turn, consumed by experts. The aim of botanical authors was therefore less the production of taxonomic expertise than of competent and reliable observers. This involved developing techniques by which readers could learn how to recognize plants and thus acquire the ability to spot rarities or other features of interest when botanizing in their own localities. The social organization of early nineteenth-century British botany, consisting largely of private individuals scattered across the country, provides an unusually good opportunity to explore the means by which the requisite observational skills were transmitted and attained in order to ensure accurate knowledge. I here explore several genres of publication to show how expert botanists attempted to shape the visual practices of their readers. As the needs and interests of experts changed, so too did the means by which their audiences were enabled to develop the observational techniques necessary for participation in botany. Learning how to see involved the ability to recognize plants both in their growing forms and as decontextualized specimens, a process for which, in both its exposition *and* its practice, we can find evidence in the space of the book.

The circuit of knowledge production is especially open to scrutiny in botany because of a spatial transformation unique to the science. Drying and pressing, the processes of botanical preservation most commonly employed, result in dramatic loss of form, as three-dimensional plants become two-dimensional specimens. One consequence of the flatness of specimens was that it rendered them capable of being incorporated into books. Their inclusion in botanical texts not only provides evidence that specimens, normally seen as the objects of study in the cabinet, were crucial to field practice, but also reveals traces of how independent collectors had to negotiate the different ways of seeing plants in order to become useful contributors of knowledge. Specimens in the space of the book challenge the assumptions that their usual contexts entail.[6] Their presence here also alerts us to the fact that although a botanical specimen is the "real thing,"[7] the radical alteration undergone by plants through drying and pressing (which also often results in loss of color) could present botanists with peculiar difficulties when honing their visual skills. Even that keen observer of nature Charles Darwin complained to botanist Asa Gray, "It is dreadful work making out anything about dried flowers; I never look at one without feeling profound pity for all botanists, but I suppose you are used to it like eels to be skinned alive."[8]

The difficulty that this transformation in plants presented to botanists, however, offers opportunities to the historian. Not only do we find in the spaces of the book instances when the quest to produce accurate observers resulted in specimens being considered preferable to illustrations, but also examples of the ways in which local collectors used specimens to mark their own progress in learning how to observe. Moreover, guides and audience responses encompassing specimens cut across class and gender lines, thus indicating that these textual spaces played an important role in creating a wide and diverse community of botanical observers. Historians have long urged that a fuller understanding of nineteenth-century science will be gained if we broaden our sense of what constitutes scientific activity. In order to achieve this, Roger Cooter and Stephen Pumfrey argue that we must stop privileging "ideas and texts over practice and object."[9] By bringing together recent work on books and their manipulation by readers, I hope to press this further by showing how botanical texts themselves can offer ways of investigating practices and objects, and thereby function as signifiers of the primacy of observational skill in the science. In order for them to do so, however, we need to revise our view that representations in books are to be interpreted only at a conceptual level rather than through their readers' practical actions.

FILLING IN SPACES

One of the clearest ways in which books acted as guides to observation was in a genre of publications that explicitly encouraged readers to record what they had seen in an orderly fashion. The practice of leaving spaces for the users of books to fill in was part of a long tradition,[10] and became a standard method of record-keeping in eighteenth-century natural history with the advent of published journals consisting of labeled columns for the daily noting of the weather or seasonal comments on the local flora and fauna. In 1800, however, a variation on this theme was introduced specifically for botanists: William Mavor's *The Lady's and Gentleman's Botanical Pocket Book; Adapted to Withering's Arrangement of British Plants. Intended to Facilitate and Promote the Study of Indigenous Botany* was designed not only to encourage owners of the book to keep records, but also to shape their way of seeing the vegetable kingdom by underlining the importance of scientific arrangement. Marketed as a companion to *A Botanical Arrangement of British Plants* of William Withering (by then in its third edition), which had done so much to promote botany and Linnaeus's system of classification, Mavor's book relied less on words than space to carry out its aims. Readers were

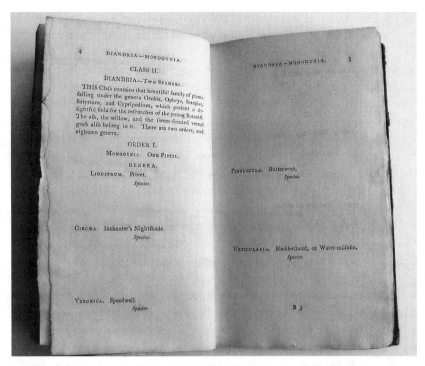

Fig. 12.1. An example of blank spaces for recording plants seen when botanizing in the field, organized according to the Linnaean classification. From William Mavor, *The Lady's and Gentleman's Botanical Pocket Book adapted to Withering's Arrangement of British Plants. Intended to Facilitate and Promote the Study of Indigenous Botany* (London: Vernor & Hood, 1800).

invited to put their learning into practice by Mavor's organization of the book according to the Linnaean classification, with "spaces left to fill up, apportioned, as far as possible, to the number of species under each genus" (fig. 12.1).[11] This collecting book exemplifies Susan Stewart's point that "it is the Linnaean system which articulates the identities of plants . . . and not the other way round,"[12] not least because the book itself was a physical representation of this classificatory arrangement. What the blank spaces of Mavor's book made clear, however, was that Linnaeus's system was given scientific meaning and the potential to elucidate the flora of a particular locality only through observational skill in the field. Cabinets of dried plants have long been regarded as sites of scientific work, but little attention has been paid to the processes by which botanists first made their reference collections. At a time when Britain's own flora was still being determined, Mavor hoped that his field-collecting book would inspire and discipline the visual habits needed to transform both neophytes and more accomplished botanists into potentially useful contributors to botanical science.

This aspiration was not confined to introductory books like Mavor's: well into the nineteenth century, it was applied to some of the most difficult areas of botany. The filling in of blank spaces in books that were, in effect, cabinets of pressed specimens could also stimulate field collecting. In 1836, the plant collector George Gardner produced his *Musci Britannici, or Pocket Herbarium of British Mosses; Named and Arranged According to Dr. Hooker's "British Flora,"* a book in which every other leaf "has a page ruled off into compartments, suited to the size of the species . . . & the names printed in lithography." Gardner had collected sufficient mosses himself to produce several books containing two hundred specimens, but the difficulty of finding some species meant that only a few copies contained significantly more than this number, the price varying accordingly. Since all the books, even the fullest, were incomplete, it was up to the possessor to "go on filling the empty spaces himself from time to time."[13] Thus books, as much as cabinets, through the provision of space and the fact that plant specimens were flat, possessed the potential to act both as works of reference and as records of their owner's field experience. The visual acuity required to detect tiny mosses in the field was further honed by placing their pressed forms in the framework of a classificatory system, itself quite literally represented in the labeled blank spaces of a book.

It was, of course, possible that specimens added to Gardner's book were not collected by the owner but obtained from other collectors to complete a reference collection. Nonetheless, the aim of such works, as we will see in later examples of books dealing with the nonflowering plants like mosses and seaweeds, was to stimulate knowledge of growing plants. It is also the case that some collectors may have eschewed dried specimens altogether, recording instead only the forms of living plants, as a popular guide to botany from the second half of the century reveals. George Bentham's *Handbook of the British Flora,* first published in 1858, was explicitly aimed at "beginners and amateurs." Despite Bentham's own botanical expertise, he styled himself an amateur and was well aware of the widespread interest in botany among those who lacked experience and had no scientific standing.[14] Writing for this audience, Bentham worked at conveying the basics of the more complex natural classification in addition to providing the most efficient means by which his readers could identify the plants they found during rambles. Clearly setting out to be a guide at a time when botany was beginning to be dominated by scientific professionals, Bentham's handbook also provides evidence of its audience by the way in which its owners made their presence known within its covers. Like most floras of the time, especially those carried into the field, the print used for the species' descriptions was small, and there were no illustrations. By making use of the margins, however, owners of Bentham's book began to make paintings of the plants they found alongside their descriptions in the text. The book thus doubled up as guide and personal record of first-ever finds. Most notable is the 1858 edition owned by Elizabeth Hood on the Isle of Wight, which contains several watercolor depictions, including that of a field gentian with a dated record of the only locality in which it grows on the island.[15] Both the sighting of the plant in the field and the painting in the margin bear witness to Hood's observational skill.

One measure of the observational skill botany required is provided by the very effort expended in producing those guides, which aimed to insure, as far as possible, that the accuracy of scientific knowledge was not compromised by individuals who failed to recognize the limits of their expertise. Perhaps surprisingly, given its popularity, botanical collecting was hard, or at least, hard to do well. Bentham's handbook, like Mavor's collecting book, got around this to some extent by omitting the most difficult groups of plants in order to capture the attention of beginners. But this similarity is deceptive. By 1858, the amateur was defined in relation to the albeit still fluid, but emerging, category of the professional man of science. In the early nineteenth century, however, not even expert botanists were "professed" in this sense, and most, regardless of social class, could be labeled "self-made" in the science of botany. In shaping the observation of field collectors according to the Linnaean system, Mavor left out the nonflowering plants, or cryptogams, which included mosses, lichens, seaweeds, fungi, and ferns, because, as Linnaeus himself had conceded, they were not susceptible to easy analysis using this system. Moreover, in 1800 very few of even the most learned botanists had turned their attention to British cryptogams. Yet, as we have already seen, by 1836 Gardner could "publish" specimens of mosses, and encourage his readers to collect the missing species by leaving blanks in the book. Clearly such a work would not have been produced had there not been a market for it.

The fact that there was any interest in nonflowering plants like mosses, which were often difficult to find and harder to classify, was largely due to the publication in 1805 of *The Botanist's Guide through England and Wales*. This pocket-sized work, produced by the Yarmouth banker Dawson Turner and Swansea porcelain manufacturer Lewis Weston Dillwyn, introduced readers to the need to consider the variety and vastness of the space of the field. By listing all the plants of note under the counties in which they were found, and by scrupulously naming the discoverers of the sites of rare plants as well as censoring those whose "botanical accuracy" they suspected (fig. 12.2), the *Botanist's Guide* aimed not only to encourage novices but also to direct "the Traveller" in "his researches."[16] Among these travelers, in addition to "the experienced Botanist," Turner and Dillwyn included the increasing number of British tourists, who, owing to the Napoleonic Wars, could no longer travel to the Continent. As the first botanical work of its kind, Turner and Dillwyn's *Guide* specifically catered to this clientele, intending to enhance the intellectual pleasure of tourists while at the same time attempting to insure that the science of botany would benefit from any discoveries they might make. The importance of contributions from private individuals is underscored by the fact that both Turner and Dillwyn, for all their expertise, were not professional botanists but a banker and porcelain manufacturer respectively, whose scientific interests had to be fitted around their business activities.

Given that Turner and Dillwyn's *Guide* is in part a record of the active participation of local collectors, stressing this as an aim of their publication might seem redundant. It was necessary, however, in the one area of botany lacking active engagement: cryptogams. Turner and Dillwyn, each of whom had made this difficult, little-studied, and unpopular area of botany their specialty, regretted that their treatment of the cryptogamia in the *Guide* was "even more faulty and imperfect than we had expected," and attributed this in part to "the small number of Botanists, from whom it was possible to procure assistance."[17] Their apology (and comments regarding cryptogams within their text) served also as an invitation to local botanists to investigate these strange and often inconspicuous plants.

Even when peace resumed, the union of polite culture and scientific botany remained explicit in Thomas Walford's *The Scientific Tourist through England, Wales, & Scotland* of 1818, also arranged by county, which directed travelers to "the principal objects of Antiquity, Art, Science, & the Picturesque, including the Minerals, Fossils, Rare Plants, and other Subjects of Natural History." Walford advised those tourists who intended to publish accounts of their travels to be well read in the sciences, "for those tours that come from the pen of scientific travellers are not only the most pleasing, but always the most instructive." To the lover of plants, he recommended "that valuable little work, the Botanist's Guide, by Dawson Turner, esq. where he will find the cryptogamous plants of each county."[18] Tourists, who had long succumbed to the charms of flowering plants, were now interested in—or, at least, were being encouraged to take an interest in—cryptogamic plants.

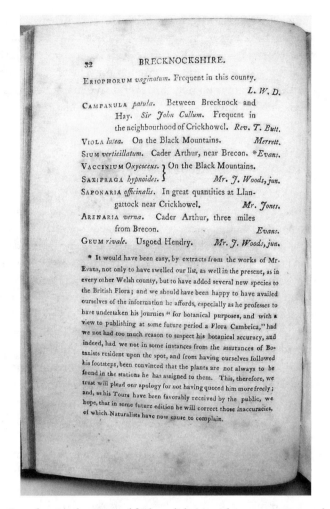

Fig. 12.2. Page showing the names of finders of plants, with a note warning readers of the botanical inaccuracies in John Evans's account of a tour through North Wales. From Dawson Turner and L. W. Dillwyn, *The Botanist's Guide through England and Wales*, 2 vols. (London: Phillips and Fardon, 1805), 1:32.

Turner and Dillwyn, by considering the local within the whole, had introduced a new aspect of accuracy into botanical observation. To know what was worth collecting, collectors had not only to possess the ability to accurately identify familiar plants in order to spot the rarities, but also to be aware that a prized rarity in one area might be extremely common elsewhere and thus of local interest only. The key issue in understanding the processes by which collectors could contribute to the production of

botanical knowledge is less the techniques by which knowledge circulated—
correspondence networks, personal contacts, publications, methods of specimen preservation and labeling—but the practices that enabled collectors to
judge which knowledge was worthy of circulation. The observational skills
that underlie this judgment do not, however, leave many historical traces
unless we consider botanical books not just as compilations of scientific
knowledge but as guides to seeing.

VISUAL IMPRESSIONS

In the "sciences of the eye," Lorraine Daston and Peter Galison stress, it is
the images that compose scientific atlases (be they folio volumes or pocket
field guides) that "make the science." Atlas images stabilize the objects of
study by showing how to distinguish the essential from the incidental, the
typical from the anomalous, and the limits of variability in nature; they
"train the eye of the novice and calibrate that of the old hand" by teaching
"how to describe, how to depict, how to see." These works "simultaneously
assume the existence of and call into being communities of observers who
see things in the same ways."[19] The ability to produce such atlases in botany, however, depended upon the state of knowledge about the plants in
question. In little-studied and difficult groups like the cryptogams, descriptions and especially illustrations—the aim of which was to make clear the
salient characters by which a plant is classified—were often less reliable
than those of flowering plants since they were subject to the biases of just
a few observers, who often disagreed about what the important classificatory features might be. Specimens and living plants thus assumed more
importance as guides to observation. Correspondence networks reveal that
botanists vied with one another to obtain specimens of hard-to-find or
difficult-to-determine plants. But evidence of the role of fresh and dried specimens in promoting the observational skills necessary for botany to progress can be found in more mundane places. In addition to the few illustrated
monographs, a range of genres developed that encompassed both book and
specimen.

Perhaps the best known of these is the "nature-printed" book, where the
impression made on a soft lead plate by an actual plant specimen was then
used to produce multiple prints for publication. This technique followed an
earlier practice of directly inking specimens themselves and pressing them
on to paper; a "recording" method that Alexander von Humboldt and Aimé
Bonpland had adopted during their travels in South America as insurance
against the destruction of their dried plant specimens by insects and hu-

midity.[20] The botanical advantage of nature-printed publications using the lead-plate technique was that the impressions so produced were regarded as contributing to observational accuracy. As botanist John Lindley made clear in 1855, in the most beautiful and best-known British example of nature printing, Thomas Moore's *The Ferns of Great Britain and Ireland*, neither "mere descriptions" nor drawings can "represent faithfully the minute peculiarities by which natural objects are often best distinguished," especially in cryptogamic botany, where "minute accuracy is of more vital importance." The "great defect of all pictorial representations," but especially those of complicated forms like ferns, Lindley warned, was that even "the most skilful and patient artist" can give only "an imperfect sketch of what he supposes to be their more important features." In short, "compared with the result of Nature-Printing, botanical drawings are often little more than indifferent diagrams." Moreover, drawings, by emphasizing the characters by which a plant was classified, illustrated the "subjects of inquiry in the cabinet" which, in his view, were better described by words alone. In contrast, although nature printing represented only what "lies upon the surface," this, Lindley pointed out, was sufficient for the "practical purposes" of "the first-sight recognition of a Fern." Nature-printing thus excelled at reproducing with "unerring truth" the features (namely, the "general manner of growth") by which the "practised eye" recognized plants "at a glance."[21]

Lindley thus intended to direct the audience for Moore's work to recognizing ferns in nature rather than studying dried specimens. He hoped that "good Nature-Printing will convey to the eye the same class of positive impressions as those which were conveyed to the mind of Gough by other organs." John Gough of Kendal, whose botanical knowledge had augmented the 1842 edition of William Wordsworth's *Guide to the Lakes*, could represent nothing but the field botanist since he lacked the requisite instrument for cabinet investigations. As Lindley explained, Gough "having become totally blind from small-pox when two years old . . . so cultivated his other senses as to recognise by touch, smell, or taste, almost every plant within twenty miles of his native place."[22] That nature-printing did succeed in promoting the recognition of plants in nature is perhaps better shown by the experience of the sculptor Robert Cockle Lucas, who found that *making* nature prints was not only a source of pleasure but also an inducement to "survey nature with a most scrutinising observation."[23]

Important though this association of plant impressions with a knowledge of their living forms was, the merging of object and image to the point of interchangeability was the limitation as well as the advantage of the process. For nature-printing produced scientifically effective images only from plants—or parts of plants such as fern fronds—that were already flat.[24] This constraint applied also to the photographic technology that emerged in the 1840s and held out the promise of replicable images of actual plants for inclusion in books. The only botanical work to be illustrated this way was Anna Atkins' *Photographs of British Algae*, which began to appear in 1843. Eventually comprising three volumes, her cyanotype impressions, with their white images of flat seaweeds against a vivid oceanlike blue, aimed at visual accuracy for scientific purposes but, for much the same reasons as nature-printing, failed to satisfy the needs of botanists. Even when, later in the century, technological advances made it easier to produce sharper images of three-dimensional plants, photography was not taken up by botanists. Given that photographs are flat and thus capable of being tucked into books, it is surprising that they were not used for capturing the particularity of local plants when first spotted in the field. Instead, the practice of "painting one's Bentham" continued well into the twentieth century,[25] perhaps because painting demanded keen observation and thus contributed to a botanist's visual skill, whereas photography did not.

While the flatness of some plants made the replication of their exact likenesses possible through nature-printing or photography, plants that were already flat could also be the cause of visual confusion. Upon receiving an album of seaweed specimens, a professor expressed his thanks for the "beautiful *sketches*" he thought done by the donor's daughters; likewise, the expert algologist Ellen Hutchins recounted watching her servant girl (whom she had trained to be a botanical assistant) trying to pluck colored drawings of seaweeds off the paper, mistaking them for real plants.[26] In these cases, the visual appearance of the pressed specimen and illustrations of the fresh form of some plants were so similar as to be almost indistinguishable. This, however, was unusual. In most cases, botanists learned to negotiate the visual differences between growing plants and pressed specimens.

Reservations about the adequacy of drawings of those "difficult divisions of the Flora" whose "collection and study" were neglected gave rise to yet another genre involving specimens in a text. Frederick Hanham's *Natural Illustrations of the British Grasses* is a good example. In this work, published in 1846, dried grass specimens took the place of the more usual colored drawings that accompanied species descriptions in monographs. "Dried Specimens," Hanham argued, were more "interesting and valuable to a Botanist" than engravings for "there is much," he claimed, "that the pencil can never show." Even the best engravings, in his view, gave but "artificial resemblances," which were "in truth a mere shadow of the substance, occult and untangible, and perhaps even calculated to mislead, or

to convey an erroneous conception of the original." For the thousand copies of the book produced, sixty-two thousand specimens were required, but
half as many again had to be "culled to obtain that number of successful
specimens."[27]

The labor and difficulties involved, as emphasized by the author, would
not have been undertaken had there been no market for such works. Botanists interested in the classification of contested and difficult groups of
plants favored specimens precisely because illustrations embodied theoretical decisions concerning what should be noticed. In learning to distinguish
grasses, however, Hanham emphasized that

> much extended observation and practice are required, in the "wild field
> of Nature, where alone a good Botanist can be formed;" for so various
> are they in appearance in their expanded and unexpanded state of flower-
> ing . . . differing also in size, depending for the most part upon the nature
> of the soil on which they are found . . . besides being variable in their
> pubescence or hairiness . . . that even the most experienced and able Bot-
> anist at times is led to fancy he has a fresh species or new variety, when
> in fact he has only the same plant differently developed in consequence
> of accidental circumstances.[28]

When exactly what was being seen was at stake, specimens allowed readers to observe and judge for themselves; they guided and trained the eye in
both the "study and collection" of plants. This point is best exemplified in
another form of publication involving specimens.

PUBLISHING SPECIMENS

Exsiccatae are published collections of labeled, dried specimens, available in
multiple copies, whose identification and taxonomic arrangement follows
that of established botanical works, regardless of whether these were illustrated or not. Two such pairings reveal the importance of specimens. As William Hooker and Thomas Taylor's 1818 illustrated monograph on mosses,
Muscologia Britannica, neared completion, Hooker encouraged Edward
Hobson, a correspondent who had supplied him with many specimens for
the work, to produce in the same year *A Collection of Specimens of British Mosses and Hepaticae, Collected in the Vicinity of Manchester, and
Systematically Arranged with Reference to the Muscologia Britannica*. Although Hooker and Taylor emphasized the "utmost care" with which their
figures of mosses had been drawn, they recommended Hobson's exsiccata to

their readers on the grounds that it "will readily be seen how much superior these works must be in point of accuracy to the best of plates."[29]

The botanical author William Henry Harvey went even further, using a previously published exsiccata in lieu of illustrations. When his *Manual of the British Algae* was published in 1841, Harvey urged his readers to consider his descriptions as a "companion" to Mary Wyatt's *Algae Danmonienses or Dried Specimens of Marine Plants, Principally Collected in Devonshire; Carefully Named According to Dr. Hooker's British Flora*, first distributed from 1834, and which, by 1841, with the production of a supplement, had extended to five volumes. Harvey claimed that readers would "experience little loss" by the lack of illustrations in his book since Wyatt's "beautifully dried and correctly named" specimens "furnish the student with a help, such as no figures, however correctly executed, can at all equal." Harvey was aware, moreover, that his readers, increasingly able to observe and collect marine plants while on seaside holidays, did not need to be taxed by the complexity of seaweed classification. Taxonomic difficulties, he explained, "are much more formidable on paper than in the field." While the "systematizer, in his study, may consume the midnight oil till his aching brains are weary with the fruitless task," the observer on the seashore "finds no difficulty whatever" in recognizing seaweeds. Recommending the dismissal of "speculations on the exact limits between the Algae and all other tribes," Harvey advocated that "we must rest satisfied with differences which we can *see*, but which we cannot *know* or *define*." In Harvey's view, although he had originally intended his work to be illustrated by figures, they would mainly have "added to the beauty of the book."[30]

These examples draw attention to the crucial role of the specimen itself in the production of observational skill in the field, and to the way in which authors attempted to utilize its potential to train readers how to observe. Neglect of this function of specimens is in part due to the more obvious scientific utility and the aesthetic quality of illustrations. Dried and pressed botanical specimens, so different in appearance from the living plant, seem to possess little appeal. Historian of art Carol Armstrong remarks upon how albums of dried plants "value the musty, tactile reality of the dead specimen's unlovely, page-buckling, rot-brown thickness, over its explanatory function."[31] This, however, misses the point of these collections. The aim was less to convey a systematic understanding of botany than to hone observational skills. These involved not only detailed comparisons of dried specimens, but also learning to see that "unlovely" pressed plants lead to the quick as well as the dead.

While the examples of Hooker and Harvey show the use of collections of specimens published as books in order to hone their readers' observational powers, we should also consider the visual skills possessed by the producers of these exsiccatae and their use of books. Both Wyatt, a servant for many years in Torquay before becoming a "dealer of shells," and Hobson, a handloom weaver and then a Manchester warehouseman, built up their collections of specimens through repeated and patient scouring of their own localities and constant observation of the habitats of the plants they collected. For such individuals, their very rootedness in particular places and their inability to travel enabled them to participate in botany. The exsiccatae formed a genre ideally suited to the observational skills of these most local of collectors. By collecting the same plants season after season, they not only learned to associate dried specimens from a previous year with living plants observed in nature, but they were also more likely to spot new or unusual plants. In these cases, the use of Hooker's books was primarily to confirm, rather than create, the observational skills of working people like Hobson and Wyatt, who could then make their collections widely available through the distribution of "carefully" named specimens arranged according to Hooker's classification. Aware that botanists would benefit from the visual acuity of these collectors only if their specimens received scientific validation, Hooker took pains to ensure the accuracy of Hobson's exsiccata, while Harvey made clear that Wyatt had been the servant of expert algologist Amelia Griffiths.

Clearly the authors of exsiccatae were excellent observers, but historians often overlook the visual skills involved in the preparation of these works by associating specimens primarily with their scientific use by experts in the cabinet. By so doing, we impoverish our understanding of why the private collector was so valued in Britain. The preeminent Victorian botanist Joseph Hooker, although notoriously impatient with local collectors who, lacking the ability to judge what they had found, fancied themselves the discoverers of new species, had nothing but praise for Hobson, appreciating that "it requires a practised eye, and some previous knowledge, thoroughly to explore a small district rich in Mosses." Moreover, pouring over a copy of Hobson's exsiccata in William Hooker's library had inspired the young Joseph to collect mosses.[32] Little wonder, given the difficulty of finding and identifying mosses, that experts encouraged workingmen like Hobson, despite being emphatic about it being "one thing to find a moss, and quite another to detect & investigate it."[33] Even though the development of moss classification depended upon the "investigation" of mosses by a few experts in the cabinet, such statements should not obscure the ways in which different observational modes had to coexist in all botanical practitioners.

As well as visual acuity in the field, the competence of a field botanist, as Withers and Finnegan point out, "was judged too by the sort of audience one could attract through the publication of one's findings and by the display—in texts and cabinets—of the knowledge gained."[34] All too often, however, such displays are regarded as the end point of fieldwork rather than its starting point. The fact that botanical books can contain both text and specimens allows us to consider the effort involved in moving from written descriptions and dried specimens to the living forms of plants in their habitats. While the accuracy of a published local flora might be measured by how the plants described in it were "easily rediscovered" by novices,[35] this does not tell us much about how the observational practices involved in the recognition of plants were transmitted to the reader. Again, it is by attending to specimens in books that this process becomes more apparent, as we will see from a modest guide to the study of mosses that appeared in 1846. This work, *Twenty Lessons on British Mosses*, by a Dundee umbrella maker, William Gardiner, was aimed at young novices.

Gardiner's "little" publication devoted to "minute" plants eschewed engravings in favor of "real specimens."[36] His *Twenty Lessons* included a brief introduction to mosses, an explanation of their structure, and eighteen descriptions from "most of the larger and more widely-distributed genera" as "examples to guide you in your farther inquiries."[37] We have already seen how affect is readily incorporated into the assessment of botanical collections, both in discussions of the appeal of drawings and in the categorization of specimens as unlovely and dead. While such views could be dismissed as anachronistic, it is, however, perilous not to historicize sentiment, for nineteenth-century botany was sprinkled with (and exploited) aesthetic judgments. Although some of the specimens in the *Twenty Lessons* may indeed appear "page-buckling" and "rot brown" (fig. 12.3), Gardiner saw in them all "a very great deal of beauty."[38] And even more than not being seen as unlovely, these dried mosses were regarded by Gardiner not as dead plants but as the best guide to the forms of living ones. The idea of illustrating his book with "real specimens, instead of engravings," he argued, "must be allowed to be more effective, for the works of Nature are always superior to the imitations of art, *and the eye can more readily recognize a plant in the growing state by this means,* than by the most careful delineations of the pencil."[39]

The difficulty of pictorial representations for encouraging beginners to look for and to recognize mosses in nature is demonstrated by comparing a specimen in Gardiner's work with the depiction of the same species in

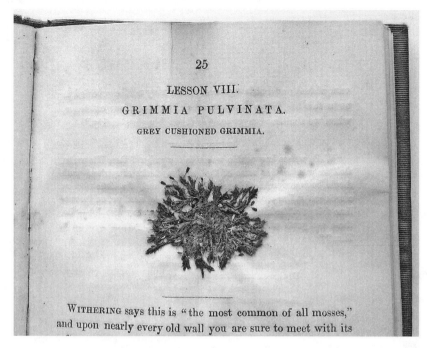

Fig. 12.3. Specimen of *Grimmia pulvinata* from the second series of William Gardiner, *Twenty Lessons on British Mosses* (London: Longman, 1849), 25.

Hooker and Taylor's illustrated monograph on mosses: *Hynum commutatum*, described by Gardiner as being distinguished by the shape and the twisting of its leaves, was illustrated in Hooker and Taylor's book by these characters alone (fig. 12.4).

While drawings do allow a magnified presentation of the classificatory characters of mosses, and thus provide the "optical consistency" required for comparative purposes, they are of more use for the scrutiny of mosses within the cabinet.[40] (Indeed, the need of the cabinet botanist for the "analysis" of cryptogamic plants "in magnified pictures" was, for Hooker, the great drawback of both nature-printing and photography.[41]) The specimen, however, does not serve to bring forth this type of discrimination only. Although in his nineteenth lesson Gardiner did explain how to identify a moss by analysis under a "good microscope,"[42] his aim was to encourage the recognition of mosses in nature. The *Twenty Lessons*, which gained the approbation of Hooker, did not, however, teach by a crude method of simply matching a flattened moss to a growing plant, since at the start of his book Gardiner explained the structure of mosses and took particular care

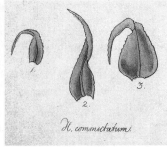

Fig. 12.4. Specimen of *Hypnum commutatum* in William Gardiner, *Twenty Lessons on British Mosses*, 2nd series (London: Longman, 1849), 57, compared to the illustration of the defining characteristic of this moss in W. J. Hooker and T. Taylor, *Muscologia Britannica*, 2nd ed. (London: Longman, 1827), plate 27.

to show the crucial characters used in their identification. Strikingly, even here he did not employ drawings but—perhaps to impress upon the reader the powers of observation required for such a study—used the minute parts of a moss itself (fig. 12.5).

SPECIMENS OF OBSERVATION

A writer in *Chambers's Edinburgh Journal* was certain that Gardiner's book would allow "any one, without the assistance of a teacher," to become acquainted with the leading tribes of mosses, and ensure "the recognition of the plants in a growing state."[43] Although Gardiner's book clearly sold well, we do not know whether the presentation of dried specimens enabled his readers to become good observers of mosses in nature. What, if any, evidence is there in the practices of botanists that shows how they might have acquired those observational skills, which, when highly honed in the keenest observers, could result in the production of exsiccatae or in significant

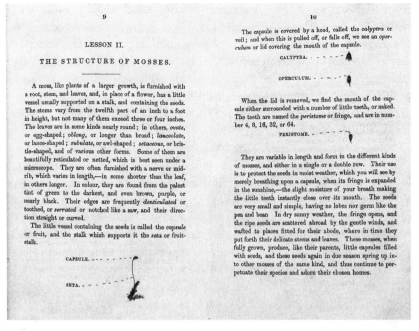

9

LESSON II.

THE STRUCTURE OF MOSSES.

A moss, like plants of a larger growth, is furnished with a root, stem, and leaves, and, in place of a flower, has a little vessel usually supported on a stalk, and containing the seeds. The stems vary from the twelfth part of an inch to a foot in height, but not many of them exceed three or four inches. The leaves are in some kinds nearly round; in others, *ovate*, or egg-shaped; *oblong*, or longer than broad; *lanceolate*, or lance-shaped; *subulate*, or awl-shaped; *setaceous*, or bristle-shaped, and of various other forms. Some of them are beautifully reticulated or netted, which is best seen under a microscope. They are often furnished with a nerve or midrib, which varies in length,—in some shorter than the leaf, in others longer. In colour, they are found from the palest tint of green to the darkest, and even brown, purple, or nearly black. Their edges are frequently *denticulated* or toothed, or *serrated* or notched like a saw, and their direction straight or curved.

The little vessel containing the seeds is called the *capsule* or fruit, and the stalk which supports it the *seta* or fruitstalk.

CAPSULE.

SETA.

10

The capsule is covered by a hood, called the *calyptra* or veil; and when this is pulled off, or falls off, we see an *operculum* or lid covering the mouth of the capsule.

CALYPTRA.

OPERCULUM.

When the lid is removed, we find the mouth of the capsule either surrounded with a number of little teeth, or naked. The teeth are named the *peristome* or fringe, and are in number 4, 8, 16, 32, or 64.

PERISTOME.

They are variable in length and form in the different kinds of mosses, and either in a single or a double row. Their use is to protect the seeds in moist weather, which you will see by merely breathing upon a capsule, when its fringe is expanded in the sunshine,—the slight moisture of your breath making the little teeth instantly close over its mouth. The seeds are very small and simple, having no lobes nor germ like the pea and bean In dry sunny weather, the fringe opens, and the ripe seeds are scattered abroad by the gentle winds, and wafted to places fitted for their abode, where in time they put forth their delicate stems and leaves. These mosses, when fully grown, produce, like their parents, little capsules filled with seeds, and these seeds again in due season spring up into other mosses of the same kind, and thus continue to perpetuate their species and adorn their chosen homes.

Fig. 12.5. The structure of moss capsules illustrated by actual specimens in William Gardiner, *Twenty Lessons on British Mosses*, 2d ed. (Edinburgh: David Mathers, 1846), 9–10. The size of the specimen on p. 9 is 2.2cm.

discoveries of rare or new plants? Once again, the unique flat-on-the-page materiality of plant specimens presents an opportunity to the historian.

For those who frequent antiquarian bookshops and fairs, it is not an uncommon experience to discover that previous owners of nineteenth-century botanical books have slipped pressed plants between the pages. Sometimes these are poor battered things lucky to have survived the opening and closing of the volume in which they remain, and giving no clue of their presence when the book is closed. In other cases, loose specimens can overwhelm a book, increasing its girth to such an extent that the covers bulge and no longer fulfill their protective function. Yet the obviously fragility of such overstuffed volumes often results in the specimens they contain having been more effectively preserved, for it is clear that they must be handled with care. Different again are those botanical books that have been dismantled and rebound with blank pages inserted between those of the printed text, so that specimens may be securely attached opposite their printed descriptions, with space for notes to be made by the collector. Even when a book bears an owner's signature, it is rare to discover much about the identity of collectors such as these. All we can know is that the author whose text has become a home for specimens has inspired an unknown collector to deposit a few found natural objects between its covers. In the absence of any other information, are the pressed plants in books "things that talk" in the sense of imparting their meaning through their particularity and context?[44] If we are to understand the significance of these traces of past scientific practice, it is to the trapped specimens that we should attend.

The site of preservation—the book itself—which acts as both stimulus to collecting and container for what is found, is potentially a revealing space of scientific practice. The meticulous annotations made by the owner of a copy of John Hull's *British Flora* of 1799 indicate some of the effort involved in honing one's observational powers. That the book was specifically adapted for such a purpose is indicated by the blank sheets bound between every page of text. There are notes beside many of the printed descriptions, and details of new plants recorded on the intercalated pages (fig. 12.6). It is

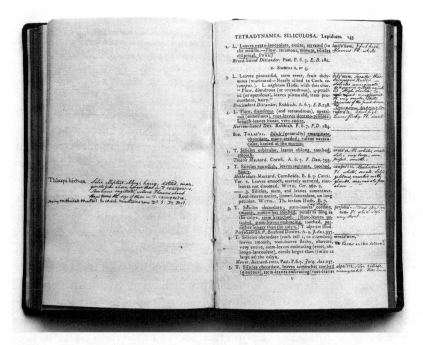

Fig. 12.6. Example of notes on the text and interleaved pages in part 1 of a copy of John Hull, *The British Flora; or, A Linnean Arrangement of British Plants*, 2 parts (Manchester: R. and W. Dean, 1799). Some of the additional notes refer to J. E. Smith's *Flora Britannica*, 3 vols. (London: J. White, 1800–1804).

not clear when these annotations were made, nor does it seem likely that they relate to field collecting rather than information gleaned from later publications. If, however, the use of this book does not bear direct witness to field practice, it perhaps does so indirectly. Mavor, when compiling his collecting book a year later and five years before the publication of Turner and Dillwyn's *Guide*, did not include the nonflowering plants. Thus, although one of Mavor's aims was to stimulate the discovery of new plants, his audience was not encouraged to pay any attention to the largely unexplored cryptogams when out in the field. The lack of interest in this class of plants is borne out by the owner of Hull's *Flora*. In contrast to the detailed notes relating to flowering plants, the intercalated pages in the part devoted to the cryptogamia remain almost completely blank.

From the evidence of the book alone, we cannot know whether the owner of Hull's *Flora*, despite the industrious notes on flowering plants, applied the knowledge so gained to collecting in the field. This contrasts with a copy of Richard Buxton's *Botanical Guide to the Flowering Plants, Ferns, Mosses, and Algae, Found Indigenous within Sixteen Miles around Manchester* (1849), also swelled to more than double its length by the binding in of blank pages. In this case, the purpose was not to make notes, but to provide space for the local plants to be preserved as specimens opposite their printed descriptions. While sections of this interleaved copy of Buxton's *Guide* remain empty, showing perhaps the difficulty of finding plants in even a restricted area, many pages register success in the field through the specimens that were glued into the book (fig. 12.7). At first glance, this practice might appear contrary to that of the book's author as revealed in an autobiographical introduction. Buxton, an impoverished Manchester shoemaker, scorned those who "attempted to learn botany from dried specimens," claiming instead that "a true idea of a plant" could be obtained only if was seen "alive and flourishing in the place where it grows, surrounded by all the conditions necessary for its growth."[45] Any apparent contradiction between the aims of the owner of the *Guide* and the author, however, stems from ignoring the fact that the presence of dried specimens was due only to their being recognized as growing plants in nature first. Just as Hobson's exsiccata confirmed his powers of observation, the different ways of seeing that a botanical collector had to possess, namely, recognizing growing and dried forms of plants and relating them to descriptions, also come together in the material evidence of the practical use of this work.

A copy of William Harvey's *Manual of the British Algae*, bulging at the seams with specimens, similarly bears witness to a reader's field collecting, although the owner's notes on the end pages of the book indicate that the

Fig. 12.7. Specimens glued into an interleaved copy of Richard Buxton's *Botanical Guide to the Flowering Plants, Ferns, Mosses, and Algae, found Indigenous within Sixteen Miles around Manchester* (London: Longman, 1849).

practical aspects of preserving seaweeds were learned from Isabella Gifford's 1848 *The Marine Botanist: An Introduction to the Study of Algology*. In this case—just as Hobson and Wyatt followed the systematic order of a recognized botanical authority—the collector seems to have used Harvey's descriptive work as an ordering device: in effect, as a cabinet, or container. The specimens were slipped in between the pages in labeled paper folders, and although they are loose within the book, they indicate care in their preparation and positioning in the text (fig. 12.8). As noted above, Harvey's *Manual* was not illustrated, and he recommended instead the use of Wyatt's exsiccata. In this case, however, Harvey's work was "illustrated" with specimens collected by the owner, who thereby also left evidence of the different ways in which field collectors had to relate to plants visually.

Fig. 12.8. Loose specimens in a folder labeled "Delesseria," opposite the start of descriptions of this genus in William Harvey, *A Manual of the British Algae* (London: John van Voorst, 1849), between pp. 54–55.

Far from signaling a lack of serious study or indicating a slovenly attitude to collections, the presence of specimens in descriptive books of botany was endorsed by expert botanists. The Reverend James Dalton, upon whose botanical knowledge the moss-mad sixteen-year-old schoolboy Henry Fox Talbot drew, emphasized that "the best memory needs frequently refreshing by overlooking the collections we make." Believing that Talbot's observation of growing mosses had advanced sufficiently that his "commencing an Herbarium is not very distant," Dalton outlined how he should proceed:

My plan with mosses, is, to attach a specimen to an interleaved Flor: Brit: opposite the description—& to draw magnified leaves &c, from microscopic observation. This facilitates investigation beyond conception. Our friend Hooker did the same, but lost his valuable Book when

returning from his Icelandic Tour. Smith's muscology is not worth four-
pence;—you had best, therefore, wait for the work now preparing by
Hooker & Doctor Taylor, have it interleaved, & proceed accordingly.[46]

Like Dalton, John Edward Smith, founder of the Linnean Society of London
and foremost proponent of Linnaean botany in Britain at the start of the nineteenth century, also used books as the means to capture most completely
the range of valuable firsthand observations upon which his work relied.
His library reveals evidence of the use of books both in the field and in preparing publications. His copy of William Hudson's *Flora Anglica* (1778) for
example, contains notes that Smith culled from other sources, new observations of his own, and pressed specimens.[47]

The book, construed as a performative space, extends our understanding of
the production, consumption, and circulation of knowledge. Evidence of the
ways in which the book can act as a signifier of scientific activity both
within and beyond the confines of a learned elite could be found across the
sciences, but it is explicit in botany because of the two-dimensional nature
of dried specimens. In emphasizing the importance of observational skill
over taxonomic expertise, the purpose of botanical guides is most clear in
the case of individuals gathering already known and often quite ordinary
plants. These particular records of practice are unlikely to be encountered
through systematic research, as they are only rarely found in libraries or
museums. Rather, such specimens and their host texts are usually serendipitous discoveries, for the context of their use and preservation is that
of private ownership. They are thus accidental remnants of past practice,
lucky survivors that prove to be rich sources of evidence. They operate as
such only if we avoid characterizations of science that devalue the "private"
and the "amateur." If, instead, we take seriously the aims of expert authors
(many of whom were amateurs) to encourage their readers to participate
in the science as articulated in their botanical guides, then the historical
traces of individuals learning how to observe are significant. Specimens in
privately owned books, like those in exsiccatae and nature-printed works,
even if now seemingly devoid of scientific import, continue to bear witness
"to practices of earlier transmission of knowledge."[48] With respect to early
nineteenth-century botanical knowledge, this involved both the means by
which authors encouraged and disciplined potential contributors to the science and the efforts of their readers to become aware of and to acquire the
observational skills that would make them worthy contributors. Although
the specimens trapped in privately owned books were not specifically des-

tined for circulation, exchange, or display, they symbolize the potential of their collectors to enter into these social networks.

Specimens gave individuals, including working men and servants their capacity to act in the private world of British botany not only by being material objects of exchange, but also because they displayed the attainment of the requisite observational skills to participate in the science. They linked the individual to the whole and tied private collecting to a notion of public good. In 1800, Mavor ensured that the users of his collecting book were aware that "the united labours and observations" of botanical observers would, in time, probably result in "a more perfect work on indigenous botany . . . than by any other method hitherto attempted."[49] We do not know how many of the collectors of the specimens placed in books went on to contribute to the researches of expert botanists. But regardless of whether their enthusiasm for botany fizzled out or their visual acuity came to be evaluated in terms of the scientific use of their collections, ultimately it is the evidence they provide of active engagement with botany that is significant.

By attending to the flat-on-the-page materiality of botanical specimens I have aimed to show that the space of the book as a site of scientific knowledge can display not only the products of observation but its very processes. The call of guides for the active involvement of collectors in the production of botanical knowledge should be seen less as an encouragement merely to wander the fields, and more as a call to develop a keen eye and the ability to translate between different ways of seeing. Focusing on the links between guides and audiences in terms of observational practices redefines the divides between private and public, amateur and professional, field and cabinet. Historicizing such categories allows us to recognize that dried specimens, even when in contexts of preservation that do not denote contributions to science, possess significance as indicators of past observational practice. These pressed plants are less important as representations of the objects described in books than of the practical experience of the reader. The traces of observational practices revealed by the presence of dried specimens in books thus show not only how plants but also how private individuals could be pressed into service.

NOTES

1. David N. Livingstone, "Text, Talk and Testimony: Geographical Reflections on Scientific Habits. An Afterword," *British Journal for the History of Science* 38 (2005): 94–95.

2. H. J. Jackson, *Marginalia: Readers Writing in Books* (New Haven, CT: Yale University Press, 2001).

3. Robert J. Mayhew, "Materialist Hermeneutics, Textuality and the History of Geography: Print Spaces in British Geography, c.1500–1900," *Journal of Historical Geography* 33 (2007): 467–68.

4. Dorinda Outram, "New Spaces in Natural History," in *Cultures of Natural History*, ed. Nicholas Jardine, James A. Secord, and Emma C. Spary (Cambridge: Cambridge University Press, 1996), 249–65; Dorinda Outram, "On Being Perseus: New Knowledge, Dislocation, and Enlightenment Exploration," in *Geography and Enlightenment*, ed. David N. Livingstone and Charles W. J. Withers (Chicago: University of Chicago Press, 1999), 281–94.

5. Charles W. J. Withers and Diarmid A. Finnegan, "Natural History Societies, Fieldwork and Local Knowledge in Nineteenth-Century Scotland: Towards a Historical Geography of Civic Science," *Cultural Geographies* 10 (2003): 336–37, 342.

6. Janet Hoskins, "Agency, Biography and Objects," in *Handbook of Material Culture*, ed. Christopher Tilley, Webb Keane, Susanne Küchler, Michael Rowlands, and Patricia Spyer (London: Sage, 2006), 82.

7. Unlike other natural history collections, living forms in botany had to be represented by wax, wood, or glass models. For the pedagogic use of such models as well as issues about the "counterfeit" replacing the real, see Lorraine Daston, "The Glass Flowers," in *Things That Talk: Object Lessons from Art and Science*, ed. Lorraine Daston (New York: Zone Books, 2004), 223–54. See also Ann Shteir, "'Fac-Similes of Nature': Victorian Wax Flower Modelling," *Victorian Literature and Culture* 35 (2007): 649–61.

8. Charles Darwin to Asa Gray, 8 March 1877, Gray Herbarium of Harvard University (117).

9. Roger Cooter and Stephen Pumfrey, "Separate Spheres and Public Places: Reflections on the History of Science Popularization and Science in Popular Culture," *History of Science* 32 (1994): 255.

10. Peter Murray Jones, "The Tabula Medicine: An Evolving Encyclopaedia," in *English Manuscript Studies, 1100–1700*, vol. 14, *Regional Manuscripts, 1200–1700*, ed. A. S. G. Edwards (London: British Library, 2008), 60–85.

11. William Mavor, *The Lady's and Gentleman's Botanical Pocket Book* (London: Vernor and Hood, 1800), viii–ix.

12. Susan Stewart, *On Longing: Narratives of the Miniature, the Gigantic, the Souvenir, the Collection* (Durham, NC: Duke University Press, 1993), 162.

13. W. J. Hooker to W. H. F. Talbot, 9 July 1836, Document 353, *The Correspondence of William Henry Fox Talbot*, ed. Larry J. Schaaf, at http://foxtalbot.dmu.ac.uk/.

14. Anne B. Shteir, "Bentham for 'Beginners and Amateurs' and Ladies: *Handbook of the British Flora*," *Archives of Natural History* 30 (2003): 237–49.

15. D. E. Allen, "An 1861 Instance of 'Painting One's Bentham,'" *Archives of Natural History* 31 (2004): 356–57.

16. Dawson Turner and L. W. Dillwyn, *The Botanist's Guide through England and Wales*, 2 vols. (London: Phillips and Fardon, 1805), 1:i, 32. Dillwyn had proposed pub-

lishing the *Guide* because he, "like many other botanists," was accustomed, when he
knew he was to leave home, to draw up from a variety of botanical works a list of the
plants he was likely to see in his travels (xii).

17. Ibid., 1:vii.

18. Thomas Walford, *The Scientific Tourist through England, Wales, and Scotland*,
2 vols. (London: John Booth, 1818), 1:i, iv–v, vi.

19. Lorraine Daston and Peter Galison, *Objectivity* (New York: Zone Books, 2007),
22, 26, 27.

20. H. Walter Lack, "The Plant Self Impressions Prepared by Humboldt and Bonpland
in Tropical America," in *Objects in Transition, an Exhibition at the Max Planck Institute for the History of Science, Berlin, August 16—September 2, 2007*, ed. Gianenrico
Bernasconi, Anna Maerker, and Susanne Pickert, (Berlin: Max Planck Institut für
Wissenschaftsgeschichte, 2007), 46–49.

21. John Lindley, "Preface," in *The Ferns of Great Britain and Ireland, by Thomas
Moore, Nature-Printed by Henry Bradbury* (London: Bradbury and Evans, 1855).

22. Ibid.

23. Quoted in Elizabeth M. Harris, "Experimental Graphic Processes in England,
1800–1859," pt. 4, *Journal of the Printing History Society*, 6 (1970): 56. Lucas's scrapbook
of nature prints is in the British Library.

24. Enid Slatter, "Botanical and Zoological Illustrations of the Nineteenth Century,"
The Linnean 24 (2008): 33; Andrea DiNoto and David Winter, *The Pressed Plant: The Art
of Botanical Specimens, Nature Prints, and Sun Pictures* (New York: Stewart, Tabori &
Chang, 1999), 92–117; Harris, "Experimental Graphic Processes," 60.

25. David E. Allen, *Books and Naturalists* (London: Collins, 2010), 394.

26. David Landsborough, *A Popular History of British Sea-Weeds* (London: Reeve,
Benham, and Reeve, 1849), 355; Ellen Hutchins to Dawson Turner, 27 November 1809,
Dawson Turner Correspondence, Trinity College, Cambridge.

27. Frederick Hanham, *Natural Illustrations of the British Grasses* (Bath: Binns and
Goodwin, 1846), ix, v, viii.

28. Ibid., vii.

29. William Hooker and Thomas Taylor, *Muscologia Britannica* (London: Longman,
1818), x.

30. W. H. Harvey, *A Manual of the British Algae* (London: John van Voorst, 1841),
liv, vi, liv.

31. Carol Armstrong, "Cameraless: From Natural Illustrations and Nature Prints to
Manual and Photogenic Drawings and Other Botanographs," in *Ocean Flowers*, ed. Carol
Armstrong and Catherine de Zegher (The Drawing Center, New York, in association with
Princeton University Press, 2004), 157.

32. "Sir Joseph Hooker's Reminiscences of Manchester," *Lancashire Naturalist* 1
(1907–8): 119; Joseph D. Hooker, *Flora Novae-Zelandiae. Introductory Essay* (London:
Lovell Reeve, 1853), vi; Jim Endersby, *Imperial Nature: Joseph Hooker and the Practices
of Victorian Science* (Chicago: University of Chicago Press, 2008).

33. William Wilson to W. J. Hooker, 1 November 1828, Directors' Correspondence,
Royal Botanic Gardens Kew, vol. 1, letter 283.

34. Withers and Finnegan, "Natural History Societies," 345.

35. Ibid., 344.

36. William Gardiner, *Twenty Lessons on British Mosses*, 2nd ed. (Edinburgh: David Mathers, 1846), 3. Gardiner's book went through two editions in the same year, and four editions had been produced by 1852, while a second series of *Twenty Lessons* containing a different set of twenty-five specimens was produced in 1849.

37. Ibid., 49.

38. Ibid., 8.

39. Ibid., 3; my emphasis.

40. Bruno Latour, "Drawing Things Together," in *Representation in Scientific Practice*, ed. Michael Lynch and Steve Woolgar (Cambridge, MA: MIT Press, 1990), 27–31.

41. W. J. Hooker to W. H. F. Talbot, 11 September 1859, Document 7954, *The Correspondence of William Henry Fox Talbot*, ed. Larry J. Schaaf, online at http://foxtalbot.dmu.ac.uk/.

42. Gardiner, *Twenty Lessons*, 2nd ed., 43–44.

43. "William Gardiner the Botanist," *Chambers's Edinburgh Journal*, n.s., 7 (1847): 251.

44. Lorraine Daston, "Speechless," in *Things That Talk: Object Lessons from Art and Science*, ed. Lorraine Daston (New York: Zone Books, 2004), 9–24.

45. Richard Buxton, *A Botanical Guide to the Flowering Plants, Ferns, Mosses, and Algae, found Indigenous within Sixteen Miles of Manchester . . . Together with a Sketch of the Author's Life* (London: Longman, 1849), vi.

46. James Dalton to W. H. F. Talbot, 2 October 1816, Document 720, *The Correspondence of William Henry Fox Talbot*, ed. Larry J. Schaaf, online at http://foxtalbot.dmu.ac.uk/. Dalton refers to James Edward Smith's *Flora Britannica*, 3 vols. (London: J. White, 1800–1804), and Hooker and Taylor's *Muscologia Britannica*.

47. H. J. Jackson, *Romantic Readers: The Evidence of Marginalia* (New Haven, CT: Yale University Press, 2005), 77.

48. Anke te Heesen, "On Scientific Objects and Their Visualization," in *Objects in Transition*, ed. Bernasconi, Maerker, and Pickert, 34.

49. Mavor, *Botanical Pocket Book*, viii.

CHAPTER THIRTEEN

Science, Print, and Crossing Borders: Importing French Science Books into Britain, 1789–1815

JONATHAN R. TOPHAM

In few periods has French science been of more significance in Britain than in the quarter century following the Revolution of 1789, when France held a dominant position in several scientific fields (notably chemistry, natural philosophy, paleontology, and comparative anatomy) and British practitioners responded to French innovations in highly productive ways. Yet the two countries were at war for twenty-one years during this period, with only two brief interludes of peace. Interaction between French and British scientific practitioners was thus marked not only by substantial cultural differences and by the national prejudices and rivalries typical of wartime, but also by the practical constraints imposed on communication. Previous scholarship on the British reception of French science in this period has generally assumed that printed accounts of French scientific work were straightforwardly available to British readers. However, printed matter always circulates partially and it does so more particularly in times of war. This partial circulation often takes place through the agency of technicians of print—such as publishers, booksellers, translators, and editors—who shape the process and outcome of communication, acting in diverse ways as "guides." In seeking to understand the processes by which scientific knowledge comes to be made and re-made in transit between spatially and culturally distinct locations, the impact of these guides on the audiences for science requires consideration.

This chapter investigates the processes of scientific communication in print between France and Britain during this critical period, and examines the role of certain key technicians of print in shaping those processes. I begin by providing a preliminary sketch of some of the means by which French science books were introduced to British readers. First, I focus on the difficulties of importing French works, particularly under Napoleon's

Continental blockade (1806–13), and on the role of booksellers in obtaining and publicizing such materials. I suggest that as more personal forms of communication were impeded, scientific readers became particularly reliant on import booksellers as mediators of French science. Secondly, I emphasize the role of periodicals in providing information about and extracts from French science books. Again, reviewers and editors, sometimes working in consort with the importing booksellers, had a critical role in guiding readers to an awareness of French science. In the remaining part of the chapter, I explore these processes in more detail through an examination of the importation into Britain of four key French scientific books: Lavoisier's *Traité élémentaire de chimie* (1789), Laplace's *Traité de mécanique céleste* (1799–1805), Lamarck's *Philosophie zoologique* (1809), and Cuvier's *Recherches sur les ossemens fossiles* (1812). These books have been selected not only because they ultimately proved to be of great significance in the development of nineteenth-century British science, but also because they range across the several sciences and across the period. In analyzing them, a further important group of print technicians come to the fore—the translators of foreign-language books and their publishers.

Although necessarily exploratory, the study provides strong evidence that French science books generally continued to reach Britain reasonably quickly throughout most of this period, despite the restrictive conditions of war, often through the activities of commercially motivated technicians of print. This finding gives new depth to existing studies of the responses of British men of science to French science in the period. The surprising alacrity with which import booksellers typically supplied copies of French scientific books to British readers stands in contrast to the selective manner in which those readers appropriated them. Two of the books considered here, those of Lavoisier and Cuvier, were seen to relate directly to existing scientific interests, and were, as a result, translated within months. The act of translation, moreover, served to advance their appropriation to British scientific interests. The other two books, though reasonably rapidly imported, appeared altogether more alien, remaining untranslated and relatively unread, at least for a spell. These selective appropriations highlight the cultural differences between British and French science, but the study also emphasizes, within Britain, differences between English and Scottish science. In all four cases, it was Scots who led the way in appropriating French science, which confirms that the close ties between Scottish and French intellectual culture that were notable in the eighteenth century continued to be important in this period. Here the complex, multi-factorial nature of the geography of nineteenth-century science comes to the fore, manifesting the importance of combining study of the mechanisms of communication between dispersed localities with the cultural conditions of those localities.

THE FRANCO-BRITISH BOOK TRADE AND SCIENCE

The Importation of French Science Books

Despite a recent turn toward the transnational history of the book, the study of the importation of French books into Britain in this period remains relatively undeveloped. According to Giles Barber, French books were "readily and promptly available" in Britain as "current stock items" from the 1740s and, despite heavy disruption as a result of war and economic measures, the growth of Franco-British book importation meant that France rose to rival Holland as the leading proximate source of Britain's Continental books by the later part of the century.[1] While a 1781 trade listing identified ten out of seventy-two London booksellers as dealing in French books, importation increasingly became a specialist business. In particular, a number of French-speaking booksellers settled in London during the war years, probably in part to serve the émigré community.[2] One émigré later recalled that in 1798 there were only three booksellers involved in the "French bookselling trade," the promising newcomer Arnaud Dulau (1762/3–1813), formerly a Benedictine mathematics teacher; the "mere retailer," Sarah Faraday's uncle Thomas Boosey; and the largest, Joseph De Boffe (1749/50–1807), on whom we shall concentrate here.[3] In fact, a number of other importing booksellers were also active in these years, several of them émigrés, but it was only after the wars that larger Continental bookselling concerns, such as Treuttel and Würtz and the Baillière family, established branches in London.[4]

Little research has been done on the practices of the eighteenth-century import trade, but Warren McDougall's study of Charles Elliot (1748–90) reveals that this prominent Edinburgh bookseller's Continental trade was based on a somewhat informal and makeshift network of correspondents and agents, including many medical practitioners.[5] More generally, networks of communication among men of science were mutually supportive of the commercial networks of booksellers. Such arrangements were significantly disrupted by the revolutionary and Napoleonic wars. While Gavin de Beer's studies of the attempts made by Joseph Banks to maintain a republic of letters under conditions of war certainly reveal the continuance of institutional and personal cooperation, they also demonstrate that there was considerable, if sporadic, disruption to personal travels, correspondence, and official communication.[6] War thus increased the reliance of scientific practitioners on those in the book trade who had a commercial motive for maintaining the importation of Continental books in increasingly difficult conditions.

While the economic effects of the wars have received extensive historical treatment, the effects on the book trade remain largely unexamined.[7] Even before declaring war in 1793, France had denounced the unpopular Anglo-French commercial treaty of 1786, and both nations soon had in place significant measures designed to curtail each others' export trade. Customs records for this period are intermittent, but official import statistics suggest a drastic reduction of book imports from France between 1792 and 1800 (see fig. 13.1). In part, this perhaps reflects the crisis in the French book trade following the revolution, but the disruption of the French export trade was one of the contributory causes in that crisis.[8]

Trade with the Netherlands, the German states, and other nations certainly provided alternative sources of supply (see fig. 13.2), and the very extensive smuggling of French goods also generally undermined official policies. Nevertheless, the importation of French books was significantly disrupted by the revolutionary war. The commercial war between France and Britain continued even during the Peace of Amiens, and in 1806 Na-

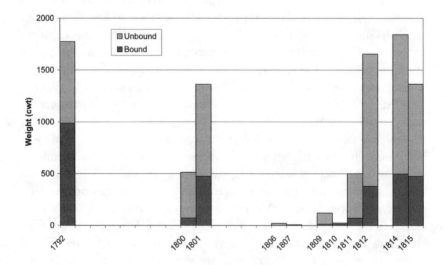

Fig. 13.1. Weight of books imported from France, 1792–1815.

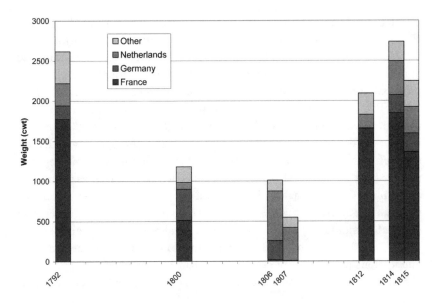

Fig. 13.2. Total weight of books imported from continental Europe, 1792–1815.

poleon sought to impose a rigorous blockade of the Continent, in order to conquer Britain by glutting its markets and to stimulate the Continental market for French imperial goods. The British counterblockade, initiated in 1807, portended increased difficulty. Official imports of French books fell away almost to nothing (fig. 13.1), and, although French goods continued to be smuggled, Continental books were for several years imported chiefly through North Sea ports (fig. 13.2). The difficult situation in 1800 was thus further exacerbated during the first half of the Napoleonic wars. The British began granting licenses for the importation of French goods in 1808, however, and when the French began issuing licenses for the importation of British colonial products in 1810, it was on the condition that French luxury products of the same value were first exported to Britain. The surfeit of books that filled French warehouses was considered ideal for these purposes, especially after 1813, although some cargoes were dumped at sea and others ultimately returned to France.[9] In the last years of the Napoleonic wars, imports of French books were thus once again considerable.

The extent to which official blockades were subverted during the wars should focus our attention on the active role of booksellers in obtaining copies of French books during these years. The evidence here is generally fragmentary, but David Bickerton's study of the Swiss natural philosopher Marc-Auguste Pictet's monthly magazine, the *Bibliothèque Britannique* (1796–1815), provides valuable insights. Pictet's efforts were mainly directed at maintaining the importation of British publications into Geneva, but his principal agent in this was the leading London importer Joseph De Boffe. The bulk of publications reached Pictet through the book trade, rather than through private individuals, although in the *"Sciences et Arts"* a surprising 30 per cent of the journal's content still came through private hands, some of it as original communications.[10] Pictet's complex supply network involved sea captains and agents in ports at each end, as well as intermediaries on the Continental supply route and elaborate financial arrangements to reimburse all those involved. Pictet initially relied on the exiled Genevan political economist François D'Ivernois to send publications from London, but he soon began to pay a clergyman named Prévost for this service. The publications passed through Dover and Paris, but arrangements proved unsatisfactory, and Pictet had to organize a chain of agents himself on a visit to England in 1798. In 1800, the ineffectual Prévost was replaced by De Boffe, whose firm continued in the role until its sale in 1817.[11]

Joseph De Boffe was the son of a French bookseller based in Fribourg, Switzerland, who, together with his sister Madeleine Eggendorffer (1744–95), had inherited his father's business in 1769. De Boffe went bankrupt in 1771 and left for London.[12] There he ultimately established himself as a significant figure in the supply of French-language publications, finally settling in Gerard Street, Soho. His success was partly based on his role in supplying books for and publishing the works of the émigré community.[13] His trade was much wider than this, however, and Chateaubriand reported that he was highly esteemed in the English book trade.[14] A catalogue issued by De Boffe in 1794 listed more than twenty-five hundred French books, many relating to the arts, sciences, travels, and natural history, and promised regular supplements containing new titles.[15] It also boasted that he could obtain French, Swiss, and German periodicals by the quickest routes. De Boffe was not only involved in the distribution of Abbé Rozier's *Journal de Physique* but was also named as the London agent on the title-page of the early volumes of Lavoisier's *Annales de Chemie* (founded 1789).[16] His reputation as a supplier of French science books attracted customers beyond London; both James Watt in Birmingham and Robert Jameson in Edinburgh sought the firm's assistance.[17] In 1800 De Boffe became the main supplier of books to the British Museum. On his death in 1807, when the business passed to his son, Joseph Charles (d. 1828), he was described as "an eminent importer of foreign books," who, "as a man of honour, and integrity," would be "long remembered."[18] By contrast, one Swiss émigré recalled De Boffe as speaking "wretched English" and as being "an ignorant drudge" whose "knowledge of books" was "*by the square foot* [emphasis in original]," but even this critic had to concede that De Boffe "possessed that sort of common sense and *tact de métier* which is sufficient for all practical purposes, and connived by thrift and prudence, and a sort of monopoly of his trade to accumulate a handsome fortune."[19]

It was his commercial know-how rather than any scientific acumen that made De Boffe invaluable to Pictet. Indeed, in the selection of publications, Pictet was later to rely on the Anglo-Genevan physician and chemist Alexander Marcet to prompt De Boffe to send more appropriate titles in his parcels.[20] However, although he was later critical of the bookseller when, under conditions of duress, his supply of books faltered, in 1803 Pictet considered that De Boffe would always know "better than anyone how we may send scientific communications."[21] De Boffe clearly had well-established routes for importing French books. On 11 and 21 August 1797, more than six hundredweight of unbound books arrived for him in the port of Dover, followed on 1 September by another consignment "consisting entirely of works relative to arts, sciences etc."[22] Given that, in 1800, the total weight of books imported from France averaged thirty-seven hundredweight per month, De Boffe's imports evidently amounted to a significant proportion of the whole—perhaps around a third. De Boffe's prominence is also evident from the lists of imported books appearing in contemporary periodicals.

It was in the import rather than the export trade that Pictet from 1797 first employed De Boffe, D'Ivernois considering him to be the only London bookseller capable of taking charge of a depot for the *Bibliotheque Britannique*. Initially De Boffe received his supplies of the journal directly from Paris on the Calais-Dover route, where neutral shipping, according to one 1795 estimate, provided a service almost on a par with that before the war. Following the French annexation of Geneva in 1798, and as conditions deteriorated after the Peace of Amiens, publications were sent via the North Sea ports and later Sweden.[23] Particularly after 1806, however, routes became unpredictable, and supplies in both directions were problematic. Pictet found it necessary to enlist Talleyrand's assistance in sending De Boffe's parcels to Geneva via prisoner-of-war cartel ships.[24] The Royal Society's records of receipt of free copies of the *Bibliothèque* suggest that the journal was "infinitely more successful in maintaining a regular link between France and the Society than any other claimant to such a role."[25] De Boffe's prowess in importing publications is confirmed by the fact that the Institut attempted to engage him as part of its network for supplying the Royal Society with books in 1806.[26]

The skill manifested by importing booksellers in overcoming wartime conditions appears to have ensured relatively continuous supplies of French books, albeit that they were rather slower, less plentiful, and less predictable than previously. As the example of De Boffe demonstrates, however, the decisions and activities of these relatively invisible technicians of print affected the availability of French science books in Britain in ways that are not easy to recover. Despite the criticisms of contemporaries, De Boffe clearly had his own sense of the market for French science books, and in 1806 he would reportedly only accept from his Paris agent such books as he expressly ordered. This might, of course, reflect the anxiety he felt concerning the government's view of his activity. When one book he imported was considered "to contain matters not exactly agreeable to certain persons," he was so alarmed at the government hint "that he should take care what works he sold" that he "sent back the whole parcel."[27] Such evidence only confirms that the activities of such importers and their selection of books are subjects worthy of further study.

The Role of Periodicals

As the example of Pictet reminds us, there were technicians of print other than booksellers with a keen commercial interest in supplying readers with information concerning new French scientific publications, namely, the editors and publishers of periodicals. By the time of the French revolution, several leading reviews devoted considerable attention to foreign literature, and this practice continued to expand as the number and variety of review journals increased over the next two decades. The liberal *Monthly Review* issued an impressive hundred-page supplementary number three times a year, consisting exclusively of reviews of foreign literature. This procedure was also followed from 1791 by the newly liberal *Critical Review* (1756–1817), and subsequently by the reactionary *Anti-Jacobin Review* (1797–1821). The High Church *British Critic* (1793–1843) included a monthly section of shorter reviews and notices entitled "Foreign Catalogue," while the leading organ of rational dissent, Joseph Johnson's *Analytical Review* (1788–99), integrated its foreign reviews with its domestic ones as, later, did the *Edinburgh Review* (f. 1802) and *Quarterly Review* (f. 1809). The *Analytical Review* was begun with an explicit promise to improve coverage of foreign works by including "regular advertisements, generally accompanied with short characters" of all the important books published on the Continent, "pointing out to literary men here, the *sources* where they may derive information relative to the subjects of their own pursuits."[28] The "sources" in question extended to the booksellers responsible for importing particular Continental books, whose names were often given in place of a publisher—for example, "Printed at Paris, and imported by De Boffe."[29]

The naming of importers became an increasingly common practice in the reviews as book supplies became constrained, suggesting that review editors relied on booksellers for their supply of new works. Before the war, the *Monthly Review* had relied for its foreign appendix chiefly on staff resident in the Netherlands, but this ended with the French occupation at the start of 1795.[30] The reviews clearly continued to depend on information about Continental literature reaching them through the mail. A regular section of the *Analytical Review* was devoted to literary news from the Continent, including notices of new books translated from foreign periodicals, and other reviews often incorporated similar coverage. While the content and production of these deserve systematic study, it is sufficient here to note the evidence they give of the disruption of literary intelligence as a result of the war. For instance, the *Analytical Review*'s early reliance on a range of Continental journals, many in French, had by 1796 been succeeded by a much narrower range, dominated by the *Allgemeine Literaturzeitung* of Jena. Similarly, by 1797 the *Critical Review* was complaining that the war made it impossible to "procure materials" for anything more than "a very imperfect list" of recent French works, although by 1800 the journal promised an improvement in consequence of having established a new "means of communication."[31]

Similar indications of the impact of the wars on the circulation of literary information are evident from the monthly magazines. While it was not the object of these magazines to provide lengthy reviews, it was in keeping with their encyclopedic ambitions that they carried extensive coverage of literary news as well as shorter book reviews. The *Monthly Magazine*, founded in 1796 by the radical publisher Richard Phillips, was notable in this regard. It came to pride itself on contributing "to a more extensive acquaintance with the proceedings, civil and literary, of other countries," than had usually been "obtainable from English publications."[32] The *Monthly*'s regular inventory of new publications came to include lists of newly imported books supplied by the various import booksellers. Significantly, the magazine also at first listed titles published in Paris "but not yet imported, or known, in England."[33] The *Monthly* sought to provide a regular retrospect of French literature, although this proved difficult at first. When, in

November 1796, a correspondent offered some literary notices of French books made possible "by means of an intelligent correspondent abroad," he noted, "our literary communications with France have been almost wholly interrupted, during the two last years, a circumstance which has tended, not a little, to restrict our knowledge of what has been lately achieved by the learned in that country."[34]

From June 1798, the *Monthly* maintained a biannual retrospect of the literature of France and other foreign countries. The first of the *Monthly*'s French literary retrospects was explicitly based on two French periodicals—the *Decade Philosophique* (1794–1807) and the *Magasin Encyclopédique* (1792–1818)—which perhaps explains why the retrospects, and the foreign literary intelligence that the *Monthly Magazine* carried, continued to be produced with impunity at times when the lists of books imported dwindled significantly.[35] Nevertheless, in July 1807 the *Monthly* apologized that the "reciprocal interdict" between Britain and France had limited the range of the retrospect of French literature, and by the following year it had become "difficult, if not impossible, to obtain any book of a very recent date."[36] Notices were often out of date, as exemplified by the reports of Laplace's *Traité de mécanique céleste* that appeared in the retrospect of French literature in 1808 and 1810.[37] In July 1811 it was reported that the "total Interruption of Communication with GERMANY" had made it "impracticable" to continue a retrospect of German literature, and indeed thereafter the foreign retrospect ceased altogether, notwithstanding that trade with France had soon returned to prewar levels.[38] A similar pattern is found in the lists of books imported: no French book imports were listed between March 1807 and November 1811.

During the war years, Britain also witnessed the development of the first commercial scientific journals, beginning in the 1790s with the chemist William Nicholson's *Journal of Natural Philosophy, Chemistry and the Arts* (f. 1797) and the inventor Alexander Tilloch's *Philosophical Magazine* (f. 1798), and later also including the Scottish chemist Thomas Thomson's *Annals of Philosophy* (f. 1813). Like the Continental journals on which they were modeled, the new journals stated that one of their key objectives was to provide readers with access to the scientific discoveries of other countries.[39] To a large extent, this meant translating or abstracting papers contributed to Continental journals. In addition, they provided some literary intelligence and reviews of books, including French titles. Here again there is evidence that the editors suffered communicative privations as a result of the wars. In 1815, for instance, Thomas Thomson wrote with relief of the ending of an "almost total exclusion from the Continent" which had lasted "for about seven years," and of the new possibility of making readers "acquainted with the various additions which the sciences have received during this eventful period" consequent on the renewed importation of "the different foreign journals." Thomson noted, however, that "the journals and scientific works" of France had already become "in some measure known" to his readers—partly through the *Annals of Philosophy*, and partly by means of the other London scientific journals.[40]

In sum, the evidence from this preliminary sketch of the mechanisms by which French scientific books were imported into Britain during the quarter-century of warfare with France suggests that the personal and institutional networks through which scientific works were communicated were significantly disrupted. Moreover, while to a large extent book importers and journal editors found ways to work around disruptions in trade, French books were for many years in short supply, with correspondingly high prices, and at certain periods (most notably under the Continental blockade between 1806 and 1812) there were serious shortfalls in the British capacity to obtain copies of, or even information about, French books. With this picture in mind, I now turn to examine the importation of four of the most significant books of French science published during the war years and consider the extent to which the peculiarities of their reception can be attributed to the changing conditions of communication.

IMPORTING FRENCH SCIENCE BOOKS: FOUR CASE STUDIES

Lavoisier's *Traité élémentaire de chimie* (1789)

While the British engagement with the work of French chemists, and of Lavoisier in particular, in the late eighteenth century has been a subject of extensive scholarship, little attention has been paid to the underlying mechanics of scientific communication until relatively recently. In her introduction to the collaborative study *Lavoisier in European Context*, for instance, Bernadette Bensaude-Vincent observes that the remarkable universalization of the French system of chemical nomenclature demands a new focus on the "vehicles and mechanisms of diffusion." She emphasizes the importance of the "intense activity of international exchanges," which took place primarily through the translation of treatises, as well as through institutional exchanges, the circulation of journals, and personal travel and correspondence.[41] These were, as noted above, typical features of Franco-British scientific exchange in the later eighteenth century. Appearing on the eve of the French Revolution, Lavoisier's *Traité* thus provides a useful reference point in relation to which we can compare examples from the war years.

Lavoisier's two-volume *Traité élémentaire de chimie*, providing in text-book form a comprehensive exposition of his new chemistry, was published in March 1789.[42] The author's presentation copy reached the Royal Society by the first meeting of the following session, in November, but copies were imported into Britain through commercial channels almost immediately, becoming available in bound form for a relatively modest ten shillings. As early as May, a brief synoptic review had appeared in Johnson's *Analytical Review*, praising the "activity of the chemists of the new school" while remaining agnostic about their prospects of ultimate success.[43] In June, a longer synoptic review appeared in the *Critical Review*, noting the work's flattering reception in Paris. The review was offered "at the earliest oppor-tunity" on the grounds that its author's "heresies" and his "abilities and zeal" had alike "astonished and instructed the world."[44] By March 1790, the *Monthly Review* had produced a lengthy laudatory review, written anony-mously by Josiah Wedgwood's secretary and chemical assistant, Alexander Chisholme. The reviewer considered that all chemists would want a copy of the new work, while acknowledging that for some readers the review's "con-cise view of the leading principles of the new system" would be sufficient.[45]

The rapid importation and reviewing of the *Traité* is striking. It may in part reflect the preexisting interest and excitement concerning Lavoisier's chemistry in Britain. As Carleton Perrin has shown, personal contacts were critical both in introducing Lavoisier's work to leading Scotttish protago-nists and in supplying them with his publications. Perrin focuses partic-ularly on the Edinburgh chemistry professor Joseph Black, who received Lavoisier's first work, the *Opuscules physiques et chimiques* (1774), as a gift from the author, who was given a copy of Lavoisier and Laplace's *Mé-moire sur la chaleur* (1783) by Jean-André Deluc, following the latter's visit to Lavoisier in Paris, and whose copy of the *Traité* was also sent by the author.[46] In addition, some of Black's students visited and were well re-ceived by Lavoisier in Paris, returning as proselytes. James Hall's champi-oning of the new chemistry after visiting Lavoisier in 1785–86 particularly helped make new converts in Edinburgh. Among these were Thomas Hope, who, as the new chemistry lecturer at Glasgow, became the first to teach the new chemistry in Britain, himself visiting Lavoisier in 1788, and the Edinburgh surgeon Robert Kerr, who, as we shall see, translated the *Traité*.[47]

These personal connections again came into play on the publication of the *Traité*. Hope had discussed the work with Black prior to publication, and on 14 July 1789 his presentation copy of the work reached him through Lord Daer, who had just returned from Paris. Hope had promised Black a sight of the volume as soon as it arrived, but thought it probable that he too had received a copy through Daer.[48] In fact, Black did not receive his own presentation copy until late September 1789, when it was sent via a student traveling to Edinburgh to complete his studies.[49] By the middle of September, a copy of the *Traité* had "fallen accidentally into [the] hands" of Robert Kerr, who set about translating it.[50] Kerr was a nephew of the prominent Edinburgh bookseller Alexander Kincaid, and he took his translation to Kincaid's protégé and successor, William Creech, then Edinburgh's leading bookseller, who had a strong commitment to producing the learned and literary works of the Scottish Enlightenment.[51] Kerr's decision to translate Lavoisier's work seems to have been actuated in part by a sense that there existed a sizeable market for the new work among Edinburgh's students of chemistry. Creech thus advised him that the translation should be "ready by the commencement of the University Session at the end of October" and Kerr accordingly "hazarded a large Edition."[52]

In the event, the translation was published in December 1789 in a single volume, undercutting the imported original at 7s. 6d.[53] Undertaken in around five weeks, Kerr's translation was largely literal, with few notes or alterations. The most significant modifications were in anglicizing the units of weight and temperature, and, given limited time for this, Kerr felt that he had to omit certain of Lavoisier's numerical tables. While he generally followed Lavoisier's nomenclature, he made one error in translation, which he noted in the preface, but he also sought to correct an error in one of the original plates.[54] The translation was reviewed in the *Critical Review* and the *Monthly Review*. As the reviewer in the former explained, the earlier review of Lavoisier's *Traité* as "a foreign work" had been "chiefly in an historical view, to point out the additions made to the stock of chemical knowledge"; in reviewing the translation, the reviewer felt able to take a more evaluative, less synoptic approach.[55] In the *Monthly Review*, Chisholme praised Kerr's translation, including his occasional pruning of Lavoisier's verbal excesses, and congratulated "the English reader on this acquisition of a complete system of the new philosophy."[56] By January 1791, Kerr's translation had almost sold out.[57] Over the next dozen years four more editions appeared. The French edition did not appear in De Boffe's catalogue of 1794, suggesting that the English editions had effectively supplanted it. Complaining in 1807 that John Murray had charged him for advertising some of the remaining copies of the translation, William Creech observed "Lavoisier's Chemistry is as well known as any book whatever."[58]

The rapid importation of Lavoisier's *Traité* through both commercial
and personal channels, and the rapid reviewing and translation of it, were
not atypical of the response to French scientific, and particularly chemical,
books in late eighteenth-century Britain. Lavoisier's earlier books had also
been swiftly translated into English. Thomas Henry's 1783 translation of a
series of Lavoisier's key papers from the *Mémoires* of the Académie provided
"the first work in *any* language to bring [them] together in a single volume."
Moreover, Henry's translations were not uncritical: he sought to advance
the debate by adding commentary on the divergences between Lavoisier and
Priestley and by sending copies to Lavoisier.[59] While Kerr did not share these
preoccupations, he did send a copy of his translation to Lavoisier in January
1791, outlining his own conversion to the latter's "luminous theory" and
expressing a desire to take into account any alterations or additions that
the author might suggest in preparing the second edition. In particular, he
requested that, if Lavoisier had a second French edition in preparation, he
should send a copy to Kerr through De Boffe. Lavoisier replied that he intended to republish the work in "an entirely new form, composing a Complete System of Philosophical Chemistry" and proposed to send it to Kerr
"sheet by sheet as it should come from the press."[60]

Kerr's letter was one of seventeen seized by the local revolutionary committee in 1793 as part of the process that led to Lavoisier's execution for
conspiracy with the enemies of France against the people. Lavoisier had
been anonymously accused of corresponding with an émigré, and, while De
Boffe was not a French émigré, Douglas McKie is surely right to argue that
"correspondence with him would certainly have been suspect during the
Terror."[61] The French chemist's untimely end thus gives dramatic form to
the disruption of an effective system of literary replication that had relied
heavily on personal contact. Trade was now also disrupted by war, and it is
symptomatic that in March 1793 Kerr told John Murray that he would send
him the next part of his translation of Linnaeus's *Systema naturae* "as soon
as a regular convoy [was] established between Leith and London," since,
much as he respected the French, he was not disposed to let them have his
work for nothing.[62]

LAPLACE'S *TRAITÉ DE MÉCANIQUE CÉLESTE* (1799–1805)

The huge, and later notorious, gulf that, at the turn of the eighteenth century,
separated natural philosophy in Britain (steeped in Newtonian geometry)
from "la physique" in France (with its reliance on analytical mathematics)
stands in sharp contrast to the extensive engagement that had taken place

between British and French chemists just a few years previously. Maurice Crosland and Smith noted in their groundbreaking study of the "transmission of physics from France to Britain" that British natural philosophers responded slowly to the physical astronomy of Laplace, Lagrange, and other French mechanicians. In particular, they observed that Laplace's work only began to be known in Britain around 1809, "through commentaries and brief reports."[63] Given that by 1799 both private and commercial channels for the circulation of French books had been overtaken by the events of the revolutionary war, it is clearly pertinent to ask whether the slow response of the British to Laplace's work was in part due to the difficulties of print communication.

Laplace's ground-breaking two-volume synthesis of physical astronomy shorn of mathematics and intended for a general readership, the *Exposition du système du monde* (1796), was imported by De Boffe within the year, and reviewed in both the *Critical Review* and the *Analytical Review*.[64] Supplies of it were probably limited, however, and early reviews did little to suggest the work's value in formulating a new approach to natural philosophy. Only the delayed review that appeared in the *Monthly Review* in October 1799 (written anonymously by the Cambridge-based enthusiast for mathematical analysis, Robert Woodhouse) gave a well-informed synopsis and appreciation of the book. Woodhouse noted, however, that, without the supporting mathematics, aspects of Laplace's reasoning would be unclear to readers, and he looked forward "with impatience to the appearance of the work which the author has promised on physical astronomy," namely, the more comprehensive and fully mathematical *Traité de mécanique céleste*.[65] Thus, while importers and reviewers brought the *Système du monde* to readers' attention, it was certainly not immediately championed in the public journals as having far-reaching consequences, and when a translation finally appeared in 1809, one reviewer noted that aspects of the work were "still in a great measure unknown" in Britain.[66]

The history of the *Mécanique céleste* also manifests both the relative effectiveness of importation mechanisms and the continuing significance of the cultural divide between "la physique" and British natural philosophy. The work's publication history is complex, with five volumes issued between 1799 and 1825. However, it was the first two volumes, published in September 1799 and sold at thirty francs, that provided the theoretical framework in which all observed astronomical phenomena could be shown, using analysis, to result from the law of gravity.[67] These quarto volumes arrived at the Royal Society in time for the new session in November, and De Boffe had imported them by December, when a lengthy synoptic review appeared in the *Critical Review*. This time, the reviewer appreciated the distinctive achievement of French physics: the work was "another remarkable instance" of the splendid and glorious prowess of French science in the post-revolutionary era. The reviewer borrowed Laplace's own description of his objective as being to treat astronomy as a "grand problem in physics," using mathematical analysis to bring together a century of research in which astronomical phenomena were reduced to the gravitational theory of Newton. Nevertheless, he regretted its "pure analytical" character "without one geometrical figure or diagram" (which he considered typical of "the modern French books upon physics") and urged instead the importance of a "happy medium" combining analysis with geometry.[68] A similarly appreciative but much more technical review of this "great and important publication" appeared in the *Monthly Review* the following April. Again written by Robert Woodhouse, the review was a lengthy and laudatory synopsis. It was, Woodhouse confessed, written with "deference and mental submission" to Laplace's "genius and acquirements," which had sometimes led him to become "entangled and bewildered in the intricacies of subtle analysis." Woodhouse considered English-speaking mathematicians so little advanced in analysis that he transliterated Laplace's word *differentiation* and explained that it meant "putting an equation or expression in fluxions."[69] His anonymous review of Silvestre Franciqs Lacroix's *Traité de calcul* (1797–1800) in the same number of the *Monthly Review* strongly urged "English" mathematicians to adopt Continental methods and notation in the calculus of fluxions.[70]

Despite these early reviews, we may infer that the number of copies imported was not large. Priced at two guineas in boards—about two-thirds as much again as the original price in France—they constituted a serious investment.[71] As previously with the *Système du monde*, while references were made to the *Mécanique céleste* in the *Philosophical Magazine* in 1800 and 1801, they appeared in translated accounts of proceedings of the Institut and contained tantalizing references to extracts from the original work published in the *Magasin Encyclopédique*.[72] Subsequent volumes and supplements were, however, also imported rapidly. The third volume was published late in 1802, during the Peace of Amiens. By April 1803, it had been imported by De Boffe and reviewed in both the *Monthly Review* and *Critical Review*. When the fourth volume appeared in May 1805, De Boffe again imported it, and it was reviewed in the *Monthly* in December.[73] The same pattern was repeated with each of the supplements issued in 1806, 1807, and 1808.[74] Yet, while in 1803 the reviewer in the *Critical Review* acknowledged the "great and universal celebrity" that the work had achieved and the "brilliant genius" of its author, he nevertheless regretted its "purely analytic nature" and warned of the need for geometrical interpretation.[75] Even Woodhouse admitted that the figures so notable by their absence from Laplace's works were "absolute necessities" to "Englishmen," and he accordingly welcomed De Boffe's importation for nine shillings of Jean Henri Hassenfratz's expository *Cours de physique céleste; ou, Leçons sur l'exposition de systême du monde* (1803).[76]

Publication of the third and fourth volumes gave John Playfair the opportunity in January 1808 to repair the earlier failure of the *Edinburgh Review* to review the work, noting that, in "the present state of Europe," it might be a long time before the fifth volume could "find its way" to Britain. To Playfair, the *Mécanique céleste* was the "most important book" of the age, and he lauded in superlative terms the results of Laplace's application of analysis to mechanics and physical astronomy. He ended, however, by bemoaning the failure of British philosophers to contribute to the post-Newtonian development of astronomy, attributing it to their "inattention to the higher mathematics." Of the *Mécanique céleste* he observed, "The number of those in this island, who can read that work with any tolerable facility, is small indeed. If we reckon two or three in London and the military schools in its vicinity, the same number at each of the two English Universities, and perhaps four in Scotland, we shall not hardly exceed a dozen; and yet we are fully persuaded that our reckoning is beyond any truth."[77] Playfair's well-known concluding criticism of the English universities and the Royal Society for failing to encourage the learning of analysis has rather obscured his observation that in Scotland, too, the understanding of Laplace's work was limited. Edinburgh had readers who early read and valued Laplace's works—not least Playfair's predecessor in the chair of natural philosophy, John Robison, who wrote appreciatively of both of Laplace's works soon after their publication, although preferring the geometrical to Laplace's analytical approach and condemning his exclusion of divine action from the universe.[78] However, the Englishman Woodhouse was also an early and consistent advocate in anonymous reviews of the importance of Laplace's work and approach, an advocacy that he also sought to bring to bear on Cambridge pedagogy.[79]

John Herschel later considered that Playfair's review gave a significant impetus to public recognition of Laplace's work in Britain.[80] Thomas Young implicitly responded to Playfair's comments a year later when, in the first issue of the *Quarterly Review*, he used the recent "paucity" of "continental publications" in Britain as an excuse for including a review of the celebrated Laplace's two "abstruse" supplements on capillary action. Agreeing with Playfair that fewer than "ten persons in the universe" had "read Laplace's

Mécanique Céleste as it ought to be read," Young nevertheless argued that British mathematicians had been preeminent in doing so. Motivated by his dispute with Laplace over capillary action, he criticized the "unnecessary abstraction" that resulted from the use of analysis, and reasserted the importance of a traditional geometric mathematical education.[81] However, Young now became part of a wider British engagement with Laplace's physics, manifested in the production of expository translations of *Mécanique céleste*. While the first two volumes of the work were rapidly translated into German, there was not even a partial English translation until 1814, when the Cambridge wrangler and Nottingham headmaster, John Toplis, privately published the first book of the *Mécanique céleste* "translated and elucidated with explanatory notes" for twelve shillings. Toplis's *A Treatise upon Analytical Mechanics* was intended to provide an introduction to analytical mechanics, which, if used with one of the Continental introductions to calculus, such as that of Lacroix, would pave the way for English readers to understand the *Mécanique céleste*.[82] Young's anonymous *Elementary Illustrations of the Celestial Mechanics of La Place*, only the first part of which was published by John Murray in 1821, also aimed, by translation and commentary, to render Laplace's work "intelligible to any person, who is conversant with the English mathematics of the old school only.[83]

Reviewing two later translations of the *Mécanique céleste* in the *Quarterly Review*, John Herschel attributed the British reluctance to become involved in the research program of which it was the crowning achievement to the country's "insular situation in the way of frequent personal communication between our mathematicians and those abroad, to the want of a widely diffused knowledge of the continental languages, and to the consequent indifference in the reading part of the public as to the direction which thought was taking . . . in other lands than our own." He also pointed to the importance of elementary works of analysis in English, most notably his own 1816 translation of Lacroix's *Traité de calcul*, in domesticating Continental notation and preparing the way for an appreciation of Laplace's work.[84] The evidence presented here demonstrates that both the importation and reviewing of Laplace's works were remarkably rapid through some of the most difficult of the war years and confirms that the reasons for the limited initial response lay elsewhere.

LAMARCK'S *PHILOSOPHIE ZOOLOGIQUE* (1809)

The delayed reaction in Britain to the extended account of Lamarck's transformist synthesis given in his *Philosophie zoologique* was, if anything, even more marked than the delayed reaction to Laplace's work. Recent scholarship has corrected an earlier perception that discussion of Lamarck's transformism was almost nonexistent before the publication of Lyell's extended criticism in the second volume of his *Principles of Geology* (1832).[85] Nevertheless, it was only in the 1820s that discussion of Lamarck's transformism became widespread in Britain, partly in consequence of the personal visits to Paris made in the postwar years by young medical men such as the radical Scots Robert Grant and Robert Knox, and by Lyell himself. It is thus valuable to examine the extent to which the British response to the *Philosophie zoologique* had earlier been affected by difficulties of print communication.

The two-volume *Philosophie zoologique* appeared in Paris in August 1809, priced at twelve francs.[86] Despite the blockade, Lamarck's presentation copy reached the Royal Society by the meeting of 1 March 1810. However, the earliest known reference to its commercial importation dates from October 1811, when the *Monthly Magazine* listed the work among the "livres nouveaux, importé par J. Deboffe, B. Dulau, et Co., et L. Deconchy," and it was priced at a hefty £1 4s., close to twice the French price.[87] This was a moment when the mutual blockade between Britain and France was having a significant impact on the importation of books, and it would not be unreasonable to conclude that the delay was at least in part a consequence of these difficulties. In addition, we must consider the possibility that the slow importation was also in part due to the work's limited reputation in France. As Pietro Corsi has shown, while Lamarck's earlier and shorter account of his transformist synthesis, the *Recherches sur l'organisation des corps vivans* (1802), excited considerable discussion in France, the new work went "virtually unnoticed" in Paris and was overshadowed by interest in Georges Cuvier and Alexandre Brongniart's 1808 paper on the geology of the Paris basin.[88]

The impression that copies of Lamarck's work did not reach Britain until the autumn of 1811 is also given by the fact that it was not reviewed in the *Monthly Review* until October of that year. Historians have hitherto taken the fact that the *Philosophie zoologique* was not reviewed in the *Edinburgh Review* or *Quarterly Review* to imply that it was not reviewed at all, but the older journals maintained a more extensive practice of reviewing than their newer rivals. The anonymous reviewer in the *Monthly Review* was Lockhart Muirhead (1765/6–1829), university librarian, lecturer, and, from 1807, first Regius Professor of Natural History at the University of Glasgow. Muirhead was well versed in French and Italian. He had traveled on the Continent on the eve of the revolution and had briefly taught both languages at the university, publishing a French grammar. He wrote extensively on natural history and Italian subjects for the *Monthly Review* and reviewed French natural history for Francis Jeffrey in the *Edinburgh Review*.[89] His lengthy two-part review of the *Philosophie zoologique* provided a detailed synopsis of the book's contents. The review was temperate in tone, devoting serious attention to Lamarck's "speculations" before concluding that "the learned and celebrated Professor" was "more ingenious than convincing." Lamarck's problem, Muirhead considered, was that, "like many of his countrymen," he was "more enamoured of bold and paradoxical discussion, than of sober observation and cool induction."[90] Nevertheless, Muirhead was somewhat persuaded of Lamarck's disproof of the "permanency of specific characters" and was eager to examine the "secondary agency" involved in producing such changes. Muirhead, the son of a Church of Scotland minister, would not allow religious objections to Lamarck's "theory of gradual evolution," pointing out that the "supreme Creator may educe his works in the way and manner which he deems most fit; that he . . . may so constitute and indue the more rude and simple forms, that they shall prove the proximate sources of gradually unfolding varieties."[91]

Other evidence from reviews seems to confirm that, while the *Philosophie zoologique* was not imported immediately, it was available to British readers relatively quickly. The *New Annual Register* for 1811, published in June 1812, gave a brief characterization of the work in its "Sketch of the Chief Productions of France, Germany, Italy, America" concerning "Physical and Mathematical" subjects.[92] Without specifically mentioning Lamarck's transmutation theory, the review reported that the book contained "much ingenuity and natural research" but that it was "in many respects loose and inconclusive." Like Muirhead, the reviewer admired Lamarck's "assemblage of facts rather than his chain of reasoning concerning them," concluding by recommending the work to all students of nature, in view of the undeveloped state of "zoological philosophy," and expressing a desire to see an English translation. A slightly longer notice of the work appeared in the *New Annual Register* two years later, which this time included a more detailed account of Lamarck's philosophy, including both his transmutation theory and his mental materialism. The reviewer's only critical observation was the opening comment that Lamarck's "fanciful principles" closely resembled "those of our ingenious but visionary countryman, Dr. Darwin."[93] A similarly synoptic but altogether more positive account of the book was contributed to the *Monthly Magazine* in June 1813 by a reader, "Verulam," whose lengthy extracts laid particular emphasis on Lamarck's transmutation theory. Such views were clearly not expected to be shocking to the magazine's liberal readers, and Verulam was able to point out that a similar "system" had been outlined by another contributor to the magazine the previous year. Lamarck's new work merited "careful translation," Verulam claimed, since it would come to "rank among our philosophical classics" and "form a proper supplement to the works of Locke, Hartley, and Reid, to which it would possess a superior claim in its more extensive application of facts, and in the greater stock of natural knowledge which the author has brought to bear upon his subject."[94]

While the general reviews and magazines gave early attention to *Philosophie zoologique,* the new scientific journals were initially heavily oriented toward natural philosophy and chemistry. It was not until 1826, once the range of magazines had diversified, that the earliest known discussion of Lamarck's work in the scientific press appeared. This was the anonymous, pro-Lamarckian article that appeared in the first volume of Robert Jameson's *Edinburgh New Philosophical Journal,* which is now thought to be by Jameson himself. Jameson kept up to date with Continental natural history better than most of his peers, publishing translations of many Continental memoirs in his journal.[95] The specialist medical press, however, had shown an earlier interest in Lamarck's work. By March 1812, London physician John Yelloly had already quoted a medical case history from the *Philosophie zoologique* in the *Medico-Chirurgical Transactions.*[96] In July 1812 the short-lived *London Medical Review* (published by Longmans and edited by London surgeon Joseph Hodgson) listed Lamarck's work among a dozen "important Works on Medical Science" published over the preceding six years, which had been imported by the medical bookseller Thomas Underwood.[97] This was followed in October by a lengthy anonymous review, which presented a brief overview of the work and a detailed exposition and criticism of the leading points of each of its three sections. As with the *Monthly's* discussion, the review was calm in tone, accorded Lamarck some respect as a man of science, but more rapidly moved to dismiss *Philosophie zoologique* as mere French speculation. It was, the reviewer observed, a "specimen of that kind of intellectual work to which the present race of scientific writers in France is so much addicted," which makes "no addition to our stock of facts" but which presents "new views," attractive to the juvenile or senile though generally "tiresome and uninteresting to grown-up minds." In particular, transmutation of species through inheritance of acquired characteristics was dismissed as being a "guess" that, since it was also the doctrine of Erasmus Darwin, had not even the "recommendation of novelty." In "these days of strict induction," it would "win over but few converts."[98]

Most of these newly discovered early reviews of *Philosophie zoologique* represented it as a speculative French curiosity, and none was unduly concerned by its doctrines, most clearly anticipating that it would not acquire British adherents.[99] As late as 1819, William MacLeay expressed the opinion that Lamarck's theory had made few "converts" in France and none in Britain.[100] By then the latter's reputation in both countries had begun to rise, however, and, partly in consequence, his transmutation theory had begun to be taken more seriously. As Pietro Corsi points out, Lamarck's *Histoire naturelle des animaux sans vertèbres* (1815–22)—published at a time when Cuvier and Brongniart's work on the Paris basin had generated fervent interest in fossil-based stratigraphy—rapidly became a standard reference work for geologists as well as zoologists, and a portion was translated into English in 1826.[101] The lengthy introduction to the *Histoire naturelle* again brought his transmutation theory to the attention of British readers, and commentators were now more concerned by its materialist import.[102] In addition, Lamarck's views were given wider currency and authority in both France and Britain by Julien-Joseph Virey's *Nouveau dictionnaire d'histoire naturelle* (2nd ed., 1816–19) and by Jean-Baptiste-Georges Bory de Saint-Vincent's *Dictionnaire classique d'histoire naturelle* (1822–31).[103]

In addition to these various forms of "literary replication," the postwar visits to Paris of British medical men and naturalists were significant in developing a sense of the importance of Lamarck's work. Both Pietro Corsi and Adrian Desmond have emphasized that British medical students in Paris in the 1820s learned to admire Lamarck from the teaching of Etienne Geoffroy Saint-Hilaire, and the political capital they made by exploiting his views only added to growing fears in Britain that they might undermine the social fabric.[104] In these circumstances, British readers such as Lyell felt a pressing need to familiarize themselves with the *Philosophie zoologique*, and French bookseller Hippolyte Baillière, who opened a London branch of the family firm in 1827, partly in order to sell some of the excess book-stock in the Paris branch, had the work in stock by July 1828, priced at twelve shillings (less than half the price of 1811).[105] In 1830, following Lamarck's death, Hippolyte's Paris-based brothers Germer and Jean-Baptiste bought up the remaining stock of *Philosophie zoologique* at auction, printing a new title page and issuing it as a second edition to meet the renewed demand.[106] Twenty years before, however, while the blockade did not delay the book's importation by more than a couple of years, the work had not seemed relevant enough to British science to be of great concern.

CUVIER'S *RECHERCHES SUR LES OSSEMENS FOSSILES* (1812)

Cuvier and Brongniart's 1808 paper on the geology of the Paris basin, published in the *Annales du Muséum national d'histoire naturelle* and the *Journal des mines*, was avidly discussed in France. As Martin Rudwick has shown, news of it reached Britain relatively quickly despite the difficulties of the blockade, with a translation appearing in the *Philosophical Magazine* in 1810.[107] By 1812, when Cuvier issued a four-volume collection of his many memoirs on fossil bones from the *Annales*, prefaced by an important preliminary discourse setting out his general conclusions in an accessible manner, the importation of French books had again become rapid and extensive. Cuvier did not donate a copy of this work to the Royal Society, but it nevertheless soon reached Britain through commercial channels. Published in December 1812, news of it reached Britain by April 1813, and it had been imported by June, priced at a monumental eight guineas. While its publication was publicly noted, it was not reviewed in any of the leading journals.[108] This is perhaps in part accounted for by the technicality of the work and also by its size and expense. Moreover, the bulk of the volume was made up of previously published papers, some of which had already been reviewed by the *Edinburgh Review*.[109] Perhaps, in addition, editors were aware that a translation of the work's lengthy and original introduction was in preparation. This had been published by mid-November 1813, and for many purposes it soon came to stand in place of the French original.[110]

The translation was carried out by Robert Kerr, the translator of Lavoisier's *Traité*, who, however, died on 11 October 1813. Given how soon afterwards the translation was in print, it seems likely that its editor, Robert Jameson, had been involved with Kerr from the outset. It was Jameson who wrote the preface, along with a number of notes and a "short account of Cuvier's Geological Discoveries" for those without "the opportunity of consulting the great work."[111] The translation was priced at a modest eight shillings, less than one-twentieth of the price for the imported original. In place of the original four heavily illustrated quarto volumes, the translation ran to 278 barely illustrated octavo pages.[112] On financial grounds alone, it is thus hardly surprising that it was in translation that Cuvier's work became generally known to an English-speaking audience, with reviews appearing across a wide range of journals. Over the next fourteen years, four more British editions appeared, undergoing significant modifications (including retranslation from later editions of Cuvier's original) at the hands of Jameson, for whom it continued to be a significant source of income.[113] Meanwhile, no complete translation of the work appeared. Leading publisher George

Byrom Whittaker estimated that such an edition, including the reengraving of the more than three hundred plates, would cost readers £20–£25 per set. Even Whittaker's 1826 attempt to publish an abridged translation in eight quarterly parts at twenty-four shillings each failed after the first issue, although he was subsequently more successful with a splendid translation of Cuvier's *Le règne animal* (1817), the plates for which cost seven thousand pounds to produce.[114]

There can be little doubt, as Martin Rudwick has argued, that the mediation of Cuvier's work through Kerr and Jameson's edition of the preliminary discourse significantly affected British readings of Cuvier's work. The choice of title for the translation—*Essay on the Theory of the Earth*—reflected Jameson's preoccupation with answering his Edinburgh rival John Playfair, the champion of James Hutton's *Theory of the Earth* (1795). While Cuvier had used the phrase "theory of the earth" *en passant* to describe his discourse, Rudwick contends that Jameson's designation ignored Cuvier's opposition to geotheory.[115] More fundamentally, Jameson's preface and notes sought to link the most recent geological revolution in Cuvier's scheme with Noah's flood, giving the latter a scientific credibility somewhat out of keeping with the intention of Cuvier's original.[116] When the book was advertised in the newspapers, it appeared with an announcement that reduced the force of the titular innovation while reinforcing Jameson's religious emphasis:

> To the Student of Geology this work will be extremely useful, as containing some very important and original views on the subject, and as being more occupied with the detail of interesting facts and observations than the establishment of an ingenious theory. The Christian may furnish himself from this production of a Parisian philosopher, with armour to defend his faith against those writers who have endeavoured to overturn it by objections against the Mosaic account of the deluge, and the age of the human race.[117]

While Jameson's version of the preliminary discourse certainly had a transformative intent that undoubtedly affected British readings of Cuvier's work, several of the contemporary reviewers were not only conscious of Jameson's intent, but resistant to it. Given that the review in the January 1814 *Edinburgh Review* was probably by Playfair, it is unsurprising that, while moderate in tone, it criticized the use of the word *theory* in the title and repudiated the connection of geological revolutions with the Noachian flood.[118] Other critics joined the chorus. The High Church *British Critic* went to some lengths to show that Jameson's injudicious choice of title was entirely at odds with Cuvier's careful distancing of himself from speculative theory and his patient fact-gathering. It also disapproved of "the pious solicitude of the English editor of Cuvier's Essay to corroborate the statements of Moses by a reference to geological discoveries" (although the reviewer exhibited a similar concern despite himself).[119] In the *Philosophical Magazine* one correspondent wrote that, having been led by Jameson's preface to expect geological evidence of the scriptural account of creation, he had been disappointed not to "discover the promised agreement." Another correspondent, an opponent of geotheory who had hoped to find it suitable as a target for his anti-Wernerian attacks, wrote disappointedly of its being "nothing like a *Theory of the Earth.*"[120] We should thus be aware that, while Jameson's rapidly produced translation was certainly responsible in part for the distinctive British reading of Cuvier's discourse, it was not exclusively so. Moreover, the translation was utterly insufficient for technical purposes, and some naturalists certainly used the expensive original work.[121]

This chapter has used four brief case studies to investigate the relevance of the practicalities of print communication in understanding the appropriation of scientific work from France to Britain during the revolutionary and Napoleonic wars. The case of Lavoisier's *Traité* clearly underlines both the importance of personal communication in the rapid appropriation of Continental scientific work in Britain on the eve of the Revolution and the rapid disruption of such communication networks from 1793. As customs records and contemporary testimony indicate, the commercial channels for obtaining French science books were put under serious strain by the restrictions on both communication and commerce during the war years. What is striking, however, is how effectively the strategies of import booksellers and periodical editors maintained the supply of information about and copies of French books under these conditions. As the case of Laplace's *Mécanique céleste* illustrates, copies of key scientific books continued to reach Britain surprisingly quickly, although perhaps at higher prices and in smaller quantities than before, and the review journals continued to provide British readers with details of their contents. Lamarck's *Philosophie zoologique* appeared at the nadir of French book importation, when the mutual blockade of Britain and France was at its height, but while its importation apparently suffered a significant delay, it is difficult to be entirely sure whether the blockade or the book's low profile in France was the key factor. Even here, however, the mechanisms of print communication worked more effectively than historians have suspected in making the work's contents available to British readers. Yet it was only after the end of the war, when more personal forms of communication were again possible, and the growing reputation and new uses of Lamarck's work in France became known, that the book began to seem particularly relevant or troubling. The case of Cuvier's *Ossemens Fossiles* brings us full circle, not least through the involvement of Robert Kerr. Benefiting from the renewed vigor of Franco-British trade toward the end of the war, the work was rapidly imported, but it was in English translation that the work came to be distinctively known in a British context.

The case studies considered here manifest the importance of taking seriously the mechanisms by which scientific publications traverse physical and cultural space. Far from being automatic, such mechanisms require the agency of a wide range of people, including not only scientific practitioners but also technicians of scientific print, often motivated by financial considerations. These actions shape the movement of knowledge. Those who carry them out become, to a greater or lesser extent, guides to the knowledge of a wider world. One of the more revisionist aspects of the account given here is the emphasis on the role played by import booksellers, notably Joseph De Boffe, in maintaining supplies of French scientific books in Britain. De Boffe's primary motivation was clearly profit, and, while this profit motive doubtless generated principles of selection, a far more extensive study would be required to identify them. A more familiar group of guides in the appropriation of French scientific publications were the reviewers in periodical publications—frequently scientific practitioners—whose synopses and evaluations often stood in place of the original works. Yet, while reviewers are more familiar to historians than book importers, they continue to be underexamined, as demonstrated here by the unearthing of previously unknown early reviews of Lamarck's *Philosophie zoologique*.

One of the most striking features of this study is the prominence of Scottish reviewers in introducing French science to British readers more generally, thus nicely highlighting the complex overlapping geographies of nation, state, language, and print distribution. Scots were also prominent as translators of French science. As the case of Jameson's editions of Cuvier's *Discours* suggest, translations often significantly transformed meaning, forming a crucial step in the process of appropriation. In the case studies considered here, the two books that were most rapidly and effectively appropriated by British men of science were those for which translations soon appeared (although it is difficult to establish the extent to which this was the cause or the effect of translation).

My purpose more generally has been to explore the impact on the local appropriation of knowledge claims of the processes by which such claims circulate in print. The point of such studies is not so much that they provide evidence of how certain knowledge claims come to achieve global acceptance but that they show how all knowledge-making takes place through such processes. While scientific work in Edinburgh, London, or Paris might be shaped by context—by local physical, social, and cultural geography—it is also shaped by the larger regional, national, and international geography in which it is situated by dint of the multifarious mechanisms of communication. As James Secord has it, it is "not so much a question of seeing how knowledge transcends the local circumstances of its production but instead of seeing how every local situation has within it connections with and possibilities for interaction with other settings."[122] By combining an understanding of local geography with greater awareness of the processes by which wider geography bears upon it, historians of science may hope to give a more multifaceted and ultimately a more satisfactory history of knowledge-making.

NOTES

I would like to acknowledge the helpful comments and criticisms of Geoffrey Cantor, Jack Morrell, Anne Secord, the editors of this volume, and all those who attended the conference from which this volume springs.

1. Giles Barber, "Treuttel and Würtz: Some Aspects of the Importation of Books from France, c. 1825," The Library, 5th ser., 23 (1968): 125; Giles Barber, "Books from the Old World and for the New: The British International Trade in Books in the Eighteenth Century," in Studies in the Book Trade of the European Enlightenment, ed. Giles Barber (London: Pindar Press, 1994), 225–64; Giles Barber, "Pendred Abroad: A View of the Late Eighteenth-Century Book Trade in Europe," in Studies in the Book Trade in Honour of Graham Pollard, ed. R. W. Hunt, I. G. Philip, and R. J. Roberts (Oxford: Oxford Bibliographical Society, 1975), 238. On the commercial hostilities between France and Britain at this period, see Eli F. Heckscher, The Continental System: An Economic Interpretation (Oxford: Clarendon Press, 1922).

2. Barber, "Pendred Abroad," 234, 253–55; Barber, "Treuttel and Würtz," 125; David Shaw, "French Émigrés in the London Book Trade to 1850," in The London Book Trade: Topographies of Print in the Metropolis from the Sixteenth Century, ed. Robin Myers, Michael Harris, and Giles Mandelbrote (New Castle, DE: Oak Knoll Press, 2003; London: British Library, 2003), 127–43.

3. John Lewis Mallet, *John Lewis Mallet: An Autobiographical Retrospect of the First Twenty-Five Years of His Life* (Windsor: Thomas E. Luff, 1890), 209; *Monthly Magazine* 36 (1813): 359.

4. Barber, "Treuttel and Würtz," 125; see also Josep Simon, "The Baillières: The Franco-British Book Trade and the Transit of Knowledge," in *Franco-British Interactions in Science since the Seventeenth Century*, ed. Robert Fox and Bernard Joly, 243–62 (London: College Publications, 2010).

5. Warren McDougall, "Charles Elliot's Medical Publications and the International Book Trade," in *Science and Medicine in the Scottish Enlightenment*, ed. Charles W. J. Withers and Paul Wood (East Linton: Tuckwell Press, 2002), 215–54.

6. Gavin de Beer, *The Sciences Were Never at War* (Edinburgh: Nelson, 1960).

7. See the classic accounts in Heckscher, *Continental System*, and François Crouzet, *L'économie Britannique et le blocus continental (1806–1813)* (Paris: Presses Universitaires de France, 1958); for a more recent introduction, see Geoffrey Ellis, *The Napoleonic Empire*, 2nd ed. (Basingstoke: Palgrave Macmillan, 2003), 109–18.

8. Carla Hesse, *Publishing and Cultural Politics in Revolutionary Paris, 1789–1810* (Berkeley and Los Angeles: University of California Press, 1991), 126–28, 136–37.

9. "Some Account of the Commerce between England and France during the Existence of the Continental System; Particularly with Respect to the Book Trade," *New Monthly Magazine* 14 (1820): 50–54; see also Barber, "Treuttel and Würtz," 123–24.

10. David Bickerton, *Marc-Auguste and Charles Pictet, the "Bibliothèque Britannique" (1796–1815), and the Dissemination of British Literature and Science on the Continent* (Geneva: Slatkine Reprints, 1986), 313–16.

11. Ibid., 333–47.

12. Georges Andrey, "Madeleine Eggendorffer, libraire à Fribourg et la Société Typographique de Neuchâtel (1769–1788): Livre, commerce et lecture dans la Suisse des Lumières," in *Aspects du livre neuchâtelois: Études réunies à l'occasion du 450ᵉ anniversaire de l'imprimerie neuchâteloise*, ed. Jacques Rychner and Michel Schlup (Neuchâtel: Bibliothèque Publique et Universitaire, 1986), 118–22. When De Boffe became a British denizen in 1806, he reported that he had been resident in the United Kingdom for more than thirty-five years; The National Archives (hereafter TNA), HO 44/46, ff. 88–89, 94.

13. Simon Burrows, *French Exile Journalism and European Politics, 1792–1814* (Woodbridge, Suffolk: Royal Historical Society and Bodyell Press, 1992), esp. 61–62, 66–68; David J. Shaw, "French-Language Publishing in London in 1900," in *Foreign-Language Printing in London, 1500–1900*, ed. Barry Taylor (Boston Spa and London: British Library, 2002), 101–22.

14. François de Chateaubriand, *Mémoires d'outre-tombe*, 4 vols. (Paris: Bordas, 1989–98), 1:563.

15. [Joseph De Boffe], *Catalogue des livres Français de J. De Boffe, libraire, Gerrard-Street, Soho, à Londres* (London: J. De Boffe, 1794).

16. Maurice Crosland, *In the Shadow of Lavoisier: The "Annales de Chimie" and the Establishment of a New Science* ([n.p.]: British Society for the History of Science, 1994), 88–89.

17. See the letters between De Boffe and Watt in the Birmingham City Archives,

James Watt Papers, 4/28 and 4/66; and the letters from J. C. De Boffe to Robert Jameson, 14 May and 1 December 1816, Edinburgh University Library, Ms. Gen. 122.73–74.

18. P. R. Harris, *A History of the British Museum Library, 1753–1973* (London: British Library, 1998), 39; *Gentleman's Magazine* 77, pt. 2 (1807), 785. In 1817 the business was taken over by Treuttel and Würtz; see Barber, "Treuttel and Würtz," 128. For wills for father and son, see TNA, PROB 11/1465 and 1748.

19. Quoted in Burrows, *French Exile Journalism*, 61.

20. Bickerton, *Marc-Auguste*, 364–67.

21. Quoted in ibid., 347 (my translation).

22. Records of the Board of Customs, Excise, Dover Letter Books, TNA, CUST 54/10, 231 and 246.

23. Bickerton, *Marc-Auguste*, 338–39, 348.

24. Ibid., 349–58.

25. Ibid., 351–52.

26. de Beer, *Sciences*, 174.

27. Ibid., *Sciences*, 174; *Hansard*, 1st ser., 38 (1818): 1022.

28. [Joseph Johnson], *Prospectus of the "Analytical Review," or a New Literary Journal, on an Enlarged Plan* ([London: Joseph Johnson], 1788), ii.

29. [Review of *Système du monde*, by P. S. Laplace], *Analytical Review* 24 (1796): 610.

30. Benjamin Christie Nangle, *The Monthly Review, Second Series, 1790–1815: Indexes of Contributors and Articles* (Oxford: Clarendon Press, 1955), 15 and 64–65.

31. *Critical Review*, 2nd ser., 20 (1797): 551; 2nd ser., 24 (1798): 561; 2nd ser., 28 (1800): 545.

32. *Monthly Magazine* 5 (1798): [iii].

33. *Monthly Magazine* 2 (1796): 813 and 899; 6 (1798): 136.

34. Z, "Present State of French Literature," *Monthly Magazine* 2 (1796): 471.

35. "Retrospect of the Present State of French Literature," *Monthly Magazine* 5 (1798): 532.

36. *Monthly Magazine* 24 (1807): 634; 26 (1808): 640.

37. *Monthly Magazine* 26 (1808): 650–51; 29 (1810): 659–60.

38. *Monthly Magazine* 31 (1811): 897.

39. James E. McClellan III, "The Scientific Press in Transition: Rozier's Journal and the Scientific Societies in the 1770s," *Annals of Science* 36 (1979): 433; Crosland, *In the Shadow*, 76.

40. Thomas Thomson, "Sketch of the Latest Improvements in the Physical Sciences," *Annals of Philosophy* 5 (1815): 1.

41. Bernadette Bensaude-Vincent, "Introductory Essay: A Geographical History of Eighteenth-Century Chemistry," in *Lavoisier in European Context: Negotiating a New Language for Chemistry*, ed. Bernadette Bensaude-Vincent and Ferdinando Abbri (Canton, MA: Science History Publications, 1995), 8–9. See also Crosland, *In the Shadow*.

42. Douglas McKie, "Introduction to Dover Edition," in *Elements of Chemistry in a New Systematic Order, Containing All the Modern Discoveries*, by Antoine-Laurent Lavoisier (New York: Dover Publications, 1965), v.

43. [Review of *Traité élémentaire de chimie*, by A.-L. Lavoisier], *Analytical Review* 4 (1789): 52–53.

44. "Lavoisier's *Traité élémentaire de chimie*," *Critical Review* 67 (1789), 531.

45. [Alexander Chisholme and Thomas Cogan], "Lavoisier's Elementary Treatise on Chemistry," *Monthly Review*, 2nd ser., 2 (1790): 311, 316.

46. C. E. Perrin, "A Reluctant Catalyst: Joseph Black and the Edinburgh Reception of Lavoisier's Chemistry," *Ambix* 29 (1982): 145 and n. 24; Arthur Donovan, "Scottish Responses to the New Chemistry of Lavoisier," *Studies in Eighteenth-Century Culture* 1 (1979): 241.

47. Perrin, "Reluctant Catalyst," 154–56.

48. Edinburgh University Library, Ms. Gen 873/III, ff. 153–54.

49. "Letters from M. Lavoisier to Dr. Black," *Report of the Fortieth Annual Meeting of the British Association for the Advancement of Science* (London: John Murray, 1871), 189.

50. Antoine-Laurent Lavoisier, *Elements of Chemistry, in a New Systematic Order, Containing All the Modern Discoveries* (Edinburgh: Printed for William Creech, and sold in London by G. G. and J. J. Robinsons, 1790), vi; Antoine-Laurent Lavoisier, *Elements of Chemistry, in a New Systematic Order, Containing All the Modern Discoveries*, 2nd ed. (Edinburgh: Printed for William Creech, and sold in London by G. G. and J. J. Robinsons, 1793), vi.

51. On Kerr, see *Oxford Dictionary of National Biography* and McKie, "Introduction," v–vi. On Kincaid and Creech, see Richard B. Sher, *The Enlightenment and the Book: Scottish Authors and Their Publishers in Eighteenth-Century Britain, Ireland, and America* (Chicago: University of Chicago Press, 2006), 311–18, 401–40.

52. Lavoisier, *Elements*, 1st ed., vi; Douglas McKie, "Antoine Laurent Lavoisier, F.R.S., 1743–1794," *Notes and Records of the Royal Society of London* 7 (1949): 13.

53. The title page gives a publication date of 1790, but the translator's preface was dated 23 October 1789, and the work was advertised as published in the London press in December. See *London Chronicle*, 22 December 1789, 604.

54. Lavoisier, *Elements*, 1st ed., vi–xii; Denis I. Duveen and Herbert S. Klickstein, *A Bibliography of the Works of Antoine-Laurent Lavoisier, 1743–1794* (London: Wm. Dawson and Sons and E. Weil, 1954), 181 and 183; Maurice P. Crosland, *Historical Studies in the Language of Chemistry* (London: Heinemann, 1962), 195.

55. "Elements of Chemistry," *Critical Review* 69 (1790), 407.

56. [Alexander Chisholme], "Kerr's Translation of Lavoisier's Elements of Chemistry," *Monthly Review*, 2nd ser., 4 (1790): 161.

57. McKie, "Lavoisier," 13.

58. Edinburgh Public Library, William Creech Papers, Item 120.

59. Frank Greenaway, introduction to *Essays Physical and Chemical*, by Antoine-Laurent Lavoisier (London: Frank Cass, 1970), xxix–xxxiii; Perrin, "Reluctant Catalyst," 146.

60. Antoine-Laurent Lavoisier, *Elements of Chemistry, in a New Systematic Order, Containing All the Modern Discoveries*, 3rd ed. (Edinburgh: Printed for William Creech, and sold in London by G. G. and J. Robinsons and T. Kay, 1796), xi; McKie, "Lavoisier," 13–14.

61. McKie, "Lavoisier," 33.

62. National Library of Scotland, MS, Acc. 12604/1644.

63. Maurice Crosland and Crosbie Smith, "The Transmission of Physics from France to Britain: 1800–1840," *Historical Studies in the Physical Sciences* 9 (1978): 11 and n. 30.

64. "La Place's Exposition of the System of the World," *Critical Review*, 2nd ser., 18 (1796): 504–11; and [Review of *Système du monde*, by P. S. Laplace], *Analytical Review* 24 (1796): 610–11.

65. [Robert Woodhouse], "Laplace on the System of the World," *Monthly Review*, 2nd ser., 29 (1799): 505–6.

66. "Laplace's System of the World," *Eclectic Review* 5 (1809): 882–83.

67. Roger Hahn, *Pierre Simon Laplace, 1749–1827: A Determined Scientist* (Cambridge, MA: Harvard University Press, 2005), 140; Jean Dhombres, "Books: Reshaping Science," in *Revolution in Print: The Press in France, 1775–1800*, ed. Robert Darnton and Daniel Roche (Berkeley and Los Angeles: University of California Press, 1989), 184.

68. "Laplace on the Mechanism of the Celestial Bodies," *Critical Review*, 2nd ser., 27 (1799): 513–14.

69. [Robert Woodhouse], "La Place on Celestial Mechanics," *Monthly Review*, 2nd ser., 31 (1800): 465 and 478; 32 (1800): 478.

70. [Robert Woodhouse], "La Croix on the Differential and Integral Calculus," *Monthly Review*, 2nd ser., 31 (1800): 493–505; 32 (1800): 485–95.

71. At 1802 exchange rates, 30 francs equated to £1 5s. See Brian R. Mitchell, *British Historical Statistics* (Cambridge: Cambridge University Press, 1988), 702.

72. "French National Institute," *Philosophical Magazine* 7 (1800): 185; Jerome de Lelande, "History of Astronomy for the Year 1800," *Philosophical Magazine* 9 (1801): 13.

73. *Journal général de la littérature de France* 6 (1803): 36; [Robert Woodhouse], "La Place on Celestial Mechanics, Vol. III," *Monthly Review*, 2nd ser., 40 (1803): 491–96; "La Place on Celestial Mechanics," *Critical Review*, 3rd ser., 1 (1804): 531–41; *Journal général de la littérature de France* 4 (1805): 141; [Robert Woodhouse], "La Place on Celestial Mechanics, Vol. 4," *Monthly Review*, 2nd ser., 48 (1805): 477–79.

74. See [Robert Woodhouse], "La Place's Supplement to His Tenth Book," *Monthly Review*, 2nd ser., 53 (1807): 483–89; [Robert Woodhouse], "Theory of Capillary Action," *Monthly Review*, 2nd ser., 60 (1809), 526–29; [Robert Woodhouse], "La Place's Supplement to His 3d Vol. of Mécanique Céleste," *Monthly Review*, 2nd ser., 60 (1809), 529–31.

75. [Woodhouse], "La Place on Celestial Mechanics," 540–41.

76. [Robert Woodhouse], "Hassenfratz's Course of Celestial Physics," *Monthly Review*, 42 (1803): 543.

77. [John Playfair], "La Place, *Traité de Mécanique Céleste*," *Edinburgh Review* 11 (1807–8): 250 and 281.

78. Crosland and Smith, "Transmission," 11–12; Jack B. Morrell, "Professors Robison and Playfair, and the 'Theophobia Gallica': Natural Philosophy, Religion and Politics in Edinburgh, 1789–1815," *Notes and Records of the Royal Society of London* 26 (1971): 43–63.

79. Harvey Becher, "William Whewell and Cambridge Mathematics," *Historical Studies in the Physical Sciences* 11 (1980–81): 8–10.

80. [John Herschel], "Mrs. Somerville's *Mechanism of the Heavens*," *Quarterly Review* 47 (1832): 544.

81. [Thomas Young], "Laplace's Supplement to the Mécanique Céleste," *Quarterly Review* 1 (1809), 107–9 and 112; Hahn, *Pierre Simon Laplace*, 163–64.

82. John Toplis, *A Treatise upon Analytical Mechanics, Being the First Book of the Mechanique Celeste of P. S. Laplace* (Nottingham: printed by H. Barnett. Sold by Longman, Hurst, Rees, Orme, and Brown, and Craddock and Joy, London; and J. Deighton, Cambridge, 1814), iii–iv.

83. [Thomas Young], *Elementary Illustrations of the Celestial Mechanics of Laplace. Part the First, Comprehending the First Book* (London: John Murray, 1821), iii.

84. [Herschel], "Mrs. Somerville," 541–42, 545.

85. See especially Pietro Corsi, "The Importance of French Transformist Ideas for the Second Volume of Lyell's *Principles of Geology*," *British Journal for the History of Science* 11 (1978): 221–44; and Pietro Corsi, *Science and Religion: Baden Powell and the Anglican Debate, 1800–1860* (Cambridge: Cambridge University Press, 1988), chap. 15.

86. *Journal général de la littérature de France* 12 (1809), 225–27.

87. *Monthly Magazine* 32 (1811): 263. At 1811 exchange rates, 12 francs equated to 13s. 5d.; see Mitchell, *British Historical Statistics*, 702.

88. Pietro Corsi, *The Age of Lamarck: Evolutionary Theories in France, 1790–1830* (Berkeley and Los Angeles: University of California Press, 1988), 207.

89. Nangle, *Monthly Review*, 47–48; James Coutts, *History of the University of Glasgow from Its Foundation in 1451 to 1909* (Glasgow: James Maclehose and Sons, 1909), 324, 328, 337–43, 512–15.

90. [Lockhart Muirhead], "Lamarck's *Zoological Philosophy*," *Monthly Review*, 2nd ser., 70 (1813): 475 and 490.

91. Ibid., 65 (1811): 479, 481, and 482. This early use of the term *evolution* in something like the modern sense predates by twenty years Charles Lyell's use of the term in his discussion of Lamarck in the *Principles of Geology*: see Peter J. Bowler, "The Changing Meaning of 'Evolution,'" *Journal of the History of Ideas* 36 (1975): 102. Similar usage also appears in the review cited in n. 98 below.

92. "Foreign Literature of the Year 1810," *New Annual Register* [32] (1811): "Literary Selections and Retrospect," 380; *Morning Chronicle*, 27 June 1812, 2.

93. "Foreign Literature of the Year 1812," *New Annual Register* [34] (1813): "Literary Selections and Retrospect," 418.

94. Verulam, "Lamarck's System of Philosophy," *Monthly Magazine* 36 (1813): 500.

95. [Robert Jameson], "Observations on the Nature and Importance of Geology," *Edinburgh New Philosophical Journal* 1 (1826): 293–302; James A. Secord, "Edinburgh Lamarckians: Robert Jameson and Robert E. Grant," *Journal of the History of Biology* 24 (1991): 14 and 15–16.

96. John Yelloly, "History of a Case of Anæsthesia," *Medico-Chirurgical Transactions* 3 (1812): 99–100.

97. *London Medical Review* 5 (1812): 264.

98. "Lamarck *Philosophie Zoologique*," *London Medical Review* 5 (1812): 313 and 316.

99. Richard Burkhardt suggests that Continental commentators at this time also

felt "little need to elaborate on Lamarck's errors." See Richard Burkhardt, *The Spirit of System: Lamarck and Evolutionary Biology* (Cambridge, MA: Harvard University Press, 1977), 190. See, by contrast, Norton Garfinkle, "Science and Religion in England, 1790–1800: The Critical Response to the Work of Erasmus Darwin," *Journal of the History of Ideas* 16 (1955): 376–88.

100. Quoted in Corsi, "Importance of French Transformist Ideas," 223.

101. Corsi, *Age of Lamarck*, 207–11; Edmund A. Crouch, *An Illustrated Introduction to Lamarck's Conchology Contained in His "Histoire naturelle des animaux sans vertèbres": Being a Literal Translation of the Descriptions of the Recent and Fossil Genera* (London: Longman, Rees, Orme, Brown, and Green, and J. Mawe, 1826).

102. "Lamarck on Invertebrate Animals," *Edinburgh Monthly Review* 3 (1820): 403–18. Compare, however, [Lockhart Muirhead?], "Lamarck's *Nat. Hist. of Animals without Vertebræ*," *Monthly Review* 90 (1819): 485–98; and 91 (1820): 512–20 (esp. 485–87), in which the reviewer excuses Lamarck's speculative theories in the light of his important contribution to invertebrate zoology. See also "[Review of *An Introduction to the Study of Conchology*, by Samuel Brookes]," *Monthly Review* 81 (1816), 328–29; and "Lamarck's *Natural History of Invertebral Animals*," *Monthly Review* 99 (1822): 485–92.

103. Corsi, *Science and Religion*, chapter 15.

104. Ibid., chap. 15; Adrian Desmond, *The Politics of Evolution: Morphology, Medicine, and Reform in Radical London* (Chicago: University of Chicago Press, 1989), esp. chap. 2.

105. Simon, "The Baillières"; [J.-B. Baillière], *A Catalogue of Books in Medicine, Surgery, Anatomy, Physiology, Natural History, Botany, Chemistry, Pharmacy, &c., &c., &c.* (Paris: J.-B. Baillière, 1828), 46. I am grateful to Josep Simon for this reference.

106. Corsi, *Age of Lamarck*, 266.

107. Martin J. S. Rudwick, *Bursting the Limits of Time: The Reconstruction of Geohistory in the Age of Revolution* (Chicago: University of Chicago Press, 2005), 494.

108. Jean Chandler Smith, *Georges Cuvier: An Annotated Bibliography of His Published Works* (Washington, DC: Smithsonian Institution Press, 1993), 181; *Caledonian Mercury*, 22 April 1813, 3c; *Monthly Magazine* 35 (1813): 438, 443; *Quarterly Review* 9 (1813): 506; *Scots Magazine* 75 (1813): 445. The literary notices in the two magazines used exactly the same wording.

109. [John Playfair], "Cuvier on Fossil Bones," *Edinburgh Review* 18 (1811): 214–30.

110. *Caledonian Mercury*, 15 November 1814, 1. For an accurate modern translation of the introduction, see Martin J. S. Rudwick, *Georges Cuvier, Fossil Bones, and Geological Catastrophes: New Translations and Interpretations of Primary Texts* (Chicago: University of Chicago Press, 1997), 173–252.

111. Robert Kerr, trans., *Essay on the Theory of the Earth, Translated from the French of M. Cuvier, with Mineralogical Notes and an Account of Cuvier's Geological Discoveries, by Professor Jameson* (Edinburgh: William Blackwood, 1813), ix; 2nd ed. (London: John Murray and Robert Baldwin, 1815).

112. Rudwick, *Bursting the Limits of Time*, 510.

113. Smith, *Georges Cuvier*, 150–59.

114. Georges Cuvier, *Researches into Fossil Osteology: Partially Abridged and*

Re-Arranged from the French (London: G. B. Whittaker, 1826), cover; Georges Cuvier, *The Animal Kingdom Arranged in Conformity with Its Organization*, 16 vols. (London: G. B. Whittaker, 1827–35); *Oxford Dictionary of National Biography*, s.v. "Whittaker, George Byrom."

115. Rudwick, *Bursting the Limits of Time*, 506, 510–11. This having been said, a later, separately published French edition of the discourse bore the title *Discours sur la théorie de la terre*; see Smith, *Georges Cuvier*, 153.

116. Rudwick, *Bursting the Limits of Time*, 511, 596–98.

117. *Caledonian Mercury*, 15 November 1813, 1; *Aberdeen Journal*, 8 December 1813, 1.

118. [John Playfair?], "Cuvier on the Theory of the Earth," *Edinburgh Review* 22 (1814): 454–55, 469. On the Continental replication of this review, see Rudwick, *Bursting the Limits of Time*, 598.

119. "Jameson's Translation of Cuvier's Theory of the Earth," *British Critic*, n.s., 1 (1814): 491.

120. Homo, "On Jameson's Preface to Cuvier's Theory of the Earth," *Philosophical Magazine* 46 (1815): 225; John Farey, "A Letter from Dr William Richardson to the Countess of Gosford (Occasioned by the Perusal of Cuvier's 'Geological Essay')," *Philosophical Magazine* 47 (1816): 358.

121. See, for example, James Burns, "John Fleming and the Geological Deluge," *British Journal for the History of Science* 40 (2007): 206 and n. 5.

122. James A. Secord, "Knowledge in Transit," *Isis* 95 (2004): 664.

CHAPTER FOURTEEN

Geological Mapping and the Geographies of Proprietorship in Nineteenth-Century Cornwall

SIMON NAYLOR

On 11 February 1814, at a meeting held at the Union Hotel, Penzance, a geological society was established.[1] Not only was it the first scientific society in the county of Cornwall; it was the first provincial geological society in Britain. The Geological Society of London, the first of its kind in the world, had been founded only a few years previously, in 1807. That a small town at the far edge of the nation should be near the forefront of geological science in particular and nineteenth-century associational science in general might at first seem surprising, but it is much less so when we remember that Cornwall at that time was the largest copper producer in the world. As well as hoping to aid in the discovery of new facts to enrich science, this newly formed Cornwall Geological Society (later renamed the Royal Geological Society of Cornwall; hereafter RGSC), set out to apply science to the advancement of the mining resources of the county.

As part of the purpose set out above, the society encouraged its members to map the county's geology, and they did so with alacrity. From the very first volume of the society's *Transactions* in 1818, color maps, sections, and traverses were in evidence as additions to the published articles. The first geological map of the whole of Cornwall was featured in the fourth volume of the *Transactions*, published in 1832, five years before the Geological Survey's map of the county. These were the first expressly scientific maps of the county of Cornwall and some of the earliest geological maps of regions of Britain. As such they form a valuable case study through which we can investigate the significance of maps in nineteenth-century science. Maps were increasingly used in the early years of the century to express scientific theories and so were an important part of science's complex visual economy. Perhaps unsurprisingly, given their provenance, maps were also used to express forms of proprietorship—over places and resources as much

345

as over scientific claims. Relatedly, they were central to the expression
of scientific authority. In other words, geological maps guided the reader
around an enclosed and owned landscape; they were texts that spoke to
particular audiences and excluded others. None of this was new in the
nineteenth century, but the increase in the quantity and quality of maps,
not to mention the uses to which they were put, certainly was. For these
reasons, it is vital that any study of the geographies of nineteenth-century
science consider the significance of maps.

The chapter begins by considering the place of geological maps in
nineteenth-century visual culture in relation to an already well-established
cartographic tradition of chorographical accounts, agricultural surveys, and
mining sections. It then discusses the social makeup of the RGSC and considers the role of the society and of its maps in the promotion of industrial
and agricultural improvement. The chapter then examines a number of the
RGSC's geological maps in order to highlight the way maps were employed
to express proprietorship and authority over economic and intellectual
territories, and the way in which they became the loci of tensions between
provincial and metropolitan geologists as well as between regional and national geology.[2]

GEOLOGICAL MAPS AND VISUAL CULTURE

Historians agree that the nineteenth century was marked by a pervasive
visual culture. There was a high degree of visual awareness and even artistic
skill among the leisured social classes in early nineteenth-century Europe,
fed partly by the Romantic obsession with landscape scenery, which had
"encouraged attempts to depict the wild mountain landscapes that had previously been considered unfit for serious artistic expression."[3] The increasing ease with which images could be reproduced and disseminated in the
nineteenth century helped habituate society to forms of visual display and
the very idea of visual knowledge. Visual means, Flint argues, were increasingly used to circulate ideas and to stimulate desire.[4]

Science in the nineteenth century was both a stimulus for and an early
proponent of this visual culture. Scientific treatises had for some time used
imagery to illustrate scientific specimens, views, and arguments. Keller, for
instance, provides a lucid account of the role of landscape and sectional
drawings in eighteenth-century explanations of earthquakes.[5] The difference between that and the following century was marked not by the use
of images themselves but by the rise in the sheer quantity—not to men-

tion quality and accessibility—of scientific images in circulation.[6] Anderson has drawn our attention to the "widespread experimentation with the visual presentation of scientific information" in the nineteenth century, while Lightman argues that the development of a mass visual culture corresponded with the growth of a mass market for science.[7] Scientific images became accessible to broader social groups and so helped make connections between elite and popular culture, even if there were concerns about these images' effects on the untrained eye and mind.[8]

The map was one type of image that was put to work in this period, as text to be read and as object to be displayed. Maps, Camerini claims, were "shared cultural images" by the 1830s and a "characteristic feature of the visual culture of nineteenth-century science."[9] By the midcentury, mapping was a recognized tool in the "visual technology" of natural history and was increasingly used to represent both natural and human geographical distributions.[10] Maps served in popular as well as more elite texts as an efficient way of communicating scientific observations, ideas, and values. More than simply two-dimensional geographical representations, however, scientific maps served a range of other functions. In Jacob's terms, they constructed worlds.[11] Not only did they reflect, summarize, or repeat scientific ideas expressed in textual or verbal form; they were also integral to the development and articulation of those ideas. Maps were not simply graphical representations of empirical information. They were highly theoretical in themselves. As Oldroyd notes, "maps of the same area made according to different theoretical presuppositions may look quite different."[12] Indeed, maps were often more effective at articulating scientific theories than traditional forms of scientific communication.

Scientific maps also served a sociological function. Maps use a "visual language" that employs rules and conventions, which in turn must be learned.[13] In doing so in the past, they relied on a social community that accepted these rules and shared an understanding of the conventions at play—what Jacob refers to as "map literacy."[14] Following the work of Harley, Jacob argues that maps presented "a seemingly objective and irrefutable appearance of factual and topographical information (the world as it is), but beyond this façade lies an elaborate rhetoric of power that organises the iconography, the social filtering and construction of the territory and the discourse of place names."[15] Maps did not simply reflect social values and prejudices; they actually *reinforced* entrenched social and political interests. These points are relevant in analysis of the nineteenth-century geological map. As this chapter will show, such maps were critical to arguments about the age and formation of Britain's geology; to the determination of scientific authority; and to the use of geology to justify elite, landed desires to implement the economic and moral ideas of improvement.

In his paper on the visual language of geology, Rudwick notes the close associations between the increase in geological maps in the early nineteenth century and the establishment of geological societies and, in particular, the Geological Society of London. The Society's *Transactions,* established in 1811, were from the outset generously illustrated with geological maps, sections, and views. While this self-conscious community of geologists helped construct an increasingly sophisticated map literacy among its members, this is not to claim that they were previously unconversant with the visual language of maps and other geographical illustrations. Rudwick argues that the gentlemanly character of the society insured that its members appreciated the topographical tradition of the eighteenth-century traveler-naturalist, on the basis of which "they developed a more formalized style that enabled landscape to bear a greater weight of structural meaning."[16] At the same time, the society's interests in utilitarian geology and its members' links with the world of mining "made them willing to adopt and develop the overtly structural maps and sections that were emerging from the work of mineral surveyors and mining engineers."[17] Not only did the society take its inspiration from the landscapes of eighteenth-century travel accounts, it was also inspired by the maps and sections of engineers, mineral surveyors, and mine adventurers.

Morrell disagrees with Rudwick's claims, however, at least in terms of the motivational forces for the development of English geology, arguing that geology's foundation as a form of polite learning meant that its relations with and contributions to mining were limited, even in industrial areas: "Given the occupational composition of the Geological Society's oligarchy, its romantic wanderlust, its love of elevated and elevating scenery, it is not surprising that this metropolitan coterie generally neglected coal formations and mining areas."[18] That said, Morrell concedes that there were exceptions to this rule: the northeast of England on the one hand, and Cornwall on the other. Of Cornwall, he notes that the RGSC brought together mine owners and the gentry to an extent otherwise unknown. Morrell claims, moreover, that the plans of the RGSC to produce a geological map, a mining school, and a mining record office failed; thus, even there, geology conformed to the more general pattern across the country, one where geological inquiry and practical mining remained practices apart. The evidence presented here, at least in terms of the production of geological maps, develops and complicates Morrell's conclusions.

CORNISH GEOLOGY AND REGIONAL IMPROVEMENT

The RGSC was founded by Dr. John Ayrton Paris, aided by Cornwall's local gentry and its wealthy industrialists.[19] Notable local members included Henry Boase, the first treasurer of the society; Joseph Carne, a wealthy local banker; Robert Were Fox, inventor of the deflector dipping-needle; Charles Fox, a Quaker shipping broker and cofounder of the Royal Cornwall Polytechnic Society; and Humphrey and John Davy. Paris himself was a local medic and later president of the Royal College of Physicians. Other notables included politicians such as William Rashleigh, Davies Gilbert, and Francis Bassett, the Lord de Dunstanville. In 1821 a new class of honorary members was instituted so that the society could assure for itself contacts and recognition from across Britain and the wider world. The society duly counted the presidents of the Royal, Linnean, and Geological Societies amongst its members, as well as professors from Oxford, Cambridge, Glasgow, and Edinburgh, such as Sir Joseph Banks, John Playfair, George Greenough, and James Edward Smith.

The society was socially elitist. Of the eight trustees of the museum in 1818, one was an earl, one a lord, and five were baronets; of the eight, three were MPs, and the one commoner, Davies Gilbert, went on to become president of the Royal Society of London. In 1834 more than a third of ordinary members (thirty-five) were from the local gentry. Many of these members were also intimately connected to industry and mining. Joseph Carne, for instance, was a partner in the Cornish Copper Company; Charles Fox was General Manager of the Perran Foundry and had investments in copper mining and smelting; and Francis Bassett was the founder of the Cornish Metal Company and owner of mines and extensive mining land. Despite this social constituency, claims were made locally that geological science was "a noble freemasonry" that brought together a wide range of constituents, including miners, farmers, and artisans who could easily collect specimens in the field that would improve knowledge of Cornwall's geology.[20] The RGSC did encourage mining professionals to join by keeping the price of membership and the cost of its *Transactions* low. Few mining professionals took up this opportunity. Although the society was more open than its London equivalent, it was by no means as inclusive as it liked to appear. It was also by far the most socially exclusive of Cornwall's scientific societies.

The first librarian of the RGSC, the Rev. Charles Val Le Grice, asserted that Penzance was an ideal center for the study of geology, as it was a "theatre in which geology displayed all her powers." "Cornubia," he warned, should not "recline in supreme indolence and ignorance amongst these

glorious monuments of Nature as the Turk does at Athens."[21] As it did for
the Geological Society of London, for the RGSC this meant collecting and
displaying geological data and specimens. The RGSC followed the "mineral
resource centre model" of the London society, which favored the collection
of facts over theoretical speculation—an effective way of avoiding awkward
theoretical and religious topics.[22] This model of science put considerable
emphasis on the collection of specimens in the field—in the Cornish case,
particularly from farmland and around and in its copper and tin mines. The
society established a museum in Penzance to house these specimens, to
which was added a public library, a newspaper reading room, and a geological
laboratory.[23] Also included was an economical department, "containing
specimens in illustration of the various changes which the ores of Tin,
Copper, &c. undergo in the processes of dressing and smelting."[24]

To its members, the society and its museum were of considerable benefit to wider society, intellectually and economically. Alongside its aim to
discover new facts to enrich science, the society also endeavored to apply
science "to improve art."[25] Paris claimed that the museum collections afforded "a most desirable and solid system of instruction; indeed it has already excited such a spirit of inquiry among the miners, as to have led to the
discovery of several minerals, before unknown in Cornwall."[26] When Davies
Gilbert said that the society would be the "sweet road to improvement," he
was referring not only to improvements to scientific knowledge, or indeed
to the improvement of miners, but also to mining and to agriculture more
generally.[27] With reference to mining, Humphry Davy, in a letter to the society in 1818, stated that "Cornwall may be regarded . . . as the Country of
Veins . . . and they are equally important to the practical miner, and to the
mineralogical philosopher."[28] In relation to agriculture, Paris argued that
"there is certainly no district in the British Empire where the natural relations between the varieties of soil and the subjacent rocks can be more easily discovered and traced, or more effectually investigated, than the county
of Cornwall; and nowhere can the information, which such an enquiry can
afford, be more immediately and successfully applied for the improvement
of waste lands, and the general advancement of agricultural science."[29]

One of the ways in which the society went about collecting facts about
Cornish geology and contributing to the applied fields of mining and agriculture was through the production of geological maps (yet another way in
which the RGSC followed the lead of the London society).[30] Such a project
was one of the founding aims of the society; Paris noted in 1818 that "should
the labours of this society terminate with the completion of this great desideratum, it will have to boast that it has presented one of the greatest

gifts which agriculture can receive from science." This was because, he continued, such a map "would not only point out the connexion between the varieties of soil and the subjacent rocks, but it would explain the local circumstances which might be friendly or hostile to their improvement."[31] This project benefited from a longer history of scientific mapping in the county of Cornwall and the country more generally.

The Ordnance Survey produced its first map of Cornwall on the scale of one inch to a mile in 1813—four years after its map of Devon—due largely to the southwest's strategic importance to Britain during the Napoleonic Wars. This is not to say that the mapping of Cornwall should be seen as only one part of a much larger *national* endeavor. Harley notes that:

> Notwithstanding the technical improvements which they embodied, the first Ordnance maps of Devon and Cornwall . . . perpetuated some of the limitations of eighteenth-century county maps. In one respect they *were* county maps: the sheets for each county, although part of a national sequence, were issued separately; . . . the borders were omitted from the inside edges of the sheets so that the gentry could mount them as one and, for those who still preferred the format of a county atlas, a separate title page was engraved. . . . And thus, owing to the strength of carto-graphic tradition, the *county* became a unit of early geological as well as of Ordnance mapping.[32]

Another contemporary national survey also chose to emphasize the county as the basis for its study: the county surveys of the Board of Ag-riculture (established in 1793). Agricultural surveys were not new. Early modern chorographical maps included agricultural information alongside their surveys of local estates, the genealogies of local families of note, and descriptions of local attractions and commodities—in short, as part of the display of elite identities and their property in the landscape. From the 1750s onward, however, these surveys' geographical basis was increasingly defined not as a cultural entity steeped in local traditions, but as a func-tional unit "whose farming practices were finely adjusted to physical and economic constraints."[33] The late eighteenth-century surveys of the Board of Agriculture examined and delimited regions by natural features as much as by political boundaries, and these "natural regions," as they were labeled, were seen as the basis upon which regional economies were examined.[34] They were also, of course, an important prop in the emerging concept of private property in land, especially in the Parliamentary Enclosure Acts of the period.[35]

Together with climate, drainage, terrain, and soil, geology was regarded as a key factor in the shaping of agricultural regions, because it was deemed to have a strong relationship with the nature and quality of the soil. The Board of Agriculture claimed that its agricultural surveys constituted the first geological maps of any part of England, although this was true only in the broadest sense.[36] In reality, influence came more strongly from the other direction—Darby notes that "the arrival of the geological map round about 1800 greatly facilitated the classification of land into what were sometimes called 'natural districts.'"[37] This was particularly the case in the southern, drift-free counties of England, where observation was easier, as opposed to the previously glaciated areas north of the Thames.

Cornwall lagged behind the rest of southwest England in agricultural innovation, in large part because its industrial energies went into mining and fishing, with agriculture often relegated to a part-time, subsistence occupation.[38] For early nineteenth-century commentators, this was a source in equal measure of frustration and optimism. In his description of West Penwith, the upland area to the west of the town of Penzance, Paris said that there the traveler plunges into a "rough, wild, and unsheltered" countryside, where farming was "in general slovenly." He continued that "the agriculturalist may, perhaps, view the district with somewhat different sensations, for the downs are certainly improvable, and those portions which have been brought into tillage have amply rewarded the labour of the adventurer."[39]

Cornwall, like its county neighbors Devon, Somerset, and Dorset, had been the subject of several agricultural surveys, including one by the Board of Agriculture in 1794.[40] For authors like Paris, rational understanding of farming was the answer to the county's agricultural malaise. Henry De la Beche, for instance, examined the relations between geology and agriculture in Devon and Cornwall, and claimed that differences in subsoil rocks were associated with differences in agricultural productivity.[41] Paris wrote papers on a similar theme, while Charles Lemon took a statistical approach to agricultural productivity.[42] The Penwith Agricultural Society also met in Penzance and had strong connections with the RGSC. Paris spoke there in 1815, arguing that the "liberal and well informed farmer" should be aware of the geology under his land because of its connection with the "agricultural economy."[43]

Before moving on to discuss the RGSC geological maps, it is worth noting that its members' cartographic literacy was informed not only by a familiarity with agricultural surveys but also by knowledge of sections and maps of mine workings, quarries, water-wells, canals, stream workings, and cliff sections. While formal geological maps were new to the nineteenth century, sections and mining surveys were not: one early example of a geological cross-section was of the coal-mines of Somerset, drawn in 1719.[44] The use of sections was certainly routine in the earliest volumes of the RGSC, which included, amongst other diagrams, two plates of mineral lodes in Cornish mines in Joseph Carne's 1822 paper; a section of the sandbanks of Mount's Bay by Henry Boase in 1827; and a diagram of a tin stream-works by John Colenso in 1832.[45]

Rudwick argues that a familiarity with the engineering style of sections and plans used in mining actually "pre-adapted individuals to the three-dimensional visualizing that structural geology required," and, by extension, that "a fully structural approach to the interpretation of the complex phenomena of geology was most readily attained within a social context of practical mining and mineral surveying."[46] Perhaps the most important consequence of the vertical, or columnar, section was the geological traverse, which in Freeman's words, "continued in a horizontal plane and on

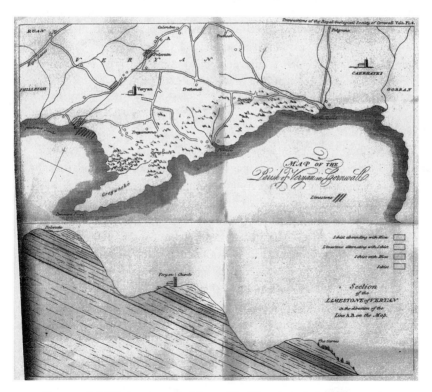

Fig. 14.1. S. J. Trist's 1818 map of the parish of Veryan in Cornwall. From the *Transactions of the Royal Geological Society of Cornwall* 1 (1818): facing page 113.

a much larger scale what the mine shaft or well-sinking revealed about an area's stratification."[47] Unlike cliff sections, but in similar manner to the maps they often accompanied, these traverses were exercises in extrapolation and interpretation—the visualization of abstract theorizing about the invisible structure of the earth.[48] In the Cornish case, the use of these visual technologies was again much in evidence—S. J. Trist's 1818 paper on the limestone rocks in the Parish of Veryan included a topographic map of the area with a section marked; and immediately below, with its vertical axis heavily exaggerated, was a section showing the rock strata and their exposure (fig. 14.1).[49]

All this is to say that chorographical accounts, agricultural surveys, and mining sections were important precursors to the production of nineteenth-century geological maps, and that agricultural and mineralogical improvement were key motivators in the production of those maps. Conversely, the resultant geological maps were valuable tools in the enhancement of the capital value of land.[50] Many members of the RGSC were landowners and industrialists, and it was in the service of these interests that the maps operated. These maps were scientific documents too, of course, but even agricultural surveys and mining sections served as important precursory tools, helping to equip readers of later geological maps with the complex visual conventions that were required to understand them.[51]

MAPPING CORNWALL'S GEOLOGY

The first map to be published by the RGSC was of the Lizard district on the south coast, as part of a paper on the geology of the area by Ashhurst Majendie, one of the society's founding members (fig. 14.2).[52] The map was based on a traverse around the entire coast and upon numerous transects across the peninsula. Majendie nonetheless complained that there was no map that distinguished elevations on which to construct his own, and he resorted to Thomas Martyn's 1748 one-inch-to-one-mile *Map of Cornwall*. Majendie's map was presented to the society first as a paper in November 1815 and then in the first volume of the society's *Transactions*. Despite its recourse to Martyn's mid-eighteenth-century map, Adam Sedgwick thought the later work good enough to refer to it positively in his own paper on the Lizard to the Cambridge Philosophical Society in 1822.[53] This was not unusual. Rudwick notes that geologists routinely relied on topographic maps as a base for their own maps. William Smith, for instance, used one-inch-to-one-mile maps of Somerset as the basis of his geological map of Bath in 1799 and continued to make use of a variety of maps, atlases, sections, and

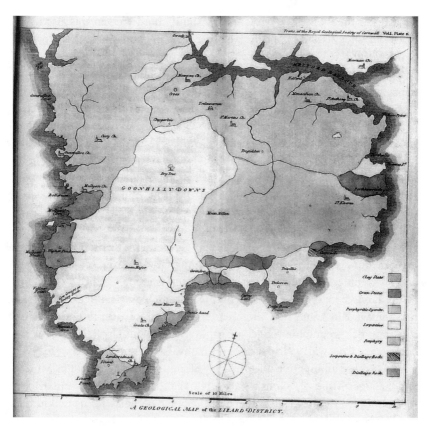

Fig. 14.2. Ashhurst Majendie's 1818 geological map of the Lizard District. From the *Transactions of the Royal Geological Society of Cornwall* 1 (1818): facing page 37.

tables in his work on English geology. Later, his own geological maps, along with those by George Greenough, were used as the basis for other geological maps of England. It is therefore interesting to note that Majendie's was not the only geological map to be presented to the RGSC in November of 1815—in that month Davies Gilbert presented a copy of William Smith's map of the geology of England and Wales to the Society.

Majendie, in map and paper, chose to limit himself to the observation of surface geology and to refrain from theoretical speculation. Rock types were colored and coded using a key so that their distribution was easily discernable to the viewer. His map included not only the areas on the Lizard where particular strata were exposed, but also the names of headlands, the churches of the district, antiquarian objects, and gentlemen's seats: his map included Trelowarren, the seat of the Vyvyan family. Sir Lyell Vyvyan was

a trustee of the RGSC Museum, and his son Richard Vyvyan was also interested in geology—and for a time was suspected of being the author *Vestiges
of the Natural History of Creation*.[54] Majendie therefore established the precedent for the display of a wide range of features on Cornwall's geological
maps; what we might see as the socialization of geological knowledge.

The next map to be printed in the society's *Transactions* took such concerns even further. The frontispiece of volume 2, published in 1822, featured a map of the Land's End District, which was accompanied by an essay
by John Forbes, an Edinburgh-trained physician at the Penzance Public
Dispensary (fig. 14.3).[55] Forbes's map of the Land's End District highlighted
the dominance of granite and included the hills that the hard rock produced when it outcropped. His map—being, he claimed, "a faithful study
of facts"—also showed the local villages and towns, the parish churches,
the tin and copper mines, and the local gentlemen's seats. Forbes's map
combined its geological themes with antiquarian ones, and set both within
a chorographical framework that emphasized the landed elite's centrality
in the landscape. Given the elite nature of the Geological Society's membership, we can argue that these early maps were concerned not only to
illustrate the geographies of geological knowledge but also to trace out the
contours of Cornwall's elite social geography. Gentlemen's seats were given
a map symbol and included in the key, alongside "Villages," "Tin Mines"
and "Copper Mines," "Granite," "Slate" and "Porphyry." Particular locations included Acton Castle, owned by the noted botanist, John Stackhouse;
Rose Price's house at Tregwainton; Michael Williams's house at Tregenna;
and John Hawkins's at Trewinnard. Price, Williams and Hawkins were all
leading figures in the Royal Geological Society of Cornwall.

This map was more than simply a symbolic projection of landed proprietorship. It also provided landowners with potentially valuable agricultural
and mining information. The mapping of some of west Cornwall's tin and
copper mines in Forbes's map was evidence of this, but it is even more explicit in a map of the parish of St. Just by Joseph Carne, published in the
same issue as Forbes's.[56] The two maps were meant to complement one
another. Carne's map included a greater wealth of economic information,
however, and was the first map by the society to do so. Included were mineral lodes and cross courses, as well as mine and stream works, alongside
topographical features, villages, roads, and antiquarian remains, and all laid
on top of geological strata. The map was, together with the accompanying
paper, entirely free from theoretical speculation on the production of the
veins it represented.[57]

In the same year in which Forbes's and Carne's maps were published,

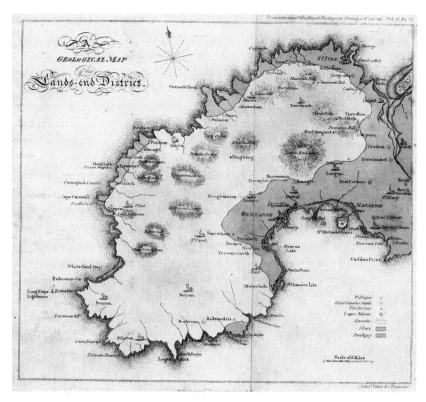

Fig. 14.3. John Forbes's 1822 map of the Land's End District. From the *Transactions of the Royal Geological Society of Cornwall* 2 (1822): frontispiece.

the RGSC apologized for not having produced a single geological map of the whole of Cornwall, but excused itself by saying that "it is an undertaking of immense extent and labour; and the map of the lodes in one Parish (St. Just) will at once shew [*sic*] the nature and importance of the plan, as well as the time requisite for the completion of such an undertaking."[58] Despite their geographical limitations then, the two published maps in the 1822 volume established early precedents and important examples for planned future work, and they did so in a number of ways. They asserted a strongly empirical and field-based approach to geological mapping. Inquiry on the ground, rather than upon theoretical speculation, functioned as the intellectual foundation upon which these maps rested. This was a fuller second precedent—geological mapping was an exercise in the gathering and displaying of data and not in philosophizing. They promoted a visual language involving the summation and abstraction of data through the use of keys and colors.

In these senses, Forbes's and Carne's maps were part of a broader mapping tradition. Secord notes that with respect to the works of Roderick Murchison, William Buckland, and others, geology was essentially stratigraphy, a taxonomic enterprise that sought order in the chaos of strata. The coloring of maps was central to this; the use of color-washes and color keys extended, in the words of William Buckland, "the progressive operations of a general inclosure act over the great common field of geology."[59] The implied link between geological mapping and agricultural enclosure was presumably not missed by Buckland's readers.

Perhaps the most striking example of this use of color was William Smith's map of the geology of England and Wales, a copy of which, as noted, had been donated to the RGSC and was hung in one of the stairwells of its building in Penzance. Smith argued that color was useful for generalizing information and summarizing the localities of thousands of specimens: "By strong lines of colour, the principal ranges of strata are rendered conspicuous, and naturally formed into classes, which may be seen and understood at a distance from the map, without distressing the eye to search for small characters."[60] Many of the ideas and arguments embodied in the maps of Majendie, Forbes, and Carne were to be developed in the work of Henry S. Boase.

HENRY S. BOASE'S GEOLOGICAL MAP OF CORNWALL

Henry Boase was born in London in 1799, the son of a Cornishman. Like Forbes, he had been trained as a medic in Edinburgh, graduating with an M.D. in 1821. He then returned to his family home and in February 1822 was appointed physician to the Public Dispensary at Penzance. With his father, he also became a partner in the Penzance Union Bank and in a company of tin smelters. In 1822 he was appointed secretary to the RGSC, a post he held until 1827 and again from 1833 to 1837. During that time he delivered two series of lectures on chemistry and read twenty-five papers on geological subjects, four of which were published in the society's *Transactions*. He also served on the committee of the Royal Cornwall Polytechnic Society, and as president of the Penzance Literary and Scientific Institution.[61]

In 1827, Boase resigned his position as secretary of the RGSC in order to devote time to an extensive study of Cornish geology. In particular, he set out to complete a geological map of the county that would help the society achieve one of its founding ambitions and act as a corrective to previous attempts that were "deficient in those minute details which a Map, proceeding from this Society, might be expected to supply."[62] Boase conducted the survey on which the map was based through extensive fieldwork, claiming to have spent two years and to have walked more than twelve hundred miles in his endeavor to "gain a more perfect knowledge of the geology of Cornwall, than could be obtained by strangers, however qualified to the task, in their hasty and partial excursions."[63] He set out with several aims: to "record the various phenomena as they present themselves in detail whilst traversing the county; avoiding, as much as possible, any theoretical observations thereon: and, in the next place, to arrange the rocks in geological order; and to discuss the various theoretical subjects suggested by the facts previously recorded."[64] The rock specimens collected during fieldwork were numbered and deposited in the society's museum in Penzance. The resulting map was supported by eight geological traverses and an extensive 308-page paper on the geology of Cornwall, a work that took up the majority of the volume in which it was published (fig. 14.4).

Boase's interpretation of Cornish geology was controversial for a number of reasons. He argued that the large granite masses, shown in pink on his map, were formed at the same time as the surrounding rocks and that, as a consequence, the majority of Cornish rocks could be considered primary rocks.[65] He went on to suggest that the mineral veins were contemporary with the rocks in which they were found and, following Werner, that they were formed by the action of water. No information on mineral deposits was included on the map. He also refused to acknowledge the presence of fossils in Cornish rocks; he believed that granite was stratified. He put forward a new and more expansive nomenclature for the primary rocks, which he followed when coloring his geological map and sections.[66] In the conclusion of his paper, Boase apologized for the "numerous innovations" he had put forward.

> As regards the new names of rocks, I have no expectation that they will be generally adopted: indeed, I merely used them to point out, in a more marked manner, those rocks which appeared to me to form distinct genera, not altogether indescribed, but which have not hitherto been accurately discriminated. . . . Should, however, any of the deductions by which I have endeavoured to disprove some received doctrines, be admitted by geologists, I shall feel gratified by such a token of the approbation of my labours.[67]

Boase's fellow geologists were largely unwilling to accept any of his ideas. When he presented his work and map to the British Association meeting in Cambridge in 1833, a number of prominent geologists, including

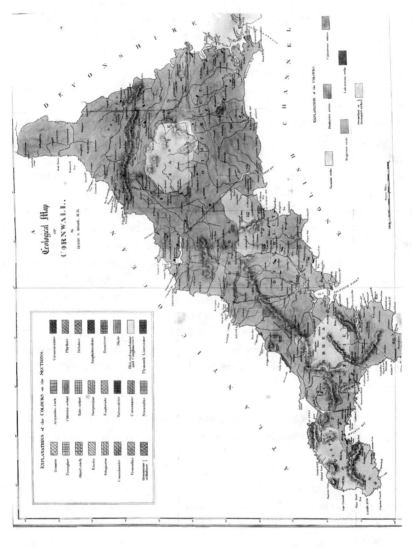

Fig. 14.4. Henry S. Boase's 1832 geological map of Cornwall. From the *Transactions of the Royal Geological Society of Cornwall* 4 (1832): facing page 361.

Adam Sedgwick, William Buckland, and John Phillips, were highly critical. By the 1830s, James Hutton's ideas about igneous intrusions had, thanks to Charles Lyell, been revived, and the division of the killas and graywacke was explained by the metamorphic alteration of the graywacke where it was near granite.[68] This interpretation was very much in contradistinction to Boase's theories, which were criticized at the association meeting where Boase was dismissed as a "provincial dissenter."[69] Despite the fact that Boase's ideas stood in the face of a powerful geological consensus, he insisted that they be discussed further at the 1834 British Association meeting in Edinburgh, by which time his *Treatise on Primary Geology* had been published, which further promoted his ideas on the formation of granite. Boase's ideas were duly attacked again by Sedgwick and Lyell on the grounds that he had no understanding of stratification, and the basis of his observations was thrown into question. Boase's only major supporter was the Wernerian Robert Jameson, editor of the *Edinburgh New Philosophical Journal* and Boase's teacher at Edinburgh. He wrote a favorable review of Boase's *Primary Geology* as it was being attacked elsewhere.[70]

The reception that Boase's map and arguments received inside Cornwall was mixed. At the meeting of the RGSC in October 1830, John Hawkins attacked Boase's work. Hawkins was from a wealthy Cornish family whose estates, at Trewithen, near Probus, included mining property. He had been educated at Trinity College Cambridge (and graduated with an M.A. in 1789), had studied under Werner, traveled with John Sibthorp, and in 1791 was elected FRS for his mineralogical and metallurgical contributions. Although he rarely attended RGSC meetings and spent much of his time at his homes in Sussex and London, he contributed papers to the society on mineral veins, mining technology, and the primitive strata.[71]

Theoretically Hawkins and Boase were both Neptunians: it was over nomenclature and the "disposition of the slate formation" that Hawkins questioned Boase's work.[72] In a barely concealed attack on Boase's claims, Hawkins noted in a paper, which was positioned at the start of the same volume of the *Transactions* in which Boase's paper appeared, that

> while I feel anxious that the objects of our combined investigation should be chiefly attained by the exertions of native geologists, for they alone have the means of revising their labours, I cannot help observing that our success will very greatly depend upon the accuracy of our conception of new facts, and on the full, clear, and candid manner in which they are repeated. It will depend too upon the scientific precision which we may apply to the nomenclature both of rocks and of single minerals.[73]

That Boase, as editor of volume 4 of the *Transactions* in which his map and paper appeared, excised two of Hawkins's papers to make way for his own did not help matters. Hawkins complained that Boase kept adding to his first communication as fresh materials were collected; thus, as he saw it, Boase violated the rule of scientific priority. The clash might also be seen as the expression of social difference—as Hawkins's disdain for the ambitions of the middle-class doctor whose ideas were eclipsing those of the landed Cornishman savant.

FIELDWORK, AUTHORITY, AND GEOLOGICAL TERRITORY

The maps discussed here help us reflect on more general points relating to the way geological maps functioned in the early nineteenth century. The first connects with Rudwick's observation that it "was only around 1820 that the enterprise [of geological surveying] became manifest in increasing numbers of standardized local 'memoirs'" and that, after this, geologists began to cope with more complex structures, and with wider regions.[74] This was obviously the case with the map produced by Boase, and it is particularly apparent when we compare it to maps such as Forbes's of the Land's End, or Majendie's of the Lizard. By the 1830s Cornwall's geologists, like those elsewhere, were confident enough to tackle complex geological structures that occupied extended areas and to express their understandings of them in textual and cartographic representations. This certainly continued in Cornwall and can be seen in the work of Charles Peach.

Peach was a member of Cornwall's Revenue Coastguard Service and previously served at Lyme Regis in Dorset, where he had developed his interest in fossils. Despite the claims of local geologists like Boase and more distant commentators like William Conybeare and William Phillips, Peach argued that he had identified fossils along the Cornish coastline. He presented his work at the British Association meetings in Plymouth in 1841 and Cork in 1843, where his work came to the attention of Roderick Murchison, amongst others.

Peach went on to publish a tabular synopsis of Cornwall's fossils in volume 6 of the *Transactions* of the RGSC—where he claimed to have detected six species of fossil fish—and to represent them on a map, also published in the same volume in 1846. In the accompanying paper and in reference to his map, he declared that "we are therefore warranted in colouring this part of the county as fossiliferous (of course excepting the granite ranges) from coast to coast; and thus a county formerly called primitive, is now shown to be, to the extent of perhaps four-fifths of it, fossiliferous."[75] This was presumably meant to be read as yet another rebuttal of Boase's ideas.

In an 1868 paper summarizing the history of research into Cornwall's fossil fish, the archaeologist William Pengelly noted that over the course of the 1850s and 1860s, many of Peach's claims had been questioned (by Adam Sedgwick among others), and his fish were presumed instead to be fossil sponges, but that by the end of the 1860s, scientific opinion had swung back in support of Peach's identifications.[76] Claiming that Peach "has done very much to elucidate the geological history of the district," he also highlighted the role he had played in showing the Devonian and the Old Red Sandstone systems to be of the same age.[77] Murchison in particular became a strong supporter of Peach and his work, describing him as "ingenious, modest and highly deserving."[78]

A second point to be considered here relates to the use of maps in the assertion of intellectual proprietorship. Both Secord and Freeman note that geological mapping was a form of territorial acquisition and that the surveying of a geographical region or the spatial expression of a geological epoch was a way of claiming it as one's own, arguing that geologists developed "powerful senses of what was their own geological territory."[79] For Cornish geologists, like their contemporaries working elsewhere in Britain, this form of territorial possession was earned through labor in the field—an opportunity that conveniently favored the local geologist over the traveler.[80] Travelers, it was argued, were at best capable of producing only a very general picture of the places journeyed through. At worst, they were guilty of shoddy fieldwork practices and hasty theorizations. In John Hawkins's opinion, "travellers glance superficially over every object, and too often from this . . . form very incorrect judgements," and such practices could not be tolerated from "native observers."[81] Boase likewise positioned his own fieldwork practices against those of the "hasty stranger," noting while he was on fieldwork, "I received many and urgent invitations, even from strangers: but I soon learnt that such intercourse, though very agreeable, interfered too much with the object in view. I accordingly avoided society altogether; and applying to no one for information, contented myself with examining such objects as came in my way, and which, in a country like Cornwall, could not fail of being both numerous and interesting."[82] Charles Peach subscribed to this value system and placed great emphasis on fieldwork and extended familiarity with place. He began his paper on fossil geology with the following assertion: "I have confined my observations on the fossiliferous strata of Cornwall to the spots I have myself visited."[83]

Mapping geological structures thus had the potential to confer authority and an enhanced reputation on the mapmaker. For Carne and Forbes,
this was of a very local form; one that did not really extend beyond its
own county boundaries. This changed from the 1830s, when the British
Association meetings facilitated the dissemination of regional geology to
national scientific audiences. It also meant that research was exposed to
a diversity of opinions and theories, and particularly to those of the powerful metropolitan elite. Within this national scientific economy, reputations could either be greatly enhanced or entirely destroyed. Both Boase
and Peach had the temerity to propose quite radical and far-reaching ideas
at the BAAS meetings—ideas founded on fieldwork in Cornwall and represented in maps. The observations made by both men were considered
by influential geologists. Peach's reputation flourished, while Boase's was
irreparably damaged by these encounters. This can be explained largely by
their willingness, or otherwise, to conform to an increasingly prevalent idea
of the limits of a provincial geologist's capabilities. Peach clearly recognized
this and was willing to perform his role as first and foremost a supplier of
observations and specimens to his intellectual superiors. In doing so he was
always careful to exhibit the right degree of humility to his audience. In
Boase's case, we might argue that the level of criticism he received at the
British Association meetings was due to his overstepping the limits of his
scientific position as a provincial geologist and being unwilling to exhibit
the right degree of subservience in presenting his work. That Boase's labors
were eclipsed by the work of Henry De la Beche, John Phillips, and the Geological Survey at the end of the 1830s supports this sense of a trend toward
the nationalization of scientific endeavor and the marginalization of local
expertise and idiosyncratic information.

This chapter has shown how important the map was to geology in the first
half of the nineteenth century. Maps were eloquent expressions of geological ideas. Yet, as we have seen, they did not confine themselves to that
subject alone—maps included other information, which helped them to
speak to other audiences: to mine prospectors, investors, landowners, agri-
culturalists, and antiquarians. Whether they were concerned with the communication of scientific or economic information, maps might all be read
as texts that expressed forms of proprietorship. All of the maps considered
here either exerted a scientific claim over geographical space or illustrated
more prosaic forms of ownership.

Because of their abilities to speak to diverse audiences, these maps
could be employed by a range of social groups, from industrialists to anti-

quarians, as much as by geologists. While this chapter has highlighted the role of maps in the claiming of surface and subterranean landscapes, it is important to remember that maps could speak to other audiences, such as tourists, albeit in much more inclusive terms. Delano-Smith has argued that maps for tourists were an invention of the nineteenth century, and that by the midcentury, guidebooks had become sophisticated navigational tools as well as reliable information resources on what localities could offer the traveler.[84] This was certainly the case in Cornwall, a region that had become increasingly popular as a tourist destination and a location for convalescence. A host of guidebooks and maps appeared shortly after the opening of the Tamar Bridge in 1859, which connected Penzance to London by railway. Such works included John Blight's *A Week at the Land's End* (1861), James Halliwell's *Rambles in Western Cornwall by the Footsteps of the Giants* (1861), Richard Edmonds's *The Land's End District* (1862), the *Routebook of Cornwall* (1863), Thomas Mills's *A Week's Wanderings in Cornwall and Devon* (1863), and R. J. King's *Handbook for Travellers in Devon and Cornwall* (1865). All of these sought to convey lessons in science, topography, and antiquities to the educated Victorian tourist. Blight's book is a good example of the genre. It was aimed at the visitor to Cornwall who harbored a wide-ranging interest in geology, geography, natural history, and antiquities. The map that accompanied the book showed, therefore, not only the churches and castles, stone circles, burial chambers, and ancient settlements of the district, along with the names of hills, headlands, and bays, but was also colored to indicate the underlying geology.

It is interesting to reflect on the similarities and differences between these tourist maps and the early nineteenth-century maps of the RGSC. The maps of Majendie, Forbes, and Carne were visually very similar to the tourist maps of the 1860s. But these earlier works were designed to be read within an eighteenth-century moral and natural-philosophical discourse of improvement, one that placed great emphasis on the role of the landowner as improver of land and of culture. While much of this was lost in the increasingly rational language of scientific specialization that Boase's and Peach's maps exhibited (where details of local antiquities were entirely absent), tourist maps and guides nonetheless also placed a moral weight on their readers to consume local geography correctly. Like other forms of popular science writing of the period, they narrated how nature, landscape, and history were to be consumed. In this sense, tourist maps, like the earlier geology maps, encouraged a close and careful interaction with place and, if followed correctly, conferred on the user their own form of moral and intellectual authority over geography.

NOTES

I am grateful to David Harvey, David Livingstone, James Ryan, Anne and Jim Secord, Leucha Veneer, Charlie Withers, and the audiences at the Edinburgh conference and at the History and Philosophy of Science Department, University of Cambridge, for valuable comments on earlier versions of this chapter. I thank the Royal Geological Society of Cornwall for allowing me access to its archives; the staff at the Morrab Library, Penzance, for their help; and Mark Pollard at Pickering and Chatto for permission to reproduce material from my book.

1. John A. Paris, "Minutes of the Proceedings of a Meeting Held at the Union Hotel, Penzance," Minute Book (Penzance: Royal Geological Society of Cornwall Archive, 1814), unpaginated; Anon., "Preface," *Transactions of the Royal Geological Society of Cornwall* 2 (1822): vi.

2. This argument is developed further in my study of the historical geographies of science in nineteenth-century Cornwall; see Simon Naylor, *Regionalizing Science: Placing Knowledges in Victorian England* (London: Pickering and Chatto, 2010).

3. Martin Rudwick, "The Emergence of a Visual Language for Geological Science, 1760–1840," *History of Science* 14 (1976): 173.

4. K. Flint, *The Victorians and the Visual Imagination* (Cambridge: Cambridge University Press, 2000), 4.

5. Suzanne B. Keller, "Sections and Views: Visual Representations in Eighteenth-Century Earthquake Studies," *British Journal for the History of Science* 31 (1998): 129; see also Alasdair Kennedy, "In Search of the 'True Prospect': Making and Knowing the Giant's Causeway as a Field Site in the Seventeenth Century," *British Journal for the History of Science* 41 (2008): 19–42.

6. Bernard Lightman, "The Visual Theology of Victorian Popularizers of Science: From Reverent Eye to Chemical Retina," *Isis* 91 (2000): 653; Rudwick, "Visual Language." See also Mary Sponberg Pedley, *The Commerce of Cartography* (Chicago: University of Chicago Press, 2005).

7. Katherine Anderson, "Mapping Meteorology," in *Intimate Universality*, ed. James Fleming, Vladimir Janković, and Deborah Cohen (Sagamore Beach, MA: Science History Publications, 2006), 70; Lightman, "Visual Theology."

8. Jennifer Tucker, *Nature Exposed: Photography as Eyewitness in Victorian Science* (Baltimore, MD: Johns Hopkins University Press, 2005), 10; Anne Secord, "Botany on a Plate: Pleasure and the Power of Pictures in Promoting Early Nineteenth-Century Scientific Knowledge," *History of Science* 93 (2002): 28–57.

9. Jane Camerini, "Evolution, Biogeography, and Maps: An Early History of Wallace's Line," *Isis* 84 (1993): 705.

10. Ibid., 708.

11. Christian Jacob, "Theoretical Aspects of the History of Cartography: Toward a Cultural History of Cartography," *Imago Mundi* 48 (1996): 192.

12. David R. Oldroyd, *The Highlands Controversy: Constructing Geological Knowledge Through Fieldwork in Nineteenth-Century Britain* (Chicago: University of Chicago Press, 1990), 343.

13. Jacob, "Theoretical Aspects of the History of Cartography," 192.

14. Ibid.

15. Ibid., 194; Brian Harley, "Maps, Knowledge, Power," in *The Iconography of Landscape*, ed. Denis Cosgrove and Stephen Daniels (Cambridge: Cambridge University Press, 1988), 277–312.

16. Rudwick, "Visual Language," 181; see also Roy Porter, "Gentlemen and Geology: The Emergence of a Scientific Career, 1660–1920," *Historical Journal* 21 (1978): 809–36.

17. Rudwick, "Visual Language," 181.

18. Jack Morrell, "Economic and Ornamental Geology: The Geological and Polytechnic Society of the West Riding of Yorkshire, 1837–53," in *Metropolis and Province: Science in British Culture, 1780–1850*, ed. Ian Inkster and Jack Morrell (Philadelphia: University of Pennsylvania Press, 1983), 232.

19. John A. Paris, *A Guide to the Mount's Bay and the Land's End* (London: Thomas and George Underwood, 1828), 26–27; Denise Crook, "The Early History of the Royal Geological Society of Cornwall, 1814–1850" (PhD diss., Open University, 1990).

20. Quoted in A. C. Todd, "The Royal Geological Society of Cornwall: Its Origins and History," in *Present Views of Some Aspects of the Geology of Cornwall and Devon*, ed. K. Hosking and G. Shrimpton (Cornwall: Royal Geological Society of Cornwall, 1964), 5.

21. Quoted in ibid., 3.

22. Martin Rudwick, *The Great Devonian Controversy: The Shaping of Scientific Knowledge Among Gentlemanly Specialists* (Chicago: University of Chicago Press, 1985).

23. Paris, *Guide to the Mount's Bay*, 29.

24. Ibid., 27–8.

25. "Preface," vi.

26. Ibid., 27.

27. Quoted in Todd, "Royal Geological Society of Cornwall," 3.

28. Humphry Davy, "Hints on the Geology of Cornwall," *Transactions of the Royal Geological Society of Cornwall* 1 (1818): 38.

29. John A. Paris, "Observations on the Geological Structure of Cornwall, with a View to Trace its Connexion with, and Influence upon Its Agricultural Economy, and to Establish a Rational System of Improvement by the Scientific Application of Mineral Manure," *Transactions of the Royal Geological Society of Cornwall* 1 (1818): 169–70.

30. Rudwick, *Great Devonian Controversy*, 20.

31. Paris, "Geological Structure of Cornwall," 170–71.

32. Brian Harley, "The Ordnance Survey and the Origins of Official Geological Mapping in Devon and Cornwall," in *Exeter Essays in Geography*, ed. Kenneth J. Gregory and William L. D. Ravenhill (Exeter: Exeter University Press, 1971), 115; my emphases.

33. Hugh Prince, "The Changing Rural Landscape, 1750–1850," in *The Agrarian History of England and Wales*, vol. 6, ed. G. E. Mingay (Cambridge: University of Cambridge Press, 1989), 10.

34. Joan Thirsk, *The Agrarian History of England and Wales*, vol. 1 (Cambridge: Cambridge University Press, 1967).

35. Tom Williamson, *The Transformation of Rural England: Farming and the Landscape, 1700–1870* (Exeter: Exeter University Press, 2002).

36. Darby notes that the "maps indicate not the stratigraphical relations of the

different rocks, but the differences in their texture and utilization." See Harold C. Darby, "Some Early Ideas on the Agricultural Regions of England," *Agricultural History Review* 2 (1954): 34.

37. Ibid., 47.

38. Thirsk, *Agrarian History of England and Wales.*

39. Paris, *Guide to the Mount's Bay,* 76–77, 80.

40. Sarah Wilmot, "The Scientific Gaze: Agricultural Improvers and the Topography of South-West England," in *Topographical Writers in South-West England,* ed. Mark Brayshay (Exeter: University of Exeter Press, 1996), 105–35; Rosalind Mitchison, "The Old Board of Agriculture (1793–1822)," *English Historical Review* 74 (1959): 41–69.

41. Prince, "Changing Rural Landscape," 80.

42. John A. Paris, *Notes on the Soils of Cornwall, with a View to Form a Rational System of Improvement by the Judicious Application of Mineral Manure* (Penzance: T. Vigurs, 1815); Paris, "Geological Structure of Cornwall"; Charles Lemon, "Notes on the Agricultural Produce of Cornwall," *Journal of the Statistical Society of London* 4 (1841): 197–208.

43. Paris, *Notes on Soils of Cornwall,* 6.

44. Keller, "Sections and Views," 146.

45. Joseph Carne, "On the Relative Age of the Veins of Cornwall," *Transactions of the Royal Geological Society of Cornwall* 2 (1822): 49–128; Henry S. Boase, "On the Sand-Banks of the Northern Shores of Mount's Bay," *Transactions of the Royal Geological Society of Cornwall* 3 (1827): 166–91; J. W. Colenso, "A Description of Happy-Union Tin Stream Work at Pentuan," *Transactions of the Royal Geological Society of Cornwall* 4 (1832): 29–39.

46. Rudwick, "Visual Language," 169.

47. Michael Freeman, *Victorians and the Prehistoric: Tracks to a Lost World* (New Haven, CT: Yale University Press, 2004), 127.

48. Ibid.

49. S. J. Trist, "Notes on the Limestone Rocks in the Parish of Veryan," *Transactions of the Royal Geological Society of Cornwall* 1 (1818): 107–13.

50. Freeman, *Victorians and the Prehistoric,* 123.

51. Keller, "Sections and Views," 142.

52. Ashhurst Majendie, "A Sketch of the Geology of the Lizard District," *Transactions of the Royal Geological Society of Cornwall* 1 (1818): 2–7.

53. Rudwick, "Visual Language," 159, 163; Crook, "Early History," 140.

54. Vyvyan was a Tory aristocrat MP who opposed a range of liberal measures. He was the author of several works, including the two-volume *Harmony of the Comprehensible World* (1842–45). See James A. Secord, *Victorian Sensation* (Chicago: University of Chicago Press, 2000), 180–83; B. T. Bradfield, "Sir Richard Vyvyan and the Country Gentlemen, 1830–1834," *English Historical Review* 83 (1968): 729–43.

55. John Forbes, "On the Geology of the Land's-End District," *Transactions of the Royal Geological Society of Cornwall* 2 (1822): 242–80.

56. Joseph Carne, "On the Mineral Productions, and the Geology of the Parish of St. Just," *Transactions of the Royal Geological Society of Cornwall* 2 (1822): 290–358.

57. Crook, "Early History," 149.

58. "Preface," viii.

59. Quoted in James A. Secord, "King of Siluria: Roderick Murchison and the Imperial Theme in Nineteenth-Century British Geology," *Victorian Studies* 25 (1982): 415. Color keys were also used in natural history as an appendage to illustrations of plants. See Luciana Martins and Felix Driver, "'The Struggle for Luxuriance': William Burchell Collects Tropical Nature," in *Tropical Visions in an Age of Empire,* ed. Felix Driver and Luciana Martins (Chicago: University of Chicago Press, 2005), 59–76.

60. Smith, quoted in D. A. Bassett, "Sheets of Many Colours or Maps of Geological Ideas: A Review," *Imago Mundi* 37 (1985): 103.

61. Denise Crook, "Boase, Henry Samuel (1799–1883)," *Oxford Dictionary of National Biography*; available online at http://www.oxforddnb.com/view/article/2738?docPos=6 (accessed 17 March 2008).

62. Henry Boase, "Contributions Towards a Knowledge of the Geology of Cornwall," *Transactions of the Royal Geological Society of Cornwall* 4 (1832): 166.

63. Ibid., 167.

64. Ibid.

65. In this treatise, Boase justified his claim thus: "It is proposed . . . to consider as primary rocks the various kinds of granite, and all those crystalline and non-fossiliferous masses, both compact and schistose, which are usually associated together under different arrangements, and intimately connected by frequent mineral transitions." See Henry S. Boase, *A Treatise of Primary Geology: Being an Examination, Both Practical and Theoretical, of the Older Formations* (London: Longman, Orme, Brown, Green, and Longman, 1834), 7–8.

66. Boase, "Contributions," 439, 446, and 451; Crook, "Boase;" Crook, "Early History."

67. Boase, "Contributions," 473 and 474.

68. Bakewell defined the graywacke in 1833 as "a coarse slate containing particles or fragments of other rocks or minerals, becoming either slate-clay or sandstone depending on the size of the particles"; and defined the killas as follows: "What is called the schist or killas in Cornwall, in the places where I have observed it in immediate junction with granite, . . . appears to have been changed by the junction: it has no appearance of slate." See R. Bakewell, *An Introduction to Geology . . . Greatly Enlarged,* 4th ed. (London: Longman, Orme, Brown, Green, and Longman, 1833), 86 and 64 respectively.

69. Jack Morrell and Arnold Thackray, *Gentlemen of Science: Early Years of the British Association for the Advancement of Science* (Oxford: Clarendon, 1981), 461, 173; Rudwick, *Great Devonian Controversy.*

70. Robert Jameson, "Reviews," *Edinburgh New Philosophical Journal* 17 (1834): 46; Crook, "Early History," 160.

71. For instance, see John Hawkins, "On a Process of Refining Tin," *Transactions of the Royal Geological Society of Cornwall* 1 (1818): 201–11; John Hawkins, "On Some Advantages Which Cornwall Possesses for the Study of Geology, and on the Use Which May Be Made of Them," *Transactions of the Royal Geological Society of Cornwall* 2 (1822): 1–13; John Hawkins, "On the Nomenclature of the Cornish Rocks," *Transactions of the Royal Geological Society of Cornwall* 2 (1822): 145–58.

72. Crook, "Early History," 158.

73. Hawkins, "Some General Observations on the Structure and Composition of the

Cornish Peninsula," *Transactions of the Royal Geological Society of Cornwall* 4 (1832):3. In turn, Boase publicly criticized Hawkins's theory about the formation of china clay; see Boase, "Contributions."

74. Rudwick, *Great Devonian Controversy*, 49.

75. Charles W. Peach, "On the Fossil Geology of Cornwall," *Transactions of the Royal Geological Society of Cornwall* 6 (1846): 181–85. See also Richard Q. Couch, "On the Silurian Remains of the Strata of the South-East Coast of Cornwall," *Transactions of the Royal Geological Society of Cornwall* 6 (1846): 147–49.

76. William Pengelly, "The History of the Discovery of Fossil Fish in the Devonian Rocks of Devon and Cornwall," *Transactions of the Devonshire Association for the Advancement of Science, Literature, and Art* 2 (1868): 423–42.

77. Ibid., 434.

78. Oldroyd, *Highlands Controversy*, 49.

79. Freeman, *Victorians and the Prehistoric*, 123; Secord, "King of Siluria."

80. Oldroyd, *Highlands Controversy.* For a recent history of geological fieldwork, see P. N. Wyse Jackson, ed., *Four Centuries of Geological Travel: The Search for Knowledge on Foot, Bicycle, Sledge and Camel* (London: Geological Society, 2007).

81. Hawkins, "Some General Observations, "20.

82. Boase, "Contributions," 360.

83. Peach, "Fossil Geology of Cornwall," 181.

84. Catherine Delano-Smith, "Milieus of Mobility: Itineraries, Route Maps and Road Maps," in *Cartographies of Travel and Navigation,* ed. James R. Akerman (Chicago: University of Chicago Press, 2006), 16–68.

CHAPTER FIFTEEN

Natural History and the Victorian Tourist: From Landscapes to Rock-Pools

AILEEN FYFE

O ver the last decade, historians of science have become fascinated by the links between science and travel. It is now widely recognized that traveling to other geographical locations can be useful, or even essential, to the development of new scientific knowledge. Much of the literature focuses on natural history, and the role of imperial, or global, travel in the work of men such as Joseph Banks, Alexander von Humboldt, Charles Darwin, and Joseph Hooker, although travel closer to home could also be crucial for geologists trying to complete their geological charts and botanists seeking to create accurate species and distribution lists.[1] In the physical sciences, the importance of global expeditions for observing certain astronomical phenomena and measuring magnetic declination has been noted.[2] Sometimes, it was other people, and specimens, which did the traveling. Eminent natural historians in London and Paris relied upon networks of correspondents to send back dried flowers, animal skins, pickled reptiles, or drawings and descriptions.[3] Without these unofficial assistants—diplomats, missionaries, and naval officers—the metropolitan men of science would have had far fewer materials to work with. The British Admiralty welcomed the incorporation of its officers into the networks of science, famously commissioning John Herschel to edit *A Manual of Scientific Enquiry: Prepared for the Use of Her Majesty's Navy, and Adapted for Travellers in General* (1849).

Yet, if we now know something of the significance of travel for the growth of scientific knowledge, we still know very little about how travel and the experience of new locations and sites might have influenced ordinary people's understanding of the natural world. During the nineteenth century, the majority of Britons who traveled did not spend years traversing oceans, jungles, and mountain ranges, nor did they send their specimens to

Kew. They were far more likely to tour the Lake District or make excursions to Margate or Morecambe, and to keep their specimens as souvenirs to adorn their domestic interiors. Thanks to the extension of the railway network and cheap excursion tickets—and the gradual development of Saturday half-holidays and annual holiday entitlements—travel had become an option for almost all sectors of the population by the second half of the nineteenth century. As tourists from Britain's industrialized cities encountered the countryside and the seaside, how did they understand what they were seeing, and what knowledge of nature did they gain?

I would like to suggest that tourism is an area that we could fruitfully investigate for traces of the public consumption of the sciences.[4] I would not be so outrageous as to claim that the sciences were ever a major part of holidays for the masses, but the continuing importance of ideas of rational recreation makes it plausible that some nineteenth-century tourists— particularly those among the educated classes—sought an instructive and entertaining angle to their holidays. While some may have found it in studying architecture, sketching views, or distributing religious tracts on the promenade, others may have turned to the natural sciences.

This chapter does not pretend to be a complete history of the links between science and tourism—a topic that surely deserves a full-length book— but I hope it will draw attention to some fascinating possibilities. One of the things that awaits a full study is a thorough examination of actual tourists' experiences, through their letters, diaries, and published travel accounts. So far, I have focused on those who provided services to tourists—especially in the form of published advice—since these agents are easier to locate in the published record. This chapter is therefore an investigation of the aspects of the sciences that were promoted to tourists in Victorian Britain, and not a study of tourist experiences.

Before continuing, I wish to draw attention to two themes that I hope the study of science and Victorian tourism might illuminate in an interesting way, namely *participation* and *observational skills*. In the older historiography, the audiences for popular science were routinely presented as passive recipients of knowledge. More recent studies have emphasized the active role of audiences both in choosing their sources of information and in creating meaning from them.[5] There is, however, another aspect to the question of audience activity, which revolves around what such audiences are expected to *do* during or as a consequence of their encounter with the sciences. Examining the popular science books of the first half of the nineteenth century, for instance, reveals an interesting contrast between the way the sciences were presented to adults and to children. For adults, the sciences were usually presented as a body of knowledge to be studied in the armchair or library and stored for future use in conversation. In contrast, children's books often emphasized practical activities, such as collecting flowers, observing the stars, or trying simple chemistry experiments.[6] Since tourism is essentially about going to new places and undertaking a different range of activities from the usual daily routine, it is possible that tourism could provide an attractive context for adults to encounter the sciences as practical activities.

It has been suggested too that guidebooks played a substantial role in forming Victorian concepts of architectural taste, encouraging their readers to form opinions on the age, original purpose, and aesthetic qualities of architectural details.[7] Forming such opinions required a combination of knowledge, most likely gained from prior reading, and the ability to know what to look at, and how, when standing in a ruined cathedral or castle. Similar observational skills are important in the field sciences—and, in a different sense, in laboratory science. It would, therefore, be interesting to see whether the guidebook's role in training an informed vision applies to natural history as well as to architecture.

I begin with an overview of domestic tourism and the emergence of a variety of services and products for tourists during the nineteenth century. My evidence for scientific tourism is based upon the appearance of the sciences within those most ubiquitous of tourist commodities, the standard guidebooks and handbooks. These, of course, were aimed at the literate, though there were guides suited to the purses of different classes of readers. I suggest that geology's links with the appreciation of scenery and the excitement of mine descents made it the earliest and most prevalent form of science to be encountered by nineteenth-century tourists, though it was joined by other branches of natural history from the middle of the century. I also suggest that geology, in its appreciation-of-scenery incarnation, encouraged the development of observational skills similar to those used in architectural appreciation, but that, from midcentury onward, both geology and natural history more generally came to be linked with a different form of tourist activity, namely, souvenir collecting.

DOMESTIC TOURISM

Until the early nineteenth century, travel and tourism were the preserve of the aristocracy—nobody else had the requisite combination of money and leisure time.[8] The classic example of eighteenth-century tourism was the Grand Tour, several months or even years of travel undertaken by young men to complete their education and broaden their horizons. The usual destination of such tours was Italy or Greece, to enable the young men to see the sites associated with their classical education.[9]

The rise of domestic tourism within Britain is usually traced to the outbreak of hostilities with France, which abruptly curtailed opportunities for Grand Tours. For the next twenty five years, would-be tourists had to settle for investigating their own country. Favorite destinations were the Lake District and the Peak District, where admiring scenery and visiting country houses were the key attractions. Such tours also had the advantage of being more practical for members of the successful upper-middle classes: a business or profession could not be left for several months, but could perhaps be handed over to a partner or trusted manager for two or three weeks.

Jane Austen gives us an account of such a tour in *Pride and Prejudice* (1813), when Elizabeth Bennett travels to Derbyshire for three weeks with her aunt and uncle Gardiner. They had originally hoped to reach the Lake District, but Mr. Gardiner's business could not spare him long enough to go so far, much to Elizabeth's disappointment. During the tour, by private carriage, the party visited Blenheim Palace, Warwick Castle, and Kenilworth Castle, with the result that, by the time they reached Mr. Darcy's seat at Pemberley, Elizabeth claimed to be "tired of seeing great houses; after going over so many, she really had no pleasure in fine carpets or satin curtains." She was persuaded, however, by the lure of Pemberley's "delightful" grounds. Thus, before entering the house itself, the group drove through the grounds, "saw and admired every remarkable spot and point of view" and appreciated the view of the house from "the top of a considerable eminence."[10] Elizabeth Bennett's experience exemplifies the turn-of-the-century tour, with its combination of country house visits and appreciation of scenery. The appeal of country houses was the combination of architecture and gardens with the interior furnishings and art collections. Since many of the collections included pieces acquired on the owner's Grand Tour, the English country house enabled tourists to encounter classical or Renaissance art without undertaking foreign travel. Although Elizabeth's party was shown around Pemberley by the housekeeper, guidebooks for various country houses were already available by the 1810s, usually identifying the items in the art collections.[11]

The appeal of scenery had much to do with the enthusiasm for the picturesque, usually associated with William Gilpin's works in the 1780s and 1790s.[12] Adherents of the picturesque learned a whole new way of looking at and talking about landscapes—something which Austen would satirize in an episode in *Northanger Abbey* (1818), when Henry Tilney teaches Catherine

Morley how to appreciate the view of Bath properly.[13] The admiration of scenery was closely associated with the activities of sketching and painting, since a view was "picturesque" precisely because it would look good in a frame. Guidebooks to the Lake District even began to recommend specific locations for would-be artists to sketch the best-known views.[14] Since these locations were not necessarily accessible by coach road, admiring and sketching the scenery could also be associated with a modest amount of walking and hill-climbing. As we will see, it could also encourage an interest in geology, as a way of better understanding landscape formations.

This sort of domestic tourism, with its emphasis on country houses and scenery, was firmly associated with the affluent upper-middle classes. Although young, single men might occasionally undertake pedestrian tours, as William Wordsworth did with a student friend in the 1790s, tourism could not easily be done cheaply.[15] Even if the expense of a private carriage could be avoided, there remained the significant cost of lodgings, food, and tips for the servants together with the requirement of sufficient leisure time. Nevertheless, domestic tourism was undoubtedly more affordable and achievable than continental travel, which explains its ongoing popularity after Waterloo in 1815. The reopening of the Continent made Paris, Florence, and Rome accessible again, but for the middle classes, such a tour was likely to be a once-in-a-lifetime opportunity.

The real transformation in domestic tourism came in the second half of the nineteenth century, once the working classes had gained more leisure time, and the railways had expanded the horizons of travelers of all classes. By the 1850s, middle- and lower middle-class employees could expect a half-holiday on Saturdays, while some employers were starting to allow a week or a fortnight's annual holiday leave, usually unpaid. Equally, some factory owners and other large employers of manual workers were coming to realize that even the working classes should have the opportunity for leisure and relaxation. The Saturday half-holiday and the unpaid annual leave allowance would become commonplace by the end of the century.[16]

For most of these new potential tourists, the traditional tour through an area of natural beauty remained inaccessible, and the day trip, or excursion, took precedence. Instructive and entertaining magazines aimed at the lower middle and working classes—such as *Chambers's Journal* and the *Leisure Hour*—routinely carried descriptions of places that might be suitable for a visit, such as Bath, Margate, and the Isle of Wight.[17] Publisher Charles Knight optimistically thought that there was a market for a new part-work devoted entirely to the needs of the new tourists: he issued *The World We Live In* from 1847 to 1849, with each sixteen-page part describing a different

town. His selection was generously broad, including obvious historic attractions (Windsor, Richmond, Hampton Court Palace) and areas of scenic
interest (Windermere), while acknowledging the truly limited options available to many of these new travelers by including Manchester, Glasgow, and
Birkenhead.[18] Later in the century, working-class opportunities for travel
expanded from day excursions to a week's summer holiday, leading to the
rapid expansion of seaside towns with easy access from the large industrial
cities, such as Colwyn Bay, Rothesay, and Scarborough.[19]

While the increased availability of leisure time was certainly crucial in
extending access to tourism, the expansion of the railway network was
equally important. After all, leisure time was of little use if there was no
quick and cheap way to get somewhere interesting within the limited
hours available. Railway travel was not necessarily cheap, but competition
helped bring down fares, and once the companies had realized the potential
profits, middle-class holiday options were transformed by special offers of
period-return tickets to the most attractive destinations on each network.
Working-class options were expanded by parliamentary insistence upon provision for third-class passengers, which resulted in the creation of the penny-
a-mile fare from 1844, and by the peculiarly British development of special
excursion trains. From the early 1840s, philanthropic employers, voluntary
societies, and excursion agents began to charter trains for group trips to
particular events, from Sunday school outings to flower festivals. Through
agreement with the railway companies, such groups were able to enjoy unusually low prices. By the 1850s, the railway companies began to run such
trains themselves, offering day returns to selected destinations at far cheaper
prices.[20]

The range of opportunities for travelers of all classes is readily apparent from the advertisements carried by the *Times* every summer. Thus, in
August 1855, a single column carried advertisements for day-return tickets
to Brighton with the South Western Railway (3s. 6d. in third class, Sundays
and Mondays only); weekly return tickets to Lowestoft or Yarmouth with
the Eastern Counties Railway (10s. in third class, Wednesday and Saturday
afternoons only); fortnightly return tickets to Paris for the 1855 Exhibition
(32s. in second class, not Thursdays or Sundays); and three-week return
tickets to the Lake District with the London and North Western Railway
(£2. 10s. in second class) or to Scarborough with the Great Northern Railway (35s. in second class, if at least four family members travel together).[21]
This was also the decade in which Thomas Cook firmly established himself
as the leader of package tours, within Britain and on the Continent.[22]

With the growth of domestic tourism, the range of activities undertaken by tourists changed and expanded. Scenery remained popular, but the attractiveness of country houses came under threat, from the 1830s onward, from the growing interest in what was known as "the Olden Time." Medieval or early modern buildings, the more ruined and romantic the better, received huge numbers of visits from lower middle- and working-class tourists.[23] Charles Knight issued another of his illustrated part-works, *Old England: A Pictorial Museum of Regal, Ecclesiastical, Baronial, Municipal and Popular Antiquities* (1845) to cash in on this interest.[24] By the second half of the century, outdoor holidays involving moderate activity were becoming more common, from the piers and promenades of the seaside resorts to the golf courses and grouse moors of the Scottish Highlands or the high peaks of the Alps. This enthusiasm for engaging in outdoor activity would open up more opportunities for engaging with and learning about the natural world.

GUIDEBOOKS

For those not taking one of Thomas Cook's tours, a guidebook was an influential source of information when planning a holiday. There is, of course, no guarantee that any tourist did any of the things recommended by a guidebook, but the guidebooks provide an opportunity to consider the likely ways in which tourists might have encountered and interpreted the natural world.

By the end of the nineteenth century, guidebooks had become indispensable to travelers. E. M. Forster poked fun at this in *A Room with a View* (1908), where Lucy Honeychurch's bewilderment arises from being lost "in Santa Croce with no Baedeker."[25] Yet, at the start of the nineteenth century, guidebooks had been a relatively new phenomenon. Generations of travelers, from medieval pilgrims onward, had used accounts written by previous travelers as a source of tips and advice, but it was not until the late eighteenth century that a distinct genre of intentional practical advice emerged.[26] By the 1780s, there were books calling themselves "guides" to Bath and Cheltenham, and the term *guidebook* came into use in the early nineteenth century.[27] Most of the earliest guidebooks were small affairs, written and published as a venture by local printers, most often for specific towns, country houses, or beauty spots.

By the 1820s and 1830s, the rise of both domestic and foreign tourism encouraged major national publishers to consider the possibilities of tourists as a market, and by the 1840s, there were several series of guidebooks offering a standardized approach to a wide range of destinations. These books were far more than simply guides to particular attractions: they typically covered wide geographical areas, offered practical advice on how to get there, where to stay, and what to see, and had a large number of detailed entries dealing with particular attractions. These were books for advance planning and consultation, in contrast to the guides to specific attractions that could be bought on the site.

The most famous of the national series was John Murray's *Handbooks for Travellers*, which, according to Samuel Smiles, had become the very "badge of the British traveller" by the end of the century.[28] The series began in 1836 with the *Handbook for Travellers on the Continent*, followed by volumes dealing with specific countries. Murray did not enter the British guidebook market until the 1850s, with a series of county volumes eventually completed in the 1890s.[29] His major competitor was the Edinburgh firm of Adam and Charles Black. The Blacks had started with guides to Edinburgh and Glasgow in 1839, but soon began treating larger areas: Scotland in 1840, the English Lakes in 1841, and England and Wales in 1843. Their distinctive feature was an emphasis on scenery: the standard series title was *Picturesque Tourist of . . .* , and the volumes usually contained several metal-engraved views. The firm continued to issue guides with "picturesque" in the title until the end of the 1850s, long after the original cult of the picturesque had faded. The Blacks produced guides to foreign destinations later in the century, but their key focus was always on domestic tourists, especially in north Britain.

As their price tags reveal, these major national series were a response to the expansion of middle-class tourism in the first half of the nineteenth century. In 1858, Black's guides to individual cities retailed at around 2s., but their *Lakes* was 5s. and *England and Wales* cost 10s. 6d.[30] Murray's volumes similarly ranged from 5s. (for *London*) to at least 15s. for the county and country guides, depending on length.[31] To put such prices in context, the third edition of Black's *England and Wales* (1853) provided a list of typical traveling expenses: a night's accommodation could cost from 1s. 6d. in a small town to 4s. in a good hotel in a major town. Dinner would be a further 3s., and breakfast 2s.[32] Guidebooks that cost more than dinner, bed, and breakfast were clearly not aimed at the masses.

The changing nature of the tourist market after midcentury did not go unnoticed by publishers. For instance, the Blacks launched a series of abridgements under the title *Shilling Guides* or *Economical Guides.* Their full guide to the Lake District ran to 249 pages by 1857, but the *Economical Guide* of the same year was just 74 pages. Moreover, new competitors entered the market. George Bradshaw was a Manchester engraver whose name had become synonymous with that most essential publication, *Bradshaw's Railway and Steam Navigation Guide*, the monthly compilation of timetables. By the late 1840s, Bradshaw's company had branched out into maps and tourist handbooks for Britain and the continent.[33] By 1848, you could purchase *Bradshaw's Descriptive Guides* for the South Western Railway, the London and South Coast Railway, the South Eastern Railway, and the Great Western Railway.[34] George Measom, also an engraver by training, launched a series of guidebooks similarly defined by railway company regions in 1852, starting with the Great Western Railway. His lavish use of wood engravings may have helped him secure the position of publisher of the *Official Illustrated Guides* to all the major railway companies in the 1850s.[35] Both Measom and Bradshaw charged just a shilling for their guidebooks, clearly indicating that they were aiming at the new tourists of the midcentury period rather than the affluent customers of Murray and the Blacks.

Guidebooks typically contained two sorts of material: general introductory remarks and detailed descriptions of particular places. The latter made up the bulk of a guidebook, offering short descriptions of towns and villages, with recommendations for visits to abbeys, manor houses, caverns, or hills with beautiful views. These entries were often arranged around recommended itineraries of various lengths. Murray's 1858 handbook to *Surrey, Hampshire and the Isle of Wight* opened by informing its readers that seven weeks would enable a tour to all the principal places in the region; but it also suggested shorter, more specialized, tours, including one focusing on antiquities and a pedestrian tour for those keen on walking and scenery.[36]

In the descriptions of locations, space was of the essence, so the entries were fairly brief and offered none of the wider context of, say, the history or geology of the region. If such contextualizing material did appear in a guidebook, it was usually in the introductory remarks. The extent and content of the introductory material varied dramatically between publishers, illustrating different conceptions of the basic information that all travelers needed to know. The Blacks offered practical advice on such matters as traveling expenses and using the railway system. Measom's early guidebooks featured a similar "Gossip Introductory," which disappeared from later editions once readers could be assumed to be familiar with the workings of the railway system.[37] In contrast, Murray's introductions provided extended essays on the geology, history, and economy of the region, to equip travelers to profit intellectually from their experiences.

Guidebooks have significant influence over travelers' decisions about where to go, what to see, and what to do. This is why they can shape travelers' experiences and knowledge of architecture and antiquities—and, perhaps, of the sciences. Their extended introductory essays made Murray's handbooks unusual in that they explicitly presented a range of general knowledge as necessary for the traveler, but their recommendation of certain types of places and activities were typical. My impression from studying a range of these nineteenth-century guidebooks is that, in the first half of the century, the sciences connected with tourism in only a limited way: the eighteenth-century traditions of visits to country houses (and to industrial sites) could have a scientific angle, while the love of scenery could certainly foster an interest in geology. But from the 1850s, there is more evidence of the sciences in the tourist literature. A growing enthusiasm for the out-of-doors helped inspire popular interest in natural history, especially but not only at the seaside. It also appears that the tourists of midcentury and after were more likely to be encouraged to see the sciences as linked to practical activities—such as collecting fossils, flowers, or seashells—rather than as a body of knowledge to be learned.

COUNTRY HOUSES, LANDSCAPES, AND MINES

We can get some idea of the sorts of places that appealed to tourists in the first half of the century by looking at the recommended "chief points of interest" in one of Murray's handbooks. The first Murray handbooks of Britain were entitled *London* and *Devon and Cornwall*, and, despite their 1851 publication date, like all Murray's handbooks, they represent the viewpoint of the upper-class or upper middle-class tourist. Figure 15.1 is part of the list of attractions in Cornwall, with asterisks marking the "most remarkable" places.[38] It is immediately apparent that historic houses and scenery remain popular. River or coast scenery is mentioned at Looe, Polperro, Truro, and Newquay, while the bleak landscapes of Bodmin Moor are listed under Jamaica Inn (Rowtor and Brown Willy are hills). The seat of the local baronet is mentioned, as are a number of antiquities from the "Olden Time," including castles at Saltash, Pendennis, and Tintagel, and the buried church of St. Enodoc, near Wadebridge. One of the striking features of the list is the inclusion of another, less famous, aspect of eighteenth- and early nineteenth-century tourism: the industrial visit. There are copper mines, tin stream-works, china-clay works, and any number of mines and quarries. Like admiration of the scenery, visits to mines and quarries could provide opportunities for tourists to encounter geology; country houses had less po-

Introd.   *Skeleton Tours.*   ix

No. III.—CORNWALL.

ROUTE.	CHIEF POINTS OF INTEREST.
Plymouth	.. p. viii.
Saltash	.. Trematon Castle.
St. Germans	.. Church. Port Eliot.
Looe	.. Scenery of the estuary and coast.
Polperro	.. Romantic coast.
Fowey	.. Place House. Scenery of the estuary.
Lostwithiel	.. Restormel Castle. Lanhydrock House. Glynn. Boconnoc.
St. Blazey	.. Treffry Viaduct*. Fowey Consols and Par Consols copper-mines.
St. Austell	.. Tin stream-works. Carclaze Mine*. China-clay works. Mevagissey. Roche Rocks*.
Grampound.	
Probus	.. Church tower.
Truro	.. Mines. Scenery of the river. St. Piran's church. Perran Round*.
Perran Wharf	.. Gardens of Carclew, seat of Sir Charles Lemon, Bart.
Falmouth	.. Pendennis Castle. Falmouth Harbour. Mabe Quarries. Tolmên.
Helston	.. Loe Pool. Kinance Cove*. Lizard Point*. Devil's Frying Pan.
Penzance	.. Museum of the Geolog. Society. St. Michael's Mount*. Land's End*. Tol Pedn Penwith*. Logan Rock*. Botallack Mine (submarine)*. Lamorna Cove. Druidic antiquities. Isles of Scilly.
Hayle	.. Iron-foundries. St. Ives and its bay*.
Redruth	.. Mines. Carn-brea Hill.
Newquay	.. Coast scenery.
St. Columb	.. Vale of Mawgan. Lanherne.
Wadebridge	.. Padstow. Church of St. Enodock.
Bodmin	.. Glynn valley. Hanter-Gantick*.
Liskeard	.. St. Keyne's Well. Clicker Tor. St. Cleer's Well. Trevethy Stone. Cheesewring*. Kilmarth Tor.
Jamaica Inn	.. Dozmare Pool. Brown Willy*. Rowtor*.
Camelford	.. Rowtor*. Devil's Jump. Hanter-Gantick*. Delabole Quarries. Tintagel*. St. Nighton's Keeve. Boscastle*.
Launceston	.. Castle. Church of St. Mary. Endsleigh*.
Callington	.. Dupath Well. Cothele*. View from Kit Hill*
Tavistock	.. p. viii.
Plymouth	.. p. viii.

No. IV.—DEVON AND E. CORNWALL.

A *walk* of 9 weeks taken by T. C. P. It comprehends the chief points of interest in Devonshire, and in Cornwall, E. of a line through Liskeard.

Note.—The best arrangement for a pedestrian tour in England is to send

a 3

Fig. 15.1. The chief points of interest in Cornwall, as listed in [Thomas Clifton Paris], *Handbook to Devon and Cornwall*, 3rd ed. (London: Murray, 1856). Reproduced courtesy of the Trustees of the National Library of Scotland.

tential as scientific tourist attractions, although there were some specific exceptions.

Owners of classical mansions continued to receive visits from the upper and upper middle classes, but those who wished to attract large parties of excursionists had to find ways to woo them away from ruined castles and monasteries.[39] Houses such as Belvoir, Chatsworth, Eaton Hall, and Alton Towers succeeded in transforming themselves into what Peter Mandler has dubbed "excursion centres," where the draw was less the house than the modern attractions in the grounds: conservatories, fountains, musical clock towers, or boating lakes.[40]

Some of these new attractions had connections to science and technology, as is particularly obvious at Chatsworth House, the family seat of the dukes of Devonshire.[41] Chatsworth had been a staple feature of any tour of the Peak District since the late eighteenth century, but it became a major tourist attraction in the mid–nineteenth century, under the aegis of William Cavendish, the sixth duke, and his gardener, Joseph Paxton. Paxton's talents also encompassed design and engineering, and he was largely responsible for turning the grounds at Chatsworth into a successful attraction. The duke encouraged Paxton to incorporate exotic species into the Chatsworth gardens, and Paxton designed two great glass houses to protect them. His Great Conservatory (1841) was the largest glass house in the world at that time, and it gave him invaluable experience for designing the Crystal Palace ten years later. The striking glass-and-iron architectural construction amazed visitors to Chatsworth, as did the birds that flew through the foliage. Paxton was also responsible for Chatsworth's impressive fountains, including the 260-foot tall Emperor Fountain (1844), built in expectation of a visit from Tsar Nicholas I.[42] The duke's arrangement with Thomas Cook resulted in visitor figures of around eighty thousand a year.[43] A visit to Chatsworth combined a mixture of forms of tourism: there was the house itself, its art collection, and its carefully landscaped grounds, in addition to its unrivaled botanical collection, its aquarium, and its fountains.

Areas of outstanding natural beauty also continued to draw tourists well into the nineteenth century, even though the enthusiasm for the picturesque was fading, and few of the new class of tourists were likely to have the skills to commit a landscape to watercolors. The growing visitor numbers led to tensions between those who sought to profit from them and those who deplored their effect upon the tranquillity of the area.[44] Wordsworth's opposition to the Kendal and Windermere railway line is perhaps the best-known example of such opposition, while the very fact that the railway line did open in 1847 demonstrates how the Lake District was being transformed from a region for upper middle-class private tours to one accessible to mass tourism.[45]

Throughout the first half of the nineteenth century, writers of tourist guidebooks routinely assured their readers that an understanding of geology was essential to understand—and thus fully to appreciate—landscape. The geologist could reveal where sheer cliffs, impressive cavern systems or picturesque valleys were to be found, and this information enabled the tourist to look with a trained eye at the landscape and to understand the forces behind its appearance. The most famous example of scenic tourism combined with geology was the guide to the Lake District created from an essay on scenery by Wordsworth and a set of letters on geology by his friend, the Cambridge geologist Adam Sedgwick. Wordsworth's essay had originally been written to introduce a volume of landscape views, but it was issued independently as a guidebook in 1822 and regularly revised and expanded for the next forty years, under variants of the title *A Complete Guide to the Lakes, Comprising Minute Directions for the Tourist.*[46] Sedgwick's contribution expanded from three to five lengthy letters.[47] The success of this format led the Blacks to commission an essay on geology from John Phillips, geology professor at King's College, London, for the second edition of their *Picturesque Guide to the English Lakes* (1844).

By the 1850s, essays for tourists by expert geologists had become rarer, but the inclusion of at least a paragraph on geology remained standard.[48] Murray's handbooks are good examples. The volume on *Devon and Cornwall* (1851) was written by Thomas Clifton Paris, the son of a geologist, and included a nine-page essay on geology plus a separate thirteen-page essay on mining, reflecting the importance of geology both to the stunning coastal scenery of the region and to its economy.[49] As Paris said, the coasts of Devon and Cornwall "display a variety of instructive sections" and provide "abundant evidence of physical convulsions which have modified the surface."[50] More commonly, geology was integrated into a more wide-ranging overview of the region. Richard John King was best known as an antiquarian and a writer on cathedrals, and he wrote or contributed to at least seven of Murray's county handbooks, including the volume on *Kent and Sussex* (1858).[51] King's introductory essays took the title "Travellers' View," and were essentially general descriptions of the countryside, including geological formations and the native vegetation and fauna. With no claims to geological expertise himself, King drew heavily (often in direct quotation) from the works of Charles Lyell and Gideon Mantell, among others.[52] King's discussion of geology and vegetation cover was intended to inform the reader about the type of views that could be expected in the area. Thus, his discussion of the clay of the Weald and the chalk of the uplands, with their oak and beech trees respectively, was followed immediately by a list of hills worth climbing for their views.[53]

The amount of actual geological knowledge a tourist could gain from reading these guidebooks varied dramatically with the author and the region. Paris's *Devon and Cornwall* essay introduced the seven mineral formations to be found in the region, from hornblende and graywacke to red sandstone and chalk, and explained their appearance and formation with considerable technical terminology. Some of the deposits associated with graywacke, for instance, were described as "both vesicular and compact, and formed of volcanic ashes and lava . . . now consolidated into slates, sandstones and conglomerates."[54] In contrast, King did little more than announce that "five parallel geological belts, of varying widths and outlines, extend throughout the county of Kent in a direction ranging from NW to SE," and used no complicated language.[55] Even the shortest discussions of geology encouraged tourists to think about the ground under their feet and the processes by which rivers and caverns had come into existence. At the very least, the routine inclusion of geology in travelers' handbooks alerted tourists to its existence as a subject of potential interest and relevance.

Geology also impinged upon nineteenth-century tourism through another eighteenth-century tradition, namely, the industrial visit.[56] This was arguably one of the earliest examples of a form of specifically scientific tourism. Wedgwood's factory at Etruria and Boulton and Watt's Soho engine works were the most famous factories to be visited by tourists with an interest in modern machinery and the production of the new generation of consumer goods, but they were not unique. In Derby in 1809, the young John Herschel was taken by his father to visit a water-powered silk mill, a steam-powered cotton mill, and a porcelain manufactory.[57] The industrial visit was supposed to be a rational inquiry into modern economic processes and technologies, but it could combine more visceral attractions: the heat, light, and smoke of blast furnaces could be seen as a sublime spectacle, while a descent into a Cornish tin mine provided a thrill of excitement.

Historians of tourism have tended to assume that industrial tourism died with the eighteenth century.[58] While not the complete truth—there was a spate of published accounts of visits to factories in the years between Charles Babbage's On the Economy of Machinery and Manufactures (1832) and the passing of factory legislation in the mid-1840s[59]—it does seem clear that factory visits never became part of the mass tourism of the middle and late nineteenth century. Not only did factory owners fear industrial espionage if large numbers of unknown visitors were admitted, but the increasing tendency toward efficiency and rationalization mitigated against disrupting factory processes by allowing visitors. And, of course, for most of the working-class excursionists, a trip to a factory was hardly an exotic prospect.

Local industry did appear in midcentury guidebooks, but it had been transformed from an object for rational inquiry to a source of consumer products. An unusually large number of local businesses were mentioned by name (and incorporated into the illustrations) in George Measom's guides, which may explain how Measom was able to sell his bulky guidebooks for just one shilling.[60] Some of the businesses mentioned by Measom were noteworthy for their technical innovation, such as the Messrs. Ellis, watchmakers and jewelers of Exeter, whose establishment had the "peculiar feature" of being in "communication, by galvanic wires, with the clock at the Guildhall, by which means the good people of Exeter are enabled to keep their appointments punctually."[61] But Measom mentioned many other Exeter businesses with fewer claims to innovation, including a "Music and Musical Emporium" selling "pianofortes, harmoniums &c" and a showroom displaying "a very large and fine collection of glass and Ceramic productions."[62] It seems as though, for Measom's midcentury tourists, local industries were principally sources of commodities for purchase.

The exception to both trends—the decline in industrial visits, and a shift towards seeing industry as a source of commodities—demonstrates another way in which geology was connected to nineteenth-century tourism. Visits to mines and quarries remained part of the tourist itinerary well after factory visits had disappeared. The attraction was the chance to descend into the depths of the earth.[63] For the geology enthusiast, this promised rational instruction, but for all tourists, it could be a slightly terrifying, sublime, experience. In 1851, for instance, the introduction to Murray's *Devon and Cornwall* recommended a mine visit for the opportunity "of descending through the crust of the earth, and examining its structure."[64] One of its recommended locations was the mine at Botallack, outside Penzance, in a breathtaking location amid "gloomy precipices of slate" rising out of the sea. The proximity of the sea was important, since "the traveller who should venture to descend into its dreary recesses may be gratified by hearing the booming of the waves and the grating of the stones as they are rolled to and fro over his head."[65] No wonder Paris had warned in his introduction that such "sublime but portentous sounds" would require great "strength of nerve in the visitor."[66] In the eighteenth and early nineteenth centuries, visits of this sort were presumably made by personal arrangement with the mine-owners, in much the same way as country house visits. The fact that handbooks like Murray's would recommend specific mines by name to their readers suggests that, by midcentury, it was perfectly possible for individuals and family groups to arrive at a mine in confident expectation of being allowed to visit.

In the first half of the nineteenth century, geology was easily the science that was most likely to be encountered by tourists. Certain country houses boasted collections of stuffed or living animals, exotic plants, or awe-inspiring fountains, but these attractions tended to be unique to specific houses. Geology, however, would routinely be encountered by tourists in search of outstanding scenery, as well as by those who thrilled at the idea of descending into the darkness of a mine. And while the lengthy geological essays offered by Sedgwick or Phillips (or even by Paris) certainly attempted to educate their readers in the details of geological theory, the subordination of geological understanding to the appreciation of scenery meant that geology was intrinsically linked to the activities of mild walking and active, careful, detailed observation. With mine visits, too, geology was providing the stimulus for a particular form of activity. After the middle of the century, this connection between tourism and scientifically grounded or motivated activities appears to have become more widespread, thanks to the growing enthusiasm for outdoor recreation.

FOSSILS, FLOWERS, AND SEA ANEMONES

In contrast to geology, other forms of natural history received very little attention in early nineteenth-century guidebooks. Early editions of Wordsworth's *Complete Guide to the Lakes* had little to say about natural history, although by midcentury, gestures were made toward those with an interest in botany and zoology. Yet in many ways the inclusion of six pages of species lists of flowers and mollusks served to emphasize the significance of the hundred pages devoted to Sedgwick's geological contributions.[67] Vegetation cover might be mentioned in passing in discussing the scenery, but as late as 1851, Murray's *Devon and Cornwall* mentioned only one fish (the pilchard) and two birds (the cormorant and the chough). The pilchard merited its three pages of discussion only because of the economic significance of its fisheries.[68]

By the mid-1850s, however, railways were enabling travelers to leave the city for cleaner air and open spaces. As one writer expressed it, "What a pleasant change it is, for those who are able to escape from endless brick and mortar, to run down a three or four hours' journey by rail, and find themselves amidst the scenery, the quietness, and the breezes of the seashore!"[69]

This opening sentence of a book about the seaside nicely sets up the contrast between the "endless bricks and mortar" of the city and the "scenery," "quietness," and "breezes" of the seaside, as well as indicating the importance of the railway as the link between them. The seaside became the favored destination of all those who lived within striking distance of the coast, although lakes and moors could play an equivalent role.[70] Bathing in (or drinking) the sea had been promoted for health purposes from the end of the eighteenth century onwards: Weymouth and Brighton offered coastal alternatives to the sort of fashionable amusement-cum-recuperation offered by such inland spa towns as Bath or Tunbridge Wells. The typical activities on offer at a seaside resort in the first half of the century included indoor and outdoor bathing, reading rooms, concerts, assemblies, and promenading. With the arrival of new classes of railway visitors, these activities were joined by donkeys and sandcastles on the beach, telescopes and mechanical toys on the piers, and a trend toward private, family entertainment.[71]

In addition, a seaside holiday offered numerous opportunities for investigating the natural world, something that appealed to those who sought rational recreation rather than mere hedonism.[72] The coast could, of course, be appreciated as scenery, as generations of travelers had already done, but there was a new interest in its flora and fauna. This was promoted by a group of successful writers of popular natural history works, including Philip Henry Gosse, who urged tourists to engage in close observation of living plants and animals, and whose works were then drawn upon by the compilers of travel guidebooks.

There are some hints to suggest that a love for collecting souvenirs may have prevented the development of the trained observational skills that the natural history writers were trying to promote.[73] This is most apparent in the tourist literature relating to natural history, but geology could also be turned into an occasion for souvenir-collecting. In Murray's *Devon and Cornwall,* for instance, the description of the romantic fishing village of Polperro is followed by speculation about the role of "a mighty torrent" acting "upon the strata of the earth" in forming the cove. Then, the tourist is informed that "the remains of fossil fish, considered characteristic of certain formations in the Silurian system of Murchison, but so imperfect that it requires an observant eye to identify them, were discovered some years ago by Mr Couch of this town. . . . Under the signal station they occur in the greatest abundance."[74] The fact that the reader is not merely told about the fish, but told precisely where to find them, may imply an expectation that some tourists were hoping to find a memento to take home. There are also references to good fossil locations and to fossil hunting in other handbooks in Murray's series (and not only in the one written by the geological enthusiast). In Devon, tourists were pointed to the impressive stretch of "fossiliferous limestone" at Pencarrow Head.[75] On the Isle of Wight, they were encouraged to read Gideon Mantell's *Geological Excursion round the Isle of Wight* (1847) and directed to places where fossil mollusks and mammals were found.[76] In the Kent handbook, the suggested route for the

Isle of Sheppey gives a detailed description of the fossils in which the area "abounds," along with "directions for the collector."[77] In contrast to scenery and mines, fossil hunting appears to be a more recent addition to the geological activities of tourists, but it is in keeping with contemporary developments in botany and zoology as tourist activities.

Two of the natural history writers who encouraged their readers to pay attention to the natural world were Philip Henry Gosse and John George Wood. Gosse was the author of the *Sea-Side Pleasures* quoted above, as well as *A Naturalist's Rambles on the Devonshire Coast* (1853) and a book about the aquarium.[78] Wood wrote many books, including *Common Objects of the Sea Shore* (1857) and *Common Objects of the Country* (1858).[79] Both writers emphasized the importance of close observation, especially of small or common creatures, and they urged their readers to go outside and learn for themselves about birds, butterflies, newts, or bats. Rather than teaching the basic anatomical facts about an animal, Gosse and Wood encouraged their readers to watch behavior and habits—things that had to be studied in the field rather than the museum. Gosse was the more respected writer in natural historical circles, but it was Wood who reached large numbers of travelers. Wood's books were published by George Routledge, who specialized in marketing cheap books to travelers on railway platforms.[80] With its colorful wrapper and shilling price-tag, Wood's *Sea Shore* (fig. 15.2) sold 15,000 copies in its first year and 14,000 the following year, while the *Country* sold 31,000 copies in its first year alone.[81] Gosse's books typically sold 1,500–2,000 copies.[82] Routledge had guessed that Wood's style of natural history writing would appeal to those travelers escaping the confines of the city, and the sales figures vindicated his instinct.

I suggest that it is more than a coincidence that this new style of writing popular natural history emerged in the middle of the nineteenth century, at the same time as railway excursions were making it easier than ever before to visit the seaside and countryside. I find support for this in the fact that handbooks for tourists were quick to incorporate it into their own works. From the middle of the century, the existing geological tourism was joined by a growing amount of natural historical activities, some of them derived directly from Gosse's writings.

When Wordsworth's *Guide to the Lakes* was abridged in the late 1850s, the editors removed Sedgwick's geology entirely but kept the short lists of plant and mollusk species. Revealingly, the preface promoted the plant list to those seeking rare flowers, but also to those wanting simply "a memento of . . . a pleasant summer's holiday."[83] Like fossils, pressed flowers could be souvenirs. Similarly, Murray's handbooks of the later 1850s contain

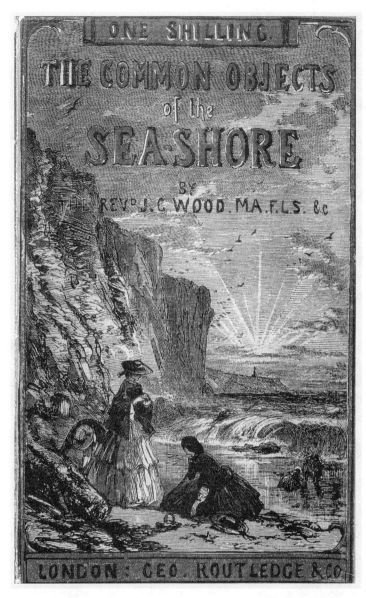

Fig. 15.2. The cover of the shilling edition of *Common Objects of the Sea Shore*, by
John George Wood (London: Routledge, 1857), reproduced by permission of the Bodleian
Library, University of Oxford (shelfmark: Johnson f.308).

references to botanical as well as geological souvenir-hunting. On the Isle of Wight, visitors are encouraged to consult Dr. Bromfield's *Flora Vectensis* (1856) for advice on the "rich and varied" species,[84] and advised that

> the *Morio, Mascula* and *Maculata* [orchids] are abundant everywhere. The downs about Bonchurch and Ventnor are covered with the *bee, dwarf* and *pyramidal Orchises,* while the *fly* is found at Quarr and Westover, the *butterfly* about Ryde and the Landslip, and the *spider* near the Boniface Hotel. . . . Off the road from Newport to Niton, a little beyond the Stag Inn at Rookley, a place which should by all means be visited by the botanist: the specimens of Osmunda [fern] are truly magnificent.[85]

The level of detail is striking: although some species are "everywhere," others are in specific localities (Bonchurch, Quarr), and a few are in extremely specific places (the Boniface Hotel, the Stag Inn). Such descriptions insured that even tourists who consulted nothing other than Murray's handbook would have a reasonable chance of locating the most notable species of orchids and ferns—the latter being something of a mania in the 1850s.[86]

Easily the most striking evidence of the increased tourist interest in natural history comes from Murray's *Devon and Cornwall* handbook. As we have already seen, its first edition featured plenty of geology but very little natural history. By the third edition, in 1856, the balance had changed completely, largely in response to Gosse's popularization of the marine natural history of Devon (though the change affected all forms of natural history, not just marine creatures). The transformation is best illustrated through a detailed example. Ilfracombe is a small coastal town in north Devon, with a population of about four thousand persons in the 1850s. Visitors from London would take the Great Western Railway from Paddington station and transfer to the Bristol & Exeter Railway; from 1851, they could then take the North Devon Railway to Barnstaple, which brought them within eleven miles of Ilfracombe.[87] The first edition of Murray's *Handbook* had little to say about the town, beyond the fact that "its principal attraction is the coast, which, stamped with a peculiar character by the irregularity of its outline, presents a front of huge dark rocks and chasms." It also had a church, and public baths.[88] The 1856 edition, however, added a completely new paragraph to its description of Ilfracombe. Now we learn that "this rocky shore has also interest in another aspect. It is a favourite haunt of those wonderful and beautiful forms of life so recently brought to our notice by such men as Gosse, who at Ilfracombe found . . . his madrepore . . . his polype . . . and his anemone."[89] Although the handbook still mentioned the rugged coast and the church, it now urged travelers to look out for the tiny creatures; Ilfracombe was a particularly good place for this sort of activity because "the tides are very favourable."

Collecting anemones and microscopic marine creatures was rather more complicated than collecting flowers, and Murray's handbook did not offer any advice on the subject, directing its readers instead to Gosse's *Naturalist's Rambles*. Gosse provided a graphic picture of the discomfort to be endured by the true enthusiast: "It is rare that the position of the hole is such as to allow of both arms working with any ease; the rock is under water, and often, if your chisel is short . . . every blow which the hammer strikes . . . has to fall upon a stratum of water, which splashes forcibly into your eyes and over your clothes."[90] Perhaps aware that most tourists were unlikely to spend hours up to their chests in cold water, Wood's technique for collecting anemones involved "slipping the thumb-nail or an ivory paper-knife under the base, and so gradually peeling them away from the support."[91] For both writers, the reason for collecting anemones was to keep them alive afterwards, in jars of saltwater or a proper aquarium, for close observation. Yet the 1850s enthusiasm for aquaria suggests that many tourists regarded a collection of living anemones as the latest fashionable souvenir, rather than an object of scientific study.[92]

Gosse was not the only natural history writer mentioned in the new edition of *Devon and Cornwall*—William Henry Harvey, an expert on seaweed, was also recommended[93]—but Gosse was the most-frequently mentioned. The pattern of references suggests that Paris referred to recently published books about the seaside when searching for new material to add to his revised edition. Ilfracombe was certainly not the only location to acquire new reasons for a visit: for instance, the headland at Braunton Burrows, which had previously been mentioned only for its lighthouse, was now commended for the "many curious plants [that] find a congenial soil among these sandhills, particularly the *round-headed club-rush,* one of the rarest in Britain. . . . Mr Gosse also mentions the *small buglos,* the rare *musky stork's-bill.*"[94] And new locations appeared for the very first time, solely because of their natural historical interest. The beach at Barricane was one such new inclusion, described as "a delightful spot, where the beach almost entirely consists of shells, many beautiful and curious."[95]

Like their competitors, Murray's *Handbooks* were aimed at general travelers and tourists, for whom the sciences were unlikely to be of more than passing interest. Nevertheless, the inclusion of information about geology, botany, and marine zoology indicates that guidebook authors and editors regarded the sciences in roughly the same light as they regarded history or architecture—as something that tourists might take an interest in and that could form the basis of particular excursions or activities. Murray's authors did not claim authority for themselves in the sciences, but were happy to recommend interested readers to more detailed accounts, such as those by Gosse and Harvey, or Lyell and Mantell. Gosse himself appreciated the role that his writings might play, suggesting that readers of his *Naturalist's Rambles* treat it as "your 'Hand-book' to the sea-side."[96] By the end of the nineteenth century, a new genre of specialized handbook would emerge to cater to those tourists and travelers with more than a passing interest. For instance, Murray launched a series of *Handbooks to the Cathedrals of England* (1861–69), and in the sciences we see the precursors of the modern "field guide" in works such as J. G. Wood's *Field Naturalist's Handbook* (1880).

Apart from travel and accommodation, the most common thing for tourists to spend money on was advice—whether a human guide hired for a particular excursion, a guidebook to a specific site, or a handbook to a larger region. Those guides were likely to play an important role in influencing both the tourist's choice of activities and the way those activities were understood. It is surely the case that areas of natural beauty and historical sites (especially if romantically ruined) attracted far more tourists than sites of scientific interest. But the evidence of the guidebooks does show that certain aspects of the sciences were presented to tourists. Geology has the longest-standing connection, offering a means to appreciate particularly spectacular coastlines, a justification for thrilling trips down mines, and the opportunity for fossil souvenirs. Other forms of natural history seem to have become part of tourist activities somewhat later, especially with the opening-up of the seaside and countryside from the 1850s. Marine zoology could justify (or stimulate) happy hours spent on the beach, while an interest in botany could lead to long afternoon walks through the countryside; both offered prospects for souvenirs to take home.

I began this chapter by drawing attention to two themes in the history of science that I hoped the study of science and tourism might illuminate: observation skills and active participation. There is no doubt that the more specialized natural history guides, such as those by Gosse or Wood, hoped to train their readers in how and what to observe, but there is rather less evidence of this in the general travel handbooks, where fossils, flowers, anemones, and seashells appear all too often as things to be taken home rather than observed in situ. For ordinary tourists, the souvenir-collecting impulse may have taken precedence over learning to observe. If there is an exception, it is in the use of geological knowledge to appreciate scenery.

Like architectural sites, landscapes must be admired and examined in situ, and this can best be done by an informed observer who knows what to look for and at. The early nineteenth-century guidebooks, which emphasized the picturesque and included lengthy essays on geology in relation to scenery, were surely attempting to train their readers to observe.

In its relation to scenery, geology might seem like a rather passive form of tourist knowledge, but it could be a stimulus to particular activities, from visits to mines to the hunt for fossils. The natural history examples discussed here were also clearly linked to an active pursuit of specimens, whether in roaming the lanes of the Isle of Wight, or by standing chest-high in seawater at Ilfracombe. This evidence supports my initial contention that people encountering the sciences as tourists were more likely to see the sciences as providing opportunities for active, participatory pursuits. Studying tourism, therefore, offers us an important perspective on the ways in which the movements of people to new, unfamiliar locations affected their understanding of and engagement with the natural world—not just for those interested in the making of scientific knowledge, but for a wide cross-section of the population.

NOTES

1. Janet Browne, *Charles Darwin: Voyaging* (London: Cape, 1995); David Philip Miller and Peter Hans Reill, eds., *Visions of Empire: Voyages, Botany, and Representations of Nature* (Cambridge: Cambridge University Press, 1996); Peter Raby, *Bright Paradise: Victorian Scientific Travellers* (London: Pimlico, 1996); Michael T. Bravo, "Precision and Curiosity in Scientific Travel: James Rennell and the Orientalist Geography of the New Imperial Age (1760–1830)," in *Voyages and Visions: Towards a Cultural History of Travel*, ed. Jas Elsner and Joan-Pau Rubies (London: Reaktion, 1999), 162–83; Felix Driver, *Geography Militant: Cultures of Exploration and Empire* (Oxford: Blackwell, 2001); Jim Endersby, *Imperial Nature: Joseph Hooker and the Practices of Victorian Science* (Chicago: University of Chicago Press, 2008). On more local travel, see Martin Rudwick, *Bursting the Limits of Time: The Reconstruction of Geohistory in the Age of Revolution* (Chicago: University of Chicago Press, 2005), 41–44, 71–75; Lisbet Koerner, *Linnaeus: Nature and Nation* (Cambridge, MA: Harvard University Press, 1999).

2. John Cawood, "The Magnetic Crusade: Science and Politics in Early Victorian Britain," *Isis* 70 (1979): 492–518; Alex Soojung-Kim Pang, "The Social Event of the Season: Solar Eclipse Expeditions and Victorian Culture," *Isis* 84 (1993): 252–57; Richard Sorrenson, "The Ship as a Scientific Instrument in the Eighteenth Century," *Osiris* 11 (1996): 221–36.

3. E. C. Spary, *Utopia's Garden: French Natural History from Old Regime to Revolution* (Chicago: University of Chicago Press, 2000), chap. 2; Jim Endersby, "'From Having No Herbarium.' Local Knowledge vs. Metropolitan Expertise: Joseph Hooker's Australasian Correspondence with William Colenso and Ronald Gunn," *Pacific Science* 55 (2001): 343–58.

4. Such an investigation would continue recent work on the history of popular science, which focuses upon the appearance of science and nature in general culture. See Aileen Fyfe and Bernard Lightman, eds., *Science in the Marketplace: Nineteenth-Century Sites and Experiences* (Chicago: University of Chicago Press, 2007); for a demonstration of the continuing importance of the sciences in general culture, see Geoffrey Cantor et al., *Science in the Nineteenth-Century Periodical* (Cambridge: Cambridge University Press, 2004), introduction, esp. 24–25. For a discussion of the state of popular science research, see the "Focus" section in *Isis* 100 (2009): 310–68.

5. For a critique of older approaches, see Stephen Hilgartner, "The Dominant View of Popularization: Conceptual Problems, Political Uses," *Social Studies of Science* 20 (1990): 519–39. For an example of a newer approach, see James A. Secord, *Victorian Sensation: The Extraordinary Publication, Reception and Secret Authorship of "Vestiges of the Natural History of Creation"* (Chicago: University of Chicago Press, 2000).

6. On children's science books, see Aileen Fyfe, "Young Readers and the Sciences," in *Books and the Sciences in History*, ed. Marina Frasca-Spada and Nicholas Jardine (Cambridge: Cambridge University Press, 2000), 276–90; James A. Secord, "Introduction," in *The Boy's Playbook of Science*, ed. John Henry Pepper, Science for Children (Bristol: Thoemmes Press, 2003). For popular science for adults, see Secord, *Victorian Sensation*, pt. 1; Aileen Fyfe, *Science and Salvation: Evangelicals and Popular Science Publishing in Victorian Britain* (Chicago: University of Chicago Press, 2004), chap. 3.

7. Jack Simmons, "Introduction," in *Murray's Handbook for Travellers in Switzerland*, ed. John Murray and William Brockeden (Leicester: Leicester University Press, 1970), 26–7. On early modern travel literature and observation, see Joan-Pau Rubiés, "Instructions for Travellers: Teaching the Eye to See," *History and Anthropology* 9 (1996): 139–90.

8. Esther A. L. Moir, *The Discovery of Britain: The English Tourist, 1540–1840* (London: Routledge, 1964); Maxine Feifer, *Going Places: The Ways of the Tourist from Imperial Rome to the Present Day* (London: Macmillan, 1985); Judith Adler, "Origins of Sightseeing," *Annals of Tourism Research* 16 (1989): 7–29; Ian Ousby, *The Englishman's England: Taste, Travel and the Rise of Tourism* (Cambridge: Cambridge University Press, 1990).

9. James Buzard, *The Beaten Track: European Tourism, Literature and the Ways to "Culture," 1800–1918* (Oxford: Clarendon Press, 1993), chap. 2.

10. Jane Austen, *Pride and Prejudice*, 2 vols. (London: J. M. Dent, 1892), ii, 58, 60 (chaps. 42–43).

11. John Harris, "English Country House Guides, 1740–1840," in *Concerning Architecture: Essays on Architectural Writers and Writing, Presented to Nikolaus Pevsner*, ed. John Summerson (London: Allen Lane, 1968), 58–74.

12. Malcolm Andrews, *The Search for the Picturesque: Landscape Aesthetics and Tourism in Britain, 1760–1800* (Aldershot: Scolar, 1989), chaps. 2–3.

13. Jane Austen, *Northanger Abbey* (London: J. M. Dent, 1892), 103–4.

14. Andrews, *Search for the Picturesque*, 158.

15. Stephen Gill, *William Wordsworth: A Life* (Oxford: Clarendon Press, 1989), 44–46, 50.

16. J. A. R. Pimlott, *The Englishman's Holiday: A Social History* (Hassocks, Sussex: Harvester Press, 1976), chaps. 5–9.

17. These places were all described in the *Leisure Hour* between July and September 1852.

18. Charles Knight, ed., *The Land We Live In: A Pictorial and Literary Sketch-Book of the British Empire*, 4 vols. (London: Knight, 1847–49).

19. Pimlott, *Englishman's Holiday*, chaps. 9–10; Alastair J. Durie, "Tourism and the Railways in Scotland: The Victorian and Edwardian Experience," in *The Impact of the Railway on Society in Britain: Essays in Honour of Jack Simmons*, ed. A. K. B. Evans and J. V. Gough (Aldershot: Ashgate, 2003), 199–209.

20. Jack Simmons, *The Victorian Railway* (London: Thames & Hudson, 1991), chap. 12.

21. All these advertisements appear in column A of the *Times* (London), 31 August 1855, 4.

22. Feifer, *Going Places*, 163–200; Buzard, *Beaten Track*, 49–65.

23. Peter Mandler, "'The Wand of Fancy': The Historical Imagination of the Victorian Tourist," in *Material Memories*, ed. Marius Kwint, Christopher Breward, and Jeremy Aynsley (Oxford: Berg, 1999), 125–42.

24. Charles Knight, *Old England: A Pictorial Museum of Regal, Ecclesiastical, Baronial, Municipal and Popular Antiquities*, 2 vols. (London: Knight, 1845).

25. E. M. Forster, *A Room with a View* (London: Softback Preview, 1995), chap. 2.

26. E. S. de Beer, "The Development of the Guidebook until the Early Nineteenth Century," *Journal of the British Archaeological Association* 15 (1952): 35–46; Harris, "Country House Guides"; John E. Vaughan, *The English Guide Book, c. 1780–1870: An Illustrated History* (Newton Abbot: David & Charles, 1974); Giles Barber, "The English Language Guide Book to Europe up to 1870," in *Journeys through the Market: Travel, Travellers and the Book Trade*, ed. Robin Myers and Michael Harris, Publishing Pathways Book Series (London: St. Paul's Bibliographies, 1999), 93–106; Nicholas Parsons, *Worth the Detour: A History of the Guidebook* (London: Sutton, 2007).

27. The *OED* gives the first use of *guidebook* as 1814, followed by a Byron quotation from 1823, and many quotations from the 1840s. *Guide*, with reference to a book, first appears in 1759.

28. Smiles 1891, cited in Simmons, "Introduction," 27.

29. Ibid.; Vaughan, *English Guide Book*, 44–47; Buzard, *Beaten Track*, 66–73.

30. Advertisement for Black's guides, in *Murray's Handbook Advertiser* (1858), 15.

31. The Cambridge University Library copy of Murray's *London* (1851) has its price on its spine.

32. *Black's Picturesque Tourist, and Road and Railway Guide Book through England and Wales*, 3rd ed. (Edinburgh: A. & C. Black, 1853), "Travelling Expenses," vii.

33. The monthly publication had been preceded by an occasional publication from 1839; see G. C. Boase, "Bradshaw, George (1801–1853)," rev. Philip S. Bagwell, *Oxford*

Dictionary of National Biography (Oxford University Press, 2004); available online at http://www.oxforddnb.com/view/article/3195 (accessed 22 February 2008).

34. See advertisement in *Bradshaw's Railway Almanack, Directory, Shareholder's Guide and Manual* (London: Adams, 1848).

35. J. D. Bennett, "The Railway Guidebooks of George Measom," *BackTrack* 15 (2001): 379–81; G. H. Martin, "Sir George Samuel Measom (1818–1901) and His Railway Guides," in *The Impact of the Railway on Society in Britain; Essays in Honour of Jack Simmons*, ed. A. K. B. Evans and J. V. Gough (Aldershot: Ashgate, 2003), 225–40.

36. Richard John King, *A Handbook for Travellers in Surrey, Hampshire and the Isle of Wight* (London: Murray, 1858).

37. Martin, "Measom and His Railway Guides," 232.

38. [Thomas Clifton Paris], *A Handbook for Travellers in Devon and Cornwall*, 3rd ed. (London: Murray, 1856), 12.

39. Peter Mandler, *The Fall and Rise of the Stately Home* (New Haven, CT: Yale University Press, 1997), chaps. 1–2.

40. Ibid., 85–97.

41. One could also mention the stuffed and living animals at Charles Waterton's Walton Hall, as discussed by Vicky Carroll, "The Natural History of Visiting: Responses to Charles Waterton and Walton Hall," *Studies in History and Philosophy of Biological and Biomedical Sciences* 35 (2004): 31–64.

42. John Kenworthy-Browne, "Paxton, Sir Joseph (1803–1865)," *Oxford Dictionary of National Biography* (Oxford University Press, 2004); available online at http://www.oxforddnb.com/view/article/21634 (accessed 19 February 2008).

43. Mandler, *Fall and Rise of the Stately Home*, 93–96.

44. Ousby, *Englishman's England*, chap. 4.

45. Gill, *William Wordsworth*, 413–14.

46. Ibid., 284–5; Vaughan, *English Guide Book*, 119–20.

47. Geology occupied 96 of the 272 pages of the 1859 edition: see John Hudson, ed., *A Complete Guide to the English Lakes, with Minute Directions for Tourists; and Mr Wordsworth's Description of the Scenery of the Country, Etc; Also, Five Letters on the Geology of the Lake District, by the Rev. Professor Sedgwick*, 5th ed. (Kendal: Hudson; London: Longman; and London: Whittaker, 1859).

48. See, for instance, John Marius Wilson, *Hand-Book to the English Lakes* (Edinburgh: T. Nelson & Sons, 1859), preface, xxxv–xxxvi.

49. Paris was the son of John Ayrton Paris.

50. Thomas Clifton Paris, *A Handbook for Travellers in Devon and Cornwall* (London: Murray, 1851), xiv–xxiii and xxxi–xliv, quotation at xiv.

51. W. P. Courtney, "King, Richard John (1818–1879)," rev. Ian Maxted, *Oxford Dictionary of National Biography* (Oxford University Press, 2004); available online at http://www.oxforddnb.com/view/article/15591 (accessed 22 February 2008).

52. For instance, the geological section in "Sussex" runs to three pages, largely due to the quotations from Lyell and Mantell; see [Richard John King], *A Handbook for Travellers in Kent and Sussex* (London: Murray, 1858), xxviii–xxxi.

53. Ibid., xvii–xix.

54. [Paris], *Devon and Cornwall*, 1st ed., xv.

55. [King], *Kent and Sussex*, xvii.

56. The only extended account of such tourism seems to be Moir, *Discovery of Britain*, chap. 8.

57. W. J. Ashworth, "Memory, Efficiency, and Symbolic Analysis: Charles Babbage, John Herschel, and the Industrial Mind," *Isis* 87 (1996): 630.

58. Moir, *Discovery of Britain*, 107.

59. For instance, George Head, *Home Tour through the Manufacturing Districts of England, in the Summer of 1835* (London: Murray, 1836); George Dodd, *Days at the Factories; or, the Manufacturing Industry of Great Britain Described and Illustrated by Numerous Engravings of Machines and Processes* (London: Knight, 1843).

60. Murray's handbooks did not accept payments in return for favorable mentions (though they did carry a separate advertising supplement), according to Simmons, "Introduction," 23–24. The only evidence regarding Measom's practice is internal, although see Martin, "Measom and His Railway Guides," 234.

61. George Measom, *The Official Illustrated Guide to the Bristol and Exeter, North and South Devon, Cornwall, and South Wales Railways*, 2nd ed. (London: Griffin; and London: Bohn, 1861), 153.

62. Ibid., 153–54.

63. On eighteenth-century mine descents, see Moir, *Discovery of Britain*, 92–96. On visits to caves, see Ousby, *Englishman's England*, chap. 4.

64. [Paris], *Devon and Cornwall*, 1st ed., xiv.

65. Ibid., 182.

66. Ibid., xxxix.

67. Geology occupies pp.168–264, while botany and zoology occupy pp. 265–70; see Hudson, *Complete Guide to the English Lakes*.

68. Paris, *Devon and Cornwall*, 1st ed., l–liii.

69. Philip Henry Gosse and Emily Gosse, *Sea-Side Pleasures* (London: Society for Promoting Christian Knowledge, 1853), 1.

70. Pimlott, *Englishman's Holiday*, chaps. 5–7; Simmons, *Victorian Railway*, chap. 13; Durie, "Tourism and the Railways."

71. Pimlott, *Englishman's Holiday*, chaps. 3 and 7.

72. David Allen, *The Naturalist in Britain: A Social History*, 2nd ed. (Princeton, NJ: Princeton University Press, 1994), chap. 6.

73. Susan Stewart, *On Longing: Narratives of the Miniature, the Gigantic, the Souvenir, the Collection* (Durham, NC: Duke University Press, 1993), 132–51.

74. [Paris], *Devon and Cornwall*, 3rd ed., 236.

75. [Paris], *Devon and Cornwall*, 1st ed., xvii.

76. King, *Surrey and Hampshire*, 276, 289.

77. [King], *Kent and Sussex*, xviii.

78. On Gosse, see Ann Thwaite, *Glimpses of the Wonderful: The Life of Philip Henry Gosse, 1810–1888* (London: Faber, 2002); Richard Broke Freeman and Douglas Wertheimer, eds., *Philip Henry Gosse: A Bibliography* (Folkestone: Dawson, 1980).

79. On Wood, see Theodore Wood, *The Rev. J. G. Wood: His Life and Work* (London: Cassell, 1890); Bernard Lightman, *Victorian Popularizers of Science: Designing Nature for New Audiences* (Chicago: University of Chicago Press, 2007), chap. 4.

80. On Routledge's publishing strategy, see F. A. Mumby, *The House of Routledge, 1834–1934, with a History of Kegan Paul, Trench, Trubner and Other Associated Firms* (London: Routledge, 1934), chap. 2.

81. For *Sea Shore*, see Publication Books 1850–58, Routledge Archives, vol. 2, f. 373, f. 426; vol. 3, f. 26. For *Country*, see Publication Books 1850–58, Routledge Archives, vol. 2, f. 424; vol. 3, f. 87 and f. 96. Note that Wood's son claimed that an edition of one hundred thousand of *Country* was sold out within a week. The claim has often been repeated (for example, by David Allen), but it is not supported by the Routledge archives; see Wood, *J. G. Wood*, 61.

82. Gosse's sales figures are mentioned in Thwaite, *Glimpses of the Wonderful*, 255, and given in detail in Freeman and Wertheimer, *Gosse Bibliography*. The *Rambles on the Devonshire Coast* sold fifteen hundred copies.

83. T. Wilson, ed. *Wilson's Hand-Book for Visitors to the English Lakes, with an Introduction by the Late W. Wordsworth, Esq., and a New Map of the Lake District, to Which Is Appended a Copious List of Plants Found in the Adjacent Country* (Kendal: Wilson; London: Longman; and London: Whittaker, n.d.), iv.

84. King, *Surrey and Hampshire*, 276.

85. Ibid., 276–77.

86. Allen, *Naturalist in Britain*, 121; David Allen, "Tastes and Crazes," in *Cultures of Natural History*, ed. Nicholas Jardine, James A. Secord, and Emma Spary (Cambridge: Cambridge University Press, 1996), 394–407.

87. Measom, *Bristol and Exeter*, 156–9.

88. [Paris], *Devon and Cornwall*, 1st ed., 117.

89. [Paris], *Devon and Cornwall*, 3rd ed., 127.

90. Philip Henry Gosse, *A Naturalist's Rambles on the Devonshire Coast* (London: Van Voorst, 1853), 25.

91. John George Wood, *The Common Objects of the Sea Shore: Including Hints for an Aquarium* (London: Routledge, 1857), 62.

92. On aquaria, see Allen, *Naturalist in Britain*, chap. 6; Thwaite, *Glimpses of the Wonderful*, 178–82; Rebecca Stott, *Theatres of Glass: The Woman Who Brought the Sea to the City* (London: Short Books, 2003).

93. [Paris], *Devon and Cornwall*, 3rd ed., 126.

94. Ibid., 129.

95. Ibid., 128.

96. Gosse, *Naturalist's Rambles*, v.

CHAPTER SIXTEEN

Place and Museum Space:
The Smithsonian Institution, National Identity,
and the American West, 1846–1896

SALLY GREGORY KOHLSTEDT

Scientific exploration of the American West and the establishment of the Smithsonian Institution occurred in tandem, each contributing to the reputation and identity of the other in the last half of the nineteenth century. Through its compelling and substantial natural history accessions, publications, and displays, the new institution described and defined the trans-Mississippi in ways that contributed to American fascination with the expansive territories acquired between 1803 and 1867. Prominent and public, the Smithsonian provided dedicated space for many of the mineralogical, zoological, and anthropological artifacts and specimens obtained during two periods of intensive exploration sponsored largely by the federal government immediately before and just after the Civil War. These material holdings gave significant visibility and importance to the striking red sandstone "castle" on the Mall strategically situated midway between the White House and the Capitol. Equally important, the museum artifacts contributed to a very material and intriguing portrayal of the American West, providing a fresh focus for Americans wrestling with a troubled self-identity as the victorious but maimed North and the devastated South were forcibly reunited.

The rapidly expanding young museum within the Smithsonian underscored the geographical provenance of western objects even as their reorganization allowed for new and malleable descriptions and meanings of the West and indeed of its very nature.[1] In the process, it helped affirm the American West as a projection of the entire country. The construction of the research and exhibition space in the Smithsonian and the public perceptions of the place designated as the American West demonstrate ways in which a geographic sensibility may be multiply constituted in imagination, through displaced artifacts, and on physical sites.

The phrase *new world* had resonated early with European colonists, and the distinctiveness of its natural history became central to their emerging national identity.[2] Over three centuries, European explorers, fur traders, the military, farmers, and families pushed Native Americans westward until the demographic edge of the frontier finally disappeared.[3] As the nation set its constitution in 1789 and expanded its territory, there was considerable interest in what lay beyond the Mississippi River. Even as its topography was being mapped in detail, the West evoked multiple images with its exotic landscapes, its potential for settlement, its evident mineral resources, its scientific surprises, and its aesthetic resonance of wilderness. Magazines, landscape paintings, travel literature, newspaper reports, geological monographs, and popular touring shows brought literary, visual, and scientific descriptions of this still somewhat mysterious region into uneasy alignment.[4] Most often by the 1850s, reference was to "western lands" or specific territories, but by the end of the century "the American West" gained currency as a phrase by juxtaposing its multiple landscapes—plains, mountains, and deserts—as elements of a distinct region.[5] The West was something of a mirage where natural history and Native Americans flickered against imagined landscapes, attractive but elusive projections on contested sites of occupation.

The Smithsonian Institution was just one space where these elements were combined to present a kind of composite place, the West, and to establish a convenient focus for national attention. Museums were particularly well positioned in the nineteenth century to be, as one scholar has framed it, "a potent force" for promoting national self-consciousness as part of the political process of democratization.[6] Their contribution was a material one, providing accessible and visible materials that had the imprimatur of science. Somewhat serendipitously, a bequest from an Englishman, James Smithson, had established the new public institution in a trust to be managed by the Congress of the young United States, with a mandate to increase and diffuse useful knowledge. Despite the reluctance of its first secretary, Joseph Henry, the Smithsonian Institution was poised to follow the model of the British Museum and others in Europe, gathering natural history materials as the American commercial and military empire spread.[7]

Washington was not yet a major city, but the United States capital slowly concentrated federal authority, especially after the Civil War as the federal government gained greater access to financial resources.[8] The Smithsonian seemed a promising symbol for national accomplishment, and Henry envisioned a research institution where scholars could pursue and then quickly publish and disseminate new knowledge. He was uncomfortable taking re-

Fig. 16.1. In the mid-1850s, Washington, D.C., was still a relatively young but
ambitious capital city. The dramatic new Smithsonian Institution (in the upper
left corner) was a prominent feature that held promise of the intellectual
leadership its residents and legislators hoped to attain. Smithsonian
Institution Archives, Record Unit 95, image no. 2004-10647.

sponsibility for permanent holdings that required maintenance, whether
books or specimens. Nonetheless, its public charter positioned the Smithsonian Institution in the thinking of many legislators as the appropriate site
for materials acquired by government expeditions and agencies and as a way
to demonstrate the resources of the expanding nation.[9] The geographic and
political site of the fledgling institution in Washington, as well as its inde
pendent resources, thus gave it a significant imprimatur of authority in the
intellectual community and in public perception (fig. 16.1).[10]

Henry's scientific vision was also coincident with national ambition. His
ideal reflected a growing aspiration for participatory and civic science as he
supported basic research on meteorology, archaeology, zoology, and other
topics of importance to North Americans.[11] His generation of scientists reflected what might be identified as "scientific nationalism," because they
were determined to gain intellectual parity with Europe even as they collaborated in the pursuit of basic scientific knowledge that clearly knew no
national boundaries. Deeply committed to the international exchange of
ideas, the Smithsonian leaders were also pushed by the pragmatic realities of their museum's awkward spaces and limited resources to frame and
constrain their holdings largely in terms of the nation. They extended this

parameter somewhat to include South as well as North America. Thus, the early character of the museum was significantly shaped by the contingent expansionist activity occurring alongside sectional strife, civil war, and literal reconstruction. As Susan Schulten has noted, "The territorial growth of the United States in the nineteenth century solidified an enduring myth about expansion as the destiny of the nation. This in turn placed a premium on knowledge of the trans-Mississippi region."[12] It also reinforced the exceptionalism too often identified with American history.[13]

Although sciences were broadly represented at the early Smithsonian, under the leadership of Assistant Secretary Spencer F. Baird, its natural history activities became particularly visible and significant. Individual objects held significance in terms of the general location in which they had been found, and initially the goal of taxonomists was to assign specific names, using an accepted classification system. But by midcentury, naturalists were increasingly interested in issues of species variation as they considered acclimatization, climate measurements, and modes of species transport from one region to another. Taxonomic data became ever more precise as issues of bio-distribution made curators more attentive to the precise provenance of specimens under their supervision. The rapid growth of the Smithsonian's (and indeed most museums') collections pushed naturalists to think in different ways about apparently related species and the relationships among them. Natural history was the original project in major European-oriented museums, and this umbrella term expanded to include anthropology and artifacts from peoples around the globe; these human data were managed in similar ways, while raising even more complex questions about their relative meaning.[14]

This chapter begins by considering the importance of nature to Americans as they forged and defined a nation. It then suggests how the Smithsonian Institution, formally established in 1846, was uniquely positioned to have its museum destiny intimately linked to the nation's so-called manifest destiny of the 1840s and 1850s. The essay argues that the Smithsonian absorbed, reflected, and enhanced the self-perception of the nation as the United States assumed, politically and intellectually, a continent-wide identity.[15] This outcome was evident in the pattern of acquisitions and the organization of natural history exhibits from the early 1850s into the mid-1870s, when the Smithsonian began planning for a dedicated National Museum building. The Smithsonian, in arguably its most influential period, thus provided a material dimension to both scientific and public perceptions of western territories. Two major figures—Spencer F. Baird, the assistant secretary in charge of the museum from 1850 to 1878 and then secretary until his death in 1887, and Otis T. Mason, who went from voluntary curator to curator of anthropology from the 1870s through the 1890s—played central roles. Both reflected long-standing natural history traditions, but they were challenged to understand the significance of geographic location in their own work, to respond to the materials they acquired, and to address public interests in their exhibits. In this process, they contributed to the image of the American West as a significant place, one that represented the country more generally.

NATURAL HISTORY IN NORTH AMERICA

Natural history assumed prominence during the period of European discovery and exploration, concentrating investigations in particular centers while encouraging voluntary efforts in sundry locations. Naturalists who traveled and even those who collected at home gained reputations by describing and cataloguing both marvelous and mundane specimens of flora and fauna.[16] Their international quest for species raised questions about the variability of natural kinds and issues of geographic distribution. What could account for outcomes that maintained some species apparently unique to a single locality, while others were found in multiple places? Theoretical schemes about migration proposed by such prominent figures as Carl Linnaeus and the Comte de Buffon did not coincide with empirical evidence, and the issues remained tantalizingly unresolved through the eighteenth century.[17] Ongoing discoveries and ever more data on distribution meant that none of the published animal and plant geographies remained stable and authoritative. Perhaps for that reason, naturalists in the young American republic concentrated on identifying the location and variation of individual species with ever greater precision but only casually suggested species range and geographical distribution, using what Jane Camerini has described as not only graphic but also mental maps as part of their classification activities.[18] Their open agenda resulted in "multiple-entry bookkeeping" that recorded layers of detail about individual specimens, including life histories and location, along with morphological description.[19]

By the mid–nineteenth century, natural history was much more than a taxonomic project. Earlier European accounts of the Americas suggested that the overlay of climate, geology, living beings, and other factors marked the characteristics, indeed the very nature, of specific places.[20] North American naturalists, self-consciously working to establish their credibility in the international community, absorbed, expanded, and sometimes challenged European constructs as these were applied to the case they knew firsthand.

Their work contributed to the complex layers of political, demographic, and natural descriptions as they collected and organized ever more empirical evidence. Such local activity made them very aware of the rich array of natural life in North America. The aggregation of these materials in private collections and then in urban museums helped to produce, as Benedict Anderson has suggested, a narrative that worked to create "imagined communities" and shared identity.[21] Historian of anthropology Curtis Hinsley, also reflecting on the relationship between collecting and nation-state identity, pointed out that natural history was particularly important in America because the status of regions, territories, and protectorates was confused and incomplete. Going further, he argued that the relocation and redefinition of objects from multiple but contiguous areas into museums provided "material, metonymic proofs of conquest, proprietorship, and ultimately incorporation."[22] It is also important to recognize that these material objects became an exercise for the national imagination as landscape, resources, and human inhabitants raised questions that destabilized any simplistic characterizations.[23]

The project of taking charge of American natural history began in Philadelphia when local naturalists emphasized their capacity to detect errors in the natural history monographs of those who had no direct experience on the American continent. Alexander Wilson recorded in the prospectus (1807) for his *American Ornithology* that "it is a mortifying truth to our literary pride that by foreigners alone, has not only this, but almost every other branch of our natural history been illustrated."[24] Indigenous naturalists were self-conscious about the flow of specimens abroad and sought to staunch it by acquiring the books and collections necessary to identify native flora and fauna. Philadelphia, Boston, New York, and Charleston naturalists established local societies as a kind of "declaration of independence." In this way, local societies could share books, compare specimens, provide secure facilities for study materials, do their own identifications, maintain type specimens locally, and become authorities on North American species. By the 1820s, private publications of colored plates and monographs were underway, typically describing species of one branch of natural history, with Alexander Wilson's and Charles Lucien Bonaparte's *American Ornithology; or, The Natural History of Birds of the United States* being an early, notable example. Philadelphia became a center for illustrated books of natural history, dependent on and a counterpoint to the specimen collections so important for the handful of naturalists who contemplated, for a short time, devising an American classification and nomenclature for native plants and animals.[25]

The Philadelphia naturalists also shared a distinctive outlook. As Juan Ilerbaig has observed, "While many European naturalists looked at American species as exotic information to be incorporated in their global systems, Americans took pride in surveying their backyard at a smaller scale. While Europeans were moved by encyclopedic concerns, Americans were driven by an impulse to know their own home and its natural productions."[26] Americans' ability to pursue that agenda was enhanced as naturalists participated in extended exploration of the West, sometimes disdaining the "closet naturalists" who remained behind in urban scientific centers; but, in fact, field and closet naturalists were typically locked in mutual dependency.[27] The bulk of the work of collecting before 1850 was done by individuals who examined the particularities of their own localities even as they relied on books and study collections for comparison with specimens acquired by others.

When the Swiss émigré Louis Agassiz arrived at Harvard in 1846, he very quickly established himself as a dynamic spokesperson for natural history. He set to work on the questions of faunal regions in North America, identifying what he called "true forms" of life and following what he envisioned as the Humboltian tradition.[28] As Philip Pauly has pointed out, zoologist Agassiz specifically decided to make "the animal creation of this continent . . . the central subject of his work." Agassiz's counterpart at Harvard, the prominent botanist Asa Gray, similarly concentrated on North America and demonstrated that scientifically there was a "natural unity of the American continent."[29] By the 1840s, naturalists had the capacity and multiple reasons to focus their attention on North America.

CONCENTRATING AUTHORITY

The expertise of Agassiz and Gray allowed them to build substantial and focused collections at Harvard, but the resources to further advance the natural sciences became best aligned in Washington. Although he was a physicist, Smithsonian's Secretary Joseph Henry also encouraged several fields where Americans were making significant contributions, including geology, zoology, botany, and archaeology. He knew firsthand the low status of American science abroad,[30] so he sought out and encouraged local authors in his Contributions to Knowledge series.[31] To make American publications known, he introduced a system of periodical exchanges with foreign societies that exceeded that of all "other societies in this country put together," and he collaborated with smaller societies to help distribute their publications as well.[32] At the same time, Henry remained uncertain about how best to promote the natural sciences, having watched the financial struggles of the Academy of Natural Sciences in Philadelphia while he taught at Princeton. He was well aware of the costs of maintaining collections, even as he acknowledged their value in advancing scientific knowledge. As a result he agreed, reluctantly, to take natural history materials into the Smithsonian's imposing structure, while fretting that "the tendency of an Institution in which collections form a prominent object" moves inexorably toward "a stationary position," encumbered by its holdings. When he hired a naturalist to oversee natural history, he sought prior agreement that acquisitions would be made only to "solve questions of scientific interest."[33] Henry also assured the Smithsonian regents that the scope would not be global; rather, "Collections illustrating the general Natural History of North America [will] become . . . an object of primary importance."[34]

Precedent and timing were significant in the pattern of accessions. Poised in the wings in the Patent Office were the results of the Wilkes Expedition of 1838–1842, the first round-the-world expedition sponsored by the United States. Its collections were deteriorating under the oversight of the private National Institute in Washington, and they were folded into the Smithsonian's collections in 1857 under an agreement with Congress to pay for their maintenance.[35] This happened about the same time that the United States significantly enlarged its territory—primarily with land from Mexico and the British Canadians—with some acquisitions acquired more peacefully than others. Territorial acquisition, in turn, led to several surveys into territories beyond the Mississippi River as entrepreneurs and politicians surveyed railroad routes across the continent to the Pacific Ocean.[36] In the absence of the Smithsonian, another venue might have emerged for the plants, animals, and minerals acquired on these western surveys, or they might have been arbitrarily distributed (and duplicates were, in fact, shared with regional natural history societies, academies, and colleges). The young Smithsonian Institution proved to be perfectly timed and well positioned, however, as natural history specimens began to flood back to the nation's capital. Its staff helped oversee the published reports that described the geology, botany, zoology, and climate conditions of the expanding nation.

As a result, and despite the hesitation of Henry, the dramatic turreted castle quickly became a highly visible cultural establishment in Washington, D.C., which was still, in 1850, an underdeveloped city of forty thousand people. Situated between the Capitol and the White House, the imposing, Romanesque building was symbolic of national aspirations. In this small town under the authority of itinerant legislators, the scientists, military officers, and agency officials became civic leaders and established ties among themselves that facilitated support and cooperation for its most cultural center, the Smithsonian Institution.

Pragmatic about the potential value of science and with little pretense of disinterestedness, Washington politicians, particularly those from the North and Midwest, agreed on common projects, such as the railroad surveys of the 1850s, which could have multiple political, financial, and military outcomes. Building on a European tradition of sending naturalists on exploring expeditions, advocates were able to persuade funders that it was important to record the natural resources of the country they governed. Substantial expedition maps, official reports, detailed illustrations, and physical specimens attest to their success.[37] The Smithsonian became a regional center by holding public lectures, hosting meetings of local scientific and historical societies, and publishing the results of significant scientific work, even as it drew in legislators from across the country who found the dramatically housed museum the best place to entertain visiting constituents. Simultaneously, the published scientific monographs and articles, using government printing services, allowed readers well beyond the United States to learn of its natural history.[38]

TAXONOMY MEETS GEOGRAPHY

Critical to the success of the Smithsonian in terms of natural history was Henry's decision to hire Spencer F. Baird as his assistant to oversee collections. By choice and by necessity, most field naturalists acquired their early expertise based on specimens found in their home vicinity.[39] Spencer Baird was no exception, and, starting when he was about seventeen, he went on extensive, often solitary, collecting trips, circling ever farther from Carlisle, Pennsylvania. As a result of these outdoor tours and his exchange with naturalists elsewhere, he had amassed perhaps the finest personal natural history collection in the country. Five hundred species of birds, more than five hundred containers of reptiles and fishes, some six hundred osteological specimens, a large collection of mammals, and a sizeable number of fossil bones filled two boxcars that went with Baird when he became assistant secretary in 1850; these specimens formed the nucleus of what eventually became the National Museum.[40] Just twenty-seven years old, Baird already had respect among naturalists, but this appointment in Washington positioned him to train and support American naturalists for the next three decades.

Baird had more than collecting experience. He was personally acquainted with leading naturalists in Philadelphia as well as Louis Agassiz in

Cambridge and James Dwight Dana in New Haven, both of whom recommended him for the Smithsonian position. By doing bibliographical work for Agassiz's monumental bibliography, *Bibliographia Zoologiae et Geologiae* (four volumes published between 1848 and 1852), and producing a translation of volume 4 (zoology) of Heck-Brockhaus's *Bilder Atlas zu Konversations Lexicon* for a New York publisher, Baird acquainted himself with European theory and practice as well. Baird's introduction to the resulting *Outline of General Zoology* explained that his translation was significantly "rewritten, with special reference to adaptation to this country" and contained "much original matter never before published."[41] It reviewed the history of classificatory systems and commented on the importance of studying zoogeography, the geographical distribution of animals. He identified certain factors, including temperature, elevation, and natural barriers as crucial to explaining animal distribution. The introduction also pointed out that humans had exercised "considerable influence" through domestication of animals, cultivation of plants, removal of forests, and the effects of stream management. In his own publications, Baird demonstrated meticulous taxonomy, but he increasingly emphasized geography, climate, and other environmental factors in his descriptions as well. As federal commissioner of fish and fisheries, a position he held simultaneously with his Smithsonian appointment in the 1870s and 1880s, Baird addressed the problem of declining fish catches in a comprehensive program that included geophysical data on oceans and streams, population studies of particular species, information on fishing techniques, and the impact of particular technologies used by the fishing industry. A major component of this research was analyzing human behavior as well as fish morphology, life history, habitat, and migration in order to explain the dynamics of fish populations, particularly in the North Atlantic.[42]

Baird thus carefully balanced the need for economically useful information with his strong incentive to add to basic knowledge, both of which required that he be strategic about the geography of natural science. From the outset, the young naturalist showed unusual aptitude for taking advantage of the real and symbolic position of the Smithsonian. He also became increasingly interested in the distribution of species, and he knit together the scientific corps whose efforts built the Smithsonian's collections, officially designated as the United States National Museum in 1879. Lynn Nyhart and others suggest that the impetus for nineteenth-century science was often intimately connected to public and political realms; she uses the term *civic zoology* to describe the ways in which Hamburg naturalist Karl Mobius projected the public value of science in his studies of living oysters and sea creatures.[43] Baird, like Mobius, served in a public institution rather than a university, a situation that offered the advantage of resources for research but also required direct accountability to public officials who needed to understand and explain their projects in practical as well as scientific terms.

Congenial and well-respected, Baird proved to be very adept at persuading political leaders, a myriad of military and other federally paid employees, and amateurs around the country that collections provided fundamental data on the material characteristics of the nation—not to mention its resources. Unlike his supervisor, Joseph Henry, Baird found it relatively easy to build working relationships with politicians who could be potential benefactors and believed that scientists had an obligation to convey (in Smithson's term, *diffuse*) their knowledge, especially in a democratic society.[44] Some of his advantage came from personal networks, including the influence of his father-in-law, Brigadier General Sylvester Churchill, inspector general of the army in the 1850s. Many of Baird's collectors were in the military or traveled under its auspices, using its food supplies, transportation networks, and protection while in Indian territories; occasionally naturalists were even deliberately posted to locations where Baird sought specimens.[45] Hearing that two Germans were planning a private collecting trip to the South Dakota badlands, for example, he quickly arranged with his colleague Joseph Leidy to send a collector with the northern railroad survey party in 1853 to circumvent the "outside barbarians."[46]

Baird's systematic and coherent accession program displaced the serendipitous system that characterized earlier natural history cabinets. Following models well established in France and England, Baird wrote numerous letters directing collectors to particularly desired specimens and produced pamphlets to describe how to collect, preserve, and ship them.[47] The exacting standards that he set helped define what one scholar has called a "Bairdian School" in ornithology, creating rules for systematics that he could impose throughout the museum.[48] An example suggests both his technique and the multiple incentives of the collaborators who shared Baird's belief that North American flora and fauna should be documented by Americans. Correspondence with Alexander Winchell, then teaching at a women's college in Alabama in the early 1850s, reveals subtle negotiation on both sides. Winchell sent more than six hundred specimens (primarily fish, birds, and plants) to Baird in 1853. He noted that he did this despite advice from colleagues who told him he would get more support and recognition from the museum in Paris. Still, he wrote, "I do not desire to divert to a foreign depository the collections that belong naturally to my own country."[49]

Winchell went on to discuss his isolation and hint at his interest in joining an expedition or visiting Washington. The reply from Baird to his southern correspondent assured the ambitious young naturalist that the entire dispatch was "one of the most important contributions the institution has ever received" from the South in terms of its variety. Baird then forwarded supplies to aid Winchell in collecting and shipping future submissions and promised to help with the publication of anything Winchell might write.[50] Baird thus nurtured a network of collectors by providing them with supplies of containers and preservatives, and he responded even to their needs for personal necessities by sending his wife Mary and daughter Lucy to get toiletries and clothing for those spending literally months or years in remote locations. Long letters encouraged young men in the field, and their reward included public recognition in annual reports and specimens named for them.[51] Baird continuously extended his network, enthusiastic about the discoveries that were defining North America, although he never ventured into the West himself.[52]

Already by 1855, Henry reported that no other collection in the world could, as he put it, "now pretend to rival the richness of the museum of the Smithsonian Institution in specimens which tend to illustrate the natural history of the continent of North America."[53] Baird continued to build both depth and breadth in the museum collections, expanding from the base of his own original holdings on the eastern side of the continent to encompass the Southwest, the plains and mountain territories, and moving northward along the Pacific Coast toward present-day Alaska. The stated goal of the Smithsonian was to acquire species that were not well studied and from regions that were under or were anticipated soon to be under the authority of the United States. Within this scope, he wanted the museum to be comprehensive. He wrote in his *Directions for Collecting*, "As the object of the Institution in making its collections is not merely to possess the different species, but also to determine their geographical distribution, it becomes important to have as full series as practicable from each locality." He added, as a reminder, "It is a fact well known in the history of museums, that the species which from their abundance would be the first expected, are the last to be received."[54] Baird's position and the need of Washington's leaders to know the country's physical character and potential resources built a network of individual, military, and corporate connections. The *Annual Reports* of the Smithsonian often recorded new accessions by the expedition that generated them and before they were analyzed in scientific detail. In 1878, the Smithsonian *Annual Report* summarized the major exploration activities that undergirded its holdings and made clear how much the emphasis had been on the trans-Mississippi West, the Alaskan territory, and North America more generally.[55] This did not mean, however, that materials from throughout the Americas and indeed the world were excluded, and Baird was quite aware of the importance of type specimens for comparison with those from Europe and Asia as well as from quite isolated locations.[56]

Expansionist thinking was an unquestioned aspect of the natural history enterprise concentrated in Washington, and Baird was deliberately entangled with essentially all federal projects that explored United States territory, often in advance of acquisition. Perhaps the most dramatic example of the political, commercial, and scientific nexus was Baird's involvement in the subarctic area, well-documented by Debra Lindsay.[57] From 1859, when the young Illinois naturalist Robert Kennicutt attached himself to the Hudson Bay Company (and later the Russian-American Telegraph Expedition) and began sending back specimens to Baird, the Smithsonian acquired considerable material from "Russian America." Lindsay downplays the impact that the naturalist's collecting had on the decision to acquire Alaska and suggests that Baird "was neither an imperialist nor an expansionist, except perhaps insofar as those terms apply to curatorial acquisitiveness; to the advancement of scientific interests, broadly defined; and to the advancement of institutional interests, narrowly defined (as Smithsonian interests)."[58] Yet there was clearly a convergence of attention toward the subarctic among certain Washington politicians, business entrepreneurs, and scientists that resulted in the acquisition of about twelve thousand Alaskan specimens added to the Smithsonian well before the territory was acquired by purchase from Russia in 1867. The unique Alaskan materials became important for exchange with European museums and a growing number of Asian museums.[59] By that date, too, the Smithsonian had arranged the bulk of its collections in eighty-six upright and a hundred and four table and window cases for public viewing.[60]

From 1867 to 1879, four major geological surveys of the western United States focused public attention on areas previously not mapped in detail.[61] By now the Smithsonian was eager for accessions and acknowledged the incoming material in its annual reports. Railroad companies and American Express facilitated shipping, sometimes gratis, showing the common interests of government and commerce in building the "national museum."[62] While new types were actively sought, a shifting interest toward species variation was spurred both by increasingly sophisticated research techniques and by evolutionary theory.[63]

North America was the prominent study site for Smithsonian curators. The incoming specimens provoked excitement when they included new species, but even familiar material encouraged reflection on geographic distribution and analysis of varieties within species. In turn, specific questions stimulated further fieldwork and collecting. Multiple specimens of species could be used for exchange of duplicates with other institutions, but increasingly the American naturalists used additional acquisitions to understand seasonal and historical migrations and to investigate regional distribution. From his earliest studies in ornithology, Baird had shown an interest in such issues, as he collected birds from along the East Coast and documented bird migration. Late in his career, renewed interest in his own early fieldwork and the large Smithsonian reference collection resulted in the monumental, three-volume *A History of North American Birds: Land Birds* (1874).[64] His research in ichthyology as head of the Commission on Fish and Fisheries, a prime example of civic zoology, emphasized biological distribution, primarily along the Atlantic Coast and extending up to Greenland and well into the Gulf of Mexico; the specimens acquired during this fieldwork, too, came to the Smithsonian.[65] Baird encouraged Joel E. Allen's work on bird distribution that established a geographical framework for the study of evolution and suggested that there was a "center of distribution" for each species.[66] The results led the editor of the *American Naturalist* to make the claim in 1876 that Baird and his colleagues, by their writing "on the laws of geographical distribution and climatic variation in mammals, and birds . . . have revolutionized our nomenclature in these classes, and bear directly on the evolution hypothesis."[67] The groundwork was in place for a particularly American view on evolution that grew from serial collections, greater attention to variation (sometimes connected to trinomialism), and a closely scrutinized conception of geographical distribution.[68]

Detailed knowledge of North America allowed younger naturalists like C. Hart Merriam to challenge even well-established international figures like Alfred Russel Wallace.[69] As chair of a Migration Committee in the 1880s, Merriam spearheaded a data-gathering campaign by distributing six thousand circulars with support of Allen and Baird that led, eventually in 1903, to a permanent Biological Survey in the Department of Agriculture.[70] A particularly significant result was what is arguably the first biogeographical map (he introduced the term), certainly the first of North America, to identify biological regions in relationship to geography (fig. 16.2).[71] Merriam noted that his result was different from those who sought to identify stations because he was interested in areas in which "regardless of local peculiarities, a general change takes place in the fauna and flora in passing from one region to another, or from low valleys or plains to high mountains—[namely] geographic differences."[72] Merriam's program sought to map pat-

The map contains the following legend and text:

Boreal.
Transition.
Upper Sonoran.
Lower Sonoran.
Lower Californian.
Tropical.

SECOND PROVISIONAL
BIO-GEOGRAPHIC MAP
OF NORTH AMERICA
Showing the Principal Life Areas
BY
DR. C. HART MERRIAM.
March, 1892.

The increase in intensity of color east of the Great Plains indicates the extent of the humid divisions of the Transition, Upper Sonoran, and Lower Sonoran Zones, known respectively as the *Alleghanian, Carolinian,* and *Austroriparian Faunas.*
The upper border of the green indicates the northern limit of trees.

GULF OF MEXICO

Fig. 16.2. C. Hart Merriam's "life zone" map was presented in his 1892 plenary address before the Biological Society of Washington and published by the Smithsonian Institution in the *Proceedings of the Biological Society of Washington* 7 (1892), inserted among pages 1–64. Courtesy of the American Geographical Society Library, University of Wisconsin at Madison Libraries.

terns based on "the relative numbers of distinctive types of mammals, birds, reptiles, and plants they contain, with due reference to the steady multiplication of species, genera, and higher groups from the poles toward the tropics."[73] While his generalizations did not always hold, his enterprise brought into sharper focus the distinctions between the eastern and western United States and required a much finer descriptive grid than had been provided by the survey maps produced over the previous three decades. Although much important work was done in universities, state survey, and other agencies, the Smithsonian and its affiliates in Washington remained key players in describing the nation and its physical continuities and distinctive areas, particularly those of the West, to the end of the nineteenth century.

Baird became secretary of the Smithsonian in 1878 and oversaw the opening of a new Natural History museum building in 1881.[74] By the time of his death in 1887, the institution had nearly twenty honorary and voluntary curators, had oversight of the Bureau of [American] Ethnology, and housed the Fish Commission, the Army's Signal Service, the U.S. Geological Survey, and the Department of Agriculture. Although Joseph Henry initially resisted acquiring a museum, he had relented when the government agreed to an annual appropriation to pay for management of its collections acquired in the course of survey expeditions. The result was that the Smithsonian had become deeply integrated into federal programs that exercised both formal and informal geographical, geological, botanical, zoological, and (as we shall see) ethnological as well as political authority.[75] Henry's fears about the materials overwhelming his space proved prescient, and the extensive collections became defining elements of the Smithsonian Institution.

Under Baird, collections contributed to an expansionist agenda and benefited from it, but the natural history information, collected with such precision, was only gradually translated into public exhibits at the Smithsonian. The geography that it represented was subtle, and many of the early exhibits were based on like types, although, as Mary P. Windsor observed, because of the nature of fieldwork and the realities of storing unopened boxes and kegs, the contents of many museums were "taxonomically heterogeneous (various families of animals, unsorted) but geographically homogeneous (all collected in one locality)."[76] The Smithsonian would be continuously refashioned even as Washington intellectuals built a local culture for themselves that was professional and vibrant. The Smithsonian opened its facilities to meetings by the new American Historical Association and the Washington Anthropological Society among other organizations with both national and local memberships.[77]

Museum display was originally driven by collections and by taxonomic intellectual schemes, with audiences expected to be receptive to the ideas of the patrons and experts.[78] But public museum curators were increasingly aware of visitors' interests and responses, and of the techniques being used by commercial enterprises and at international exhibitions. If early nineteenth-century museums were mirrored in publications, and publications reflected museum holdings, that relationship was changing. By midcentury, museums assumed more active roles as arbiters of natural history questions and were expected to facilitate public education.[79] One of the pressing issues presented by the American West involved the peoples who had lived and were living there.

EMPIRICAL ANTHROPOLOGY

Although Baird concentrated on zoology, archeology and anthropology had been a modest part of the Smithsonian Institution's purview at the outset. The first monograph in Henry's publication series had been an extensive study of *Ancient Monuments of the Mississippi Valley.* Moreover, archeological and anthropological objects, including a reasonably large collection from the South Seas, came as part of the transfer of the Wilkes's Exploring Expedition collection. The railroad surveys added some materials, and increasing public interest in anthropology led Henry to insert a request into his annual report in 1861: "The Smithsonian Institution being desirous of adding to its collections in archaeology all such material as bears upon the physical type, the arts and manufactures of the original inhabitants of America, solicits the co-operation of officers of the army and navy, missionaries, superintendents, and agents of the Indian department, residents of the Indian country, and travelers to that end."[80] After the Civil War, and as four additional major surveys under Clarence King, John Wesley Powell, Ferdinand V. Hayden, and George Wheeler pushed through the Great Plains and Rocky Mountains (much of this area originally designated as Indian Territories), political and military attention turned to indigenous inhabitants, often in aggressive ways.[81] At the same time, more sympathetic public attention, shaped by worries about cultural and population extinction, motivated private and foreign collectors to seek out whatever "authentic" objects remained from relatively isolated indigenous peoples.[82] The potential loss of presumably unique material pushed Baird to add to the emphasis on archeological materials related to the question of origins and on systematic attention to contemporary artifacts.

Thus when a local teacher, Otis T. Mason, trained in classics of the Eastern Mediterranean, visited the Smithsonian shortly after the end of the Civil War to translate some Semitic inscriptions, Baird suggested that the young man shift his attention. He suggested tantalizingly, "If you devote your life to such a subject as this, you will have to take the leavings of European workers. It will not be possible for you here in America to obtain the material for important researches; but—I give you the two Americas."[83] Mason took the suggestion and joined several other volunteer curators who used their spare time to preserve, organize, and increase the Smithsonian's collections.[84] Essentially self-taught, he took his initial direction from the ideas of Dresden collector Gustav Klemm, who sought to organize ethnological material scientifically and used a classification system loosely derived from natural history. When Klemm's collections were taken to Leipzig in the 1870s, they were arranged so that "like specimens" were grouped together.[85] Such arrangements, also evident in the young Pitt Rivers Museum at Oxford, were intended to demonstrate the growing complexity of artifacts over time and the intellectual and social progress of the human species.[86] Otis Mason, surrounded by the natural history expertise in the National Museum, was drawn to this approach. He published a translation of Klemm's project, with additional comments, in the Smithsonian Institution's annual report for 1875.[87] The material objects of human art and industry were to him the "models of [man's] earliest manufactures" and reflected his "means of subsistence and action."[88] Mason applied a natural science taxonomic approach in his personal research program, studying baskets, particularly those made by the Native Americans from the Southwest. His intensive study of *Aboriginal Indian Basketry* is a classic and is still available in paperback.[89] He borrowed techniques as well as language explicitly from natural history as he "dissected" artifacts in order to understand, for example, the technology of complex basket weaving or unusual patterning. He also discussed baskets in terms of their morphology and phylogeny.[90] His colleague, artist and anthropologist William Henry Holmes, complemented Mason's work by concentrating on ceramics during the 1880s. Like the rest of the nation, they were particularly attentive to the American Indians of the West, and those of the Southwest and Pacific Coast held particular fascination.

In 1875 Mason was put in charge of gathering anthropological materials for the Centennial Exposition in Philadelphia the following year, with the goal of showing "the history of culture among the aborigines of America, including the tribes now in existence, and those which are nearly or quite extinct."[91] A pamphlet to solicit objects provided detailed instructions and was primarily directed at agents of the Indian Bureau and others interested in ethnology. Mason summarized his approach: "Considering the whole human race in space and time as a single group, and all the arts and industries of man in the light of genera and species, the arrangement shall be such as to show the natural history of the objects. All of the lines of investigation pursued by naturalists in their respective fields may here be followed."[92] This taxonomic organizational principle, however, was never fully implemented.

The exhibit at the Philadelphia Exposition in 1876 was large and impressive. It made the Smithsonian Institution a household name and meant that for the next thirty years the Smithsonian would be responsible for significant displays at subsequent national and international fairs.[93] Charles Rau, a staff member who later became curator of archaeology, maintained the display on site. Because he had family in town, Baird stayed for an extended period of time as well, becoming acquainted with foreign exhibitors. When the fair concluded, these visitors had donated so much of their material that it required forty boxcars for the trip to Washington. This windfall helped persuade Congress to establish a permanent National Museum in its own building, under the Smithsonian's auspices, which was completed in 1881. By that time, Anthropology Hall had been reorganized on the second floor of the original castle, and a visitor's guide boasted that it was "especially rich in North American articles," as it was actively acquiring weapons, utensils, textiles, and ceramics from Asia, Africa, Australia and Pacific Islands.[94] The recovered portraits of Indians and scenes of Indian life on the plains by George Catlin were hung in the museum's lecture hall, and another gallery presented an elaborate model of cliff dwellings from the Four Corners region.[95] Finally, in 1884, Mason was hired as a full-time, permanent curator at the Smithsonian. Using techniques from the Philadelphia Centennial Exposition, Mason and others shaped their exhibits by tribal groups even as they arranged "like with like" specimens in glass-fronted cases. As a naturalist anthropologist, he assumed that there were distinguishable and distinguishing linguistic, technological, and social categories that could be identified and correlated, even as he and his colleagues traced mobility and development over time.[96]

By the late 1880s, Mason had become ambivalent about his taxonomic science method, especially for public display. Like fellow curators in natural history, he wanted to stretch his analysis to include environmental and other factors. John Wesley Powell had been named director of the new Bureau of Ethnology in 1879, under the auspices of the Smithsonian, and his legislative mandate was to collect only materials from within the United States.[97] Powell shared Mason's interest in the unity of humankind, but as a geologist and linguist, he was understandably interested in environmental factors.[98] Mason modified his natural history approach, but refused to relinquish the comparative approach in his well-publicized debate with cultural anthropologist Franz Boas in 1887.[99] The American Museum in New York, however, moved toward organizing its materials by ethnographic area, as in the Northwest Coast Hall.[100] Ideas about how to display museum anthropology were unsettled. In 1889, Mason had an opportunity to visit the Paris Exposition, attend European anthropological meetings, and take a two-month tour of museums in England, France, Germany, Sweden, Denmark, and the Low Countries. Here he encountered a variety of theoretical ideas and carefully scrutinized public displays in ethnology for approaches to use in Washington. Having seen the public enthusiasm for display maps in the Dresden Museum, for example, Mason adopted maps into his exhibits and continued to look for ways to correlate geographic distribution with exhibit series as he wrestled with ways to incorporate environmental factors.[101] His labels also became more detailed, always including the increasingly specific place where a specimen or artifact had been collected.[102]

In the 1880s, a more sophisticated taxidermy had created public interest in realistic representation of animals, placed in what appeared to be their natural habitat.[103] At the Chicago Centennial Exposition in 1893, Mason used a linguistic classification for indigenous people but concluded that geographical or culture area displays were more effective for the public and began to use those in the museum.[104] In 1895 Mason produced a chart with twelve "ethnic environments" in North America that shared some elements of Merriam's approach to bio-geography, although there is no evidence that this set of categories was much used.[105] His final exhibition effort, implemented by his subordinates because of Mason's ill health, was the Smithsonian's anthropological display at the 1901 Pan American Exhibition in Buffalo; a third of the exhibit was devoted to ethnology. Still experimenting, they presented material to illustrate that local environments shape devices, choosing fire-starting implements as an example. Geography seemed more evident in a display of harpoons, which varied considerably from the southern Straits of Magellan north to the Arctic. The exhibit case showed the simple barbed harpoon through enhancements that ended with a sophisticated toggle harpoon used by Eskimos because it worked best for large prey.[106] But Mason also created life-history groups, echoing the habitat groups used in natural history that reflected an environmental and cultural sensibility.[107]

Mason's ideas had changed over time, influenced by Powell, Boas, and his trip abroad as well as in response to the challenges of trying to use oversimplified categories for dealing with human artifacts. While his earliest research focused on recent historical objects or contemporary examples from the American West deemed "authentic," he found that as he acquired contemporary objects, he needed to take adaptation and modification into account. Simply dealing with location or linguistics had been challenging for those who had hoped to "disentangle this confusion and to establish the identity of all these tribes." Linguistics seemed a possible tool, and Mason observed, "Before this [extensive language study] was done it was useless to attempt the making of an aboriginal map of our continent. But the task did not end with the straightening out of the names. The old maps are defective, writers were mistaken and reported a tribe to be in areas that they never visited, other omitted to mention stocks living in certain regions. . . . Moreover, some tribes were migratory and none were absolutely settled."[108] As ethnological data poured into the Smithsonian during the heyday of American anthropology, the provenance of humans and their artifacts seemed significantly more complex than that of other living creatures.

The lack of precise categories and the malleability of all aspects of culture, including place names, influenced Mason's thinking about instabilities in his own community. For more than a decade, starting in 1890, he sat on a newly constituted United States Board on Geographic Names. Here the challenge was to decide among competing names and spellings for cities, geographic features, and specific locations (including railroad stops), and to choose names that would become standard within federal government agencies.[109] This exercise involved making important choices about priority, proximity, persistence, and politics. It was an assignment that made Mason aware of the impossibility of trying to situate any human activity as static. Places named by indigenous peoples, by visiting miners or trappers, by local settlers, or by legislators vied for priority, and, as it turned out, even an official decision did not inevitably change local usage.[110]

During the 1880s, the Smithsonian expanded into contemporary technology, reflecting the scope and public enthusiasm for major industrial and international expositions where the staff members were frequently involved.[111] Under Baird's successor, George Brown Goode, a new Division of Arts and Industries encompassed "all ethnological material except that belonging to archeological prehistory."[112] Exhibits suggested the measure of "progress" of humanity into the nineteenth century, but Mason did not subscribe wholly to that sensibility. His exhibition of the Eskimo collection in 1886 in its new gallery in the National Museum building, for example, emphasized "types and materials," but also suggested the importance of a culture as multifaceted and coherent. To make the point, the exhibit was arranged so that a visitor walked along one axis of the exhibit and viewed a range of inventions by Eskimo people that featured clothing, architecture, and handiwork—a view of a specific culture—but moving along the other axis, followed particular a particular invention as expressed in other cultures.[113] His Eskimo axis integrated geographic location and climate, domestic and hunting technology, and significant icons. His comparative axis was taxonomic, maintaining the importance of the thing itself, significant in a framework larger than that of particular groups and suggesting the cultural evolution of humans. Here the anthropologist paid attention to fine detail, recording essential details of function, material, morphology, ornamentation, and other elements that made it possible to think about levels of skill and thus of development.[114] Recognizing the unsettled state of theory, Mason put displays of particular kinds of artifacts by a linguistic group in a case on wheels, allowing him to cluster cases either by group or by like objects for comparative study. Anthropologists were debating the complex mix of environmental, cultural, and hereditary influences at play, and Mason sought to acknowledge important theories as he presented materials representing the peoples of the Southwest and the Northwest, including Eskimos, in his exhibits. Never a field anthropologist, Mason was influential through his exhibits, his publications, and his bibliographies of recent research published annually in the reports of the Smithsonian Institution.[115]

TIME AND PLACE

The Smithsonian's visibility was intimately tied to its pivotal location in the nation's capital at a particular moment in time. Its authority in natural history had grown based on the richness of its holdings and their presentation to the public in publications and display. Historical circumstances conspired to provide the resources for exhibits, publications, and public activities that highlighted the "nature" of a West that extended up to Alaska, also called Russian America. Although not contiguous (except in imagination), with its aesthetic drama, Alaska offered distinctive fauna and flora, as well as indigenous peoples. The United States National Museum was built to represent the entire country, even the continent, but its western acquisitions arrived in disproportionate numbers and attracted the most public attention. Moreover, its displays reflected the political and economic interests of legislators, business leaders, and a country avidly following accounts of discovery on the western frontiers. In the 1880s, the general collecting in the West tapered off even as materials gathered earlier were catalogued and put on display. Some special expeditions led to features like the dramatic buffalo group of William Hornaday (fig. 16.3).[116] But while there had been

Fig. 16.3. Among the most iconic images of the West were bison (or buffalo).
Taxidermist and later zoo director William Hornaday mounted this group in
the 1880s when worries about extinction of various animals became pressing.
Smithsonian Institution Archives, Record Unit 95, image no. 5470.

two hundred and fourteen named expeditions from 1850 to 1876, there were
only twenty-nine in the last quarter-century, and only four modest ones in
the West.[117] The era of continental expansion was over, and the natural history of the West had redefined the nation.

In the late nineteenth century, the Smithsonian remained a powerful
national institution and had established a prominent place internationally,
but it also faced significant competition. It exchanged specimens around the
globe, often distributing duplicates of the species that had come in from the
territories. Equally important was its management of periodicals distributed
and received, which numbered more than ten thousand; the outgoing journals and reports related to Smithsonian holdings, which typically underscored the natural history and anthropology of North America (fig. 16.4).[118]
The goal of a comprehensive museum of American natural history, however,
had proven illusive. The annual report for 1900 pointed out that exploration
and gifts had left gaps in the holdings, even though more than 4,819,836
items had been acquired by then. The emphasis on bio-geography meant

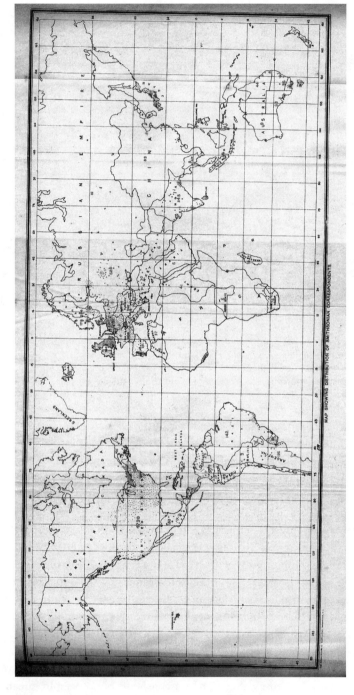

Fig. 16.4. The Smithsonian Institution not only exercised local influence through its extensive publications and distribution network, but positioned itself as an international representative of American scientific activity in the last half of the nineteenth century, distributing literally hundreds of its annual reports and monographs annually. Smithsonian Institution, *Annual Report for 1894*, insert.

that similar specimens from multiple, even contingent locations could have scientific value, but this elasticity prescribed no limit to collections. Smithsonian holdings were so extensive that, as Henry had feared, they sometimes seemed to overwhelm the staff with their breadth and depth.[119] Thorough classification of the entire resources of North America had been the Smithsonian's original goal, but the complexity of that task was now far more evident. In an effort to make better sense of the objects, geography had become more important, even as it became clear that geography was itself a malleable subject. This was nowhere more evident than in the human collections, which did not have a settled place in the museum space.

Another constraining factor came as the Smithsonian's prominence in natural history was eroded by federal agencies with their own scientific agendas, professional societies, research universities, and other major urban museums. Some of the exclusive arrangements regarding western information and materials that had made the Smithsonian so powerful were lost as agencies like the Department of the Interior, the Department of Agriculture, and even the Bureau of Education extended their influence westward.[120] Moreover, a generation of major philanthropists built new natural history (and art) museums in New York, Chicago, and Pittsburgh, which now competed for western materials as well as more global acquisitions.

As the late nineteenth century ushered in new foreign policies, and the United States military and commercial leaders reoriented their attention, the Smithsonian reflected and reinforced emerging public interests. Significantly, Mason pondered the possibility of creating an "appendix" to the ethnological exhibition of North America that would show "the primitive industries of that apportion [sic] of Africa from whose tribes were taken the parents of our negro population."[121] In the imperial mood of the period, characterization of American Indians seemed to echo language associated with colonization elsewhere, but their materials were already becoming familiar. Mason turned his own attention from North America toward the Pacific and Asia for the last decade of his life.[122] As new objects came in from the Pacific region, the aging Mason told a colleague that his heart was "now in Malaysia, with my proto-Americans." His final publication focused once again on the subject of his important early research, and his book *The Vocabulary of Malaysian Basketwork* was issued on the day of his funeral in 1908.[123]

In the nineteenth century, however, as Steven Conn has pointed out, anthropology in the United States was American Indian anthropology.[124] Mason and his colleagues had added another significant dimension to the representations of western America found in books, magazines fine arts, and popular

illustrations. The result, however, was no more singular than those portray-
als, although the anthropological theories Mason and others had acquired
stressed a progressive framework of cultural development that underscored
the distinctness of the inhabitants encountered in the American West.

It is a truism that museums move objects out of time and place. In the last
half of the nineteenth century, natural specimens and artifacts from the
West as well as the rest of the United States were reordered in scientific
terms, but these terms also proved to have their own temporality. The West
as a place, and the place of the West in the nation became at once more
specific and more unstable as curators integrated objects and collections,
defined in part by their location of origin, into social, taxonomic, and exhibit spaces.[125] Museum staff members, like the public they addressed, were
thinking hard and not uniformly about humans and their natural habitation. The Smithsonian's museum activity provided only one among many
ways that the West was being defined—in some sense a sampler of what an
explorer might find but intensely packed rather than in wide open spaces.[126]
Nonetheless, it was influential in and beyond the capital. The fact that its
scope was deliberately national and its exhibits concentrated on the West
as a place of particular interest contributed to the conflation of identities
between the West and the reuniting nation. Moreover, the Smithsonian's
mandate to disseminate knowledge led to prodigious publication and distribution efforts in its first half-century, making it arguably the largest scientific publisher in the nation.[127]

    Western expeditions, represented in official reports, detailed illustra-
tions, public lectures, and museum display, shaped taxonomical knowledge
into new and persuasive descriptions of the places they traversed. Such scientific work helped establish several distinct "Wests" on C. Hart Merriam's
map. Indeed maps—some graphic, some textual, and some mental—were
overlaid in ways that filled in the topographical, geological, botanical, zoological, demographic, climatological, and other features of the continental
United States in the second half of the nineteenth century. The material
objects provided apparent authenticity, documentation that was visual and
tactile, and they became, as Henry fretted, a well-maintained "set" that represented the United States, inevitably over-representing the West. That outcome was coincident with the interests, even preoccupation, of the citizens
gaining real and imaginary access through settlement and travel along the
railroads, as well as magazines, books, and art exhibits and the popular Wild
West shows of Buffalo Bill. Moreover, the emphasis on ethnology, which
identified Native Americans almost exclusively with the West, simulta-

neously put these people into the category of subjects to be studied rather than peoples with whom to negotiate.[128]

In the 1890s, however, Washington's leaders were turning their attention beyond the country's continental borders, engaged in new forms of expansion in the Pacific and Caribbean. The Smithsonian staff also began to shift its orientation, turning both inward to the nation's human history and outward in gathering natural history and ethnology well beyond the Americas. Segments of the American West were named national parks and wilderness areas in an effort to preserve the best of a geographical place that had so much symbolic resonance. At the same time, the identification of the desert Southwest, the Rockies, and the Pacific Northwest as specific and distinctive regions, as well as the growth of regional cities eroded much of the symbolism that had generalized the West as nature and nation.[129]

Located in the national capital, the Smithsonian Institution was a visible and influential enterprise in its first fifty years.[130] Its scientific collections, publications, and display furthered a tradition that identified the nation with nature but reshaped that identification with systematic science and in particular with new knowledge about the American West.[131] This was a subtle nationalism linked to the natural topography of an expanding country, promoted among its citizens and extending the profile of America abroad through exchanges of specimens and illustrated publications. The ambiguity of dealing with the West as both emblematic of the nation and as quite distinctive in its own right was never resolved, and the two outlooks were juxtaposed within the exhibits of the Smithsonian. The implied empiricism of museum objects as presented by Baird, Mason, and their curatorial colleagues meant that the museum's exhibit space seemed an authentic, realistic representation of a West captured by memory and giving shape to a community. It was also true that these scientists, whether explorers in the West or curators in the museum, were also in some sense captured by the political and social realities that came with being affiliated with the United States National Museum at the time when the West was "won" and before the nation moved on. In the historic moment when the Smithsonian established itself as preeminent in the natural history of North America, the American West provided resources sufficient to give the nation a positive perspective on itself. Nature on display provided the icons of identity.[132]

NOTES

Participants at the Edinburgh conference at the Institute of Geography provided thought-ful questions and comments that helped shape this essay. Also helpful were Nathan Crowe of the University of Minnesota, who tabulated data on Smithsonian accessions; Pamela Henson, historian for the Smithsonian Institution, who read an early version and helped acquire illustrations; and John Findlay, historian of the American West and fellow visitor at the University of Auckland in 2008.

1. Museums served multiple functions, and recent historiography has emphasized both their civic function and the ways they negotiated among active constituents in-volved in museum activities. See Benedict Anderson, *Imagined Communities: Reflec-tions on the Origin and Spread of Nationalism*, rev. ed. (London: Verso, 2006), esp. chap. 4; and Susan Leigh Star and James Griesemer, "Institutional Ecology, 'Translations,' and Boundary Objects: Amateurs and Professionals in Berkeley's Museum of Comparative Zoology, 1907–1939," *Social Studies of Science* 19 (1989): 387–420.

2. See Susan Scott Parrish, *American Curiosity: Cultures of Natural History in the Colonial British Atlantic World* (Chapel Hill: University of North Carolina Press, 2006). Also see Steven Conn, *Museums and American Intellectual Life, 1876–1926* (Chicago: University of Chicago Press, 1998), 34.

3. Alan Bogue's biography, *Frederick Jackson Turner: Strange Roads Going Down* (Norman: University of Oklahoma Press, 1998), argues that Turner's frontier thesis reflected popular thinking about the "loss" of the frontier.

4. David M. Wrobel, *Promised Lands: Promotion, Memory, and the Creation of the American West* (Lawrence: University of Kansas Press, 2002). On representations of sci-ence in the West, see Rebecca Bedell, *The Anatomy of Nature: Geology and American Landscape Painting, 1825–1875* (Princeton, NJ: Princeton University Press, 2001), esp. chap. 6.

5. Donald Meinig borrows the contemporary term *new West* as he disassembles its components in his classic *The Shaping of America: A Geographic Perspective on Five Hundred Years of History*, vol. 3, *Transcontinental America, 1850–1915* (New Haven, CT: Yale University Press, 1998), 31–36. The literature on the West as place and symbol is large; see Henry Nash Smith, *Virgin Land: The American West as Symbol and Myth* (Cambridge, MA: Harvard University Press, 1978); Annette Kolodny, *The Lay of the Land: Metaphor as Experience and History in American Life and Letters* (Chapel Hill: University of North Carolina Press, 1975); and Jules David Prown, *Discovered Lands, Invented Pasts: Transforming Visions of the American West* (New Haven, CT: Yale University Press, 1992).

6. Flora S. Kaplan, ed., *Museums and the Making of "Ourselves": The Role of Objects in National Identity* (London: Leicester University Press, 1994), 1. In 1991 the Smithsonian launched an exhibit entitled "The West as America," but its challenge to well-established, often mainstream myths proved controversial; see William H. Truettner and Nancy Adams, eds., *The West as America: Reinterpreting Images of the Frontier, 1829–1920* (Washington, DC: Smithsonian Institution Press, 1991).

7. Imperialist activities are deeply implicated in museum development and vice versa, as suggested in David Livingstone, *Putting Science in Its Place: Geographies of*

Scientific Knowledge (Chicago: University of Chicago Press, 2003). On museums in the colonies, see Susan Sheets-Pyenson, *Cathedrals of Science: The Development of Colonial Natural History Museums during the Late Nineteenth Century* (Kingston, Ontario: McGill-Queens University Press, 1988); and Sally Gregory Kohlstedt, "Australian Museums of Natural History: Public Priorities and Scientific Initiatives in the Nineteenth Century," *Historical Records of Australian Science* 5 (1983): 1–29.

8. J. Kirkpatrick Flack, *Desideratum in Washington: The Intellectual Community in the Capitol City, 1870–1900* (Cambridge: Schenkman, 1975).

9. The Smithsonian gradually divested itself of earlier holdings: its extensive library eventually became the Library of Congress; entomology and botany went to the Department of Agriculture; the medical materials were transferred to the Army Medical Museum (but were returned in 1904); and patents remained secured in their own agency, while the Smithsonian acquired natural history and, less extensively, art. In a personal communication, Pam Henson clarified this divestment, presented here in abbreviated form.

10. The scientific community understood in the period before research universities were established that the Smithsonian's resources and publications were for common use. Henry strategically initiated projects that relied on voluntary networks throughout the country and created broad visibility; see Daniel Goldstein, "'Yours for Science': The Smithsonian Institution's Correspondence and the Shape of the Scientific Community in the Nineteenth Century," *Isis* 85 (1994): 573–99.

11. For the ways in which voluntarism in natural history related to civic society, see Diarmid A. Finnegan, *Natural History Societies and Civic Culture in Victorian Scotland* (London: Pickering and Chatto, 2009).

12. Susan Schulten, *The Geographical Imagination in America, 1880–1950* (Chicago: University of Chicago Press, 2001), 69.

13. See Daniel T. Rodgers, "Exceptionalism," in *Imagined Histories: American Historians Interpret the Past*, ed. Anthony Molho and Gordon S. Wood (Princeton, NJ: Princeton University Press, 1998), 3–20.

14. After considerable controversy about whether anthropology belonged in museums of natural history in the 1990s, accommodation was reached at most institutions, giving more voice to the peoples represented in them.

15. American politicians argued against Europeans that this was a "domestic and therefore inevitable policy of territorial expansion across the continent," according to Charles Vevier; see Arthur Power Dudden, ed., *American Empire in the Pacific: From Trade to Strategic Balance, 1700–1922* (Aldershot: Ashgate, 2004), 91–103; quotation on 91.

16. Brian W. Ogilvie, *The Science of Describing: Natural History of Renaissance Europe* (Chicago: University of Chicago Press, 2006), 5–8.

17. Prominent naturalists of the eighteenth century had various competing theories about the distribution of plant and animal life: see Lisbet Koerner, *Linnaeus: Nature and Nation* (Cambridge, MA: Harvard University Press, 1999); and James B. Larson, *Interpreting Nature: The Science of Living Form from Linnaeus to Kant* (Baltimore, MD: Johns Hopkins University Press, 1994).

18. Jane Camerini points out that mental mapping was part of an epistemological shift that occurred in the nineteenth century. A new language of maps grew from the

need to make order out of complex bio-geographical data, as geological, botanical, and zoological distinctions alone were inadequate. See Jane Camerini, "Darwin, Wallace, and Maps" (PhD diss., University of Wisconsin–Madison, 1987); and Alfred Wallace, *The Geographical Distribution of Animals* (New York: Hafner, 1876), "Introduction."

19. Juan Ilerbaig, "Pride in Place: Fieldwork, Geography, and American Field Zoology, 1850–1920" (PhD diss., University of Minnesota, 2002), 8.

20. The classic account of these early natural histories is Raymond P. Stearns, *Science in the British Colonies of America* (Urbana: University of Illinois Press, 1970). Count Buffon had commented on the nature of North America that accounted for the smaller and less healthy species of the New World; see Lee Alan Dugatkin, *Mr. Jefferson and the Giant Moose: Natural History in Early America* (Chicago: University of Chicago Press, 2009).

21. Anderson, *Imagined Communities*, esp. chap. 4.

22. Curtis M. Hinsley, "Collecting Culture and the Cultures of Collecting: The Lure of the American Southwest, 1880–1915," *Museum Anthropology* 16 (1992): 12–20.

23. On the way in which maps played a role in this envisioning identity, see John Rennie Short, "A New Mode of Thinking: Creating a National Geography in the Early Republic," in *Surveying the Record: North American Scientific Exploration to 1930*, ed. Edward C. Carter II, Memoirs of the American Philosophical Society (Philadelphia, PA: American Philosophical Society, 1999), 19–50.

24. Quoted in Margaret Welsh, *The Book of Nature: Natural History in the United States, 1825–1875* (Boston: Northeastern University Press, 1998), 11.

25. Charlotte Porter, *The Eagle's Nest: Natural History and American Ideas, 1812–1842* (Tuscaloosa: University of Alabama Press, 1986), 51. The early Philadelphia Academy of Natural History Society's "rooms" held the cabinets of individual members, a practice that gave way to an organizational scheme that grouped objects by species. See also Pamela M. Henson, "Spencer Baird's Dream: A U.S. National Museum," in *Cultures and Institutions of Natural History: Essays in the History and Philosophy of Science*, ed. Michael T. Ghiselin and Alan E. Leviton (San Francisco: California Academy of Sciences, 2000), 101–26.

26. Ilerbaig, "Pride in Place," 16.

27. Many, of course, engaged in both fieldwork and research at the museum bench or laboratory. For discussion of this tension, see Livingstone, *Putting Science in Its Place*, 40–42.

28. Larson, *Interpreting Nature*, 1994, 131. Studies of the New World heightened tension around the issue of including animals in landscape illustrations and whether to provide stylized or very accurate representations, according to Victoria Dickinson, *Drawn from Life: Science and Art in the Portrayal of the New World* (Toronto: University of Toronto Press, 1998), 230.

29. Philip J. Pauly, *Biologists and the Promise of American Life: From Meriwether Lewis to Alfred Kinsey* (Princeton, NJ: Princeton University Press, 2000), 37. Pauly notes that Gray also shifted his emphasis over time to point out the continuities between Asian and American Pacific region flora.

30. Henry had worried about the status of American scientists since the 1830s. His goal was to promote good work that would establish their reputation abroad, but he did

not want such work to appear to be provincial; see Nathan Reingold, ed., *The Papers of Joseph Henry*, vol. 3 (Washington, DC: Smithsonian Institution Press, 1979): 3:xvii–xix. Original data and new interpretations led him to make the first volume in the Smithsonian's Contributions to Knowledge series an important account entitled *Ancient Monuments of the Mississippi Valley*, by Ephriham G. Squier and Edwin Davis, in 1848. See Terry A. Barnhart, *Ephriham George Squier and the Development of American Anthropology* (Lincoln: University of Nebraska Press, 2005), 51–52.

31. Baird immediately initiated a national and international exchange system with museums, learned societies, and colleges with federal support that included not only shipping and payment for Smithsonian publications but also those of other learned societies. See Goldstein, "'Yours for Science,'" 573–99; and Nancy E. Gwinn, "The Origins and Development of International Publication Exchange in Nineteenth-Century America" (PhD diss., George Washington University, 1996).

32. Gwinn, "International Publication Exchange," 220. The Smithsonian's *Annual Report* was distributed widely, but the Contributions to Knowledge series more sparingly.

33. Smithsonian Institution (hereafter SI), *Annual Report . . . 1850* (Washington, DC: Government Printing Office, 1851), 22. For Henry the goal was to acquire "undescribed species" from "unexplored parts of the country" that could "furnish materials for an interesting series of memoirs on physiology, embryology, and comparative anatomy" (22). Besides contributing to knowledge, Henry noted, the data provided practical help to American students of physiology and comparative anatomy who needed illustrations of native animals that they were dissecting.

34. Ibid., 45. Baird's report went on to point out that next in importance were specimens from Europe because of a close natural history relationship with the same or closely aligned species.

35. Sally Gregory Kohlstedt, "A Step toward Scientific Self-Identity in the United States: The Failure of the National Institute, 1844," *Isis* 62 (1971): 342–48.

36. A series of memos from Baird regarding the Upper Yellowstone Expedition of 1873 revealed that the railroad would allow transport of people and materials, the army quartermaster would provide supplies, and the army engineers would pay salaries. See Baird to War Department, 6 May 1873, Assistant Secretary of the Smithsonian, Record Unit 189 (hereafter Baird MSS), Smithsonian Institution Archives, Washington, DC (hereafter SIA).

37. The central goal of the extensive literature on western exploration of the period was to create reliable maps and document resources; see Clifford M. Nelson, "Toward a Reliable Geologic Map of the United States, 1803–1993," in Carter, *Surveying the Record*, 51–76.

38. Steven Shapin elaborates on the relationship in "Science and the Public," in *Companion to the History of Modern Science*, ed. R. C. Olby, G. N. Canton, J. R. R. Christie, and M. J. S. Hodge (London: Routledge, 1990), 990–1007.

39. Welch, *The Book of Nature*, esp. chaps. 1 and 2. Welch makes the point that young naturalists, who were often isolated, relied heavily on print culture, both text and illustrations, to help them gain requisite skills for identifications.

40. William A. Deiss, "Spencer F. Baird and His Collectors," *Journal of the Society of Natural History* 9 (1980): 635–36.

41. Offered this opportunity through his mentor, George P. Marsh, Baird produced the sections on reptiles and fishes, with S. S. Haldeman on invertebrates, John Cassin on birds, and Charles Girard on mammals. See Spencer F. Baird, ed., *Outlines of General Zoology* (New York: Rudolph Garrigue, 1851), "Notice."

42. Dean Allard, *Spencer Fullerton Baird and the U. S. Fish Commission* (New York: Arno Press, 1978). It is ironic that as location became more important, museum curators were often less able to spend time in the field.

43. Lynn Nyhart, "Civic and Economic Zoology in Nineteenth-Century Germany: The 'Living Communities' of Karl Mobius," *Isis* 89 (1990): esp. 635–40.

44. Alpheus Hyatt, "Sketch of the Life and Services to Science of Spencer F. Baird," *Boston Society of Natural History, Proceedings* 24 (1888): 560. Baird worked well with legislators, avoiding partisan politics even as he cultivated important political connections.

45. President of the National Geographic Society Gardiner Greene Hubbard elaborated on the Smithsonian's participation in government expeditions in his "Geography," in *The Smithsonian, 1846–1896: The History of the First Half Century*, ed. George Brown Goode (Washington, DC: Government Printing Office, 1897), 773–84.

46. Geologist James Hall had also written to Leidy about these foreign explorers: "I feel a little patriotic just now and am anxious to prevent all these things from going to Europe." See Mike Foster, *Strange Genius: The Life of Ferdinand Vandeveer Hayden* (Niwot, CO: Roberts Rinehart, 1994), 40.

47. Baird produced a handbook and often reprinted instructions to inexperienced collectors that made clear the requisite preparation, *General Directions for Collecting, Preserving, and Transporting Specimens of Natural History* (Washington, DC: Government Printing Office, 1857); see also Goldstein, "'Yours for Science," 573–99.

48. Several of the "Biographical Memoirs" of Baird in the SI *Annual Report . . . for 1888*, 703–44 make this point. See esp. 707–8, by Robert Ridgway; 732–33, by William Dahl; and 740–41, by John Wesley Powell.

49. The correspondence from Winchell to Baird, 6 January 1853, is in William H. Dall, *Spencer Fullerton Baird: A Biography* (Philadelphia, PA: Lippincott, 1915), 287–91. For another detailed example of Baird's mentorship, see Mathew Godfrey, "Traversing the Fortieth Parallel: The Experiences and Letters of Robert Ridgway, Teenage Ornithologist" (MA thesis, Utah State University, 1997).

50. The correspondence from Baird to Winchell, 19 March 1853, is reproduced in Dall, *Spencer Fullerton Baird*, 291–93. Baird sympathized with Winchell's "narrow bonds" while assuring him that his "circumstances [would] change" and that this apprenticeship would make him "brighter and more capable one day hereafter." Indeed Winchell went on to teach geology at the University of Michigan and then served as president of Syracuse and Vanderbilt universities.

51. Debra Lindsay, "Intimate Inmates: Wives, Households, and Science in Nineteenth-Century America," *Isis* 89 (1998): 635–40; and Deiss, "Spencer F. Baird," 629–43.

52. Robert Kohler, *All Creatures: Naturalists, Collectors, and Biodiversity, 1850–1950* (Princeton, NJ: Princeton University Press, 2006), 7; Jeremy Vetter, "Science along the Railroad: Expanding Field Work in the United States Central West," *Annals of Science* 61 (2004): 187–211.

53. SI, *Annual Report . . . 1855*, 31.

54. Baird, *General Directions*, 5–6.

55. In 1867, the designated categories of origin were British and Russian America, Western America, Interior Mountain region, Eastern and Southern states, West Indies, Mexico, Central America, and South America. See SI, *Annual Report . . . 1867*, 41–55.

56. See Louis Agassiz to Baird, 4 March 1861, Baird MSS, SIA; and Baird's response in Elmer Charles Herber, ed., *Correspondence between Spencer Fullerton Baird and Louis Agassiz: Two Pioneer American Naturalists* (Washington, DC: Smithsonian Institution Press, 1963), 167–69. Baird was aware, for example, of the similarity between North American plants and those of Japan but observed that only the Dutch had managed to explore that country; as a result, acquisitions would need to come indirectly through exchange with museums in Holland. See SI, *Annual Report . . . 1850*, 45.

57. Debra Lindsay, *Science in the Subarctic: Trappers, Traders, and the Smithsonian Institution* (Washington, DC: Smithsonian Institution Press, 1993).

58. Lindsay, *Science in the Subarctic*, 124–25. She does point out (note 11), that the annual appropriations for the care of government collections jumped from a rather steady $4,000 to $10,000 in 1867.

59. Sally Gregory Kohlstedt, "Otis T. Mason's Museum Tour: Observation, Exchange, and Standardization in Public Museums, 1889," *Museum History Journal* 2 (2008): 181–208.

60. *Guide to the Smithsonian Institution and National Museum* (Philadelphia, PA: Collins, 1863). Most of the zoological material was from North America, although there were European and South Pacific specimens and artifacts, the latter primarily from the Wilkes Expedition.

61. William H. Goetzmann, *Exploration and Empire: The Explorer and the Scientist in the Winning of the American West* (Austin: Texas State Historical Association 1993); and Richard A. Bartlett, *Great Surveys of the American West* (Norman: University of Oklahoma Press, 1962).

62. There were no tax incentives for this cooperation, but Baird garnered cooperation based on the position of the Smithsonian and its powerful connections to political and thus economic leaders. He also used his connections to solicit volunteers, such as transit agents, to distribute his circulars to local naturalists who might collect for the Smithsonian; see Baird to James C. Fargo, 4 and 18 June 1878, Baird MSS, SIA. I thank Ellen Alers for bringing this material to my attention.

63. Robert Kohler makes a strong case for the ongoing refinements and documents the identification of species in *All Creatures.*

64. S. F. Baird, T. M. Brewer, and R. Ridgway, *A History of North American Birds*, 3 vols. (Boston: Little, Brown, 1875). Published with colleagues and with lithographed plates hand-colored by Lavinia Bowen, these volumes were among the last in the era of high-quality monographs; specialized journals with limited illustrations became the mode of specialists, and cheaper productions were marketed to the general public.

65. Allard, *Spencer Fullerton Baird*. His Fish Commission research, while concentrated on the eastern seaboard where Baird typically spent his summers (away from the heat of Washington), eventually extended to inland lakes and rivers.

66. Ilerbaig, "Pride in Place," 75. Allen was particularly attentive to breeding seasons and temperature variation along with geography.

67. Alpheus S. Packard Jr., "A Century's Progress in American Zoology," *American Naturalist* 10 (1876): 597. This assessment was homegrown, however, and only slowly and reluctantly accepted by Europeans, who were not enthusiastic about the trinomial system that resulted from species collections; see Mark Barrow Jr., *A Passion for Birds: American Ornithology after Audubon* (Princeton, NJ: Princeton University Press, 1998), chap. 4.

68. Ilerbaig, "Pride in Place," 75. Ilerbaig suggests that this emerging tradition was a hybrid of indigenous and cosmopolitan influences represented by Baird and Agassiz.

69. Merriam's challenge, which suggested inaccuracies as well as failure to incorporate historical explanation in his discussion of North America, is reported in Ilerbaig, "Pride in Place," 131.

70. Initially the project was called the Division of Economic Ornithology and Mammalogy within the Department of Agriculture, and one of its original goals was to identify the boundary between eastern and western faunas of North America. See Ilerbaig, "Pride in Place," 126; and Keir B. Sterling, *Last of the Naturalists: The Career of C. Hart Merriam* (New York: Arno Press, 1974).

71. But see the alternate claim by Malte C. Ebach and Daniel F. Goujet, "The First Biogeographical Map" *Journal of Biogeography* 33 (2006): 761–69.

72. C. H. Merriam, "The Geographical Distribution of Life in North America with Special Reference to Mammalia," *Proceedings of the Biological Society of Washington* 7 (1892): 1–64: 3. This paper explains the distinctive qualities of this map, which relies on empirical data on mammals (thus omitting plants, insects, and other life forms) and uses historical/paleontological as well as climatological (temperature and humidity) and geological explanations. Originally the survey was under the Department of Agriculture, but in 1886 a new Division of Economic Ornithology and Mammalogy was established under Merriam as first chief, and its collections were to belong to the Smithsonian; see Jenks Cameron, *The Bureau of Biological Survey: Its History, Activities, and Organization* (Baltimore, MD: Johns Hopkins University Press, 1929).

73. Merriam, "Geological Distribution," 64.

74. Pamela M. Henson, "A National Science and a National Museum," *Proceedings of the California Academy of Sciences* 55, supp. 1 (2004): 334–57.

75. Allard, *Spencer Fullerton Baird*, 244–47.

76. Mary P. Winsor, "Agassiz's Notions of a Museum: The Vision and the Myth," in *The Cultures and Institutions of Natural History: Essays in the History and Philosophy of Science*, ed. Michael T. Ghiselin and Alan E. Leviton (San Francisco: California Academy of Sciences, 2000), 262.

77. A useful survey of federal agencies related to scientific activities and voluntary or professional societies is found in Otis T. Mason, "The Scientific Growth of Washington" ms. 7239 in the National Anthropological Archives (hereafter NAA), Smithsonian Institution (hereafter SI).

78. Disciplinary boundaries within the natural sciences were in transition and would, by the twentieth century, be transformed into ecology, systematics, ethology, genetics, and other areas. See the essays in Ronald Rainger, Keith Benson, and Jane Maienschein, eds., *The American Development of Biology* (Philadelphia, PA: University of Pennsylvania Press, 1988).

79. Alfred R. Wallace, "Museums for the People," *Macmillan's Magazine* 19 (1869): 244–50.

80. George Gibbs, "Instructions for the Archaeological Investigations in the U. States," SI, *Annual Report . . . 1861*, 392. The "big four" geological surveys after 1867 were intended to assess resources and provide maps, but they were all attentive to native populations as well.

81. Joseph Henry was alert to changing public tastes and noted, "As ethnology is a branch of study which, at this time, is occupying popular attention, it may be proper to give a more detailed account than usual." See SI, *Annual Report . . . 1869*, 31. See also Stephen Williams, *Fantastic Archeology: The Wild Side of North American Prehistory* (Philadelphia: University of Pennsylvania Press, 1991), chap. 4.

82. On the international flow and exchange of materials, see H. Glenn Penny, *Objects of Culture: Ethnology and Ethnographic Museums in Imperial Germany* (Chapel Hill: University of North Carolina Press, 2002), esp. 2–7. The influence of German research in the shift away from the Indian mounds to still-living Indian groups and their ancestor relationships needs further study, but see John Kelly, "Charles Rau: Developments in the Career of a Nineteenth-Century German-American Archaeologist," in *New Perspectives on the Origins of American Archeology*, ed. David L. Browman and Stephen Williams (Tuscaloosa: University of Alabama Press, 2002), 117–32. The broad public concern about extinction is traced in Mark Barrow, *Nature's Ghosts: Confronting Extinction from the Age of Jefferson to the Age of Ecology* (Chicago: University of Chicago Press, 2009).

83. Quoted in Curtis Hinsley, *Savages and Scientists: The Smithsonian Institution and the Development of American Anthropology, 1846–1910* (Washington, DC: Smithsonian Institution Press, 1981), 85; another edition, with a new preface, has been published as *The Smithsonian and the American Indian: Making a Moral Anthropology in Victorian America* (Washington, DC: Smithsonian Institution Press, 1994).

84. The emergence of European prehistorical work in the 1840s and 1840s gave impetus to Americans; see Don D. Fowler and David R. Wilcox, eds., *Philadelphia and the Development of Americanist Archaeology* (Tuscaloosa: University of Alabama Press, 2003), xvi–xvii.

85. This arrangement was apparently only in place from 1874 to 1878; see Penny, *Objects of Culture*, 167–69.

86. The Pitt Rivers Museum had from its origins been organized in typological and developmental terms, although always with clear identification by location, to show evolution. Its founder Augustus Henry Lane-Fox Pitt-Rivers's insistence that his particular arrangement be maintained led the British Museum to turn down the collection, which went instead to Oxford. See George W. Stocking Jr., *Victorian Anthropology* (New York: Free Press, 1987), 263–65.

87. Otis T. Mason, *Ethnological Directions Relative to the Indian Tribes of the United States* (Washington, DC: Government Printing Office, 1875). He requested not only specimens but also models of large objects, photographs, drawings, and descriptions.

88. Otis T. Mason, "Leipsic Museum of Ethnology," SI, *Annual Report . . . 1873*, 396. Gustav Klemm had studied in Leipzig, and from 1852 until his death in 1867 he was the librarian in Dresden. His concept of culture history was based on the theory that human

intellectual development goes through historically specific stages that can be identified with material objects.

89. Otis T. Mason, *Aboriginal Indian Basketry* (Washington, DC: Government Printing Office, 1902); reprinted by Dover Books in 1989. The 1904 version of his book has the evocative subtitle *Studies in a Textile Art without Machinery.*

90. Discussion of his method is detailed in Laurel Smith, "In the Museum Case of Otis T. Mason: Natural History, Anthropology, and the Nature of Display in the United States National Museum" (MA thesis, University of Oklahoma, 1994), 25–35.

91. A considerable amount of the material came from medical doctors and was eventually handed over to the Army's medical museum. Michael G. Rhode and James H. T. Conner, "'A Repository for Bottled Monsters and Medical Curiosities': The Evolution of the Army Medical Museum," in *Defining Memory: Local Museums and the Construction of History in America's Changing Communities*, ed. Amy K. Levin (Walnut Creek, CA: Altamira Press, 2007), 177–96.

92. United States National Museum (hereafter USNM), *Annual Report . . . 1884* (Washington, DC: Government Printing Office, 1885), 63. This pamphlet is particularly notable because it outlines a stratigraphic approach to examining sites and also emphasizes site identification by longitude and latitude along with the request for photographs, artifacts, and reports of interviews and observations.

93. Robert W. Rydell, *All the World's a Fair: Visions of Empire at American International Expositions, 1876–1916* (Chicago: University of Chicago Press, 1984).

94. *Visitor's Guide to the Smithsonian Institution and National Museum, Washington, DC* (Washington: Judd and Detweiler, 1880), 63.

95. The exploring surveys produced paintings, photographs, chromolithographs, and wood-engraved illustrations that found their way into magazines, books, and major exhibitions as well as published reports. See Debora Anne Rindge, "The Painted Desert: Images of the American West from the Geological and Geographical Surveys of the Western Territories, 1867–1879" (PhD diss., University of Maryland, College Park, 1993).

96. The emphasis of Mason and the Smithsonian staff on premodern material is characterized in Zachary Androus, "An Examination of Otis T. Mason's Standard of Authenticity: Salvage Ethnology and Indian Baskets at the Smithsonian Institution" (MA thesis, University of Montana, 2002). Androus argues that anthropologists were oblivious to the illusion of uncontaminated baskets (Mason's specialty).

97. G. Brown Goode to A. J. Miller, 22 March 1889, Baird MSS, SIA. Powell was probably the staff member who most envisioned the West as representing American destiny. See Daniel H. Jones, "The Panorama from Point Sublime: John Wesley Powell's 'Religion of Science' and the Intellectual Origins of his Arid Land Reforms" (PhD diss., Indiana University, 1997).

98. Powell also had a mandate to study American Indians in order to determine scientifically how public policy influenced the indigenous peoples; see Regna Darnell, "Daniel Garrison Brinton and the View from Philadelphia," in Fowler and Wilcox, *Philadelphia*, 22.

99. Boas is typically presented as "modern" in this debate, but a major figure has challenged perceptions of Mason and sees him as essentially pragmatic; see George W. Stocking Jr., "Footnotes for the History of Anthropology: Dogmatism, Pragmatism, Essen-

tialism, Relativism: The Boas/Mason Museum Debate Revisited," *History of Anthropology Newsletter* 21 (1994): 3–12.

100. Ira Jacknis, "Franz Boas and Exhibits: On the Limitations of the Museum Method of Anthropology," in *Objects and Others: Essays on Museums and Material Culture*, ed. George W. Stocking Jr. (Madison: University of Wisconsin Press, 1985), 75–111.

101. Kohlstedt, "Otis T. Mason's Tour of Europe." This public enthusiasm was fueled by the fact that geography had become a common school subject. See Anne Baker, "Geography, Pedagogy, and Race: Schoolbooks and Ideology in the Antebellum United States," *Proceedings of the American Antiquarian Society* 113 (2003): 163–90; Schulten, *Geographical Imagination*, 17–24. It may be true that geography had lost status by the early nineteenth century in Europe, but popular geography had strong advocates in North America.

102. Smith, "Museum Case of Otis Mason," 12, 154.

103. Mary Anne Andrei, "Nature's Mirror: How the Taxidermy of Ward's Natural Science Establishment Transformed Wildlife Display in American Natural History Museums and Fought to Save Endangered Species" (PhD diss., University of Minnesota, 2006).

104. USNM, *Annual Report . . . 1894*, 211.

105. SI, *Annual Report . . . 1895*, chart on 656–65. His various efforts to define fields and subfields had little impact; see Darnell, "Daniel Brinton," 31–33.

106. Smith, "Museum Case of Otis Mason," 154.

107. In this exhibit, he showed a family dynamic in which the decision of a young man in the family to use a sled to carry a small seal was the object of considerable humor. Careful analysis of group behavior would come much later to the nonhuman habitat groups. On gender in natural science exhibition, see Sally Gregory Kohlstedt, "Nature by Design: Masculinity and Animal Display in Nineteenth-Century America," in *Figuring It Out: Science, Gender, and Visual Culture*, ed. Ann B. Shteir and Bernard Lightman (Hanover, NH: Dartmouth College Press, 2006), 110–39.

108. Otis Mason, "History of the Division of Ethnology, 1849–1906," MSS and Pamphlet File, box 2, #20, NAA, SI.

109. *Second Report of the United States Board on Geographic Names, 1890–1899* (Washington, DC: Government Printing Office, 1901).

110. For a useful insight into contemporary naming practices elsewhere, see Giselle Byrnes, *Boundary Markers: Land Surveying and the Colonisation of New Zealand* (Wellington: Bridget Williams Books, 2001).

111. Although emphasizing the Department of Arts and Industry, under Goode's supervision in 1881, as a "backdoor" way to include technology, Linda Eickmeier Endersby documents the acquisitions and interest within the new United States National Museum in her "Expositions, Museums, and Technological Display: Building Cultural Institutions for the 'Citizen Inventor' in the Late Nineteenth-Century United States" (PhD diss., Massachusetts Institute of Technology, 1999).

112. Goode, "Review of the Work of the Scientific Departments," *USNM Report* (1893), 115.

113. Simon J. Bronner, ed., *Folklife Studies from the Gilded Age: Object, Rite, and Custom in Victorian America* (Ann Arbor: University of Michigan Research Press, 1987), 117–18.

114. Otis T. Mason, "Traps of the Amerinds—A Study in Psychology and Invention," *American Anthropologist* 2 (1900): 657–75. Mason grew ever more interested in the variety within and among societies in ways that constrained his enthusiasm for the "unity of culture" themes of contemporaries like Franz Boas; and he was in this instance interested in how the psychology of animals was part of the psychology of the humans trapping them.

115. David L. Browman, "Origins of Stratigraphic Excavation in North America: The Peabody Museum Method and the Chicago Method," in *New Perspectives on the Origins of Americanist Archaeology*, ed. David L. Browman and Stephen Williams (Tuscaloosa: University of Alabama Press, 2002), 248.

116. Buffalo were one of the prominent icons of the American West, represented in the enormously popular traveling show of Buffalo Bill Cody; see Juti Anne Winchester, *"All the West's a Stage: Buffalo Bill, Cody, Wyoming, and Western Heritage Presentation, 1846–1997"* (PhD diss., Northern Arizona University, 1999). The irony, of course, is that the popularity coincided with agricultural settlement of the western lands and the loss of cowboy culture.

117. See the summary list of expeditions to 1877 in SI, *Annual Report . . . 1877*, 45–51, and for the period from 1878 to 1917, see the Smithsonian's Web site at http://www.siarchives.si.edu/findingaids/faexplist.htm. Given the time lag between collecting and display, the new National Museum building became a showpiece for the American West in the last two decades of the century.

118. Othniel Charles Marsh, having spent three years in Europe, wrote to Henry in 1867 to tell him that the quality of Smithsonian publications and its system of exchange led men of science in Europe to regard it as the "fountainhead of science" in America; see Gwinn, "Origins and Development of International Publication Exchange," 378, 276.

119. On the tenuous history of this discipline, see Gareth Nelson, "From Candolle to Croizat: Comments on the History of Biogeography," *Journal of the History of Biology* 11 (1978): 269–305. Geography as a field was added to the annual bibliographical scientific record in 1882; see SI, *Annual Report . . . 1882*, 247–364.

120. SI, *Annual Report . . . 1900*, 29. For more on this level and stage of work, see Kohler, *All Creatures*.

121. Mason observed that "the opening of the Congo, the easy access to the centers of influence along the west coast make this task comparatively easy now"; see Otis T. Mason, "History of the Division of Ethnology, 1849–1906," unpublished ms., Manuscripts and Pamphlet Files, box 2, folder 20, NAA, SI. On the belated attention, see Michele Alicia Gates Moresi, "Exhibiting Race, Creating Nation: Representation of Black History and Culture at the Smithsonian Institution, 1895–1976" (PhD diss., George Washington University, 2004).

122. Roy MacLeod and Philip F. Rehbock, eds., *Nature in Its Greatest Extent: Western Science in the Pacific* (Honolulu: University of Hawaii Press, 1988).

123. Quoted in Hinsley, *Savages and Scientists*, 114–15.

124. Steven Conn, *History's Shadow: Native Americans and Historical Consciousness in the Nineteenth Century* (Chicago: University of Chicago Press, 2004).

125. With the frontier "closed," Americans turned to shaping the place in terms of larger American themes; see Paul Scolari, "Indian Warriors and Pioneer Mothers: Ameri-

can Identity and the Closing of the Frontier in Public Monuments, 1890–1930" (PhD diss., University of Pittsburgh, 2005), 4.

126. Gregg Mitman makes a similar point about how documentary film compressed the experience of filming animal behavior so that, after editing, only active and interesting moments would remain; see his *Reel Nature: America's Romance with Wildlife on Film* (Cambridge, MA: Harvard University Press, 1999).

127. Michael Lacey, "The Mysteries of Earth-Making Dissolve: A Study of Washington's Intellectual Community and the Origins of American Environmentalism in the Late Nineteenth Century" (PhD diss., George Washington University, 1979).

128. There is extensive literature on the way this occurred in the United States and elsewhere. For one particularly insightful perspective, see Conal McCarthy, *Exhibiting Maori: A History of Colonial Cultures of Display* (Wellington, New Zealand: Te Papa Museum Press, 2007).

129. Richard W. Etulain, *Re-imagining the Modern American West: A Century of Fiction, History, and Art* (Tucson: University of Arizona Press, 1996), 3–30; see also William H. Goetzmann and William N. Goetzmann, *The West of the Imagination* (New York: W. W. Norton, 1986), a companion volume to the PBS series of the same name.

130. In 1896 George Brown Goode, Baird's successor in charge of the National Museum, died, marking the end of an era during which natural history dominated the Smithsonian.

131. For more discussion on how science and technology connect with national identity, see Carol E. Harrison and Ann Johnson, eds., "National Identity: The Role of Science and Technology," special issue, *Osiris* 24 (2009).

132. On the importance of focusing on time when considering what he termed *Arcadian space* (with reference to American western parks), see Denis Cosgrove, *Geography and Vision: Seeing, Imagining and Representing the World* (London: I. B. Tauris, 2008), 83.

Putting the Geography of Science in Its Place

NICOLAAS RUPKE

In bringing this collection of essays on the geography of nineteenth-century science to a close, I propose to look at the spaces of the spatialist approach itself. In other words, and borrowing from the title of David Livingstone's book, *Putting Science in Its Place* (2003), I suggest we might put the *geographical approach*, not just *science*, in its place, including the contributions to this volume. In what spaces has the geography of scientific knowledge taken shape and been operating? Which sites has it occupied? How may this approach be placed in relation to comparable projects?

More particularly, I want to draw attention to some nineteenth-century geographies of science. Some of the scientists whose work is part of the general concern of this collection of essays, were deeply convinced of the constitutive importance of location for the development of civilization, including science, and believed that its level of attainment was place-specific. What were the ideological coordinates of these nineteenth-century geographies of science? In what ways are ours different? Before turning to that question, however, some reflections on the current situation may be in order.

THE SPATIAL TURN IN THE HISTORY OF SCIENCE

Diarmid Finnegan has recently observed that the now commonplace concern with the spatiality of scientific knowledge represents a veritable "spatial turn" in the historiography of science.[1] Over the past two decades or so a project has been in development to locate past scientific knowledge in the spaces of its production and circulation so as to ascertain the role played by local context. Science—it has been stressed—is not just a collection of abstract theories and general truths but a concrete practice with spatial dimensions. Scientific knowledge is "situated knowledge,"[2] and the wide-spread acceptance it enjoys is not merely an automatic consequence of any universal validity of methods and concepts but has been contingent on vi-cissitudes of travel, on ways and means of dispersal "over space." Book his-tory, print culture, translation studies, and the like, have all been brought to bear on the geography of scientific knowledge. The geographical approach to reading, moreover, has underscored the fact that scientific texts may be read differently in different settings.[3] A further refinement in the historical geography of knowledge has come from the analysis of metaphor and rheto-ric in scientific discourse, establishing that a relationship exists between "location and locution."[4] The introduction to our volume succinctly pre-sents us with the full panoply of theoretical options for projecting science onto the screen of historical geography, and I need not elaborate here.

What Finnegan calls the "spatial turn" follows from previous fashion changes in the historiography of science going back to the 1960s. Then the traditional history of ideas, which had treated science as an autonomous phenomenon based on rational and philosophical principles, was to a sig-nificant extent abandoned and supplanted by a belief that science is sig-nificantly shaped by its social institutions. Through the 1970s and 1980s, the social history of science and science studies made further inroads, and historians began taking science down from its elevated level of cognitive purity to the concrete conditions of institutions, organizations, politics, public concern, and vested interests.[5]

Context and *culture* became the buzzwords of fashionable history of sci-ence, and, initially, national context proved a favored contextual category, especially in reception studies. The comparative reception of Darwinism, for example, provided a nationally differentiated picture of the history of evolutionary biology.[6] A series of collected volumes, edited by Roy Porter and Mikuláš Teich, beginning in the early 1980s and continuing into the 1990s, refracted major European philosophies, cultural styles, and socioeco-nomic movements to show a spectrum of nationally varied manifestations.[7] Barely had the reviewing press pronounced judgment on the most recent of these, than a second series of collected volumes, this one edited by David Livingstone and Charles Withers, was started, the first volume, like that of the Porter and Teich series, dealing with the Enlightenment. This work probed questions of space more deeply, well below the level of national con-text, defining contextual space in a more discriminating and fine-grained way and paying special attention to the spaces of science.[8]

Our current volume is the latest in the Livingstone-Withers series. Col-lectively, its chapters situate a range of scientific knowledge claims in civic, metropolitan, and even colonial island sites, and in such architectural spaces as museums and laboratories. They follow science's claims on overseas expeditions and domestic tourist trips; locate them on maps, and both on and between the pages of printed books; and place them, too, in the oratory of science lectures and platform culture. Nineteenth-century scientific knowledge, it is convincingly shown, constituted a plurality of knowledges, each shaped by local customs and norms, dependent on locally generated authority and credibility, and serving partisan political purposes.

Some of these concerns had their forebears. So let me now look back at several of the nineteenth-century authors who, long before the recent spatial turn in the history of science, were developing geographies of science. Each of these scientists offered a particular take on the discussion of why science has the geography it does.

NINETEENTH-CENTURY GEOGRAPHIES OF SCIENCE

ALEXANDER VON HUMBOLDT AND THE CARTOGRAPHIC IMPERATIVE

The attempt to understand varieties of people, their social systems, and their cultural attainments on the basis of their distribution across the globe has a history that goes back at least some two hundred years, to the very beginning of the period covered by this collection of essays. As early as the beginning of the nineteenth century, a hermeneutic of spatiality was systematically applied to the study of civilization.[9] The geographical approach to the study of natural and cultural phenomena was put into place, comparing and contrasting these phenomena in the context of their various provinces of occurrence worldwide.

Associated with "the spatial turn in science" were some of the founder names of the modern study of geography, most prominently Alexander von Humboldt. What we now call "Humboldtian science" formed a major part of the nineteenth-century study of nature, characterized by a preoccupation with the measurement of environmental parameters on a global scale and a mental transposition from eighteenth-century taxonomic classifications of minerals, rocks, plants, animals, humans, and even parts of human cultures to their many and varied locations of actual occurrence.[10] Interests in spatial relations went hand in hand with visual representation, in particular with cartography and with major changes in the nature and quantity of geographical atlases.[11]

Some of these atlases were specialized, thematic volumes, of which the most successful was Heinrich Berghaus's (1797–1884) *Physikalischer Atlas*

(1849–52; individual maps had been in circulation long before).[12] The maps functioned as the unofficial illustrations to the text of Humboldt's *Cosmos*. Berghaus's cartography was widely republished, and the Edinburgh cartographer Alexander Keith Johnston (1804–1871), geographer in ordinary to Queen Victoria, put together an English-language version of Berghaus's masterpiece, entitled *The Physical Atlas* (1848–50; second enlarged edition 1856).[13] Plagiarized, popular, and school editions followed.

Humboldt's contributions to cartographic representation were highly effective if not revolutionizing. He devised the isoline or, more particularly, the isotherm, and a veritable isoline craze developed. Climate, weather types, ocean currents, land masses, rock formations, geomagnetism, plants, and animals were all studied in terms of their place on the global map: even *Homo sapiens* in its varied manifestations was plotted on isotherm charts. The study of the geographical relations of natural phenomena and the related delineation of so-called natural territories surreptitiously turned into a hierarchical elevation of Western European culture.[14] In many different ways the Humboldtian interest in global distribution served to place those who made the maps in the geographical center and thus at the hub of the world—the natural place from where the world could and should be ruled. For example, Berghaus's anthropogeographical maps showed Europe (or just northwestern Europe) as the world's pivotal region where race and bodily health, diet and clothing, mental development, systems of government, as well as religious beliefs all reached their global optimum. The Humboldtians quite consciously connected biogeography with political geography, and the distribution maps were tools for the promotion of such cultural values as Protestantism, constitutional monarchy, and the nation-state.

Eurocentricity was believed to be supported by, among other things, one of the most remarkable Berghaus maps of the period—at the same time also the most elementary—showing the distribution of land and sea: "*Erdkarte zur Übersicht der Vertheilung des Starren und Flüssigen*" (Map of the world giving a general view of the distribution of the solid and fluid) first printed in 1839.[15] It offered what was still at the time a novel representation of continental masses and oceanic expanse, plotted on two hemispheres.[16] These were, therefore, not the conventional eastern and western hemispheres, nor the northern and southern, but instead an "*Hämisphere der grössten Masse Landes*" (greatest area of dry land) and an "*Hämisphere der grössten Masse Wassers*" (greatest area of water). The feature of this division of land and water that gave it enormous contemporary interest was that the continental hemisphere had a geometrically definable center, and this center proved to be Europe —more specifically, northwestern Europe. Nationalistic debates took place over the precise location of this world center, the French claiming it for Paris, the Germans placing it in the vicinity of Berlin, and the British finding it not a million miles from London. Mary Somerville, for one, in her *Physical Geography* (1848), opined that a viewpoint high above Falmouth would reveal the largest possible hemispheric extent of dry land.[17] In any case, to all parties, the centrality of their nation's position on the global map seemed to confer natural legitimacy to their claims of political hegemony. Location was fate and a justification for imperialist domination, colonialism, and, in many instances, racism.

It perhaps merits emphasis that this map was by and large accurate—a quality product of Victorian geodesy and cartography that already at that time included much of Antarctica—and yet it was simultaneously through-and-through ideological, which indicates that science, whether good or bad, partakes of the sociopolitical values that co-constitute its space or spaces.

CARL RITTER AND A GROWING EUROCENTRISM

The person generally credited with having initiated the scientific study of the outlines of landmasses was Humboldt's fellow "Berliner" Carl Ritter (1779–1859) who gave the notion of a Eurocentric geography its sharpest definition. Already about a dozen years before the appearance of the Berghaus map, Ritter argued that the single greatest contrast on the surface of the earth is formed by a northeastern terrestrial hemisphere, with Europe at its center, and a southwestern oceanic one. He developed this observation in a set of twin papers, read in 1826 and 1828 to the Königliche Akademie der Wissenschaften zu Berlin—papers that simultaneously placed the centrality of the European continent on a scientific footing and put forward as a natural consequence of this central position the superiority and world hegemony of Europe's civilization. According to Ritter, "In the middle of the northern land-hemisphere, or this large terrestrial circle, lies the part of Europe that dominates all other areas of the earth, in consequence of its superior civilization and of the greatest many-sidedness of its contact with the large continental form of the planet, in the centre of greatest effectiveness, with the broadest acquired sphere of historical influences and developments."[18]

Ritter discussed, in addition to the global arrangement of landmasses, various other and more detailed physiographical variables—all in support of the argument that Europe had been destined to become the great theater of world traffic and of world history, the *"grossen Schauplatz des gemeinsamen Weltverkehrs"* (great theatre of general world traffic) and the

"*Verbindungsglied Aller zu Allen*" (part connecting all to everyone). One of these features was the outline of continental masses—coastal contours. Europe proved to be the most "*gegliedertes*" (indented) continent, with major peninsulas both in the south and north. Civilization had reached its greatest heights in those regions where a high ratio of length of coastline to surface area of land existed—as in Greece, Italy, or Europe generally.[19] The more complex and thus longer a continent's coastline in relation to the surface area of dry land, the more favorable to advanced human development that part of the globe was. This variable appealed to many Europeans because it seemed to invest their claims to economic, political, and religious world domination with natural justification.

Fascination with the Berghaus hemispheres of land versus water diminished through the second half of the nineteenth century, as North America, and in particular the United States, grew to become a significant new center of Western civilization. Eurocentricity appeared no longer so unquestionably apt a spatial model of history or contemporary power politics. If anything, it now seemed that a natural dominance of the northern over the southern hemisphere prevailed—of Europe over Africa, of North over South America, and of a soon to be Christianized and Europeanized—it was confidently anticipated—Asia over Australia. Humboldtian representations and Ritter's rule served to underpin also this Eurasian-American top versus bottom view of the world. The Swiss-born Arnold Guyot (1807–84)—a fervent Humboldtian and educated in the German geographical tradition of Ritter—having moved from Germany to Boston and thence to Princeton—recalculated Ritter's ratio of coastal length versus surface area, concluding that not Europe but North America had the highest number and thus appeared destined to be the place on earth where civilization was to reach its apogee.

More generally, Guyot observed that the northern continents looked like "beautiful trees with abundant spreading branches, or bodies richly articulated with useful members."[20] The southern continents, by contrast, were "trunks without branches, or bodies without members." "It is a remarkable fact that the deeply indented, well articulated continents, are, and have always been, the abode of the most highly civilized nations. The unindented ones, shut up within themselves and less accessible from without, have played no important part in the drama of history." If one were to apply this coastal contour criterion globally, the northern landmasses were significantly "higher" than the "simple" southern ones. In his earlier Boston lectures entitled "The Earth and Man" (1849) Guyot stated,

Tropical nature cannot be conquered and subdued, save by civilized men, armed with all the might of discipline, intelligence, and of skilful industry. It is, then, from the northern continents that those of the south await their deliverance; it is by the help of the civilized men of the temperate continents that it shall be vouchsafed to the men of the tropical lands to enter into the movement of universal progress and improvement, wherein mankind should share. . . . In this way, alone, will the inferior races be able to come forth from the state of torpor and debasement wherein they are plunged, and live the active life of the higher races. . . . The three northern continents, however, seem made to be the leaders; the three southern, the aids. The people of the temperate continents will always be the men of intelligence, of activity, the brain of humanity, if I may venture to say so; the people of the tropical continents will always be the hands, the workmen, the sons of toil.[21]

Charles Darwin—we may want to remind ourselves—was right in there with the "north over south" biogeographers. In the *Origin of Species* he offered struggle and selection as the explanation of the north-to-south vector of invasion and conquering migration. "I suspect that this preponderant migration from north to south is due to . . . the northern forms having existed in their own homes in greater numbers, and having consequently been advanced through Natural Selection and competition to a higher stage of perfection or dominating power, than the southern forms."[22] Darwin was referring to plants but, as James Moore has incisively argued, "Darwin's living organisms behaved like Englishmen; or rather, Englishmen to him were invasive organisms, multiplying, spreading across the earth, keeping evolution on the march."[23] The geography of plants and animals had a common context with political geography, no less in the case of Darwin than of Guyot or, for example, of Alphonse de Candolle.

Alphonse de Candolle and Religion

In 1873, the year that Guyot published his *Physical Geography*, an exceptional work appeared that marked the early days of a new turn in the geography of knowledge. It addressed the question, "What produces excellence in science?" and concluded that the answer came in the form of a study of the places that conditioned eminence in men of science. The book was written by the Humboldtian plant geographer Alphonse de Candolle (1806–1893) and carried the title *Histoire des sciences et des savants* (1873).

Already Berghaus had dealt with the distribution of such cultural phenomena as "mental development" (*Geistige Bildung*). His own location on
the map, the predominantly Protestant part of Europe, had "proved" to be
the cradle of the most advanced level. Candolle, whose *Géographie bota-
nique raisonnée* (1855) established his international reputation as a plant
geographer and who participated prominently in the Humboldtian turn, in
later life transposed the methods of his phytogeography to the study of science and scientific excellence, attempting to establish causal links with
social, political, and, above all, religious environments as well as conventional physical-geographical ones. To this end, he selected a sample of more
than three hundred eminent scientists who had attained the fairly rare distinction of election as foreign member or associate to the leading French,
British, and German scientific academies (for 1750, 1789, 1829, and 1869).[24]
Candolle found significant national differences: Switzerland, for example,
had produced more than 10 percent of the eminent men of science with less
than 1 percent of the total population of Europe. Such differences Candolle
attributed to the degree of favorableness of the scientists' environment in
fostering scientific interests. Although at a general level he believed in the
importance of both temperate climate and white race, as well as a high
standard of living, he particularly stressed favorable factors of a religious
kind—the absence of dogmatism and the promotion of free inquiry, which
predominated in Protestant countries and made them "out-produce" Catholic ones. A strong correlation existed between renown in science and a
clergymen's family background. A large number of famous scientists were
the sons of pastors and, Candolle calculated, the greater extent to which
Protestant rather than Catholic countries—or, within Switzerland, Protestant rather than Catholic cantons—had contributed to science was just
about equal to what the sons of pastors had accomplished:

> The sciences would not have progressed to the point where they are to-
> day if Linnaeus, Hartsoeker, Euler, Jenner, Wollaston, Olbers, Blumen-
> bach, Robert Brown, Berzelius, Encke, Mitscherlich, Agassiz, etc. had
> not been born. Fortunately, their fathers, although clergymen, were not
> subject to the rules of celibacy. If we remove from the lists of eminent
> scientists in the Protestant countries the sons of pastors, a near balance
> is restored between the populations of the two denominations with re-
> spect to their influence on the sciences. Thus a purely disciplinary rule,
> which had nothing to do with doctrinal matters and has not even always
> existed in the Roman Church, has been of fatal consequence for the sci-
> ences in Catholic countries.[25]

Candolle, scion of a Genevan dynasty of accomplished botanists, who himself was of French and Swiss Calvinist extraction, attributed the "pastors' sons effect" to the fact that in clergymen's families children are, more than on average, taught habits of simple hard work and unselfish intellectual pursuits.

In *The Origin of Species*, Darwin extensively referred to Candolle's botanical researches and, upon the appearance of Darwin's great work, Candolle converted to Darwinism. Its notion of the sorting of organisms by environmental conditions and the Spencerian "survival of the fittest" added a new dimension to the environmental determinism he favored. The subtitle of Candolle's book expressed his Darwinian allegiance: "*suivie d'autres études sur des sujets scientifiques en particulier sur la sélection dans l'espèce humaine*" (followed by other scientific studies especially about selection in humans). Candolle's statistical case for the environmental causation of intellectual ability differed, however, from that of another Darwinian and was to a certain extent made to take issue with it, namely, that of Francis Galton, who in *Hereditary Genius* (1869), and, later, in *English Men of Science: Their Nature and Nurture* (1874), argued for innate causes. In his 1869 book, Galton cited Candolle's family in support of his thesis of inherited intellectual ability. Candolle himself, however, prioritized environmental influences—family tradition in particular. The controversy and correspondence between the two men led to an early enunciation of the "nature-nurture" problem.[26]

When in the early twentieth century the Leipzig Nobel Laureate for chemistry, Wilhelm Ostwald (1853–1932), helped translate Candolle's book into German, he nudged it toward Galton's eugenicist position, adding the word *heredity* to the subtitle (*nebst anderen Studien über wissenschaftlichen Gegenstände insbesondere über Vererbung und Selektion beim Menschen*) and turning the book into volume two of his series "Grosse Männer: Studien zur Biologie des Genies" (1911).

Let me round off this bird's-eye view of nineteenth-century geographies of science with a brief mention of an influential figure who carried the approach well into the twentieth century.[27]

ELLSWORTH HUNTINGTON, RACE, AND CULTURE

Across the Atlantic, the Yale University geographer Ellsworth Huntington (1876–1947), known for his climatic determinism, continued to follow the primarily environmentalist line. How, he asked, is genius related to environmental factors? In his now notorious *The Character of Races* (1925) he countered the Galtonian position as follows:

Assume, for the moment, that the people in all parts of Europe are en-
dowed with exactly the same degree of hereditary ability, and that the
state of progress is everywhere the same, politically, industrially, socially,
and otherwise. Would the proportion of men who rise to eminence be everywhere the same? I doubt it. The regions around the North Sea would
probably always excel eastern and southern Europe. This is mainly
because on an average the men of genius in the North Sea countries
would be more energetic than those of other regions because they would
enjoy better health, even though the medical service were everywhere
equally good. They would be continually stimulated by their cool, brac-
ing climate, and would feel like working hard all the year, whereas their
southern and eastern colleagues in either hot weather or cold would be
subject to periods of depression which are a regular feature of the less-
favored parts of Europe. Because of their strength and energy the men of
genius in the North Sea region would cause civilization to advance and
incidentally would improve the conditions of health more than would
their comrades of the south and east, even though the degree of innate
ability was the same in all countries. Thus from whatever point of view
the matter is approached, we seem forced to conclude that the phase of
racial character which expresses itself in differences in energy, initiative,
and the power of achievement is closely correlated with differences in
the physical environment.[28]

As had Candolle, Huntington made much of the connection of scientific excellence with religion. In true Humboldtian fashion, he used isoline
maps to link geographical location, denominational preponderance, and excellence in science. Candolle had compared and contrasted Catholics and
Protestants, but Huntington produced a more fine-grained analysis, dealing separately with Presbyterians, Congregationalists, Baptists, Methodists,
Episcopalians, Unitarians, and others. A eugenicist, he attributed much to
inheritance, yet inherited predisposition and talent ultimately went back to
environmental selection. Like Candolle, he credited the clergyman's son-
effect, adding a Darwinian twist: "Taken as a whole the extent to which the
clergymen of the various denominations have been the fathers of eminent
sons is almost directly in proportion to the degree to which the denominations have suffered persecution and selection."[29] Unitarians were given
highest marks. "The eminent sons of Unitarian ministers seem to represent
the final result of a long process of natural selection."[30] The most notable
instance of environmental selection, however, was that of the Puritans, to
which he devoted much space.

THE PLACE OF TODAY'S SPATIAL TURN

This historical retrospective can easily be expanded to include other authors who stressed the importance of the physical environment for cultural development, such as Ellen Semple (1863–1932), who communicated much of the anthropogeography of Friedrich Ratzel (1844–1904) to an English-language readership.[31] My afterword is not intended, however, as a comprehensive history of these approaches. The few instances discussed above suffice to make it clear that the spaces occupied by nineteenth-century geographies of scientific knowledge were invariably infused with ideological features that served those who defined the spaces in the first place. Let us now return to our own retrospectively defined geographies of nineteenth-century science and briefly consider the ideological location of these definitions.

It is ironic that the trend of the past few decades to scrutinize science in terms of its social conditions and situations has hardly been accompanied by a twin fashion to explore history of science in the same probing way. We—the historians—have come to stress the rootedness of science in the actual circumstances and spaces of the scientists. But what about the historiography of the historians of science? Has this not equally been rooted in their—our—sites and spaces? The locations of the locators are likely to be as many and varied, and as formative of their—our—views, as those of the scientists. Similarly, the spaces of our "geography of knowledge" approach to nineteenth-century science will be no less ideological than the spaces in which Humboldt, Ritter, Guyot, Candolle, Huntington, and their fellow geographers put forward spatialist explanations of cultural and scientific excellence.

In following on from the "science in culture" and, not long after, "science as culture" approach, the spatial turn has provided board and lodging for many libertarian and left-wing values of recent decades. Accommodation has been given to protest causes that are very different from the geographical determinism discussed above—from its Eurocentrism, imperialism, racism, and Protestant religious supremacism. In line with the social history project of the 1960s, the spatial turn in the history of science has in fact subverted canonical notions of the old environmental determinism and has "created space" for egalitarian, non-elitist, postcolonialist, antiracist, and similar ideologies. Popular science, colonial science, feminist science, the amateurs, religious believers, the provincial outsiders at the periphery—these and more were given places in the pluralistic, patchwork-quilt world of postmodern history of science.

Consider just a few examples. In *The Politics of Evolution* (1989), a previously unnoticed group of non-establishmentarian London radicals who advocated evolutionary ideas in the context of the British class struggle was put on the map by Adrian Desmond.[32] Feminism was given a place when Donna Harraway and others showed that the scientific study of nature had traditionally been dressed up in men's clothes.[33] James Secord paid detailed attention not only to a select scientific readership of *Vestiges of the Natural History of Creation*, but also to such non-elite readers as apprentices in a provincial factory town.[34] And Anne Secord—to cite one of several other possible examples—drew attention to "science in the pub," conducted by common artisan botanists in early nineteenth-century Lancashire.[35] "Me too" and "we too" claims in postcolonialist studies put the spotlight on contributions to science and the architecture of science in former overseas possessions of European powers.[36] Our volume continues this trend, allocating scientific spaces to botanical tourists, small-time curators, local informants, even working-class servants; to creationists as well as Darwinists, the amateur as well as the professional, women as well as men, indigenous culture as well as colonial importation; and to printers, instrument makers, and booksellers.

In summing up, I perhaps should stress that the sites and spaces identified in our volume for nineteenth-century science are not merely realities "out there" that we discover and record; they are also *assignments* of place by us historians—assignments that reflect our place and serve to instrumentalize the prestige of science for a range of self-serving purposes. Studying spatial distribution does not merely mean determining natural realms of occurrence. The categories of spatiality in which we examine science are themselves invariably constitutively influenced by our own locations as historians of science—whether these categories be as concrete as national territories and museums of natural history or as abstract as denominational discourses and rhetorical spaces, and whether we think of them as material "container spaces" or as social coproductions of the scientific endeavor itself. We do not merely happen to find nineteenth-century science in certain locations; we *assign* science to locations that are selected and constructed by us, not just *re*constructed, to validate our purposes by conflating them with science—with that product of modern culture we so much admire and value. This is not to condemn or even just criticize such work, but to cast a "metahistorical" critical look at our own stance and avoid committing the sin of "the scholar's pretense" that one's own stance at long last is the Archimedean point that definitively settles the issues.[37] We're all in this together.

NOTES

1. Diarmid A. Finnegan, "The Spatial Turn: Geographical Approaches in the History of Science," *Journal of the History of Biology* 41 (2008): 369–88. See also David N. Livingstone, *Putting Science in Its Place: Geographies of Scientific Knowledge* (Chicago: University of Chicago Press, 2003); Charles W. J. Withers, "The Geography of Scientific Knowledge," in *Göttingen and the Development of the Natural Sciences*, ed. Nicolaas A. Rupke (Göttingen: Wallstein, 2002), 9–18.

2. Donna Harraway, "Situated Knowledge: The Science Question in Feminism as a Site of Discourse on the Privilege of Partial Perspective," *Feminist Studies* 14 (1988): 575–99.

3. The classic study is by James A. Secord, *Victorian Sensation: The Extraordinary Publication, Reception, and Secret Authorship of "Vestiges of the Natural History of Creation"* (Chicago: University of Chicago Press, 2000).

4. Simon Schaffer, "The History and Geography of the Intellectual World: Whewell's Politics of Language," in *William Whewell: A Composite Portrait*, ed. Menachem Fisch and Simon Schaffer (Oxford: Clarendon Press, 1991), 201–31; David N. Livingstone, "Science, Site, and Speech: Scientific Knowledge and the Spaces of Rhetoric," *History of the Human Sciences* 20 (2007):71–98.

5. An indication of this was the title (as well as the contents) of a Festschrift for the Oxford historian of the atomic bomb, Margaret Gowing; see Nicolaas A. Rupke, ed., *Science, Politics, and the Public Good* (London: Macmillan, 1988).

6. Thomas F. Glick, *The Comparative Reception of Darwinism*, 2nd ed. (Chicago: University of Chicago Press, 1988). For a more recent and wide ranging follow-up, see Eve-Marie Engels and Thomas F. Glick, eds., *The Reception of Charles Darwin in Europe*. 2 vols. (London: Continuum, 2008).

7. See the following works by Roy Porter and Mikuláš Teich, eds. *The Enlightenment in National Context* (Cambridge: Cambridge University Press, 1981); *Romanticism in National Context* (Cambridge: Cambridge University Press, 1988); *The Renaissance in National Context* (Cambridge: Cambridge University Press, 1992); and *The Industrial Revolution in National Context* (Cambridge: Cambridge University Press, 1996).

8. David N. Livingstone and Charles W. J. Withers, eds., *Geography and Enlightenment* (Chicago: University of Chicago Press, 1999) and *Geography and Revolution* (Chicago: University of Chicago Press, 2005).

9. To be sure, the notion that location—especially in terms of physical environment—is of constitutive importance to humans, their physical appearance, their diseases and wellness, and their moral and intellectual character has ancient roots and reemerged in the late eighteenth century in a variety of forms in studies of the geographical distribution of living beings. Climate in particular was often regarded as a primary determinant of inferior or superior physique and vigor of body and mind. Such environmental determinism is expressed in the title of a treatise by William Falconer (the English physician, not the Scottish poet): *Remarks on the Influence of Climate, Situation, Nature of Country, Population, Nature of Food, and Way of Life, on the Disposition of Temper, Manners*

and Behaviour, Intellects, Laws and Customs, Form of Government, and Religion of
Mankind (London: Dilly, 1781).

    10. See, for example, Nicolaas A. Rupke, "Humboldtian Distribution Maps: The
Spatial Ordering of Scientific Knowledge," in The Structure of Knowledge: Classifications
of Science and Learning since the Renaissance, ed. Tore Frängsmyr (Berkeley and Los
Angeles: University of California, Office for History of Science and Technology, 2001),
93–116. For other aspects, see Malcolm Nicolson, "Alexander von Humboldt, Humboldtian Science, and the Origins of the Study of Vegetation," History of Science 25 (1987):
169–94, and "Humboldtian Plant Geography after Humboldt: The Link to Ecology," British Journal for the History of Science 29 (1996): 289–310; Michael Dettelbach, "Humboldtian Science," in Cultures of Natural History, ed. Nicholas Jardine, James A. Secord, and
Emma C. Spary (Cambridge: Cambridge University Press, 1996), 287–304; Nicolaas A.
Rupke, "Humboldtian Medicine," Medical History 40 (1996): 293–310. On the nature
of eighteenth and early nineteenth-century geography and universal mapping, see Anne
Marie Claire Godlewska, Geography Unbound: French Geographic Science from Cassini
to Humboldt (Chicago: University of Chicago Press, 1999).

    11. Nicolaas A. Rupke and Karen E. Wonders, "Humboldtian Representations in
Medical Cartography," in Medical Geography in Historical Perspective, ed. Nicolaas A.
Rupke, Medical History, Supplement no. 20 (London: The Wellcome Trust Centre for the
History of Medicine at University College, London, 2000) 163–75.

    12. Jane Camerini, "The Physical Atlas of Heinrich Berghaus: Distribution Maps
as Scientific Knowledge," in Non-Verbal Communication in Science Prior to 1900, ed.
Renato Mazzolini (Florence: Leo S. Olschki, 1993): 479–512; Jane Camerini, "Heinrich
Berghaus's Map of Human Diseases," in Medical Geography in Historical Perspective, ed.
Nicolaas A. Rupke, Medical History, Supplement no. 20 (London: The Wellcome Trust
Centre for the History of Medicine at University College, London, 2000), 186–208.

    13. Alexander Keith Johnston, The Physical Atlas: A Series of Maps and Notes Il-
lustrating the Geographical Distribution of Natural Phenomena (Edinburgh: W. and A. K.
Johnston, 1848–50); see also Gerhard Engelmann, "Der Physikalische Atlas des Heinrich
Berghaus und Alexander Keith Johnston's Physical Atlas," Petermanns Geographische
Mitteilungen 108 (1964): 133–49.

    14. Nicolaas A. Rupke, "Paradise and the Notion of a World Centre, from the
Physico-Theologians to the Humboldtians," in Phantastische Lebensräume, Phantome
und Phantasmen, ed. Hans-Konrad Schmutz (Marburg: Basiliskenpresse, 1997), 77–88.
See also David Turnbull, Maps Are Territories; Science Is an Atlas (Chicago: University
of Chicago Press, 1989); Denis Wood, The Power of Maps (New York: Guilford Press,
1992); Denis Wood and John Fels, The Nature of Maps: Cartographic Constructions of the
Natural World (Chicago: University of Chicago Press, 2008); Derek Gregory, Geographical Imaginations (Oxford: Blackwell, 1994).

    15. Heinrich Berghaus, Physikalischer Atlas oder Sammlung von Karten, auf denen
die hauptsächlichsten Erscheinungen der anorganischen und organischen Natur nach
ihrer geographischen Verbreitung und Vertheilung bildlich dargestellt sind, 2 vols.
(Gotha: Justus Perthes Verlag, 1845–1848), pt. 3, no. 1.

    16. Land and water hemispheres had already been depicted in 1816 by Stieler: see
Adolf Stieler and C. G. Reichard, Hand-Atlas ueber alle Theilen der Erde nach dem

neuesten Zustande und über das Weltgebaeude (Gotha: Justus Perthes Verlag, 1827), plate 8.

17. Mary Somerville, *Physische Geographie*, vol. 1 (Leipzig: Weber, 1851), 57.

18. Translated from Carl Ritter, "Ueber geographische Stellung und horizontale Ausbreitung der Erdtheile," in *Abhandlungen der historisch-philologischen Klasse der Koeniglichen Akademie der Wissenschaften zu Berlin, 1826* (Berlin: Koenigliche Akademie der Wissenschaften, 1829), 107.

19. See also Carl Ritter, "Bemerkungen ueber Veranschaulichungsmittel raeumlicher Verhaeltnisse bei graphischen Darstellungen durch Form und Zahl," in *Abhandlungen der historisch-philologischen Klasse der Koeniglichen Akademie der Wissenschaften zu Berlin, 1828* (Berlin: Koenigliche Akademie der Wissenschaften, 1831): 213–32; see further Nicolaas A. Rupke, "Eurocentric Ideology of Continental Drift," *History of Science* 34 (1996): 251–72.

20. Arnold Guyot, *Physical Geography* (London: Sampson Low, Marston, Low, and Searle, 1873), 23.

21. Arnold Guyot, *The Earth and Man* (Boston: Gould and Lincoln, 1859), 330–31.

22. Charles Darwin, *On the Origin of Species by Means of Natural Selection* (London: John Murray, 1859), 379.

23. James Moore, "Revolution of the Space Invaders: Darwin and Wallace on the Geography of Life," in *Geography and Revolution*, ed. David N. Livingstone and Charles W. J. Withers (Chicago: University of Chicago Press, 2005), 126.

24. Raymond E. Fancher, "Alphonse de Candolle, Francis Galton, and the Early History of the Nature-Nurture Controversy," *Journal of the History of the Behavioral Sciences* 19 (1983): 341–52.

25. Translated from Alphonse de Candolle, *Histoire des sciences et des savants depuis deux siècles suivie d'autres études sur des sujets scientifiques en particulier sur la sélection dans l'espèce humaine* (Geneva: Bale; Lyon: Georg,1873), 125–26.

26. Fancher, "Alphonse de Candolle."

27. Others could be added, including Friedrich Ratzel and Ellen Churchill Semple. The latter's *Influences of Geographic Environment* (New York: Russell and Russell, 1911) made Ratzel's anthropogeography known to a North American readership. Colonialist settlement purposes continued to be supported by this. Most directly in line with Huntington was the Anglo-Australian-North American geographer Griffith Taylor (1880–1963) who, in a series of influential works, such as *Canada: A Study of Cool Continental Environments* (1947) and *Australia: A Study of Warm Environments* (1955), promoted aspects of Huntington's environmental determinism. His pioneering work produced insights and biases that have survived in popular culture to the present day. Yet, in the wake of the defeat of the Third Reich and reinforced by the cultural revolution of the 1960s, the nature-nurture pendulum swung toward environmental causation and support of new, liberal and left-wing causes. What were the conditions that produced Nobel Laureates in the United States? Why were there so many Jewish scientists among the Laureates? Or, to look back at early modern times, which spaces—national, religious, or other—had given us modern science, and why had the very early scientific-technological brilliance of the Chinese—magisterially documented by the Anglican Marxist Joseph Needham—come to nought?

28. Ellsworth Huntington, *The Character of Races as Influenced by Physical Environment, Natural Selection and Historical Development* (New York: Scribner, 1925), 233.

29. Ibid., 321.

30. Ibid., 326.

31. On Semple, see Innes M. Keighren, "Bringing Geography to the Book: Charting the Reception of *Influences of Geographic Environment*," *Transactions of the Institute of British Geographers* 31 (2006): 525–40.

32. Adrian Desmond, *The Politics of Evolution: Morphology, Medicine, and Reform in Radical London* (Chicago: University of Chicago Press, 1989).

33. Donna Harraway, *Primate Visions: Gender, Race, and Nature in the World of Modern Science* (New York: Routledge, 1989).

34. See Secord, *Victorian Sensation.*

35. Anne Secord, "Science in the Pub: Artisan Botanists in Early Nineteenth-Century Lancashire," *History of Science* 32 (1994): 269–315.

36. Susan Sheets-Pyenson, *Cathedrals of Science: The Development of Colonial Natural History Museums during the Late Nineteenth Century* (Kingston, Ontario: McGill-Queen's University Press, 1988).

37. Needless to say, this afterword itself has a subjective "location" with a partial perspective. For more on the meta-approach, see my *Alexander von Humboldt: A Metabiography* (Chicago: University of Chicago Press, 2008), 214–18.

Adler, Judith. 1989. "Origins of Sightseeing." *Annals of Tourism Research* 16:7–29.

Alberti, Samuel J. M. M. 2001. "Amateurs and Professionals in One County: Biology and Natural History in Late Victorian Yorkshire." *Journal of the History of Biology* 34:115–47.

———. 2002. "Placing Nature: Natural History Collections and Their Owners in Nineteenth-Century Provincial England." *British Journal for the History of Science* 35:291–311.

———. 2005. "Civic Cultures and Civic Colleges in Victorian England." In Daunton, *Organisation of Knowledge in Victorian Britain*, 337–56.

———. 2007. "The Museum Affect: Visiting Collections of Anatomy and Natural History." In Fyfe and Lightman, *Science in the Marketplace*, 371–403.

———. 2009. *Nature and Culture: Objects, Disciplines and the Manchester Museum.* Manchester: Manchester University Press.

Albion, Robert Greenhalgh. 1938. *Square-Riggers on Schedule: The New York Sailing Packets to England, France, and the Cotton Ports.* Princeton, NJ: Princeton University Press.

Alexander, Theodore Edmund. 1969. "Francis Maitland Balfour's Contributions to Embryology." PhD diss., University of California.

Allard, Dean Conrad. 1978. *Spencer Fullerton Baird and the U.S. Fish Commission.* New York: Arno Press.

Allen, David E. 1994. *The Naturalist in Britain: A Social History.* 2nd ed. Princeton, NJ: Princeton University Press.

———. 1996. "Tastes and Crazes." In Jardine, Secord, and Spary, *Cultures of Natural History*, 394–407.

———. 1998. "On Parallel Lines: Natural History and Biology from the Late Victorian Period." *Archives of Natural History* 25:361–71.

———. 2004. "An 1861 Instance of 'Painting One's Bentham.'" *Archives of Natural History* 31:356–57.

———. 2010. *Books and Naturalists.* London: Collins.

Allen, Garland E. 1975. *Life Science in the Twentieth Century.* New York: Wiley.

Alter, Peter. 1987. *The Reluctant Patron: Science and the State in Britain, 1850–1920.* Oxford: Berg.

Altick, Richard D. 1978. *The Shows of London: A Panoramic History of Exhibitions, 1600–1862.* Cambridge, MA: Belknap Press.

Anderson, Benedict. 2006. *Imagined Communities: Reflections on the Origin and Spread of Nationalism.* London: Verso. (Orig. pub. 1983.).

Anderson, Katherine. 2006. "Mapping Meteorology." In *Intimate Universality*, ed. James Fleming, Vladimir Janković, and Deborah Cohen, 69–92. Sagamore Beach, MA: Science History Publications.

Anderson, R. G. W. 1992. "'What Is Technology?': Education Through Museums in the Mid-Nineteenth-Century." *British Journal for the History of Science* 25:169–84.

Andrei, Mary Anne. 2006. "Nature's Mirror: How the Taxidermy of Ward's Natural Science Establishment Transformed Wildlife Display in American Natural History Museums and Fought to Save Endangered Species." PhD diss., University of Minnesota.

Andrews, Malcolm. 1989. *The Search for the Picturesque: Landscape Aesthetics and Tourism in Britain, 1760–1800.* Aldershot: Scolar.

———. 2006. *Charles Dickens and His Performing Selves.* Oxford: Oxford University Press.

Andrey, Georges. 1986. "Madeleine Eggendorfer, libraire à Fribourg et la Société Typographique de Neuchâtel (1769–1788): Livre, commerce et lecture dans la Suisse des Lumières." In *Aspects du Livre neuchâtelois: Études réunies à l'occasion du 450ᵉ anniversaire de l'imprimerie neuchâteloise*, ed. Jacques Rychner and Michel Schlup, 118–57. Neuchâtel: Bibliotheque Publique et Universaire.

Androus, Zachary T. 2002. "An Examination of Otis T. Mason's Standard of Authenticity: Salvage Ethnology and Indian Baskets at the Smithsonian Institution." PhD diss., University of Montana.

Annan, N. G. 1955. "The Intellectual Aristocracy." In *Studies in Social History: A Tribute to G. M. Trevelyan*, ed. J. H. Plumb, 243–87. London: Longmans, Green.

Armstrong, Carol. 2004. "Cameraless: From Natural Illustrations and Nature Prints to Manual and Photogenic Drawings and Other Botanographs." In *Ocean Flowers*, ed. Carol Armstrong and Catherine de Zegher, 54–165. New York: The Drawing Center, in association with Princeton University Press.

Arnold, David. 2005. *Tropics and the Traveling Gaze: India, Landscape and Science, 1800–1856.* Seattle: University of Washington Press.

Ashworth, W. J. 1996. "Memory, Efficiency, and Symbolic Analysis: Charles Babbage, John Herschel, and the Industrial Mind." *Isis* 87:629–53.

"Asiatic Intelligence—Ceylon." *Asiatic Journal*, n.s., 13 (1834): 172–73.

Atkins, Anna. 1843–53. *Photographs of British Algae: Cyanotype Impressions.* 3 vols. Halstead: privately printed.

Atkinson, Edward. 1905. "The Negro a Beast." *North American Review* 181:202.

Auerbach, Jeffrey A. 1999. *The Great Exhibition of 1851: A Nation on Display.* New Haven, CT: Yale University Press.

Austen, Jane. 1892. *Northanger Abbey.* London: J. M. Dent.

———. 1892. *Pride and Prejudice.* 2 vols. London: J. M. Dent.

Ayrton, Frederick. 1848. "Observations upon M. D'abbadie's Account of His Discovery of the Sources of the White Nile, and upon Certain Observations and Certain

Objections and Statements in Relation Thereto, by Dr. Beke." *Journal of the Royal Geographical Society* 18:48–74.

[Baillière, J.-B.]. 1828. *A Catalogue of Books in Medicine, Surgery, Anatomy, Physiology, Natural History, Botany, Chemistry, Pharmacy, &c., &c., &c.* Paris: J.-B. Baillière.

Baines, T., ed. 1904. *Annual Register: A Review of Public Events at Home and Abroad for the Year 1903.* London: Longmans, Green.

Baird, Spencer F. 1851. *Outlines of General Zoology.* New York: Rudolph Garrigue.

———. 1857. *General Directions for Collecting, Preserving, and Transporting Specimens of Natural History.* Washington, DC: Government Printing Office.

Baird, Spencer F., Thomas M. Brewer, and Robert Ridgeway. 1875. *A History of North American Birds.* 3 vols. Boston: Little, Brown.

Baker, Anne. 2003. "Geography, Pedagogy, and Race: Schoolbooks and Ideology in the Antebellum United States." *Proceedings of the American Antiquarian Society* 113:163–99.

Baker, John Norman Leonard. 1944. "Sir Richard Burton and the Nile Sources." *English Historical Review* 59:49–61.

Baker, R. A., and J. M. Edmonds. 1998. "Louis Compton Miall (1842–1921): The Origins and Development of Biology at the University of Leeds." *Linnean* 14:40–8.

Baker, Samuel W. 1855. *Eight Years' Wanderings in Ceylon.* London: John Murray; repr. Dehiwela, Sri Lanka: Tisara Prakasakayo, 1983.

———. 1866. "Account of the Discovery of the Second Great Lake of the Nile, Albert Nyanza." *Journal of the Royal Geographical Society* 36:1–18.

Bakewell, R. 1833. *An Introduction to Geology . . . Greatly Enlarged.* 4th ed. London: Longman, Orme, Brown, Green, and Longman.

Balfour, Alice Blanche. 1930. "Butterflies and Moths Found in East Lothian." *Transactions of the East Lothian Antiquarian and Field Naturalists' Society* 1:169–84.

Balfour, Arthur James (First Earl of Balfour). 1912. *Arthur James Balfour as Philosopher and Thinker: A Collection of the More Important and Interesting Passages in His Non-Political Writings, Speeches, and Addresses, 1879–1912.* Ed. Wilfrid M. Short. London: Longmans, Green.

———. 1930. *Chapters of Autobiography.* Ed. Mrs. Edgar Dugdale. London: Cassell.

Balfour, Francis M. 1873. "On the Disappearance of the Primitive Groove in the Embryo Chick." *Quarterly Journal of Microscopical Science* 13:276–80.

———. 1878. *A Monograph on the Development of Elasmobranch Fishes.* London: Macmillan.

Balfour, G. W., and F. M. Balfour. 1872. "On Some Points in the Geology of the East Lothian Coast." *Geological Magazine* 9:161–64.

Balfour Memorial: Undergraduate Meeting at the Union, Cambridge, 30th October, 1882. Cambridge: Fabb & Tyler.

Bandaranayake, Senake, and Gamini Jayasinghe. 1986. *The Rock and Wall Paintings of Sri Lanka.* Colombo, Sri Lanka: Lake House Press.

Barbellion, W. N. P. [Bruce F. Cummings]. 1919. *The Journal of a Disappointed Man.* London: Chatto and Windus.

Barber, Giles. 1968. "Treuttel and Wurtz: Some Aspects of the Importation of Books from France, c.1825." *The Library,* 5th ser., 23:118–44.

————. 1975. "Pendred Abroad: A View of the Late Eighteenth-Century Book Trade in Europe." In *Studies in the Book Trade in Honour of Graham Pollard*, ed. R. W. Hunt, I. G. Philip, and R. J. Roberts, 231–77. Oxford: Oxford Bibliographical Society.

————. 1994. "Books from the Old World and for the New: The British International Trade in Books in the Eighteenth Century." In *Studies in the Book Trade of the European Enlightenment*, ed. G. Barber, 225–64. London: Pindar Press.

————. 1999. "The English Language Guide Book to Europe up to 1870." In *Journeys through the Market: Travel, Travellers and the Book Trade*, ed. Robin Myers and Michael Harris, 93–106. Publishing Pathways Book Series. London: St. Paul's Bibliographies.

Barnes, Trevor J. 2004. "Placing Ideas: Genus Loci, Heterotopia, and Geography's Quantitative Revolution." *Progress in Human Geography* 28:565–95.

Barnett, Clive. 1998. "Impure and Worldly Geography: The Africanist Discourse of the Royal Geographical Society, 1831–73." *Transactions of the Institute of British Geographers* 23:239–51.

Barnhart, Terry A. 2005. *Ephriham George Squier and the Development of American Anthropology*. Lincoln: University of Nebraska Press.

Barringer, Tim, and Tom Flynn, eds. 1998. *Colonialism and the Object: Empire, Material Culture, and the Museum*. London: Routledge.

Barrow, Mark, Jr. 2009. *Nature's Ghosts: Confronting Extinction from the Age of Jefferson to the Age of Ecology*. Chicago: University of Chicago Press.

————. 1998. *A Passion for the Birds: American Ornithology after Audubon*. Princeton, NJ: Princeton University Press.

Bartlett, Richard A. 1962. *Great Surveys of the American West*. Norman: University of Oklahoma Press.

Barton, Ruth. 1990. "'An Influential Set of Chaps': The X-Club and Royal Society Politics, 1864–65." *British Journal for the History of Science* 23:53–81.

————. 2004. "Scientific Authority and Scientific Controversy in *Nature*: North Britain against the X Club." In *Culture and Science in the Nineteenth-Century Media*, ed. Louise Henson, Geoffrey Cantor, Gowan Dawson, Richard Noakes, Sally Shuttleworth, and Jonathan R. Topham, 223–35. Aldershot: Ashgate.

Bassett, D. A. 1985. "Sheets of Many Colours or Maps of Geological Ideas: A Review." *Imago Mundi* 37:101–5.

Bassett, Thomas J., and Phillip W. Porter. 1991. "From the Best Authorities: The Mountains of Kong in the Cartography of West Africa." *Journal of African History* 32:367–413.

Bates, Alan W. 2008. "'Indecent and Demoralising Representations': Public Anatomy Museums in Mid-Victorian England." *Medical History* 52:1–22.

Bateson, Beatrice. 1928. *William Bateson, F.R.S., Naturalist: His Essays and Addresses*. Cambridge: Cambridge University Press.

[Bateson, William]. "Discussion of Reports." *Cambridge University Reporter*, 20 February 1894, 494.

Bateson, William. 1900. "Memorandum from the Evolution Committee of the Royal Society." *Entomologist's Monthly Magazine*, 2nd series, 11:139–40.

————. 1909. "Heredity and Variation in Modern Lights." In *Darwin and Modern Sci-*

ence: Essays in Commemoration of the Centenary of the Birth of Charles Darwin and of the Fiftieth Anniversary of the Publication of the Origin of Species, ed. A.C. Seward, 85–101. Cambridge: Cambridge University Press.

———. 1979. *Problems of Genetics*, ed. G. Evelyn Hutchinson and Stan Rachootin. New Haven, CT: Yale University Press. (Reprint; orig. pub. in 1913).

———. 1894. *Materials for the Study of Variation: Treated with Especial Regard to Discontinuity*. London: Macmillan.

Bayly, Christopher A. 1996. *Empire and Information: Intelligence Gathering and Social Communication in India, 1770–1870*. Cambridge: Cambridge University Press.

Bayne, Peter. 1855. *The Christian Life: Social and Individual*. Edinburgh: James Hogg.

———. 1871. *The Life and Letters of Hugh Miller*. Boston: Gould and Lincoln.

Bazerman, Charles. 1999. *The Languages of Edison's Light*. Cambridge, MA: MIT Press.

Beauchamp, Kenneth. 1997. *Exhibiting Electricity*. London: Institute of Electrical Engineers.

Becher, Harvey. 1980–81. "William Whewell and Cambridge Mathematics." *Historical Studies in the Physical Sciences* 11:1–48.

Becker, Bernard H. 1968. *Scientific London*. London: Frank Cass.

Bedell, Rebecca. 2001. *The Anatomy of Nature: Geology and American Landscape Painting, 1825–1875*. Princeton, NJ: Princeton University Press.

Beer, Gillian. 2000. *Darwin's Plots: Evolutionary Narratives in Darwin, George Eliot and Nineteenth-Century Fiction*. 2nd ed. Cambridge: Cambridge University Press.

Beeton, Isabella. 1861. *The Book of Household Management: Comprising Information for the Mistress*. London: S. O. Beeton.

Beke, Charles Tilstone. 1843. "On the Countries South of Abyssinia." *Journal of the Royal Geographical Society* 13:254–69.

———. 1847. "On the Nile and Its Tributaries." *Journal of the Royal Geographical Society* 17:1–84.

———. 1860. *The Sources of the Nile: Being a General Survey of the Basin of That River, and of Its Head-Streams; with the History of Nilotic Discovery*. London: James Madden.

Belk, Russell W. 1988. "Possessions and the Extended Self." *Journal of Consumer Research* 15:139–68.

———. 1995. *Collecting in a Consumer Society*. London: Routledge.

Bellon, Richard. 2007. "Science at the Crystal Focus of the World." In Fyfe and Lightman, *Science in the Marketplace*, 301–35.

———. 2011. "Inspiration in the Harness of Daily Labor: Darwin, Botany and the Triumph of Evolution, 1859–1868." *Isis* 102 (forthcoming, 2011).

Bendysche, Thomas. 1863–64. "The History of Anthropology." *Memoirs Read Before the Anthropological Society of London* 1:335–420.

Bennett, J. D. 2001. "The Railway Guidebooks of George Measom." *BackTrack* 15:379–81.

Bennett, J. W. 1843. *Ceylon and Its Capabilities: An Account of Its Natural Resources, Indigenous Productions and Commercial Facilities*. London: W. H. Allen.

Bennett, Tony. 1988. "The Exhibitionary Complex." *New Formations* 4:73–102.

———. 1995. *The Birth of the Museum: History, Theory, Politics*. London: Routledge.

———. 2004. *Pasts beyond Memory: Evolution, Museums, Colonialism*. London: Routledge.

Bensaude-Vincent, Bernadette. 1995. "Introductory Essay: A Geographical History of Eighteenth-Century Chemistry." In *Lavoisier in European Context: Negotiating a*

New Language for Chemistry, ed. Bernadette Bensaude-Vincent and Ferdinando Abbri, 1–17. Canton, MA: Science History Publications.

Bentham, George. 1858. *Handbook of the British Flora.* London: Lovell Reeve.

Berghaus, Heinrich. 1845–48. *Physikalischer Atlas oder Sammlung von Karten, auf denen die hauptsächlichsten Erscheinungen der anorganischen und organischen Natur nach ihrer geographischen Verbreitung und Vertheilung bildlich dargestellt sind.* 2 vols. Gotha: Justus Perthes Verlag.

Berman, Morris. 1978. *Social Change and Scientific Organization: The Royal Institution, 1799–1844.* Ithaca, NY: Cornell University Press.

Bevis, Matthew. 2003. "Volumes of Noise." *Victorian Literature and Culture* 31:578–79.

Bickerton, David. 1986. *Marc-Auguste and Charles Pictet, the "Bibliothèque Britannique" (1796–1815) and the Dissemination of British Science and Literature on the Continent.* Geneva: Slatkine Reprints.

Bird, Isabella Lucy. 1862. "Christian Individuality." *North British Review* 37:249–84.

Blackman, Helen J. 2004. "A Spiritual Leader? Cambridge Zoology, Mountaineering and the Death of F. M. Balfour." *Studies in the History and Philosophy of Biological and Biomedical Sciences* 35:93–117.

———. 2007. "The Natural Sciences and the Development of Animal Morphology in Late-Victorian Cambridge." *Journal of the History of Biology* 40:71–108.

Black's Picturesque Tourist, and Road and Railway Guide Book through England and Wales. 1853. 3rd ed. Edinburgh: A. & C. Black.

Bloor, David. 1976. *Knowledge and Social Imagery.* Chicago: University of Chicago Press.

———. 2005. "Toward a Sociology of Epistemic Things." *Perspectives on Science* 13:285–312.

Boas, Franz. 1907. "Some Principles of Museum Administration." *Science* 25:921–33.

Boase, Frederic. 1905. "Becker, Bernard Henry." In *Modern English Biography* 4:333. Truro: Netherton and Worth.

Boase, Henry S. 1827. "On the Sand-Banks of the Northern Shores of Mount's Bay." *Transactions of the Royal Geological Society of Cornwall* 3:166–91.

———. 1832. "Contributions Towards a Knowledge of the Geology of Cornwall," *Transactions of the Royal Geological Society of Cornwall* 4:166.

———. 1834. *A Treatise of Primary Geology: Being an Examination, Both Practical and Theoretical, of the Older Formations.* London: Longman, Orme, Brown, Green, and Longman.

Bogue, Alan. 1998. *Frederick Jackson Turner: Strange Roads Going Down.* Norman: University of Oklahoma Press.

Bolas, Thomas. 1882. "The Fire Risks Incidental to Electric Lighting." *Journal of the Society of Arts* 30:663–72.

Bonsor, N. R. P. 1975–80. *North Atlantic Seaway.* 5 vols. Jersey: Brookside.

———. 1983. *South Atlantic Seaway.* Jersey: Brookside.

Bourguet, Marie-Noëlle. 1999. "The Explorer." In *Enlightenment Portraits*, ed. Michel Vovelle, 257–315. Chicago: University of Chicago Press.

Bourguet, Marie-Noëlle, Christian Licoppe, and H. Otto Sibum, eds. 2003. *Instruments, Travel and Science: Itineraries of Precision from the Seventeenth to the Twentieth Century.* London: Routledge.

Bowen, Margarita. 1981. *Empiricism and Geographical Thought: From Francis Bacon to Alexander von Humboldt.* Cambridge: Cambridge University Press.

Bowie, Henry. 1865. "Music Hall: An Annoyance." *Scotsman*, 8 November.

Bowler, Peter. 1975. "The Changing Meaning of 'Evolution.'" *Journal of the History of Ideas* 36:95–114.

———. 1992. *The Fontana History of the Environmental Sciences.* London: Fontana.

Box, Joan Fisher. 1978. *R. A. Fisher: The Life of a Scientist.* New York: John Wiley & Sons.

Bradfield, B. T. 1968. "Sir Richard Vyvyan and the Country Gentlemen, 1830–1834." *English Historical Review* 83:729–43.

Bradshaw's Railway Almanack, Directory, Shareholder's Guide and Manual (London: Adams, 1848).

Brain, Robert. 1993. *Going to the Fair: Readings in the Culture of Nineteenth-Century Exhibitions.* Cambridge: Whipple Museum of the History of Science.

Bravo, Michael T. 1999. "Precision and Curiosity in Scientific Travel: James Rennell and the Orientalist Geography of the New Imperial Age (1760–1830)." In *Voyages and Visions: Towards a Cultural History of Travel*, ed. Jas Elsner and Joan-Pau Rubiés, 162–83. London: Reaktion.

Brett, Reginald Baliol (second Viscount Esher). 1912. "Balfour Professorship of Genetics: Munificent Gift to Cambridge." *Times* (London), 13 March.

———. *Journals and Letters of Reginald, Viscount Esher.* 1938. Ed. Oliver, Viscount Esher. 4 vols. London: Ivor Nicholson & Watson.

Bridges, Roy C. 1976. "W. D. Cooley, the RGS and African Geography in the Nineteenth Century." *Geographical Journal* 142:27–47, 274–86.

———. 2008. "William Desborough Cooley (1795–1883)." *Geographers Biobibliographical Studies* 27:43–62.

"British-Grown Fruits and Vegetables: A Great Exhibition at Chiswick." 1903. *Garden.* Supplement.

Bronner, Simon J., ed. 1987. *Folklife Studies from the Gilded Age: Object, Rite, and Custom in Victorian America.* Ann Arbor: University of Michigan Research Press.

Brooke, John H. 1996. "Like Minds: The God of Hugh Miller." In *Hugh Miller and the Controversies of Victorian Science*, ed. Michael Shortland, 171–86. Oxford: Clarendon Press.

Brooke, Christopher N. L. 1993. *A History of the University of Cambridge.* Vol. 4 (1870–1990). 4 vols. Cambridge: Cambridge University Press.

Browman, David L. 2002. "Origins of Stratigraphic Excavation in North America: The Peabody Museum Method and the Chicago Method." In *New Perspectives on the Origins of Americanist Archaeology*, ed. David L. Browman and Stephen Williams, 242–64. Tuscaloosa: University of Alabama Press.

Browne, Janet. 1995. *Charles Darwin: Voyaging.* London: Cape.

Buchanan, Angus. 2002. *Brunel: The Life and Times of Isambard Kingdom Brunel.* London: Hambledon.

Burchfield, J. D. 2002. "John Tyndall at the Royal Institution." In *"The Common Purposes of Life": Science and Society at the Royal Institution of Great Britain*, ed. Frank A. J. L. James, 147–68. Aldershot: Ashgate.

———. 2004. "Tyndall, John." In *Dictionary of Nineteenth-Century British Scientists*, ed.
      Bernard Lightman. 4:2053–58. Bristol: Thoemmes Continuum Press.

Burke, Peter, ed. 1991. *New Perspectives on Historical Writing*. Cambridge: Cambridge
      University Press.

Burkhardt, Frederick, Janet Browne, Duncan M. Porter, and Marsha Richmond, eds.
      1993. *The Correspondence of Charles Darwin, 1860*. Vol. 8. Cambridge: Cambridge
      University Press.

Burkhardt, Frederick, Joy Harvey, Duncan M. Porter, and Jonathan R. Topham, eds.
      1997. *The Correspondence of Charles Darwin, 1862*. Vol. 10. Cambridge: Cambridge
      University Press.

Burkhardt, Richard W. 1977. *The Spirit of System: Lamarck and Evolutionary Biology*.
      Cambridge, MA: Harvard University Press.

———. 1979. "Closing the Door on Lord Morton's Mare: The Rise and Fall of Telegony."
      Studies in the History of Biology 3:1–21.

———. 2007. "The Leopard in the Garden: Life in Close Quarters at the Muséum
      D'histoire Naturelle." *Isis* 98:675–94.

Burmeister, Maritha Rene. 2000. "Popular Anatomical Museums in Nineteenth-Century
      England." PhD diss., Rutgers University.

Burns, James. 2007. "John Fleming and the Geological Deluge." *British Journal for the
      History of Science* 40:205–25.

Burrow, John W. 1963. "Evolution and Anthropology in the 1860s: The Anthropological
      Society of London, 1863–1871." *Victorian Studies* 7:137–54.

Burrows, Simon. 1992. *French Exile Journalism and European Politics, 1792–1814*.
      Woodbridge, Suffolk: Royal Historical Society and Boydell Press.

Burton, Richard Francis. 1859. "The Lake Regions of Central Africa, with Notices of the
      Lunar Mountains and the Sources of the White Nile." *Journal of the Royal Geograph-
      ical Society* 29:1–454.

———. 1860. *The Lake Regions of Central Africa, a Picture of Exploration*. New York:
      Harper & Brothers.

———. 1865. "On Lake Tanganyika, Ptolemy's Western Lake-Reservoir of the Nile."
      Journal of the Royal Geographical Society 35:1–15.

Buxton, Richard. 1849. *A Botanical Guide to the Flowering Plants, Ferns, Mosses, and
      Algae, Found Indigenous within Sixteen Miles of Manchester . . . Together with a
      Sketch of the Author's Life*. London: Longman.

Buzard, James. 1993. *The Beaten Track: European Tourism, Literature and the Ways to
      "Culture," 1800–1918*. Oxford: Clarendon Press.

Byrnes, Giselle. 2001. *Boundary Markers: Land Surveying and the Colonisation of New
      Zealand*. Wellington: Bridget Williams Books.

Cairns, John. 1860. "The Late Dr George Wilson of Edinburgh." *Macmillan's Magazine*
      1:199–203.

Cambridge University Association. 1906. "A Plea for Cambridge." *Quarterly Review*
      204:499–525.

Camerini, Jane. 1987. "Darwin, Wallace, and Maps." PhD diss., University of Wisconsin–
      Madison.

———. 1993. "Evolution, Biogeography, and Maps: An Early History of Wallace's Line."
    Isis 84:700–727.

———. 1993. "The Physical Atlas of Heinrich Berghaus: Distribution Maps as Scientific
    Knowledge." In *Non-Verbal Communication in Science Prior to 1900*, ed. Renato
    Mazzolini, 479–512. Florence: Leo S. Olschki.

———. 2000. "Heinrich Berghaus's Map of Human Diseases." In *Medical Geography in
    Historical Perspective*, ed. Nicolaas A. Rupke, 186–208. Medical History, Supplement
    no. 20. London: The Wellcome Trust Centre for the History of Medicine at Univer-
    sity College, London.

Cameron, Jenks. 1929. *The Bureau of Biological Survey: Its History, Activities, and Orga-
    nization.* Baltimore, MD: Johns Hopkins University Press.

Cameron, Verney Lovett. 1874–75. "Exploration of Lake Tanganyika: Letter from Lieut.
    V. L. Cameron, Describing the Discovery of an Outlet." *Proceedings of the Royal
    Geographical Society of London* 19:75–78.

Campbell, Mary Baine. 1999. *Wonder and Science: Imagining Worlds in Early Modern
    Europe.* Ithaca, NY: Cornell University Press.

Candlin, Fiona. 2010. *Art, Museums and Touch.* Manchester: Manchester University
    Press.

Candolle, Alphonse de. 1873. *Histoire des sciences et des savants depuis deux siècles
    suivie d'autres études sur des sujets scientifiques en particulier sur la sélection dans
    l'espèce humaine.* Geneva: Bale; Lyon: Georg.

Cannadine, David. 1999. *The Decline and Fall of the British Aristocracy.* Rev. ed. New
    York: Vintage Books.

Cannon, Susan Faye. 1978. *Science in Culture: The Early Victorian Period.* New York:
    Dawson and Science History Publications.

Cantor, Geoffrey, Gowan Dawson, Graeme Gooday, Richard Noakes, Sally Shuttleworth,
    and Jonathan R. Topham. 2004. *Science in the Nineteenth-Century Periodical.* Cam-
    bridge: Cambridge University Press.

Carne, Joseph. 1822. "On the Mineral Productions, and the Geology of the Parish of St.
    Just." *Transactions of the Royal Geological Society of Cornwall* 2:290–358.

———. 1822. "On the Relative Age of the Veins of Cornwall." *Transactions of the Royal
    Geological Society of Cornwall* 2:49–128.

Caron, François, and Christine Berthet. 1984. "Electrical Innovation: State Initiative or
    Private Initiative? Observations on the 1881 Paris Exhibition." *History and Technol-
    ogy* 1:307–18.

Carroll, Charles. 1900. *"The Negro a Beast"; or, "In the Image of God."* St. Louis, MO:
    American Book and Bible House.

Carroll, Vicky. 2004. "The Natural History of Visiting: Responses to Charles Waterton
    and Walton Hall." *Studies in History and Philosophy of Biological and Biomedical
    Sciences* 35:31–64.

*A Catalogue of the Valuable Collection of Natural History Belonging to the Late Mr
    Richard Rutledge Wingate.* Newcastle: Blackwell, 1859.

Cavalli-Sforza, Luigi Luca. 1990. "Recollections of Whittingehame Lodge." *Theoretical
    Population Biology* 38:301–5.

Cawood, John. 1979. "The Magnetic Crusade: Science and Politics in Early Victorian Britain." *Isis* 70:492–518.

Cecil, Lady Gwendolen. 1931. *Life of Robert, Marquis of Salisbury*. 5 vols. London: Hodder & Stoughton.

Chadarevian, Soraya de. 1996. "Laboratory Science Versus Country-House Experiments: The Controversy between Julius Sachs and Charles Darwin." *British Journal for the History of Science* 29:17–41.

Charnley, Berris, and Gregory Radick. 2009. "Plant Breeding and Intellectual Property before and after the Rise of Mendelism: The Case of Britain." In *Living Properties: Making Knowledge and Controlling Ownership in the History of Biology*, ed. Jean-Paul Gaudillière, Daniel J. Kevles, and Hans-Jörg Rheinberger. Preprint 382. Berlin: Max-Planck-Institut für Wissenschaftsgeschichte.

Chartier, Roger. 1981. *The Cultural Uses of Print in Early Modern Europe*. Princeton, NJ: Princeton University Press.

———, ed. 1989. *The Culture of Print: Power and the Uses of Print in Early Modern Europe*. Cambridge: Polity Press.

Chateaubriand, François de. 1989–98. *Mèmoires d'outre-tombe*. Paris: Bordas.

[Chisholme, Alexander]. 1790. "Kerr's Translation of Lavoisier's Elements of Chemistry." *Monthly Review*, 2nd ser., 4:159–62.

[Chisholme, Alexander, and Thomas Cogan]. 1790. "Lavoisier's Elementary Treatise on Chemistry." *Monthly Review*, 2nd ser., 2:308–17.

Chung, Yun Shun Susie. 2003. "John Britton (1771–1857): A Source for the Exploration of the Foundations of County Archaeological Society Museums." *Journal of the History of Collections* 15:113–25.

Clark, J. F. M. 2009. *Bugs and the Victorians*. New Haven, CT: Yale University Press.

Clark, John Willis. 1904. *Endowments of the University of Cambridge*. Cambridge: Cambridge University Press.

Clendinning, Anne. 2002. *Demons of Domesticity: Women and the English Gas Industry, 1889–1939*. Aldershot: Ashgate.

Clydeside Cameos: A Series of Sketches of Prominent Clydeside Men. 1885. Glasgow: Ranken.

Cock, Alan G., and Donald R. Forsdyke. 2008. *Treasure Your Exceptions: The Science and Life of William Bateson*. New York: Springer.

Cohn, Bernard. 1996. *Colonialism and Its Forms of Knowledge: The British in India*. Princeton, NJ: Princeton University Press.

Coleman, William. 1968. "Bateson Papers." *Mendel Newsletter* 2:1–3.

———. 1970. "Bateson and Chromosomes: Conservative Thought in Science." *Centaurus* 15:228–314.

———. 1971. *Biology in the Nineteenth Century: Problems of Form, Function, and Transformation*. New York: Wiley.

Colenso, J. W. 1832. "A Description of Happy-Union Tin Stream Work at Pentuan." *Transactions of the Royal Geological Society of Cornwall* 4:29–39.

Colman, Henry. 1891. "Pre-Adamites." *Methodist Review* 7:891–902.

Conn, Steven. 1998. *Museums and American Intellectual Life, 1876–1926*. Chicago: University of Chicago Press.

———. 2004. *History's Shadow: Native Americans and Historical Consciousness in the Nineteenth Century*. Chicago: University of Chicago Press.

Connerton, Paul. 1989. *How Societies Remember*. Cambridge: Cambridge University Press.

Cooley, William Desborough. 1841. *The Negroland of the Arabs Examined and Explained; or, an Inquiry into the Early History and Geography of Central Africa*. New York: J. Arrowsmith.

———. 1845. "The Geography of N'yassi, or the Great Lakes of Southern Africa, Investigated; with an Account of the Overland Route from the Quanza in Angola to the Zambezi in the Government of Mozambique." *Journal of the Royal Geographical Society* 15:185–235.

———. 1854. *Claudius Ptolemy and the Nile; or, an Inquiry into That Geographer's Real Merits and Speculative Errors, His Knowledge of Eastern Africa and the Authenticity of the Mountains of the Moon*. London: John W. Parker and Son.

———. 1855–57. "Journey of Joachim Rodriguez Graça to the Mwata Ya Nvo." *Proceedings of the Royal Geographical Society of London* 1:92.

———. 1863–64. "On the Travels of Portuguese and Others in Inner Africa." *Proceedings of the Royal Geographical Society of London* 8:255–56.

Coombes, Annie E. 1994. *Reinventing Africa: Museums, Material Culture and Popular Imagination in Late Victorian and Edwardian England*. New Haven, CT: Yale University Press.

Cooper, Malcolm. 2003. "Thomas Skinner's Castle Line." *Ships in Focus* 24:250–59.

Cooter, Roger. 1997. "The Moment of the Accident: Culture, Militarism and Modernity in Late Victorian Britain." In *Accidents in History: Injuries, Fatalities and Social Relations*, ed. Roger Cooter and Bill Luckin, 107–57. Amsterdam: Rodopi.

Cooter, Roger, and Stephen Pumfrey. 1994. "Separate Spheres and Public Places: Reflections on the History of Science Popularization and Science in Popular Culture." *History of Science* 32:237–67.

Cordiner, James. 1807. *A Description of Ceylon: Containing an Account of the Country, Inhabitants and Natural Productions*. London: Longman, Hurst, Rees, and Orme.

Corlett, Ewan. 1990. *The Iron Ship: The Story of Brunel's SS Great Britain*. London: Conway Maritime Press.

Corsi, Pietro. 1978. "The Importance of French Transformist Ideas for the Second Volume of Lyell's *Principles of Geology*." *British Journal for the History of Science* 11:221–44.

———. 1988. *The Age of Lamarck: Evolutionary Theories in France, 1790–1830*. Berkeley and Los Angeles: University of California Press.

———. 1988. *Science and Religion: Baden Powell and the Anglican Debate, 1800–1860*. Cambridge: Cambridge University Press.

Cosgrove, Denis. 2008. *Geography and Vision: Seeing, Imagining and Representing the World*. London: I. B. Tauris.

Cosgrove, Denis, and Stephen Daniels, eds. 1988. *The Iconography of Landscape: Essays on the Symbolic Representation, Design and Use of Past Environments*. Cambridge: Cambridge University Press.

Coutts, James. 1909. *History of the University of Glasgow from Its Foundation in 1451 to 1909*. Glasgow: James Maclehose and Sons.

Crane, Susan A., ed. 2000. *Museums and Memory.* Stanford, CA: Stanford University Press.

Croal, D. 1904. *Sketches of East Lothian.* 4th ed. Haddington: Haddingtonshire Courier.

Crompton, Rookes E. B. 1884. "Artificial Lighting in Relation to Health." *Journal of the Society of Telegraph Engineers and Electricians* 13:390–415.

———. 1928. *Reminiscences.* London: Constable.

Crosland, Maurice P. 1962. *Historical Studies in the Language of Chemistry.* London: Heinemann.

———. 1994. *In the Shadow of Lavoisier: The "Annales de Chimie" and the Establishment of a New Science.* N.p.: British Society for the History of Science.

Crosland, Maurice P., and Crosbie Smith. 1978. "The Transmission of Physics from France to Britain: 1800–1840." *Historical Studies in the Physical Sciences* 9:1–61.

Crouch, Edmund A. 1826. *An Illustrated Introduction to Lamarck's Conchology Contained in His "Histoire naturelle des animaux sans vertèbres": Being a Literal Translation of the Descriptions of the Recent and Fossil Genera.* London: Longman, Rees, Orme, Brown, and Green and J. Mawe.

Crouzet, François. 1958. *L'économie Britannique et le blocus continental (1806–1813).* Paris: Presses Universitaires de France.

Crozier, I. R. 1985. *William Fee McKinney of Sentry Hill: His Family and Friends.* Coleraine: Impact Printing.

Couch, Richard Q. 1846. "On the Silurian Remains of the Strata of the South-East Coast of Cornwall." *Transactions of the Royal Geological Society of Cornwall* 6:147–49.

Crook, Denise. "Boase, Henry Samuel (1799–1883)." *Oxford Dictionary of National Biography.* Available online at http://www.oxforddnb.com/view/article/2738?docPos=6 (accessed 17 March 2008).

———. 1990. "The Early History of the Royal Geological Society of Cornwall, 1814–1850." PhD diss., Open University.

Cullen, Paul Card, et al. 1874. "Pastoral Addresses of the Archbishops and Bishops of Ireland." *Irish Ecclesiastical Record* 11:49–70.

Cuvier, Georges. 1826. *Researches into Fossil Osteology: Partially Abridged and Rearranged from the French.* London: G. B. Whittaker.

———. 1827–35. *The Animal Kingdom Arranged in Conformity with Its Organization.* 16 vols. London: G. B. Whittaker.

Dalgleish, Walter. S. 1894. "Geography at the British Association, Oxford, August 1894," *Scottish Geographical Magazine* 10:463–73.

Dall, William H. 1915. *Spencer Fullerton Baird: A Biography.* Philadelphia: J. B. Lippincott.

Darby, Henry C. 1954. "Some Early Ideas on the Agricultural Regions of England." *Agricultural History Review* 2:30–47.

Darnell, Regna. 2003. "Daniel Garrison Brinton and the View from Philadelphia." In *Philadelphia and the Development of American Archaeology,* ed. Don D. Fowler and David R. Wilcox, 21–35. Tuscaloosa: University of Alabama Press.

Darwin, Charles. 1859. *On the Origin of Species by Means of Natural Selection, or the Preservation of Favoured Races in the Struggle for Life.* London: John Murray.

Darwin, Francis, ed. 1887. *The Life and Letters of Charles Darwin, Including an Autobiographical Chapter.* 3 vols. London: John Murray.

Daston, Lorraine. 1988. "The Factual Sensibility." *Isis* 79:452–70.

———. 2000. "Introduction: The Coming into Being of Scientific Objects." In *Biographies of Scientific Objects*, ed. Lorraine Daston, 1–14. Chicago: University of Chicago Press.

———. 2004. "Speechless." In *Things That Talk: Object Lessons from Art and Science*, ed. Lorraine Daston, 9–24. New York: Zone Books.

———. 2004. "The Glass Flowers." In *Things That Talk: Object Lessons from Art and Science*, ed. Lorraine Daston, 223–54. New York: Zone Books.

———. 2004. "Type Specimens and Scientific Memory." *Critical Inquiry* 31:153–82.

Daston, Lorraine, and Katherine Park. 1998. *Wonders and the Order of Nature, 1150–1750*. New York: Zone Books.

Daston, Lorraine, and Peter Galison. 2007. *Objectivity*. New York: Zone Books.

Daunton, Martin J., ed. 2005. *The Organisation of Knowledge in Victorian Britain*. Oxford: Oxford University Press.

Davy, Humphry. 1818. "Hints on the Geology of Cornwall." *Transactions of the Royal Geological Society of Cornwall* 1:38–50.

Davy, John. 1821. *An Account of the Interior of Ceylon and of Its Inhabitants*. London: Longman, Hurst, Rees, Orme & Brown; repr., Dehiwela, Sri Lanka: Tisara Prakasakayo, 1983.

de Beer, E. S. 1952. "The Development of the Guidebook until the Early Nineteenth Century." *Journal of the British Archaeological Association* 15:35–46.

de Beer, Gavin. 1960. *The Sciences Were Never at War*. Edinburgh: Nelson.

De Boffe, Joseph. 1794. *Catalogue des livres français de J. De Boffe, libraire, Gerrard-Street, Soho, à Londres*. London: J. De Boffe.

De Butts, Augustus. 1841. *Rambles in Ceylon*. London: W. H. Allen.

Deiss, William A. 1980. "Spencer F. Baird and His Collectors." *Journal of the Society of Natural History* 9:636–45.

Delano-Smith, Catherine. 2006. "Milieus of Mobility: Itineraries, Route Maps and Road Maps." In *Cartographies of Travel and Navigation*, ed. James R. Akerman, 16–68. Chicago: University of Chicago Press.

De Silva, Colvin R. 1953–1962. *Ceylon under British Occupation*. 2 vols. Colombo, Sri Lanka: Colombo Apothecaries.

De Silva, K. M. 1981. *A History of Sri Lanka*. Delhi: Oxford University Press.

De Silva, W. A. 1891. "A Contribution to Sinhalese Plant Lore." *Journal of the Royal Asiatic Society, Ceylon Branch*, 12:113–44.

———. 1919. "A Probable Origin of the Name Kushtharajagala." *Royal Asiatic Society, Ceylon Branch* 28:86.

Desmond, Adrian. 1989. *The Politics of Evolution: Morphology, Medicine, and Reform in Radical London*. Chicago: University of Chicago Press.

———. 1997. *Huxley: From Devil's Disciple to Evolution's High Priest*. Reading, MA: Perseus Books.

Desmond, R. 1992. *The European Discovery of the Indian Flora*. Oxford: Oxford University Press.

Dettelbach, Michael. 1996. "Humboldtian Science." In Jardine, Secord, and Spary, *Cultures of Natural History*, 287–304.

Dewaraja, Lorna. 1972. *A Study of the Political, Administrative and Social Structure of the Kandyan Kingdom of Ceylon, 1707–1760*. Colombo, Sri Lanka: Lake House Press.

Dhombres, Jean. 1989. "Books: Reshaping Science." In *Revolution in Print: The Press in France, 1775–1800*, ed. Robert Darnton and Daniel Roche, 177–202, 337–41. Berkeley and Los Angeles: University of California Press.

Diamond, Jared. 1997. *Guns, Germs and Steel: The Fates of Human Societies*. London: Jonathan Cape.

———. 2005. *Collapse: How Societies Choose to Fail or Survive*. London: Allen Lane.

Dickinson, Victoria. 1998. *Drawn from Life: Science and Art in the Portrayal of the New World*. Toronto: University of Toronto Press.

Dierig, S., J. Lachmund, and A. J. Mendelsohn. 2003. "Introduction: Toward an Urban History of Science." In "Science and the City," special issue, *Osiris* 18:1–19.

Dillon, Maureen. 2001. *Artificial Sunshine: A Social History of Lighting*. London: National Trust.

———. 2001. "'Like a Glow-Worm That Has Lost Its Glow': The Invention of the Electric Incandescent Lamp and the Development of Artificial Silk and Jewellery." *Costume: The Journal of the Costume Society* 35:76–81.

DiNoto, Andrea, and David Winter. 1999. *The Pressed Plant: The Art of Botanical Specimens, Nature Prints, and Sun Pictures*. New York: Stewart, Tabori & Chang.

Dodd, George. 1843. *Days at the Factories; or, the Manufacturing Industry of Great Britain Described and Illustrated by Numerous Engravings of Machines and Processes*. London: Knight.

Dodds, Klaus, and Stephen Royle. 2003. "The Historical Geography of Islands. Introduction: Rethinking Islands." *Journal of Historical Geography* 29:487–98.

Donovan, Arthur. 1979. "Scottish Responses to the New Chemistry of Lavoisier." *Studies in Eighteenth-Century Culture* 1:237–49.

Dorn, Harold. 1991. *The Geography of Science*. Baltimore, MD: Johns Hopkins University Press.

Drayton, Richard. 2000. *Nature's Government: Science, Imperial Britain, and the "Improvement" of the World*. New Haven, CT: Yale University Press.

"Dr. David Page on Man, in His Natural History Relations." 1868. *Anthropological Review* 6:109–14.

Dritsas, Lawrence. 2010. *Zambesi: David Livingstone and Expeditionary Science in Africa*. London: I. B. Tauris.

Driver, Felix. 2001. *Geography Militant: Cultures of Exploration and Empire*. Oxford: Blackwell.

Driver, Felix, and Luciana Martins, eds. 2005. *Tropical Visions in An Age of Empire*. Chicago: University of Chicago Press.

Dubow, Saul. 2000. "A Commonwealth of Science: The British Association in South Africa, 1905 and 1929." In *Science and Society in Southern Africa*, ed. Saul Dubow, 66–99. Manchester: Manchester University Press.

Duckworth, W. L. H. 1919. "Professor Alexander Macalister." *Man* 19 (November): 164–8.

Dudden, Arthur Power, ed. 2004. *American Empire in the Pacific: From Trade to Strategic Balance, 1700–1922*. Aldershot: Ashgate.

Dugatkin Alan. 2009. *Mr. Jefferson and the Giant Moose: Natural History in Early America.* Chicago: University of Chicago Press.

Duncan, Isabelle. 1860. *Pre-Adamite Man; or, the Story of Our Old Planet and Its Inhabitants, Told by Scripture and Science.* London: Saunders, Otley.

Duncan, James. 1990. *The City as Text: The Politics of Landscape Interpretation in the Kandyan Kingdom.* Cambridge: Cambridge University Press.

———. 2007. *In the Shadow of the Tropics: Climate, Race and Biopower in Nineteenth-Century Ceylon.* Aldershot: Ashgate.

Dunlop, Alexander. 1848. *Emerson's Orations to the Modern Athenians; or, Pantheism.* Edinburgh: J. Elder.

Durie, Alastair J. 2003. "Tourism and the Railways in Scotland: The Victorian and Edwardian Experience." In *The Impact of the Railway on Society in Britain: Essays in Honour of Jack Simmons,* ed. A. K. B. Evans and J. V. Gough, 199–209. Aldershot: Ashgate.

Duveen, Denis I., and Herbert S. Klickstein. 1954. *A Bibliography of the Works of Antoine Laurent Lavoisier, 1743–1794.* London: Wm. Dawson and Sons and E. Weil.

Dyer, Frank Lewis, Thomas Commerford Martin, and William Henry Meadowcroft. 1929. *Edison: His Life and Inventions.* New York: Harper & Bros.

Ebach, Malte C., and Daniel F. Goujet. 2006. "The First Biogeographical Map." *Journal of Biogeography* 33:761–9.

The Edinburgh Academy Register: A Record of All Those Who Have Entered the School since Its Foundation in 1824. 1914. Edinburgh: Constable.

"Edinburgh Dissected." 1857. *Athenaeum,* 30 May.

"Editorial." 1858. *Engineer* 6:357–58.

"Editorial." 1860. *Engineer* 9:9.

[Editorial]. 1867. *Scotsman,* 29 March, 2.

Edmond, Rod, and Vanessa, Smith, eds. 2003. *Islands in History and Representation.* London: Routledge.

Edmonds J. M., and R. A. Beardmore. 1955. "John Phillips and the Early Meetings of the British Association," *Advancement of Science* 12:97–104.

Edney, Matthew H. 1999. "Reconsidering Enlightenment Geography and Map Making: Reconnaissance, Mapping, Archive." In Livingstone and Withers, *Geography and Enlightenment,* 165–98.

"The Electric Light and Its Friends." 1881. *Journal of Gas Lighting, Water Supply and Sanitary Improvement* 38:1030–31.

"The Electric Light at the Savoy Theatre." 1882–83. *Nature* 27:418–19.

"Electricity in Stageland." 1898. *Electrician* 42:336–37.

"Elements of Chemistry." 1790. *Critical Review* 69:407–21.

Ellis, Geoffrey. 2003. *The Napoleonic Empire.* 2nd ed. Basingstoke: Palgrave Macmillan.

Elsner, John, and Roger Cardinal, eds. 1994. *The Cultures of Collecting.* London: Reaktion.

Elston, Mary Ann. 1987. "Women and Anti-Vivisection in Victorian England, 1870–1900." In Rupke, *Vivisection in Historical Perspective,* 259–94.

Elwick, James. 2007. *Styles of Reasoning in the British Life Sciences: Shared Assumptions, 1820–1858.* London: Pickering and Chatto.

Emmerson, George S. 1977. *John Scott Russell: A Great Victorian Engineer and Naval Architect.* London: John Murray.

Endersby, Jim. 2001. "'From Having No Herbarium': Local Knowledge vs. Metropolitan Expertise: Joseph Hooker's Australasian Correspondence with William Colenso and Ronald Gunn." *Pacific Science* 55:343–58.

Endersby, Jim. 2004. "Hooker, Joseph Dalton." In *Dictionary of Nineteenth-Century British Scientists*, ed. Bernard Lightman, 2:994–1001. Bristol: Thoemmes Continuum Press.

———. 2008. *Imperial Nature: Joseph Hooker and the Practices of Victorian Science.* Chicago: University of Chicago Press.

Endersby, Linda Eickmeier. 1999. "Expositions, Museums, and Technological Display: Building Cultural Institutions for the 'Citizen Inventor' in the Late Nineteenth-Century United States." PhD diss., Massachusetts Institute of Technology.

Engelmann, Gerhard. 1964. "Der Physikalische Atlas des Heinrich Berghaus und Alexander Keith Johnston's Physical Atlas." *Petermanns Geographische Mitteilungen* 108:133–49.

Engels, Eve-Marie, and Thomas F. Glick, eds. 2008. *The Reception of Charles Darwin in Europe.* 2 vols. London: Continuum.

Erhardt, James. 1855–1857. "Reports Concerning Central Africa, as Collected in Mambara and on the East Coast, with a New Map of the Country." *Proceedings of the Royal Geographical Society of London* 1:8–10.

Essig, Mark. 2003. *Edison and the Electric Chair.* Stroud: Sutton.

Etulain, Richard W. 1996. *Re-imagining the Modern American West: A Century of Fiction, History, and Art.* Tucson: University of Arizona Press.

"The Evangelical Alliance Conference." *Scotsman*, 7 July, 3.

Eve, A. S., and C. H. Creasey. 1945. *Life and Work of John Tyndall.* London: Macmillan.

Fabian, Johannes. 2000. *Out of Our Minds: Reason and Madness in the Exploration of Central Africa.* Berkeley and Los Angeles: University of California Press.

Falconer, William. 1781. *Remarks on the Influence of Climate, Situation, Nature of Country, Population, Nature of Food, and Way of Life, on the Disposition of Temper, Manners and Behaviour, Intellects, Laws and Customs, Form of Government, and Religion of Mankind.* London: Dilly.

Fan, Fa-Ti. 2004. *British Naturalists in Qing China: Science, Empire, and Cultural Encounter.* Cambridge, MA: Harvard University Press.

Fancher, Raymond E. 1983. "Alphonse de Candolle, Francis Galton, and the Early History of the Nature-Nurture Controversy." *Journal of the History of the Behavioral Sciences* 19:341–52.

Fara, Patricia. 2002. *An Entertainment for Angels: Electricity in the Enlightenment.* Cambridge: Icon.

Farber, Paul Lawrence. 2000. *Finding Order in Nature: Naturalist Tradition from Linnaeus to E. O. Wilson.* Baltimore, MD: Johns Hopkins University Press.

Farey, John. 1816. "A Letter from Dr William Richardson to the Countess of Gosford (Occasioned by the Perusal of Cuvier's 'Geological Essay')." *Philosophical Magazine* 47:354–64.

F[armer], J. B. 1927. "William Bateson, 1861–1926." *Proceedings of the Royal Society* B101:i–xli.

"Fatalities from Electric Lighting." 1890. *Telegraphic Journal and Electrical Review* 26:39.

Feifer, Maxine. 1985. *Going Places: The Ways of the Tourist from Imperial Rome to the Present Day.* London: Macmillan.

Findlay, Alexander George. 1867. "On Dr. Livingstone's Last Journey, and the Probable Ultimate Sources of the Nile." *Journal of the Royal Geographical Society* 37:193–212.

Finnegan, Diarmid A. 2008. "The Spatial Turn: Geographical Approaches in the History of Science." *Journal of the History of Biology* 41:369–88.

———. 2009. *Natural History Societies and Civic Culture in Victorian Scotland.* London: Pickering and Chatto.

"The Fire Insurance Companies and the Electrical Exhibition." 1882. *Journal of Gas Lighting* 29:509.

Fish, Stanley. 2003. *Is There a Text in This Class? The Authority of Interpretive Communities.* Cambridge, MA: Harvard University Press.

Flack, J. Kirkpatrick. 1975. *Desideratum in Washington: The Intellectual Community in the Capitol City, 1870–1900.* Cambridge: Schenkman.

Fletcher, R. A. 1910. *Steam-Ships: The Story of Their Development to the Present Day.* London: Sidgwick & Jackson.

Fletcher, T. W. 1961. "The Great Depression of English Agriculture, 1873–1896." *Economic History Review* 13:417–32.

Flint, K. 2000. *The Victorians and the Visual Imagination.* Cambridge: Cambridge University Press.

Flower, William Henry. 1898. *Essays on Museums and Other Subjects Connected with Natural History.* London: Macmillan.

Forbes, Jonathan. 1840. *Eleven Years in Ceylon, Comprising Sketches of the Field Sport and Natural History and an Account of Its History and Antiquities.* 2 vols. London: R. Bentley.

Forbes, John. 1822. "On the Geology of the Land's-End District." *Transactions of the Royal Geological Society of Cornwall* 2:242–80.

"Foreign Literature of the Year 1810." 1811. *New Annual Register* [32]: "Literary Selections and Retrospect," 367–91.

"Foreign Literature of the Year 1812." 1813. *New Annual Register* [34]: "Literary Selections and Retrospect," 411–36.

Forgan, Sophie. 1985. "Faraday–From Servant to Savant: The Institutional Context." In *Faraday Rediscovered: Essays on the Life and Work of Michael Faraday*, ed. David Gooding and Frank A. J. L. James, 51–67. Basingstoke: Macmillan.

———. 1986. "Context, Image and Function: A Preliminary Enquiry into the Architecture of Scientific Societies." *British Journal for the History of Science* 19:89–113.

———. 2005. "Building the Museum: Knowledge, Conflict, and the Power of Place." *Isis* 96:572–85.

Forster, E. M. 1995. *A Room with a View.* London: Softback Preview.

Fortey, Richard. 2008. *Dry Store Room No. 1: The Secret Life of the Natural History Museum.* London: Harper.

F[oster], M. "Francis Maitland Balfour." 1883. *Proceedings of the Royal Society of London* 35: xx–xxvii.

Foster, Mike. 1994. *Strange Genius: The Life and Services of Ferdinand Vandeveer Hayden.* Niwot, CO: Roberts Rinehart.

Foucault. Michel. 1986. "Of Other Spaces." *Diacritics* 16:22–27.

Fowler, Don D., and David R. Wilcox, eds. 2003. *Philadelphia and the Development of Americanist Archaeology.* Tuscaloosa: University of Alabama Press.

Fox, Robert. 1995. "Edison et la Presse Française à l'Exposition Internationale D'électricité de 1881." In *Science, Industry, and the Social Order in Post-Revolutionary France,* ed. Robert Fox, 223–25. Aldershot: Variorum.

Fox, Stephen. 2003. *The Ocean Railway: Isambard Kingdom Brunel, Samuel Cunard and the Revolutionary World of the Great Atlantic Steamships.* London: HarperCollins.

Frasca-Spada, Maria, and Nicholas Jardine, eds. 2000. *Books and the Sciences in History.* Cambridge: Cambridge University Press.

Freeman, Michael. 2004. *Victorians and the Prehistoric: Tracks to a Lost World.* New Haven, CT: Yale University Press.

Freeman, R. B. 1978. *Charles Darwin: A Companion.* Folkestone: Dawson.

Freeman, Richard Broke, and Douglas Wertheimer, eds. 1980. *Philip Henry Gosse: A Bibliography.* Folkestone: Dawson.

"French National Institute." 1800. *Philosophical Magazine* 7:183–89.

Frost, C. 1853. *On the Prospective Advantages of a Visit to the Town of Hull by the British Association for the Advancement of Science.* Hull: Privately printed.

Frostick, Elizabeth. 1985. "Museums in Education: A Neglected Role?" *Museums Journal* 85:67–74.

Fyfe, Aileen. 2000. "Young Readers and the Sciences." In Frasca-Spada and Jardine, *Books and the Sciences in History,* 276–90.

———. 2004. *Science and Salvation: Evangelicals and Popular Science Publishing in Victorian Britain.* Chicago: University of Chicago Press.

Fyfe, Aileen, and Bernard Lightman. 2007. "Science in the Marketplace: An Introduction." In Fyfe and Lightman, *Science in the Marketplace,* 1–19.

———, eds. 2007. *Science in the Marketplace: Nineteenth-Century Sites and Experiences.* Chicago: University of Chicago Press.

Gairdner, W. T. 1896. "The Late Dr William Smith." *Scotsman,* 30 May.

Galison, Peter. 2003. *Einstein's Clocks, Poincaré's Maps.* London: Hodder and Stoughton.

Galton, Francis. 1873. "Transactions of the Sections: Geography." *Report of the Forty-Second Annual Meeting of the British Association for the Advancement of Science,* 198–202. London: John Murray.

Gardiner, J. Stanley. 1934. *The Zoological Department, Cambridge.* Cambridge: W. Heffer and Sons.

Gardiner, William. 1846. *Twenty Lessons on British Mosses.* 2nd ed. Edinburgh: David Mathers.

———. 1849. *Twenty Lessons on British Mosses.* Second series. London: Longman.

Gardner, George. 1836. *Musci Britannici, or Pocket Herbarium of British Mosses.* Glasgow: Allan and Ferguson.

Garfinkle, Norton. 1955. "Science and Religion in England, 1790–1800: The Critical Response to the Work of Erasmus Darwin." *Journal of the History of Ideas* 16:376–88.

G[askell], W. H. 1908. "Sir Michael Foster, 1836–1907." *Proceedings of the Royal Society* 80:lxxi–lxxxi.

Gay, Albert and Charles H. Yeaman. 1906. *Central Station Electricity Supply.* 2nd ed. London: Whittaker.

Gay, Hannah. 2003. "Science and Opportunity in London, 1871–85: The Diary of Herbert McLeod." *History of Science* 41:427–58.

Geiger, William, ed. 1929–1930. *Culavamsa: Being the More Recent Part of the Maha-vamsa.* London: Pali Text Society.

Geison, Gerald L. 1978. *Michael Foster and the Cambridge School of Physiology: The Scientific Enterprise in Late-Victorian Society.* Princeton, NJ: Princeton University Press.

Giebelhausen, Michaela, ed. 2003. *The Architecture of the Museum: Symbolic Structures, Urban Contexts.* Manchester: Manchester University Press.

Gieryn, Thomas. 1999. *Cultural Boundaries of Science: Credibility on the Line.* Chicago: University of Chicago Press.

———. 1999. "Two Faces on Science: Building Identities for Molecular Biology and Biotechnology." In *Architecture of Science*, ed. Peter Galison and E. Thompson, 423–55. Cambridge, MA: MIT Press.

———. 2002. "Three Truth-Spots." *Journal of the History of the Behavioral Sciences* 38:113–32.

———. 2006. "City as Truth-Spot: Laboratories and Field-Sites in Urban Studies." *Social Studies of Science* 36:5–38.

Gifford, Isabella. 1848. *The Marine Botanist: An Introduction to the Study of Algology.* London: Darton.

Gilfillan, George. 1852. *A Second Gallery of Literary Portraits.* Edinburgh: James Hogg.

Gill, E. Leonard. 1908. *The Hancock Museum and Its History.* Newcastle: Natural History Society.

Gill, Stephen. 1989. *William Wordsworth: A Life.* Oxford: Clarendon Press.

Glick, Thomas F. 1988. *The Comparative Reception of Darwinism.* 2nd ed. Chicago: University of Chicago Press. (Orig. pub. 1974.)

Goddard, T. Russell. 1929. *History of the Natural History Society of Northumberland, Durham and Newcastle Upon Tyne, 1829–1929.* Newcastle-Upon-Tyne: Reid.

Godfrey, Mathew. 1997. "Traversing the Fortieth Parallel: The Experiences and Letters of Robert Ridgway, Teenage Ornithologist." MA thesis, Utah State University.

Godlewska, Anne Marie Claire. 1999. *Geography Unbound: French Geographic Science from Cassini to Humboldt.* Chicago: University of Chicago Press.

Goetzmann, William H. 1993. *Exploration and Empire: The Explorer and the Scientist in the Winning of the American West.* Austin: Texas State Historical Association.

Goetzmann, William H., and William N. Goetzmann. 1986. *The West of the Imagination.* New York: W. W. Norton.

Goldstein, Daniel. 1994. "'Yours for Science': The Smithsonian Institution's Correspondence and the Shape of the Scientific Community in the Nineteenth Century." *Isis* 85:573–99.

Golinski, Jan. 1998. *Making Natural Knowledge: Constructivism and the History of Science.* Cambridge: Cambridge University Press.

Gooday, Graeme J. N. 1990. "Precision Measurement and the Genesis of Physics Teaching Laboratories in Victorian Britain." *British Journal for the History of Science* 23:25–51.

———. 1991. "'Nature' in the Laboratory: Domestication and Discipline with the Microscope in Victorian Life Science." *British Journal for the History of Science* 24:307–41.

———. 1998. "The Premisses of Premises: Spatial Issues in the Historical Construction of Laboratory Credibility." In Smith and Agar, *Making Space for Science*, 216–45.

———. 2008. *Domesticating Electricity: Technology, Uncertainty and Gender, 1880–1914.* London: Pickering & Chatto.

———. 2008. "Liars, Experts and Authorities." *History of Science* 46:431–56.

———. 2008. "Placing or Replacing the Laboratory in the History of Science." *Isis* 99:783–95.

Gooding, David. 1999. *Experiment and the Making of Meaning.* Dordrecht: Kluwer Academic.

Gooding, David, R. Pinch, and Simon Schaffer, eds. 1989. *The Uses of Experiment.* Cambridge: Cambridge University Press.

Gordon, Mrs. J. E. H. [Alice M.]. 1891. *Decorative Electricity, with a Chapter on Fire Risks by J. E. H. Gordon.* London: Sampson and Low.

Gosden, Chris. 1999. *Anthropology and Archaeology: A Changing Relationship.* London: Routledge.

Gosden, Chris, and Frances Larson, with Alison Petch. 2007. *Knowing Things: Exploring the Collections at the Pitt Rivers Museum, 1884–1945.* Oxford: Oxford University Press.

Gosse, Philip Henry. 1853. *A Naturalist's Rambles on the Devonshire Coast.* London: Van Voorst.

Gosse, Philip Henry, and Emily Gosse. 1853. *Sea-Side Pleasures.* London: Society for Promoting Christian Knowledge.

Gossett, Thomas F. 1963. *Race: The History of an Idea in America.* Dallas, TX: Southern Methodist University Press.

Gould, Stephen Jay. 1999. "The Pre-Adamite in a Nutshell." *Natural History* 108:24–27, 72–77.

Graham-Smith, G. S., and D. Keilin. 1939. "George Henry Falkiner Nuttall, 1862–1937." *Obituary Notices of Fellows of the Royal Society* 2:492–9.

Gray, Asa. 1860. "Darwin on the Origin of Species." *Atlantic Monthly* 6:109–16, 229–39.

Greenaway, Frank. 1970. Introduction to *Essays Physical and Chemical*, by Antoine-Laurent Lavoisier, v–xxxiii. London: Frank Cass.

Greene, John C. 1954. "The American Debate on the Negro's Place in Nature, 1780–1815." *Journal of the History of Ideas* 15:384–96.

Greenhough, Beth. 2006. "Imagining an Island Laboratory: Representing the Field in Geography and Science Studies." *Transactions of the Institute of British Geographers* 31:224–37.

Greenwood, Thomas. 1888. *Museums and Art Galleries.* London: Simpkin, Marshall.

Gregory, Derek. 1994. *Geographical Imaginations.* Oxford: Blackwell.

Gross, David. 2000. *Lost Time: On Remembering and Forgetting in Late Modern Culture.* Amherst: University of Massachusetts Press.

Grove, Richard. 1995. *Green Imperialism: Tropical Island Edens and the Origins of Environmentalism.* Cambridge: Cambridge University Press.

Guest, Ivor. 1957. *Victorian Ballet Girl: The Tragic Story of Clara Webster.* London: Adam and Charles Black.

Guide to the Smithsonian and National Museum. 1863. Philadelphia: Collins.

Gunn, Simon. 2000. *The Public Culture of the Victorian Middle Class: Ritual and Authority in the English Industrial City, 1840–1914.* Manchester: Manchester University Press.

Gunther, Albert Everard. 1975. *A Century of Zoology at the British Museum through the Lives of Two Keepers, 1815–1914.* London: Dawsons.

Guyot, Arnold. 1859. *The Earth and Man.* Boston: Gould and Lincoln.

———. 1873. *Physical Geography.* London: Sampson Low, Marston, Low, and Searle.

Gwinn, Nancy E. 1996. "The Origins and Development of International Publication Exchange in Nineteenth-Century America." PhD diss., George Washington University.

Hahn, Roger. 2005. *Pierre Simon Laplace, 1749–1827: A Determined Scientist.* Cambridge, MA: Harvard University Press.

Halbwachs, Maurice. 1992. *On Collective Memory.* Trans. Lewis A. Coser. Chicago: University of Chicago Press.

Hall, Brian K. 2003. "Francis Maitland Balfour (1851–1882): A Founder of Evolutionary Embryology." *Journal of Experimental Zoology* 299B:3–8.

———. 2005. "Betrayed by Balanoglossus: William Bateson's Rejection of Evolutionary Embryology as the Basis for Understanding Evolution." *Journal of Experimental Zoology* 304B:1–17.

Hallam, Elizabeth. 2010. *Anatomy Museum: Death and the Body Displayed.* London: Reaktion.

Hammond, Robert. 1884. *The Electric Light in Our Homes.* London: Warne.

Harcourt, Freda. 2006. *Flagships of Imperialism: The P&O Company and the Politics of Empire from Its Origins to 1860.* Manchester: Manchester University Press.

Harley, J. Brian. 1971. "The Ordnance Survey and the Origins of Official Geological Mapping in Devon and Cornwall." In *Exeter Essays in Geography,* ed. Kenneth J. Gregory and William L. D. Ravenhill, 105–24. Exeter: Exeter University Press.

———. 1988. "Maps, Knowledge, Power." In *The Iconography of Landscape,* ed. Denis Cosgrove and Stephen Daniels, 277–312. Cambridge: Cambridge University Press.

Hanham, Frederick. 1846. *Natural Illustrations of the British Grasses.* Bath: Binns and Goodwin.

Harraway, Donna. 1988. "Situated Knowledge: The Science Question in Feminism as a Site of Discourse on the Privilege of Partial Perspective." *Feminist Studies* 14:575–99.

———. 1989. *Primate Visions: Gender, Race, and Nature in the World of Modern Science.* New York: Routledge.

Harris, Elizabeth M. 1970. "Experimental Graphic Processes in England, 1800–1859, Part 4." *Journal of the Printing History Society* 6:53–89.

Harris, John. 1968. "English Country House Guides, 1740–1840." In *Concerning Architecture: Essays on Architectural Writers and Writing, Presented to Nikolaus Pevsner,* ed. John Summerson, 58–74. London: Allen Lane.

Harris, Paul. 1989. *Life in a Scottish Country House: The Story of A. J. Balfour and Whittingehame House.* Haddington: Whittingehame House Publishing.

Harris, P. R. 1998. *A History of the British Museum Library, 1753–1973.* London: British Library.

Harris, Steven J. 1998. "Long Distance Corporations, Big Sciences, and the Geography of Knowledge." In "The Scientific Revolution as Narrative," ed. Mario Bagioli and Steven J. Harris. Special issue, *Configurations* 6:269–305.

Harrison, Carol E., and Ann Johnson, eds. 2009. "National Identity: The Role of Science and Technology." Special issue, *Osiris* 24.

Harvey, Paul. 2005. *Freedom's Coming: Religious Culture and the Shaping of the South from the Civil War through the Civil Rights Era.* Chapel Hill: University of North Carolina Press.

Harvey, R. D. 1985. "The William Bateson Letters at the John Innes Institution." *Mendel Newsletter* 25:1–11.

Harvey, W. H. 1841. *A Manual of the British Algae.* London: John van Voorst.

———. 1854. "Ceylon Botanic Gardens," In *Ceylon Almanac* [no vol. no.]: 44

Haslett, Caroline. 1931. "Electricity in the Household." *Journal of the Institution of Electrical Engineers* 69:1376–77.

Hasskarl, G. G. H. 1898. *"The Missing Link"; or, The Negro's Ethnological Status. Is He a Descendant of Adam and Eve? Is He the Progeny of Ham? Has He a Soul? What Is His Relation to the White Race? Is He a Subject of the Church or the State, Which?* Chambersburg, PA: Democratic News.

Hawkins, John. 1818. "On a Process of Refining Tin." *Transactions of the Royal Geological Society of Cornwall* 1:201–11.

———. 1822. "On some Advantages Which Cornwall Possesses for the Study of Geology, and on the Use Which May Be Made of Them." *Transactions of the Royal Geological Society of Cornwall* 2:1–13.

———. 1822. "On the Nomenclature of the Cornish Rocks." *Transactions of the Royal Geological Society of Cornwall* 2:145–58.

———. 1832. "Some General Observations on the Structure and Composition of the Cornish Peninsula." *Transactions of the Royal Geological Society of Cornwall* 4:1–20.

Haynes, Clare. 2001. "A 'Natural' Exhibitioner: Sir Ashton Lever and His *Holosphusikon*." *British Journal for Eighteenth-Century Studies* 24:1–14.

Hays, Jo N. 1983. "The London Lecturing Empire." In Inkster and Morrell, *Metropolis and Province*, 91–119.

Head, George. 1836. *Home Tour through the Manufacturing Districts of England, in the Summer of 1835.* London: Murray.

Heckscher, Eli F. 1922. *The Continental System: An Economic Interpretation.* Oxford: Clarendon Press.

Heiton, John. 1861. *The Castes of Edinburgh.* 3rd ed. Edinburgh: William P. Nimmo.

Hellrigel, Mary Ann. 1998. "The Quest to Be Modern: The Evolutionary Adoption of Electricity in the United States, 1880s to 1920s." In *Elektrizität in der Geistesgeschichte*, ed. Klaus Plitzner, 65–86. Bassum: GNT-Verlag.

Henare, Amiria. 2005. *Museums, Anthropology and Imperial Exchange.* Cambridge: Cambridge University Press.

Henke, Christopher R. 2000. "Making a Place for Science: The Field Trial." *Social Studies of Science* 30:483–511.

Henson, Pamela M. 2000. "Spencer Baird's Dream: A U.S. National Museum," In *Cultures and Institutions of Natural History: Essays in the History and Philosophy of Science*, ed. Michael T. Ghiselin and Alan E. Leviton, 101–26. San Francisco: California Academy of Sciences.

———. 2004. "A National Science and a National Museum. *Proceedings of the California Academy of Sciences* 55, supp. 1: 334–57.

Herber, Elmer Charles, ed. 1963. *Correspondence between Spencer Fullerton Baird and Louis Agassiz: Two Pioneer Naturalists.* Washington, DC: Smithsonian Institution Press.

[Herschel, John]. 1832. "Mrs. Somerville's *Mechanism of the Heavens.*" *Quarterly Review* 47:537–59.

Hesse, Carla. 1991. *Publishing and Cultural Politics in Revolutionary Paris, 1789–1810.* Berkeley and Los Angeles: University of California Press.

Hewitt, Martin. 1998. "Ralph Waldo Emerson, George Dawson, and the Control of the Lecture Platform in Mid-Nineteenth-Century Manchester." *Nineteenth-Century Prose* 25:1–23.

———. 2002. "Aspects of Platform Culture in Nineteenth-Century Britain." *Nineteenth Century Prose* 29:1–32.

Higgitt, Rebekah, and Charles W. J. Withers. 2008. "Science and Sociability: Women as Audience at the British Association for the Advancement of Science, 1831–1901." *Isis* 99:1–27.

Hilgartner, Stephen. 1990. "The Dominant View of Popularization: Conceptual Problems, Political Uses." *Social Studies of Science* 20:519–39.

Hill, Kate. 2005. *Culture and Class in English Public Museums, 1850–1914.* Aldershot: Ashgate.

Hilton, Mary, and Jill Shefrin, eds. 2009. *Educating the Child in Enlightenment Britain: Beliefs, Cultures, Practices.* Farnham: Ashgate.

Hinsley, Curtis M. 1981. *Savages and Scientists: The Smithsonian Institution and the Development of American Anthropology, 1846–1910.* Washington: Smithsonian Institution Press.

———. 1992. "Collecting Culture and the Cultures of Collecting: The Lure of the American Southwest, 1880–1915." *Museum Anthropology* 16:12–20.

———. 1994. *The Smithsonian and the American Indian: Making a Moral Anthropology in Victorian America.* Washington: Smithsonian Institution Press. (Orig. pub. 1981.)

Hobson, Edward. 1818. *A Collection of Specimens of British Mosses and Hapaticae, Systematically Arranged with Reference to the Muscologia Britannica, English Botany, &c. &c. &c.* Manchester: M. Wilson.

Hodder, Edwin. 1890. *Sir George Burns, Bart: His Times and Friends.* London: Hodder and Stoughton.

Holland, John. 1837. *The Tour of the Don.* London: Groombridge.

Holt, John Clifford. 1996. *The Religious World of Kirti Sri: Buddhism, Art and Politics in Late Medieval Sri Lanka.* Oxford: Oxford University Press.

Homo. 1815. "On Jameson's Preface to Cuvier's Theory of the Earth." *Philosophical Magazine* 46:225–29.

Hooker, Joseph D. 1853. *Flora Novae-Zelandiae. Introductory Essay.* London: Lovell Reeve.

Hooker, William J., and Thomas Taylor. 1818. *Muscologia Britannica; Containing the Mosses of Great Britain, Systematically Arranged and Described.* London: Longman.

———. 1827. *Muscologia Britannica; Containing the Mosses of Great Britain and Ireland, Systematically Arranged and Described.* 2nd ed. London: Longman.

Hoskins, Janet. 2006. "Agency, Biography and Objects." In *Handbook of Material Culture*, ed. Christopher Tilley, Webb Keane, Susanne Kuechler-Fogden, Michael Rowlands, and Patricia Spyer, 74–84. London: Sage.

Hospitalier, E. 1889. *Domestic Electricity for Amateurs.* Trans. C. J. Wharton. London: E. &. F. N. Spon.

Howard, Jill. 2004. "'Physics and Fashion': John Tyndall and His Audiences in Mid-Victorian Britain." *Studies in History and Philosophy of Science* 35:729–58.

Howarth, Elijah. 1891. "On Some Recent Museum Legislation." *Report of the Proceedings of the Museums Association* 2:121–24.

———. 1892. "Library and Museum Legislation." *Report of the Proceedings of the Museums Association* 3:87–95.

Hubbard, Gardiner Greene. 1897. "Geography." In *The Smithsonian, 1846–1896: The History of the First Half Century*, ed. George Brown Goode, 773–84. Washington, DC: Government Printing Office.

Hudson, John, ed. 1859. *A Complete Guide to the English Lakes, with Minute Directions for Tourists; and Mr Wordsworth's Description of the Scenery of the Country, Etc; Also, Five Letters on the Geology of the Lake District, by the Rev. Professor Sedgwick.* 5th ed. Kendal: Hudson; London: Longman; and London: Whittaker.

Hudson, Kenneth. 1975. *A Social History of Museums: What the Visitors Thought.* London: Macmillan.

Hudson, William. 1778. *Flora Anglica.* 2 vols. 2nd ed. London: For the Author.

———. 1874. *The Life of John Holland of Sheffield Park.* London: Longmans Green.

Hughes, Thomas Parkes. 1983. *Networks of Power: Electrification in Western Society, 1880–1930.* Baltimore, MD: Johns Hopkins University Press.

Hull, John. 1799. *The British Flora, or a Linnean Arrangement of British Plants.* 2 Parts. Manchester: R. and W. Dean.

Hunt, Bruce. 1997. "Doing Science in a Global Empire: Cable Telegraphy and Electrical Physics in Victorian Britain." In *Victorian Science in Context*, ed. Bernard Lightman, 312–33. Chicago: University of Chicago Press.

Huntington, Ellsworth. 1925. *The Character of Races as Influenced by Physical Environment, Natural Selection and Historical Development.* New York: Scribner.

Hurren, Elizabeth T. 2004. "A Pauper Dead-House: The Expansion of the Cambridge Anatomical Teaching School under the Late-Victorian Poor Law, 1870–1914." *Medical History* 48:69–94.

Huxley, Leonard. 1902. *Life and Letters of Thomas H. Huxley.* 2 vols. New York: D. Appleton & Company.

———. 1918. *Life and Letters of Sir Joseph Dalton Hooker.* 2 vols. London: John Murray.

Huxley, Thomas H. 1862. "Professor Huxley and His Critics." *Scotsman*, 24 January.

———.1869. "On the Physical Basis of Life." *Fortnightly Review* 5:129–45.

———. 1896. "Suggestions for a Proposed Natural History Museum in Manchester." *Report of the Proceedings of the Museums Association* 7:126–31.

Hyatt, Alpheus. 1888. "Sketch of the Life and Services to Science of Spencer F. Baird," *Boston Society of Natural History, Proceedings* 24:558–65.

Ilerbaig, Juan. 2002. "Pride in Place: Fieldwork, Geography, and American Field Zoology, 1850–1920." PhD diss., University of Minnesota.

Ingold, Tim. 2000. *The Perception of the Environment: Essays in Livelihood, Dwelling and Skill.* New York: Routledge.

Inkster, Ian. 1997. *Scientific Culture and Urbanisation in Industrialising Britain.* Aldershot: Ashgate.

Inkster, Ian, and Jack Morrell, eds. 1983. *Metropolis and Province: Science in British Culture, 1780–1850.* London: Hutchinson.

Irschick, Eugene. 1994. *Dialogue and History: Constructing South India, 1795–1895.* Berkeley and Los Angeles: University of California Press.

Jacknis, Ira. 1985. "Franz Boas and Exhibits: On the Limitations of the Museum Method in Anthropology." In *Objects and Others: Essays on Museums and Material Culture,* ed. George W. Stocking Jr., 75–111. Madison: University of Wisconsin Press.

Jackson, H. J. 2001. *Marginalia: Readers Writing in Books.* New Haven, CT: Yale University Press.

———. 2005. *Romantic Readers: The Evidence of Marginalia.* New Haven, CT: Yale University Press.

Jacob, Christian. 1996. "Theoretical Aspects of the History of Cartography: Toward a Cultural History of Cartography." *Imago Mundi* 48:191–8.

Jacyna, L. S. 1980. "Science and Social Order in the Thought of A. J. Balfour." *Isis* 71:11–34.

James, Frank A. J. L., ed. 1991–96. *The Correspondence of Michael Faraday.* 3 vols. London: Institution of Electrical Engineers.

———. 2002. "Introduction." In *'The Common Purposes of Life': Science and Society at the Royal Institution of Great Britain,* ed. Frank A. J. L. James, 1–16. Aldershot: Ashgate.

James, Frank A. J. L, and Anthony Peers. 2007. "Constructing Space for Science at the Royal Institution for Great Britain." *Physics in Perspective* 9:130–85.

"James Howden." 1913–14. *Marine Engineer and Naval Architect* 36:165.

[Jameson, Robert]. 1826. "Observations on the Nature and Importance of Geology." *Edinburgh New Philosophical Journal* 1:293–302.

———. 1834. "Reviews." *Edinburgh New Philosophical Journal* 17:46.

"Jameson's Translation of Cuvier's Theory of the Earth." 1814. *British Critic,* n.s., 1:479–92.

Jardine, Nicholas. 2000. "Books, Texts, and the Making of Knowledge." In Frasca-Spada and Jardine, *Books and the Sciences in History,* 393–407.

Jardine, Nicholas, James A. Secord, and Emma C. Spary, eds. 1996. *Cultures of Natural History.* Cambridge: Cambridge University Press.

Jeal, Timothy. 2007. *Stanley: The Impossible Life of Africa's Greatest Explorer.* London: Faber.

John, A. H. 1959. *A Liverpool Merchant House: Being the History of Alfred Booth and Company, 1863–1958.* London: Allen & Unwin.

Johns, Adrian. 1998. *The Nature of the Book: Print and Knowledge in the Making.* Chicago: University of Chicago Press.

Johnson, Kristin. 2005. "Type-Specimens of Birds as Sources for the History of Ornithology." *Journal of the History of Collections* 17:173–88.

[Johnson, Joseph]. 1788. *Prospectus of the "Analytical Review"; or, a New Literary Journal, on an Enlarged Plan.* [London: Joseph Johnson].

Johnston, Alexander Keith. 1848–50. *The Physical Atlas: A Series of Maps and Notes Illustrating the Geographical Distribution of Natural Phenomena.* Edinburgh: W. and A. K. Johnston.

Johnston, Harry Hamilton. 1904. *The Nile Quest: A Record of the Exploration of the Nile and Its Basin.* London: Lawrence and Bullen.

Johnston, William. 1975. "Preface." In *Science and Revelation: A Series of Lectures in Reply to the Theories of Tyndall, Huxley, Darwin, Spencer.* Belfast: William Mullan.

Jones, Daniel H. 1997. "The Panorama from Point Sublime: John Wesley Powell's 'Religion of Science' and the Intellectual Origins of His Arid Land Reforms." PhD diss., Indiana University.

Jones, Greta. 1997. "Catholicism, Nationalism and Science." *Irish Review* 20:40–61.

———. 2001. "Scientists against Home Rule." In *Defenders of the Union: A Survey of British and Irish Unionism Since 1801*, ed. D. George Boyce and Alan O'Day, 188–208. London: Routledge.

———. 2004. "Darwinism in Ireland." In *Science and Irish Culture*, ed. David Attis, 1:115–37. Dublin: Royal Dublin Society.

Jones, Peter Murray. "The *Tabula Medicine: An Evolving Encyclopaedia*." In *English Manuscript Studies, 1100–1700*, vol. 14, *Regional Manuscripts, 1200–1700*, ed. A. S. G. Edwards, 60–85. London: British Library, 2008.

Kaplan, Flora S., ed. 1994. *Museums and the Making of "Ourselves": The Role of Objects in National Identity.* London: Leicester University Press.

Kavanagh, Gaynor. 2000. *Dream Spaces: Memory and the Museum.* Leicester: Leicester University Press.

Keeling, A. M. 2007. "Charting Marine Pollution Science: Oceanography on Canada's Pacific Coast, 1938–1970." *Journal of Historical Geography* 33:403–28.

Keighren, Innes M. 2006. "Bringing Geography to the Book: Charting the Reception of *Influences of Geographic Environment*." *Transactions of the Institute of British Geographers* 31:525–40.

Keller, Suzanne B. 1998. "Sections and Views: Visual Representations in Eighteenth-Century Earthquake Studies." *British Journal for the History of Science* 31:129–59.

Kelly, John. 2002. "Charles Rau: Developments in the Career of a Nineteenth-Century German-American Archaeologist." In *New Perspectives on the Origins of American Archaeology*, ed. David L. Browman and Stephen Williams, 117–32. Tuscaloosa: University of Alabama Press.

Kemp, Martin. 1990. *The Science of Art: Optical Themes in Western Art from Brunelleschi to Seurat.* New Haven, CT: Yale University Press.

———. 2000. *Visualisations: The Nature Book of Art and Science.* Oxford: Oxford University Press.

Kennedy, Alasdair. 2008. "In Search of the 'True Prospect': Making and Knowing the Giant's Causeway as a Field Site in the Seventeenth Century." *British Journal for the History of Science* 41:19–42.

Kennedy, Dane. 1996. *Magic Mountains: Hill Stations and the British Raj*. Berkeley and Los Angeles: University of California Press.

———. 2005. *The Highly Civilized Man: Richard Burton and the Victorian World*. London: Harvard University Press.

———. 2007. "British Exploration in the Nineteenth Century: A Historiographical Survey." *History Compass* 5/6:1879–1900.

Kenny, Judith. 1995. "Climate, Race and Imperial Authority: The Symbolic Landscape of the British Hill Stations in India." *Annals of the Association of American Geographers* 85:694–714.

Keppie, Lawrence. 2007. *William Hunter and the Hunterian Museum in Glasgow, 1807–2007*. Edinburgh: Edinburgh University Press.

Kerr, Robert, trans. 1813. *Essay on the Theory of the Earth. Translated from the French of M. Cuvier, with Mineralogical Notes and an Account of Cuvier's Geological Discoveries, by Professor Jameson*. Edinburgh: William Blackwood; 2nd ed., London: John Murray and Robert Baldwin, 1815.

[King, Richard John]. 1858. *A Handbook for Travellers in Kent and Sussex*. London: Murray.

———. 1858. *A Handbook for Travellers in Surrey, Hampshire and the Isle of Wight*. London: Murray.

Kingsley, Charles. 1891. *Charles Kingsley: His Letters and Memories of His Life*. London: Kegan Paul.

Kirk, John Foster. 1899. "Becker, Bernard Henry." In *A Supplement to Allibone's Critical Dictionary of English Literature and British and American Authors*, 1:118. Philadelphia: J. B. Lippincott.

Kitteringham, Guy Stuart. 1981. "Studies in the Popularisation of Science in England, 1800–30." PhD diss., University of Kent at Canterbury.

Knell, Simon J. 2000. *The Culture of English Geology, 1815–1851: A Science Revealed Through Its Collecting*. Aldershot: Ashgate.

Knight, Charles. 1845. *Old England: A Pictorial Museum of Regal, Ecclesiastical, Baronial, Municipal and Popular Antiquities*. 2 vols. London: Knight.

———. ed. 1847–49. *The Land We Live In: A Pictorial and Literary Sketch-Book of the British Empire*. 4 vols. London: Knight.

Knight, David. 2002. "Establishing the Royal Institution: Rumford, Banks and Davy." In *'The Common Purposes of Life': Science and Society at the Royal Institution of Great Britain*, ed. Frank A. J. L. James, 97–118. Aldershot: Ashgate.

———. 2002. "Scientific Lectures: a History of Performance." *Interdisciplinary Science Reviews* 27:217–24.

Knorr-Cetina, Karin. 1999. *Epistemic Cultures: How the Sciences Make Knowledge*. Cambridge, MA: Harvard University Press.

Knorr-Cetina, Karin, and M. Mulkay, eds. 1983. *Science Observed: Perspectives on the Social Study of Science*. London: Sage.

Knox, Robert. 1681. *An Historical Relation of the Island of Ceylon*. London: R. Chiswell.

Koerner, Lisbet. 1999. *Linnaeus: Nature and Nation*. Cambridge, MA: Harvard University Press.

Kohler, Robert E. 2002. *Landscapes and Labscapes: Exploring the Lab-Field Border in Biology*. Chicago: University of Chicago Press.

———. 2006. *All Creatures: Naturalists, Collectors, and Biodiversity, 1850–1950.* Princeton, NJ: Princeton University Press.

———. 2008. "Lab History. Reflections." *Isis* 99:761–68.

Kohlstedt, Sally Gregory. 1971. "A Step toward Scientific Self-Identity in the United States: The Failure of the National Institute, 1844." *Isis* 62:339–62.

———. 1983. "Australian Museums of Natural History: Public Priorities and Scientific Initiatives in the Nineteenth Century. *Historical Records of Australian Science* 5:1–29.

———. 2006. "Nature by Design: Masculinity and Animal Display in Nineteenth-Century America." In Shteir and Lightman, *Figuring It Out*, 110–39.

———. 2008. "Otis T. Mason's Museum Tour: Observation, Exchange, and Standardization in Public Museums, 1889." *Museum History Journal* 2:181–208.

Kolodney, Annette. 1975. *The Lay of the Land: Metaphor as Experience and History in American Life and Letters.* Chapel Hill: University of North Carolina Press.

Kreilkamp, Ivan. 2005. *Voice and the Victorian Storyteller.* Cambridge: Cambridge University Press.

Lacey, Michael. 1979. "The Mysteries of Earth-Making Dissolve: A Study of Washington's Intellectual Community and the Origins of American Environmentalism in the Late Nineteenth Century." PhD diss., George Washington University.

Lack, H. Walter. 2007. "The Plant Self Impressions Prepared by Humboldt and Bonpland in Tropical America." In *Objects in Transition, An Exhibition at the Max Planck Institute for the History of Science, Berlin, August 16—September 2*, ed. Gianenrico Bernasconi, Anna Maerker, and Susanne Pickert, 46–49. Berlin : Max-Planck-Institut fur Wissenschaftsgeschichte.

"Lamarck on Invertebrate Animals." 1820. *Edinburgh Monthly Review* 3:403–18.

"Lamarck's *Natural History of Invertebral Animals*." 1822. *Monthly Review* 99:485–92.

"Lamarck *Philosophie Zoologique*." 1812. *London Medical Review* 5:312–28.

Land Use Consultants. 1987. *An Inventory of Gardens and Designed Landscapes in Scotland.* 5 vols. Perth: Countryside Commission for Scotland.

Landsborough, David. 1849. *A Popular History of British Sea-Weeds.* London: Reeve, Benham, and Reeve.

"La Place on Celestial Mechanics." 1804. *Critical Review*, 3rd ser., 1 (1804): 531–41.

"La Place's Exposition of the System of the World." 1796. *Critical Review* , 2nd. ser. 18:504–11.

"Laplace on the Mechanism of the Celestial Bodies." 1799. *Critical Review*, 2nd ser., 27:513–21.

"Laplace's System of the World." 1809. *Eclectic Review* 5:881–95.

Larson, James B. 1994. *Interpreting Nature: The Science of Living Form from Linnaeus to Kant.* Baltimore, MD: Johns Hopkins University Press.

Latour, Bruno. 1987. *Science in Action: How to Follow Scientists and Engineers Through Society.* Cambridge, MA: Harvard University Press.

———. 1990. "Drawing Things Together." In *Representation in Scientific Practice*, ed. Michael Lynch and Steve Woolgar, 19–68. Cambridge, MA: MIT Press.

———. 2005. *Reassembling the Social: An Introduction to Actor-Network Theory.* Oxford: Oxford University Press.

Lavoisier, Antoine-Laurent. 1790. *Elements of Chemistry, in a New Systematic Order, Containing All the Modern Discoveries.* Edinburgh: Printed for William Creech, and sold in London by G. G. and J. J. Robinsons.

———. 1793. *Elements of Chemistry, in a New Systematic Order, Containing All the Modern Discoveries.* 2nd ed. Edinburgh: Printed for William Creech, and sold in London by G. G. and J. J. Robinsons.

———. 1796. *Elements of Chemistry, in a New and Systematic Order, Containing All the Modern Discoveries.* 3rd ed. Edinburgh: printed for William Creech, and sold in London by G. G. and J. Robinsons and T. Kay.

"Lavoisier's *Traité élémentaire de chimie.*" 1789. *Critical Review* 67:531–35.

"Lecture on Sir George Harvey." 1868. *Scotsman,* 1 February.

Lee, Louise. 2008. "Voicing, Devoicing and Self-Silencing: Charles Kingsley's Stuttering Christian Manliness." *Journal of Victorian Culture* 13:1–17.

Leebody, John R. 1872. "The Theory of Evolution and Its Relations to Religious Thought." *British and Foreign Evangelical Review* 21:1–35.

———. [An Irish Graduate]. 1872. "The Irish University Question." *Fraser's Magazine* 10:55–64.

———. 1876. "The Scientific Doctrine of Continuity." *British and Foreign Evangelical Review* 25:742–74.

———. 1890. "Evolution." *Witness,* 10 October, 3.

———. 1915. *A Short History of McCrea Magee College, Derry, During Its First Fifty Years.* Londonderry: Printed at "Derry Standard" Office.

Lefebvre, Henri. 1991. *The Production of Space.* Trans. Donald Nicholson-Smith. Oxford: Blackwell.

Lelande, Jerome de. 1801. "History of Astronomy for the Year 1800." *Philosophical Magazine* 9:1–15.

Le Mesurier, C. J. R. 1893. *Manual of the Nuwara Eliya District of the Central Province of Ceylon.* Colombo, Sri Lanka: G. J. A. Skeen.

Lemon, Charles. 1841. "Notes on the Agricultural Produce of Cornwall." *Journal of the Statistical Society of London* 4:197–208.

Leslie, George. 1881. "The Invertebrate Fauna of the Firth of Forth." *Proceedings of the Royal Physical Society of Edinburgh* 4:68–97.

"Letters from M. Lavoisier to Dr. Black." 1871. *Report of the Fortieth Annual Meeting of the British Association for the Advancement of Science* (London: John Murray), 189–92.

Levtzion, Nehemia. 2000. "Arab Geographers, the Nile, and the History of Bilad Al-Sudan." In *The Nile: Histories, Cultures, Myths,* ed. Haggai Erlich and Israel Gershoni, 71–6. Boulder, CO: Lynne Reinner.

Lewis, Geoffrey. 1989. *For Instruction and Recreation: A Centenary History of the Museums Association.* London: Quiller.

The Life of Hugh Miller: A Sketch for Working Men. 1862. London: Samuel W. Partridge.

Lightman, Bernard. 2000. "The Visual Theology of Victorian Popularizers of Science: From Reverent Eye to Chemical Retina." *Isis* 91:651–60.

———. 2007. "Lecturing in the Spatial Economy of Science." In Fyfe and Lightman, *Science in the Marketplace,* 97–132.

———. 2007. *Victorian Popularizers of Science: Designing Nature for New Audiences.*
    Chicago: University of Chicago Press.

Lindley, John. 1855. "Preface." In *The Ferns of Great Britain and Ireland*, by Thomas
    Moore, Nature-Printed by Henry Bradbury. London: Bradbury and Evans.

Lindsay, Debra. 1993. *Science in the Subarctic: Trappers, Traders, and the Smithsonian
    Institution.* Washington, DC: Smithsonian Institution Press.

———. 1998. "Intimate Inmates: Wives, Households, and Science in Nineteenth-Century
    America." *Isis* 89:631–53.

Lindsay, W. S. 1874–76. *History of Merchant Shipping and Ancient Commerce.* 4 vols.
    London: Sampson Low, Marston, Low, and Searle.

Lipset, David. 1980. *Gregory Bateson: The Legacy of a Scientist.* Englewood Cliffs, NJ:
    Prentice-Hall.

"Liverpool Polytechnic Society." 1859. *Engineer* 7:381.

Livingstone, David. 1863–64. "Letters from the Zambesi to Sir R. I. Murchison, and
    (the Late) Admiral Washington." *Proceedings of the Royal Geographical Society of
    London* 8:256–63.

Livingstone, David, and Charles Livingstone. 1865. *Narrative of an Expedition to the
    Zambesi and Its Tributaries, and of the Discovery of the Lakes Shirwa and Nyassa,
    1858–1864.* London: John Murray.

Livingstone, David N. 1992. "Darwinism and Calvinism: The Belfast-Princeton Connec-
    tion." *Isis* 83:408–28.

———. 1992. *The Geographical Tradition: Episodes in the History of a Contested Enter-
    prise.* Oxford: Blackwell.

———. 1995. "The Spaces of Knowledge: Contributions towards a Historical Geography
    of Science." *Environment and Planning D: Society and Space* 13:5–34.

———. 1997. "Darwin in Belfast: The Evolution Debate." In *Nature in Ireland: A Scien-
    tific and Cultural History*, ed. John W. Foster, 387–408. Dublin: Lilliput Press.

———. 1999. "Tropical Climate and Moral Hygiene: Anatomy of a Victorian Debate."
    British Journal for the History of Science 32:93–100.

———. 2003. *Putting Science in Its Place: Geographies of Scientific Knowledge.* Chicago:
    University of Chicago Press.

———. 2003. "Science, Religion and the Geography of Reading: Sir William Whitla and
    the Editorial Staging of Isaac Newton's Writings on Biblical Prophecy." *British Jour-
    nal for the History of Science* 36:27–42.

———. 2005. "Science, Text and Space: Thoughts on the Geography of Reading." *Trans-
    actions of the Institute of British Geographers* 35:391–401.

———. 2005. "Text, Talk and Testimony: Geographical Reflections on Scientific Habits:
    An Afterword," *British Journal for the History of Science* 38:93–100.

———. 2007. "Science, Site and Speech: Scientific Knowledge and the Spaces of Rheto-
    ric." *History of the Human Sciences* 20:71–98.

———. 2008. *Adam's Ancestors: Race, Religion and the Politics of Human Origins.* Balti-
    more, MD : Johns Hopkins University Press.

Livingstone David N., and Charles W. J. Withers, eds. 1999. *Geography and Enlighten-
    ment.* Chicago: University of Chicago Press.

———, eds. 2005. *Geography and Revolution.* Chicago: University of Chicago Press.

Lorimer, Hayden. 2000. "Guns, Game and the Grandee: The Cultural Politics of Deer-stalking in the Scottish Highlands." *Ecumene* 7:403–31.

Loughney, Claire. 2005. "Colonialism and the Development of the English Provincial Museum, 1823–1914." PhD diss., Newcastle University.

Lowe, E. Ernest. 1928. *A Report on American Museum Work*. Edinburgh: Carnegie United Kingdom Trust.

Lowe, Phillip. 1976. "Amateurs and Professionals: The Institutional Emergence of British Plant Ecology." *Journal of the Society for the Bibliography of Natural History* 7:517–35.

———. 1981. "The British Association and the Provincial Public." In MacLeod and Collins, *Parliament of Science*, 118–44.

Lubbock, Basil. 1946. *The China Clippers*. Glasgow: Brown, Son & Ferguson.

———. 1948. *The Colonial Clippers*. Glasgow: Brown, Son & Ferguson.

Macalister, Alexander. 1870. "Review of Works on Life and Organisation." *Dublin Quarterly Journal of Medical Science* 50:131–32.

———. 1871. "The Body—The Temple of God." *Plain Words* 9 (May 1): 137–40.

———. 1871. Review of *The Descent of Man, and Selection in Relation to Sex*, by Charles Darwin. *Dublin Quarterly Journal of Medical Science* 52:133–52.

———. 1882. *Evolution in Church History*. Dublin: Hodges, Figgis.

———. 1886. *Man, Physiologically Considered*. Present-Day Tracts No. 38. London: Religious Tract Society.

McCarthy, Conal. 2007. *Exhibiting Maori: A History of Colonial Cultures of Display*. Wellington, New Zealand: Te Papa Museum Press.

McComb's Presbyterian Almanack, and Christian Remembrancer for 1875. 1875. Belfast: James Cleeland.

McClellan III, James E. 1979. "The Scientific Press in Transition: Rozier's Journal and the Scientific Societies in the 1770s." *Annals of Science* 36:425–49.

McDougall, Warren. 2002. "Charles Elliot's Medical Publications and the International Book Trade." In *Science and Medicine in the Scottish Enlightenment*, ed. Charles W. J. Withers and Paul B. Wood, 215–54. East Linton: Tuckwell Press.

MacDonald, Helen. 2006. *Human Remains: Dissection and Its Histories*. London: Yale University Press.

Macdonald, William. 1862. "The Origin of Species." *Scotsman*, 29 January.

MacGregor, Arthur. 2007. *Curiosity and Enlightenment: Collectors and Collections from the Sixteenth to the Nineteenth Century*. New Haven, CT: Yale University Press.

MacGregor, David R. 1961. *The China Bird: The History of Captain Killick and the Firm He Founded, Killick Martin and Company*. London: Chatto and Windus.

MacKenzie, John M. 1995. "The Provincial Geographical Societies in Britain, 1884–1914." In *Geography and Imperialism, 1820–1940*, ed. Morag Bell, Robin Butlin, and Michael Heffernan, 93–124. Manchester: Manchester University Press.

———. 2009. *Museums and Empire: Natural History, Human Cultures and Colonial Identities*. Manchester: Manchester University Press.

McKie, Douglas. 1949. "Antoine Laurent Lavoisier, F.R.S., 1743–1794." *Notes and Records of the Royal Society of London* 7:1–41.

———. 1965. "Introduction to Dover Edition." In *Elements of Chemistry in a New*

Systematic Order, Containing All the Modern Discoveries, by Antoine-Laurent Lavoisier, v–xxxi. New York: Dover Publications.

McKinney, S. B. G. 1886. *Disease and Sin: A New Text-Book for Medical and Divinity Students.* London: Wyman & Sons.

———. 1888. *The Science and Art of Religion.* London: Kegan Paul, Trench.

———. 1890. *The Abolition of Suffering.* London: Elliot Stock.

———. 1891. *The Revelation of the Trinity.* Edinburgh and London: Oliphant, Anderson & Ferrier.

———. 1898. *The Origin and Nature of Man.* London: Hutchinson.

McOuat, Gordon R. 1996. "Species, Rules and Meaning: The Politics of Language and the Ends of Definitions in Nineteenth-Century Natural History." *Studies in History and Philosophy of Science* 27:473–519.

MacLeod, Roy. 1983. "Whigs and Savants: Reflections on the Reform Movement in the Royal Society, 1830–48." In Inkster and Morrell, *Metropolis and Province,* 55–90.

———. 1994. "Embryology and Empire: The Balfour Students and the Quest for Intermediate Forms in the Laboratory of the Pacific." In *Darwin's Laboratory: Evolutionary Theory and Natural History in the Pacific,* ed. Roy MacLeod and Philip F. Rehbock, 140–64. Honolulu: University of Hawaii Press.

MacLeod, Roy, and Peter Collins, eds. 1981. *The Parliament of Science: The British Association for the Advancement of Science.* Northwood: Science Reviews.

MacLeod, Roy, J. R. Friday, and C. Gregor. 1975. *The Corresponding Societies of the British Association for the Advancement of Science 1883–1929: A Survey of Historical Records, Archives and Publications.* London: Mansell.

MacLeod, Roy, and Philip F. Rehbock, eds. 1988. *Nature in Its Greatest Extent: Western Science in the Pacific.* Honolulu: University of Hawaii Press.

MacLeod, Suzanne, ed. 2005. *Reshaping Museum Space: Architecture, Design, Exhibitions.* London: Routledge.

MacQueen, James. 1850. "Notes on the Present State of Geography of Some Parts of Africa." *Journal of the Royal Geographical Society* 20:235–52.

———. 1856. "Notes on the Geography of Central Africa, from the Researches of Livingstone, Monteiro, Graça, and Others." *Journal of the Royal Geographical Society* 26:109–30.

———. 1858–59. "Observations on the Geography of Central Africa." *Proceedings of the Royal Geographical Society of London* 3:208–14.

Majendie, Ashhurst. 1818. "A Sketch of the Geology of the Lizard District." *Transactions of the Royal Geological Society of Cornwall* 1:2–7.

Mallet, John Lewis. 1890. *John Lewis Mallet: An Autobiographical Retrospect of the First Twenty-Five Years of His Life.* Windsor: Thomas E. Luff.

Mandler, Peter. 1997. *The Fall and Rise of the Stately Home.* New Haven, CT: Yale University Press.

———. 1999. "'The Wand of Fancy': The Historical Imagination of the Victorian Tourist." In *Material Memories,* ed. Marius Kwint, Christopher Breward, and Jeremy Aynsley, 125–42. Oxford: Berg.

Manguel, Alberto. 1997. *A History of Reading.* London: Flamingo.

Marie, Jennifer. 2004. "The Importance of Place: A History of Genetics in 1930s Britain," PhD diss., University College London.

Markham, S. Frank. 1938. *A Report on the Museums and Art Galleries of the British Isles (Other Than the National Museums)*. Edinburgh: Constable.

Marsden, Ben. 1998. "Blowing Hot and Cold: Reports and Retorts on the Status of the Air-Engine as Success or Failure." *History of Science* 36:373–420.

———. 2008. "The Administration of the 'Engineering Science' of Naval Architecture at the British Association for the Advancement of Science, 1831–1872." *Yearbook of European Administrative History* 20:67–94.

Marsden, Ben, and Crosbie Smith. 2005. *Engineering Empires: A Cultural History of Technology in Nineteenth-Century Britain*. Basingstoke: Palgrave Macmillan.

Marshall, Henry. 1823. "Contribution to a Natural and Economical History of the Coconut Tree." In *Memoirs of the Wernerian Natural History Society* 5: 107–43.

Martin, G. H. 2003. "Sir George Samuel Measom (1818–1901) and His Railway Guides." In *The Impact of the Railway on Society in Britain: Essays in Honour of Jack Simmons*, ed. A. K. B. Evans and J. V. Gough, 225–40. Aldershot: Ashgate.

Martin, William Todd. 1887. *The Evolution Hypothesis: A Criticism of the New Cosmic Philosophy*. Edinburgh: James Gemmell.

Martins, Luciana, and Felix Driver. 2005. "'The Struggle for Luxuriance': William Burchell Collects Tropical Nature." In Driver and Martins, *Tropical Visions*, 59–76.

Marston, Sallie. 2000. "The Social Construction of Scale." *Progress in Human Geography* 24:219–42.

Marston, Sallie, John Paul Jones III, and Keith Woodward. 2005. "Human Geography without Scale." *Transactions of the Institute of British Geographers* 30:416–32.

Marvin, Carolyn. 1988. *When Old Technologies Were New: Thinking About Electric Communication in the Late Nineteenth Century*. New York: Oxford University Press.

Marvin, Charles. 1886. *The Moloch of Paraffin*. London: Privately printed. Copy in Institution of Engineering and Technology Archives, London.

Mason, Otis T. 1874. "Leipsic Museum of Ethnology." *Smithsonian Institution Annual Report for 1873*, 369–410. Washington, DC: Government Printing Office.

———. 1875. *Ethnological Directions Relative to the Indian Tribes of the United States*. Washington, DC: Government Printing Office.

———. 1900. "Traps of the Amerinds—A Study in Psychology and Invention." *American Anthropologist* 2:657–75.

———. 1902. *Aboriginal Indian Basketry*. Washington, DC: Government Printing Office.

Masson, David. 1865. "Dead Men Whom I Have Known; or, Recollections of Three Cities." *Macmillan's Magazine* 12:74–90.

Mavor, William. 1800. *The Lady's and Gentleman's Botanical Pocket Book; Adapted to Withering's Arrangement of British Plants. Intended to Facilitate and Promote the Study of Indigenous Botany*. London: Vernor and Hood.

Mayhew, Robert. 2007. "Materialist Hermeneutics, Textuality and the History of Geography: Print Spaces in British Geography, c.1500–1900." *Journal of Historical Geography* 33:466–88.

Measom, George. 1861. *The Official Illustrated Guide to the Bristol and Exeter, North and South Devon, Cornwall, and South Wales Railways.* 2nd ed. London: Griffin, Bohn.

Meinig, Donald. 1998. *The Shaping of America: A Geographic Perspective on Five Hundred Years of History.* 3 vols. New Haven, CT: Yale University Press.

Melville, Richard V. 1995. *Towards Stability in the Names of Animals: A History of the International Commission on Zoological Nomenclature 1895–1995.* London: International Trust for Zoological Nomenclature.

Merriam, C. Hart. 1892. "The Geographical Distribution of Life in North America with Special Reference to Mammalia." *Proceedings of the Biological Society of Washington* 7:1–64.

Merriman, Nick. 2008. "Museum Collections and Sustainability." *Cultural Trends* 17:3–21.

Miall, Louis Compton, and F. Greenwood. 1878. "Anatomy of the Indian Elephant." *Journal of Anatomy and Physiology* 12:261–87.

Miers, Henry. 1928. *A Report on the Public Museums of the British Isles (Other Than the National Museums).* Edinburgh: Constable.

Mill, Hugh R. 1887. "Report to Council," *Scottish Geographical Magazine* 3:521–30.

———. 1889. "Report to Council on the British Association Meeting at Newcastle, 1889," *Scottish Geographical Magazine* 5:606–8.

Miller, W. Addis. 1949. *The "Philosophical": A Short History of the Edinburgh Philosophical Institution.* Edinburgh: C. J. Cousland and Sons.

Miller, David Philip. 1983. "Between Hostile Camps: Sir Humphry Davy's Presidency of the Royal Society of London, 1820–1827." *British Journal for the History of Science* 16:1–47.

———. 1986. "The Revival of the Physical Sciences in Britain, 1815–1840." *Osiris* 2:107–34.

———. 1996. "Joseph Banks, Empire and 'Centers of Calculation' in Late Hanoverian London." In Miller and Reill, *Visions of Empire,* 21–37.

Miller, David Philip, and Peter Hans Reill, eds. 1996. *Visions of Empire: Voyages, Botany, and Representations of Nature.* Cambridge: Cambridge University Press.

Miller, Hugh. 1863. *Sketch-Book of Popular Geology: Being a Series of Lectures Delivered before the Philosophical Institution of Edinburgh.* Edinburgh: Adam and Charles Black.

———. 1857. *The Testimony of the Rocks.* Edinburgh: Thomas Constable.

Mitchell, Brian R. 1988. *British Historical Statistics.* Cambridge: Cambridge University Press.

Mitchison, Rosalind. 1959. "The Old Board of Agriculture (1793–1822)." *English Historical Review* 74:41–69.

Mitman, Gregg. 1999. *Reel Nature: America's Romance with Wildlife on Film.* Cambridge, MA: Harvard University Press.

Moir, Esther A. L. 1964. *The Discovery of Britain: The English Tourist, 1540–1840.* London: Routledge.

Molho, Anthony, and Gordon S. Wood. 1998. *Imagined Histories: American Historians Interpret the Past.* Princeton, NJ: Princeton University Press.

Moody, T. W. 1958. "The Irish University Question of the Nineteenth Century." *History* 43:90–109.

Moon, Alexander. 1821. *Catalogue of the Indigenous and Exotic Plants Growing in Cey-lon*. Colombo: Wesleyan Press.

Moore, James. 2005. "Revolution of the Space Invaders: Darwin and Wallace on the Geography of Life." In Livingstone and Withers, *Geography and Revolution*, 106–32.

Moresi, Michele Alicia Gates. 2004. "Exhibiting Race, Creating Nation: Representation of Black History and Culture at the Smithsonian Institution, 1895–1976." PhD diss., George Washington University.

Morrell, Jack B. 1971. "Professors Robison and Playfair, and the 'Theophobia Gallica': Natural Philosophy, Religion and Politics in Edinburgh, 1769–1815." *Notes and Records of the Royal Society of London* 26:43–63.

———. 1983. "Economic and Ornamental Geology: the Geological and Polytechnic Society of the West Riding of Yorkshire, 1837–53." In Inkster and Morrell, *Metropolis and Province*, 231–56.

Morrell, Jack, and Arnold Thackray. 1981. *Gentlemen of Science: Early Years of the British Association for the Advancement of Science*. Oxford, Clarendon.

Morton, Graeme. 1999. *Unionist Nationalism: Governing Urban Scotland, 1830–1860*. East Linton: Tuckwell Press.

Morus, Iwan Rhys. 1998. *Frankenstein's Children: Electricity, Exhibition and Experiment in Early Nineteenth-Century London*. Princeton, NJ: Princeton University Press.

———. 2006. "Seeing and Believing Science." *Isis* 97:101–10.

Morus, Iwan, Simon Schaffer, and James Secord. 1992. "Scientific London." In *London—World City, 1800–1840*, ed. Celina Fox, 129–42. New Haven, CT: Yale University Press.

Moseley, H. N. 1882. "Francis Maitland Balfour." *Fortnightly Review* 32:568–80.

Moser, Stephanie. 2006. *Wondrous Curiosities: Ancient Egypt at the British Museum*. Chicago: University of Chicago Press.

"Mr Lowe MP on Education." 1867. *Scotsman*, 2 November.

"Mr Moncrieff MP on Modern Scientific Speculations." 1868. *Scotsman*, 14 January.

Muensterberger, Werner. 1994. *Collecting: An Unruly Passion. Psychological Perspectives*. Princeton, NJ: Princeton University Press.

[Muirhead, Lockhart]. 1811–13. "Lamarck's *Zoological Philosophy*." *Monthly Review*, 2nd ser., 65 (1811):473–84; 70 (1813): 481–90.

[———?]. 1819–20. "Lamarck's *Nat. Hist. of Animals without Vertebræ*." *Monthly Review* 90 (1819):485–98; 91 (1820): 512–20.

Mumby, F. A. 1934. *The House of Routledge, 1834–1934, with a History of Kegan Paul, Trench, Trubner and Other Associated Firms*. London: Routledge.

Murchison, Roderick I. 1863. "Presidential Address." *Journal of the Royal Geographical Society* 33:113–192.

———. 1864. "Geography and Ethnology." *Report of the Thirty-Third Meeting of the British Association for the Advancement of Science*, 122–23. London: John Murray.

———. 1865. "Transactions of the Sections: Geography and Ethnology." *Report of the Thirty-Fourth Annual Meeting of the British Association for the Advancement of Science*, 130–35. London: John Murray.

Myers, Fred R., ed. 2001. *The Empire of Things: Regimes of Value and Material Culture*. Oxford: Currey.

Nadis, Fred. 2005. *Wonder Shows: Performing Science, Magic, and Religion in America.* New Brunswick, NJ: Rutgers University Press.

Nangle, Benjamin Christie. 1955. *The Monthly Review, Second Series, 1790–1815: Indexes of Contributors and Articles.* Oxford: Clarendon Press.

Natural History Society of Northumberland Durham and Newcastle-upon-Tyne. 1847. *Report.* Newcastle-upon-Tyne: Private circulation.

Naylor, Simon. 2002. "The Field, the Museum and the Lecture Hall: The Space of Natural History in Victorian Cornwall." *Transactions of the Institute of British Geographers* 27:494–513.

———. 2005. "Introduction: Historical Geographies of Science—Places, Contexts, Cartographies." *British Journal for the History of Science* 38:1–12.

———. 2010. *Regionalizing Science: Placing Knowledges in Victorian England.* London: Pickering and Chatto.

"The Negroland of the Arabs by W. D. Colley [Review]." 1842. *Journal of the Royal Geographical Society* 12:120–5.

Nell, Andreas. 1953. "Some Trees and Plants Mentioned in the Mahavamsa." *Ceylon Historical Journal* 2:258–64.

Nelson, Clifford M. 1999. "Toward a Reliable Geologic Map of the United States, 1803–1993." In *Surveying the Record: North American Exploration to 1930*, ed. Edward C. Carter II, 51–74. Memoirs of the American Philosophical Society. Philadelphia, PA: American Philosophical Society.

Nelson, Gareth. 1978. "From Candolle to Croizat: Comments on the History of Biogeography." *Journal of the History of Biology* 11:269–305.

"Sir George Watson, Philanthropist, Dies." 1930. *New York Times*, 14 July, 17.

Newall, Peter. 2004. *Orient Line: A Fleet History.* Preston: Ships in Focus.

Newton, Alfred. 1878. *Zoology.* London: Society for Promoting Christian Knowledge.

Nicol, Stuart. 2001. *MacQueen's Legacy: A History of the Royal Mail Line.* 2 vols. Brimscombe Port Stroud: Tempus.

Nicolson, Malcolm. 1987. "Alexander von Humboldt, Humboldtian Science, and the Origins of the Study of Vegetation." *History of Science* 25:169–94.

———. 1996. "Humboldtian Plant Geography after Humboldt: The Link to Ecology." *British Journal for the History of Science* 29:289–310.

Noll, Mark A. 1989. *Princeton and the Republic, 1768–1822: The Search for a Christian Enlightenment in the Era of Samuel Stanhope Smith.* Princeton, NJ: Princeton University Press.

———. 1995. "The Rise and Long Life of the Protestant Enlightenment in America." In *Knowledge and Belief in America: Enlightenment Traditions and Modern Religious Thought*, ed. William M. Shea and Peter A. Huff, 88–124. New York: Cambridge University Press.

"Notes." 1880. *Telegraphic Journal and Electrical Review* 8:54.

Nye, David. 1990. *Electrifying America: Social Meanings of a New Technology.* Cambridge, MA: MIT Press.

Nyhart, Lynn K. 1990. "Civic and Economic Zoology in Nineteenth-Century Germany: The 'Living Communities' of Karl Mobius." *Isis* 89:605–30.

————. 1995. *Biology Takes Form: Animal Morphology and the German Universities, 1800–1900.* Chicago: University of Chicago Press.

————. 1996. "Natural History and the 'New' Biology." In Jardine, Secord, and Spary, *Cultures of Natural History,* 426–43.

————. 2004. "Science, Art, and Authenticity in Natural History Displays." In *Models: The Third Dimension of Science,* ed. Soraya de Chadarevian and Nick Hopwood, 307–35. Stanford, CA: Stanford University Press.

Obeyesekere, Gananath. 1963. "The Great Tradition and the Little Tradition in the Perspective of Sinhalese Buddhism." *Journal of Asian Studies* 22:139–53.

O'Connor, Ralph. 2007. *The Earth on Show: Staging Prehistoric Worlds for the British Public, 1802–1856.* Chicago: University of Chicago Press.

Ogborn, Miles. 2007. *Indian Ink: Script and Print in the Making of the English East India Company.* Chicago: University of Chicago Press.

Ogilvie, Brian W. 2006. *The Science of Describing: Natural History of Renaissance Europe.* Chicago: University of Chicago Press.

Olby, Robert. 1989. "The Dimensions of Scientific Controversy: The Biometric-Mendelian Debate." *British Journal for the History of Science* 22:299–320.

————. 1989. "Scientists and Bureaucrats in the Establishment of the John Innes Horticultural Institution under William Bateson." *Annals of Science* 46:497–510.

————. 1991. "Social Imperialism and State Support for Agricultural Research in Edwardian Britain." *Annals of Science* 48:509–26.

————. 2004. "Bateson, William (1861–1926)." In *Oxford Dictionary of National Biography,* ed. Colin Matthew and Brian Harrison. Oxford: Oxford University Press.

Oldroyd, David. 1990. *The Highlands Controversy: Constructing Geological Knowledge Through Fieldwork in Nineteenth-Century Britain.* Chicago: University of Chicago Press.

[Oliphant, Laurence]. 1865. "Nile Basins and Nile Explorers." *Blackwood's Edinburgh Magazine* 97:100–17.

Ondaatje, Christopher. 1998. *Journey to the Source of the Nile.* Toronto: Harper Collins.

Ophir, Adir, and Steven Shapin. 1991. "The Place of Knowledge: A Methodological Survey," *Science in Context* 4:3–21.

Opitz, Donald Luke. 2004. "Aristocrats and Professionals: Country-House Science in Late-Victorian Britain." PhD diss., University of Minnesota.

————. 2004. "'Behind Folding Shutters in Whittingehame House': Alice Blanche Balfour (1850–1936) and Amateur Natural History." *Archives of Natural History* 31:330–48.

————. 2006. "'This House Is a Temple of Research': Country-House Centres for Late Victorian Science." In *Repositioning Victorian Sciences: Shifting Centres in Nineteenth-Century Scientific Thinking,* ed. David Clifford, Elisabeth Wadge, Alex Warwick, and Martin Willis, 143–53. London: Anthem Press.

Orange, A. D. 1971. "The British Association for the Advancement of Science: The Provincial Background," *Science Studies* 1:315–29.

"Our Little Chatter Box—'Fire-Proof Ladies Dresses.'" 1857. *Theatrical Journal* 18:276.

Ousby, Ian. 1990. *The Englishman's England: Taste, Travel and the Rise of Tourism.* Cambridge: Cambridge University Press.

Outram, Dorinda. 1996. "New Spaces in Natural History." In Jardine, Secord, and Spary, *Cultures of Natural History*, 249–65.

———. 1999. "On Being Perseus: New Knowledge, Dislocation, and Enlightenment Exploration." In Livingstone and Withers, *Geography and Enlightenment*, 281–94.

Ozer, Mark N. 1966. "The British Vivisection Controversy." *Bulletin of the History of Medicine* 40:158–67.

Packard, Alpheus S., Jr. 1876. "A Century's Progress in American Zoology." *American Naturalist* 10:591–8.

Pall Mall Gazette. 1888. "How to Become an Orator." 24 October, 1–2.

Palladino, Paolo. 1993. "Between Craft and Science: Plant Breeding, Mendelian Genetics, and British Universities, 1900–1920." *Technology and Culture* 34: 300–323.

Pang, Alex Soojung-Kim. 1993. "The Social Event of the Season: Solar Eclipse Expeditions and Victorian Culture." *Isis* 84:252–77.

Paradis, James. 1996. "The Natural Historian as Antiquary of the World." In *Hugh Miller and the Controversies of Victorian Science*, ed. Michael Shortland, 122–50. Oxford: Clarendon Press.

Paranavitana, K. D., and C. G. Uragoda. 2007. "Medicinalia Ceylonica: Specifications of Indigenous Medicines of Ceylon sent by the Dutch to Batavia in 1746." *Journal of the Royal Asiatic Society, Ceylon Branch* 53:1–55.

Paris, John A. 1814. "Minutes of the Proceedings of a Meeting held at the Union Hotel, Penzance." Minute Book, Penzance: Royal Cornwall Geological Society Archive.

———. 1815. *Notes on the Soils of Cornwall, with a View to Form a Rational System of Improvement by the Judicious Application of Mineral Manure.* Penzance: T. Vigurs.

———. 1818. "Observations on the Geological Structure of Cornwall, with a View to Trace Its Connexion with, and Influence upon Its Agricultural Economy, and to Establish a Rational System of Improvement by the Scientific Application of Mineral Manure." *Transactions of the Royal Geological Society of Cornwall* 1:169–70.

———. 1828. *A Guide to the Mount's Bay and the Land's End.* London: Thomas and George Underwood.

[Paris, Thomas Clifton]. 1851. *A Handbook for Travellers in Devon and Cornwall.* London: Murray.

———. 1856. *A Handbook for Travellers in Devon and Cornwall.* 3rd ed. London: Murray.

Parrish, Susan Scott. 2006. *American Curiosity: Cultures of Natural History in the Colonial British Atlantic World.* Chapel Hill: University of North Carolina Press.

Parsons, Nicholas. 2007. *Worth the Detour: A History of the Guidebook.* London: Sutton.

Pauly, Philip J. 2000. *Biologists and the Promise of American Life: From Meriwether Lewis to Alfred Kinsey.* Princeton, NJ: Princeton University Press.

Peach, Charles W. 1846. "On the Fossil Geology of Cornwall." *Transactions of the Royal Geological Society of Cornwall* 6:181–85.

Pearce, Susan M. 1995. *On Collecting: An Investigation into Collecting in the European Tradition.* London: Routledge.

———. 1998. *Collecting in Contemporary Practice.* London: Sage.

Pearson, Karl. 1906. "Walter Frank Raphael Weldon, 1860–1906." *Biometrika* 5:1–52.

Pearson, Nicholas M. 1982. *The State and the Visual Arts: A Discussion of State Inter-*

vention in the Visual Arts in Britain, 1760–1981. Milton Keynes: Open University Press.

Pedley, Mary Sponberg. 2005. *The Commerce of Cartography.* Chicago: University of Chicago Press.

Peebles, Patrick. 2006. *The History of Sri Lanka.* Westport, CT: Greenwood Press.

Pengelly, William. 1868. "The History of the Discovery of Fossil Fish in the Devonian Rocks of Devon and Cornwall." *Transactions of the Devonshire Association for the Advancement of Science, Literature, and Art* 2:1–20.

Penny, H. Glenn. 2002. *Objects of Culture: Ethnology and Ethnographic Museums in Imperial Germany.* Chapel Hill: University of North Carolina Press.

Percival, Robert. 2006. *An Account of the Island of Ceylon Containing Its History, Geography, Natural History.* London: C. and R. Baldwin. (Orig. pub. 1805.)

Perrin, C. E. 1982. "A Reluctant Catalyst: Joseph Black and the Edinburgh Reception of Lavoisier's Chemistry." *Ambix* 29:145.

Petch, T. 1920. "The Early History of Botanic Gardens in Ceylon with Notes on the Topography of Ceylon." *Ceylon Antiquary and Literary Register* 5:119–24.

———. 1921. "The Early History of Botanic Gardens in Ceylon." *Ceylon Antiquary and Literary Register* 7:63–73.

———. 1925. *Bibliography of Books and Papers Relating to Agriculture and Botany to the End of the Year 1915.* Colombo, Sri Lanka: H. Ross Cottle.

Pethiyagoda, Rohan. 1998. "The Family de Alwis Seneviratne of Sri Lanka: Pioneers in Biological Illustration." *Journal of South Asian Natural History* 4:99–110.

———. 2007. *Pearls, Spices and Green Gold: An Illustrated History of Biodiversity Exploration in Sri Lanka.* Colombo, Sri Lanka: W. H. T. Publications.

Philalethes. 1864. "Peyrerius and Theological Criticism." *Anthropological Review* 2:109–16.

"The Philosophical Institution." 1846. *Scotsman,* 3 October.

"Philosophical Institution." 1851. *Scotsman,* 8 February.

"Philosophical Institution." 1851. *Scotsman,* 12 November.

"Philosophical Institution." 1853. *Scotsman,* 12 February.

"Philosophical Institution: Mr H. Miller's Lectures." 1853. *Scotsman,* 14 December.

"The Philosophical Institution: Technology." 1856. *Scotsman,* 20 February.

"Philosophical Institution." 1862. *Scotsman,* 11 January.

"Philosophical Institution." 1862. *Scotsman,* 5 November.

"Philosophical Institution." 1867. *Scotsman,* 27 March.

"The Philosophical Institution." 1888. *Scotsman,* 2 April.

Picker, John M. 2000. "The Soundproof Study: Victorian Professionals, Work Space, and Urban Noise." *Victorian Studies* 42:427–53.

Pickering, Andrew, ed. 1992. *Science as Practice and Culture.* Chicago: University of Chicago Press.

Pickstone, John V. 2000. *Ways of Knowing: A New History of Science, Technology and Medicine.* Manchester: Manchester University Press.

———. 2007. "Working Knowledges before and after circa 1800: Practices and Disciplines in the History of Science, Technology, and Medicine." *Isis* 98:489–516.

Pimlott, J. A. R. 1976. *The Englishman's Holiday: A Social History*. Hassocks, Sussex: Harvester Press.

[Playfair, John]. 1807–8. "La Place, *Traité de Mécanique Céleste*." *Edinburgh Review* 11:249–84.

[———]. 1811. "Cuvier on Fossil Bones." *Edinburgh Review* 18:214–30.

[———?]. 1814. "Cuvier on the Theory of the Earth." *Edinburgh Review* 22:454–75.

Pollard, Sidney, and Paul Robertson. 1979. *The British Shipbuilding Industry, 1870–1914*. Cambridge, MA: Harvard University Press.

Poole, E. H. L. 1931. "An Early Portuguese Settlement in Northern Rhodesia." *Journal of the Royal African Society* 30:164–8.

Popkin, Richard. 1987. *Isaac la Peyrère (1596–1676): His Life, Work and Influence*. Leiden: Brill.

Porter, Charlotte. 1986. *The Eagle's Nest: Natural History and American Ideas, 1812–1842*. Tuscaloosa: University of Alabama Press.

Porter, J. L. 1874. *Science and Revelation: Their Distinctive Provinces, with a Review of the Theories of Tyndall, Huxley, Darwin, and Herbert Spencer*. Belfast: William Mullan.

———. 1874. *Theological Colleges: Their Place and Influence in the Church and in the World; with Special Reference to the Evil Tendencies of Recent Scientific Theories, Being the Opening Lecture of Assembly's College, Belfast, Session 1874–75*. Belfast: William Mullan.

Porter, Roy. 1978. "Gentlemen and Geology: The Emergence of a Scientific Career, 1660–1920." *Historical Journal* 21:809–36.

Porter, Roy, and Mikuláš Teich, eds. 1981. *The Enlightenment in National Context*. Cambridge: Cambridge University Press.

———, eds. 1988. *Romanticism in National Context*. Cambridge: Cambridge University Press.

———, eds. 1992. *The Renaissance in National Context*. Cambridge: Cambridge University Press, 1992.

Porter, Roy, and Mikuláš Teich, eds. *The Industrial Revolution in National Context*. Cambridge: Cambridge University Press, 1996.

Porter, Theodore M. 2004. *Karl Pearson: The Scientific Life in a Statistical Age*. Princeton, NJ: Princeton University Press.

Porter, William Smith. 1922. *Sheffield Literary and Philosophical Society: A Centenary Retrospect*. Sheffield: Northend.

P[oulton], E. B. 1887. "The Study of Embryology." *Nature* 36:601.

Powell, Richard. 2007. "Geographies of Science: Histories, Localities, Practices, Futures." *Progress in Human Geography* 31:309–30.

Prakash, Gyan. 1999. *Another Reason: Science and the Imagination of Modern India*. Princeton, NJ: Princeton University Press.

"Preface." 1822. *Transactions of the Royal Geological Society of Cornwall* 2:viii.

"Presentation of a Bust of Mr William Smith to the Institution." 1863. *Scotsman*, 31 July.

Preziosi, Donald. 2006. "Philosophy and the Ends of the Museum." In *Museum Philosophy for the Twenty-First Century*, ed. Hugh H. Genoways, 69–78. Lanham, MD: Altamira.

Pridham, Charles. 1849. *An Historical and Statistical Account of Ceylon and Its Dependencies.* 2 vols. London: Boone.

Prince, Hugh. 1989. "The Changing Rural Landscape, 1750–1850." In *The Agrarian History of England and Wales*, vol. 6, ed. G. E. Mingay, 7–83. Cambridge: University of Cambridge Press.

Prior, Nick. 2002. *Museums and Modernity: Art Galleries and the Making of Modern Culture.* Oxford: Berg.

———. 2003. "The Art of Space in the Space of Art: Edinburgh and Its Gallery, 1780–1860." *Museum and Society* 1:63–74.

"Professor Huxley on the Bases of Physical Life." 1868. *Scotsman*, 9 November.

"Professor Tulloch's Introductory Address." 1857. *Scotsman*, 7 November.

"Professor Winchell's 'Preadamites.'" 1880. *Appleton's Journal: A Magazine of General Literature* 9 (July): 86–90.

Prown, Jules David. 1992. *Discovered Lands, Invented Pasts: Transforming Visions of the American West.* New Haven, CT: Yale University Press.

Punnett, R. C. 1950. "Early Days of Genetics." *Heredity* 4:1–10.

Pyenson, Lewis. 2002. "An End to National Science: The Meaning and Extension of Local Knowledge." *History of Science* 40:251–90.

Pyenson, Lewis, and Susan Sheets-Pyenson. 1999. *Servants of Nature: A History of Scientific Institutions, Enterprises, and Sensibilities*, Norton History of Science Series. London: HarperCollins.

Pym, H., ed. 1882. *Memories of Old Friends, Being Extracts from the Journals and Letters of Caroline Fox.* London, Smith Elder.

Raby, Peter. 1996. *Bright Paradise: Victorian Scientific Travellers.* London: Pimlico.

Rader, Karen A., and Victoria E. M. Cain. 2008. "From Natural History to Science: Display and the Transformation of American Museums of Science and Nature." *Museum and Society* 6:152–71.

Rainbow, Phil, and Roger J. Lincoln. 2003. *Specimens: The Spirit of Zoology.* London: Natural History Museum.

Rainger, Ronald. 1978. "Race, Politics and Science: The Anthropological Society of London in the 1860s." *Victorian Studies* 22:51–70.

Rainger, Ronald, Keith Benson, and Jane Maienschein. 1988. *The American Development of Biology.* Philadelphia: University of Pennsylvania Press.

Raj, Kapil. 2006. *Relocating Modern Science: Circulation and the Construction of Knowledge in South Asia and Europe, 1650–1900.* Basingstoke: Palgrave Macmillan.

Rankine, W. J. M. 1871. *A Memoir of John Elder: Engineer and Shipbuilder.* Edinburgh: Blackwood.

Rees, Terence. 1978. *Theatre Lighting in the Age of Gas.* London: Society for Theatre Research.

Reingold, Nathan, and Marc Rothenberg, eds. 1972–2007. *The Papers of Joseph Henry.* 12 vols. Washington: Smithsonian Institution Press.

Relaño, Francesc. 1995. "Against Ptolemy: The Significance of the Lopes-Piggafetta Map of Africa." *Imago Mundi* 47:49–65.

———. 2002. *The Shaping of Africa: Cosmographic Discourse and Cartographic Science in Late Medieval and Early Modern Europe.* Aldershot: Ashgate.

"Remarks upon the Comparative Healthfulness and Other Local Advantages of Nuwera
    Eliya in the Island of Ceylon and the Neil Gerry Hills in Hindoostan." 1832. *Co-
    lombo Journal* 1:472–73.
"Retrospect of the Present State of French Literature." 1798. *Monthly Magazine*
    5:532–41.
[Review of *An Introduction to the Study of Conchology*, by Samuel Brookes]. 1816.
    Monthly Review 81:328–29.
[Review of *Système du monde*, by P. S. Laplace.] 1796. *Analytical Review* 24:610–11.
"Review of *The Journal of a Disappointed Man*, by W. N. P. Barbellion," *Nature*, 10 July
    1919, 363.
[Review of *Traité élémentaire de chemie*, by A.-L. Lavoisier]. 1789. *Analytical Review*
    4:52–53.
Rhode, Michael G., and James H. T. Conner. 2007. "'A Repository for Bottled Monsters
    and Medical Curiosities': The Evolution of the Army Medical Museum." In *Defining
    Memory: Local Museums and the Construction of History in America's Changing
    Communities*, ed. Amy K. Levin, 177–96. Walnut Creek, CA: Altimira Press.
Richmond, Marsha L. 1997. "'A Lab of One's Own': The Balfour Biological Laboratory for
    Women at Cambridge University, 1884–1914." *Isis* 88:422–55.
———. 2001. "Women in the Early History of Genetics: William Bateson and the Newn-
    ham College Mendelians, 1900–1910." *Isis* 92:55–90.
———. 2006. "The 1909 Darwin Celebration: Reexamining Evolution in the Light of
    Mendel, Mutation, and Meiosis." *Isis* 97:447–84.
———. 2006. "The 'Domestication' of Heredity: The Familial Organization of Geneticists
    at Cambridge, 1895–1910." *Journal of the History of Biology* 39:565–605.
Rindge, Debora Anne. 1993. "The Painted Desert: Images of the American West from the
    Geological and Geographical Surveys of the Western Territories, 1867–1879." PhD
    diss., University of Maryland, College Park.
Ritter, Carl. 1829. "Ueber Geographische Stellung und Horizontale Ausbreitung der
    Erdtheile." *Abhandlungen der historisch-philologischen Klasse der Koeniglichen
    Akademie der Wissenschaften zu Berlin, 1826*. Berlin: Koenigliche Akademie der
    Wissenschaften,103–27.
———. 1831. "Bemerkungen ueber Veranschaulichungsmittel raeumlicher Verhaeltnisse
    bei graphischen Darstellungen durch Form und Zahl." *Abhandlungen der Historisch-
    Philologischen Klasse der Koeniglichen Akademie der Wissenschaften zu Berlin,
    1828*. Berlin: Koenigliche Akademie der Wissenschaften, 213–32.
Ritvo, Harriet. 1987. *The Animal Estate: The English and Other Creatures in the Victo-
    rian Age*. London: Harvard University Press.
———. 1997. *The Platypus and the Mermaid and Other Figments of the Classifying
    Imagination*. Cambridge, MA: Harvard University Press.
———. 2004. "Understanding Audiences and Misunderstanding Audiences: Some Publics
    for Science." In *Science Serialized: Representations of Science in Nineteenth-
    Century Periodicals*, ed. Geoffrey Cantor and Sally Shuttleworth, 331–49. Cambridge,
    MA: MIT Press.
Robb, Johnston Fraser. 1993. "Scotts of Greenock, Shipbuilders and Engineers, 1820–1920:
    A Family Enterprise." 2 vols. PhD diss., University of Glasgow.

Roberts, Andrew. 2000. *Salisbury: Victorian Titan.* London: Phoenix.

Rodger, Richard. 2001. *The Transformation of Edinburgh.* Cambridge: Cambridge University Press.

Rodgers, Daniel T. 1998. "Exceptionalism." In Molho and Wood, *Imagined Histories,* 3–20.

"Rowan and Horton's Steam Engines and Boilers." 1860. *Engineer* 10:210.

Rose, Caroline. 2002. "Charles Kingsley Speaking in Public: Empowered or at Risk?" *Nineteenth-Century Prose* 29:133–50.

Rose, Mark. 1995. *Cities of Light and Heat: Domesticating Gas and Electricity in Urban America.* University Park: Pennsylvania State University Press.

Rowan, Frederick J. 1880. "On the Introduction of the Compound Engine, and the Economical Advantage of High Pressure Steam; with a Description of the System Introduced by the Late Mr J. M. Rowan." *Transactions of the Institution of Engineers and Shipbuilders in Scotland* 23: 51–97.

Rubiés, Joan-Pau. 1996. "Instructions for Travellers: Teaching the Eye to See." *History and Anthropology* 9:139–90.

Rubinstein, W. D. 1981. *Men of Property: The Very Wealthy in Britain since the Industrial Revolution.* New Brunswick, NJ: Rutgers University Press.

Rudwick, Martin. 1976. "The Emergence of a Visual Language for Geological Science, 1760–1840." *History of Science* 14:149–95.

———. 1985. *The Great Devonian Controversy: The Shaping of Scientific Knowledge Among Gentlemanly Specialists.* Chicago: University of Chicago Press.

———. 1997. *Georges Cuvier, Fossil Bones, and Geological Catastrophes: New Translations and Interpretations of Primary Texts.* Chicago: University of Chicago Press.

———. 2002. *Scenes from Deep Time: Early Pictorial Representations of the Prehistoric World.* Chicago: University of Chicago Press.

———. 2005. *Bursting the Limits of Time: The Reconstruction of Geohistory in the Age of Revolution.* Chicago: University of Chicago Press.

Rupke, Nicolaas A., ed. 1987. *Vivisection in Historical Perspective.* London: Croom Helm.

———, ed. 1988. *Science, Politics, and the Public Good.* London: Macmillan.

———. 1994. *Richard Owen: Victorian Naturalist.* New Haven, CT: Yale University Press.

———. 1996. "Eurocentric Ideology of Continental Drift." *History of Science* 34:251–72.

———. 1996. "Humboldtian Medicine." *Medical History* 40:293–310.

———. 1997. "Paradise and the Notion of a World Centre, from the Physico-Theologians to the Humboldtians." In *Phantastische Lebensräume, Phantome und Phantasmen,* ed. Hans-Konrad Schmutz, 77–88. Marburg: Basiliskenpresse.

———. 2001. "Humboldtian Distribution Maps: The Spatial Ordering of Scientific Knowledge." In *The Structure of Knowledge: Classifications of Science and Learning since the Renaissance,* ed. Tore Frängsmyr, 93–116. Berkeley: University of California, Office for History of Science and Technology.

———. 2005. *Alexander von Humboldt: A Metabiography.* Frankfurt am Main: Peter Lang; repr., Chicago: University of Chicago Press, 2008.

———. 2009. *Richard Owen: Biology without Darwin.* 2nd ed. Chicago: University of Chicago Press.

Rupke, Nicolaas A., and Karen E. Wonders. 2000. "Humboldtian Representations in

Medical Cartography." In *Medical Geography in Historical Perspective*, ed. Nicolaas A. Rupke, 163–75. Medical History, Supplement no. 20. London: The Wellcome Trust Centre for the History of Medicine at University College, London.

Rutnam, James T. 1957. "Ancient Nuwara Eliya, Part 1: The Ramayana Legend." *Ceylon Fortnightly Review*, 21 June, 25, 27, 36.

———. 1957. "Ancient Nuwara Eliya, Part 2: A Link with King Dutugemunu." *Ceylon Fortnightly Review*, 19 July, 19, 21, 33.

———. 1957. "Ancient Nuwara Eliya, Part 3: A Lithic Record." *Ceylon Fortnightly Review*, 23 August, 17, 19.

Ryan, James R., and Simon Naylor, eds. 2010. "Exploration and the Twentieth Century." In *New Spaces of Exploration: Geographies of Discovery in the Twentieth Century*, ed. James R. Ryan and Simon Naylor, 1–22. London: I. B. Tauris.

Rydell, Robert W. 1984. *All the World's a Fair: Visions of Empire at American International Expositions, 1876–1916*. Chicago: University of Chicago Press.

Samuel, Raphael. 1994. *Theatres of Memory: Past and Present in Contemporary Culture*. London: Verso.

Sappol, Michael. 2002. *A Traffic of Dead Bodies: Anatomy and Embodied Social Identity in Nineteenth-Century America*. Princeton, NJ: Princeton University Press.

"The Savoy Theatre." 1882. *Electrical Review* 10:10.

Schaaf, Larry J., ed. *The Correspondence of William Henry Fox Talbot*. Available online at http://foxtalbot.dmu.ac.uk/.

Schaffer, Simon. 1991. "The History and Geography of the Intellectual World: Whewell's Politics of Language." In *William Whewell: A Composite Portrait*, ed. Menachem Fisch and Simon Schaffer, 201–31. Oxford: Clarendon Press.

———. 1992. "Late Victorian Metrology and Its Instrumentation: A Manufactory of Ohms." In *Invisible Connections: Instruments, Institutions, and Science*, ed. Robert Bud and S. E. Cozzens, 23–56. Bellingham, WA: SOPIE Optical Engineering Press.

———. 1992. "Self-Evidence." *Critical Enquiry* 18:327–62.

———. 1998. "The Leviathan of Parsonstown: Literary Technology and Scientific Representation." In *Inscribing Science: Scientific Texts and the Materiality of Communication*, ed. Timothy Lenoir and Hans Ulrich Gumbrecht, 182–222. Stanford, CA: Stanford University Press.

———. 1998. "Physics Laboratories and the Victorian Country House." In Smith and Agar, *Making Space for Science*, 149–80.

Schaffer, Simon, Lissa Roberts, Kapil Raj, and James Delbourgo, eds. 2009. *The Brokered World: Go-Betweens and Global Intelligence, 1770–1820*. Sagamore Beach, MA: Science History Publications.

Schivelbusch, Wolfgang. 1988. *Disenchanted Night: The Industrialisation of Light in the Nineteenth Century*. Oxford: Berg.

Schuchert, Charles. 1897. "What Is a Type in Natural History?" *Science* 5:636–40.

Schulten, Susan. 2001. *The Geographical Imagination in America, 1880–1950*. Chicago: University of Chicago Press.

Scolari, Paul. 2005. "Indian Warriors and Pioneer Mothers: American Identity and the Closing of the Frontier in Public Monuments, 1890–1930." PhD diss., University of Pittsburgh.

"Scottish Matters." 1859. *Engineer* 7:375.

"The Screw Steamer 'Brandon.'" 1854. *Nautical Magazine and Naval Chronicle* 23:507–9.

Scrope, William. 1883. *Days of Deer Stalking in the Scottish Highlands.* London: Hamilton, Adams.

Second Report of the United States Board on Geographic Names, 1890–1899. 1901. Washington, DC: Government Printing Office.

Secord, Anne. 1994. "Science in the Pub: Artisan Botanists in Early Nineteenth-Century Lancashire." *History of Science* 32:269–315.

———. 2002. "Botany on a Plate: Pleasure and the Power of Pictures in Promoting Early Nineteenth-Century Scientific Knowledge." *History of Science* 93:28–57.

Secord, James A. 1982. "King of Siluria: Roderick Murchison and the Imperial Theme in Nineteenth-Century British Geology." *Victorian Studies* 25:413–42.

———. 1985. "Darwin and the Breeders: A Social History." In *The Darwinian Heritage*, ed. David Kohn, 519–42. Princeton, NJ: Princeton University Press.

———. 1991. "Edinburgh Lamarckians: Robert Jameson and Robert E. Grant." *Journal of the History of Biology* 24:1–18.

———. 1994. "Introduction." In *Vestiges of the Natural History of Creation and Other Evolutionary Writings*, by Robert Chambers, ix–xlv. Chicago: University of Chicago Press.

———. 1996. "The Crisis of Nature." In Jardine, Secord, and Spary, *Cultures of Natural History*, 447–59.

———. 1981. "Nature's Fancy: Charles Darwin and the Breeding of Pigeons." *Isis* 72:163–86.

———. 2000. *Victorian Sensation: The Extraordinary Publication, Reception, and Secret Authorship of "Vestiges of the Natural History of Creation."* Chicago: University of Chicago Press.

———. 2003. "Introduction." In *The Boy's Playbook of Science*, ed. John Henry Pepper, v–x. Bristol: Thoemmes Press.

———. 2004. "Knowledge in Transit." *Isis* 95:654–72.

———. 2004. "Monsters at the Crystal Palace." In *Models: The Third Dimension of Science*, ed. Soraya de Chadarevian and Nick Hopwood, 138–69. Stanford, CA: Stanford University Press.

———. 2007. "How Scientific Conversation Became Shop Talk." In Fyfe and Lightman, *Science and the Marketplace*, 23–59.

Semple, Ellen Churchill. 1911. *Influences of Geographic Environment.* New York: Russell and Russell.

Seward, A. C., ed. 1910. *Darwin and Modern Science: Essays in Commemoration of the Centenary of the Birth of Charles Darwin and the 50th Anniversary of the Publication of the Origin of Species.* Cambridge: Cambridge University Press.

Shapin, Steven. 1982. "History of Science and Its Sociological Reconstructions." *History of Science* 20:157–211.

———. 1983. "'Nibbling at the Teats of Science': Edinburgh and the Diffusion of Science in the 1830s." In Inkster and Morrell, *Metropolis and Province*, 151–78.

———. 1988. "The House of Experiment in Seventeenth-Century England." *Isis* 79:373–404.

————. 1990. "Science and the Public." In *Companion to the History of Modern Science*, ed. R. C. Olby, G. N. Canton, J. R. R. Christie, and M. J. S. Hodge, 990–1007. London: Routledge.

————. 1994. *A Social History of Truth: Civility and Science in Seventeenth-Century England*. Chicago: University of Chicago Press.

————. 1995. "Here and Everywhere: Sociology of Scientific Knowledge." *Annual Review of Sociology* 21:289–321.

————. 1998. "Placing the View from Nowhere: Historical and Sociological Problems in the Location of Science." *Transactions of the Institute of British Geographers* 23:5–12.

Shapin, Steven, and Barry Barnes. 1977. "Science, Nature and Control: Interpreting Mechanics' Institutes." *Social Studies of Science* 7:31–74.

Shapin, Steven, and Simon Schaffer. 1985. *Leviathan and the Air-Pump: Hobbes, Boyle, and the Experimental Life*. Princeton, NJ: Princeton University Press.

Shaw, David. 2002. "French-Language Publishing in London in 1900." In *Foreign-Language Printing in London, 1500–1900*, ed. Barry Taylor, 101–22. Boston Spa and London: British Library.

————. 2003. "French Émigrés in the London Book Trade to 1850." In *The London Book Trade: Topographies of Print in the Metropolis from the Sixteenth Century*, ed. Robin Myers, Michael Harris and Giles Mandelbrote, 127–43. New Castle, DE: Oak Knoll Press; London: British Library.

Sheail, John. 1987. *Seventy-Five Years in Ecology: The British Ecological Society*. Oxford: Blackwell.

Sheets-Pyenson, Susan. 1988. *Cathedrals of Science: The Development of Colonial Natural History Museums during the Late Nineteenth Century*. Kingston, Ontario: McGill-Queen's University Press.

Shelton, Anthony A. 2000. "Museum Ethnography: An Imperial Science." In *Cultural Encounters: Representing "Otherness,"* ed. Elizabeth Hallam and Brian Street, 155–93. London: Routledge.

Sheppard, Eric, and R. B. McMaster, eds. 2004. *Scale and Geographic Enquiry: Nature, Society and Method*. Oxford: Blackwell.

Sher, Richard B. 2006. *The Enlightenment and the Book: Scottish Authors and Their Publishers in Eighteenth-Century Britain, Ireland, and America*. Chicago: University of Chicago Press.

Short, John Rennie. 1999. "A New Mode of Thinking: Creating a National Geography in the Early Republic." In *Surveying the Record: North American Scientific Exploration to 1930*, ed. Edward E. Carter II, 19–50. Philadelphia, PA: American Philosophical Society.

Shteir, Ann B. 2003. "Bentham for 'Beginners and Amateurs' and Ladies: *Handbook of the British Flora*." *Archives of Natural History* 30:237–49.

————. 2007. "'Fac-Similes of Nature': Victorian Wax Flower Modelling." *Victorian Literature and Culture* 35:649–61.

Shteir, Ann B., and Bernard Lightman, eds. 2006. *Figuring It Out: Science, Gender and Visual Culture*. Hanover, NH: University Press of New England.

Sibum, H. Otto. 2003. "Exploring the Margins of Precision." In Bourget, Licoppe, and Sibum, *Instruments, Travel and Science*, 216–42.

Simmons, Jack. 1970. "Introduction." In *Murray's Handbook for Travellers in Switzerland*, ed. John Murray and William Brockeden, 9–29. Leicester: Leicester University Press.

———. 1991. *The Victorian Railway*. London: Thames & Hudson.

Simoes, A., A. Carneiro, and M. P. Diogo, eds. 2003. *Travels of Learning: A Geography of Science in Europe*. Dordrecht: Kluwer Academic.

Simon, Josep. 2010. "The Baillières: The Franco-British Book Trade and the Transit of Knowledge." In *Franco-British Interactions in Science since the Seventeenth Century*, ed. Robert Fox and Bernard Joly, 243–62. London: College Publications.

"Sir John Lubbock on Savages." 1869. *Scotsman*, 4 November.

"Sir Joseph Hooker's Reminiscences of Manchester." 1907–8. *Lancashire Naturalist* 1:118–20.

"Sir William Pearce, Bart, M.P." 1888–89. *Marine Engineer* 10:344 (obituary of Pearce).

Sivasundaram, Sujit. 2005. *Nature and the Godly Empire: Science and Evangelical Mission in the Pacific, 1795–1850*. Cambridge: Cambridge University Press.

Sivasundaram, Sujit, ed. Forthcoming (2010). "Global Histories of Science." *Isis* 101: n.p.

———. Forthcoming. "Ethnicity, Indigeneity and Migration in the Advent to British Rule in Sri Lanka." *American Historical Review*.

Slatter, Enid. 2008. "Botanical and Zoological Illustrations of the Nineteenth Century: A Survey of the Changes in Printing Techniques with Examples taken from Our Archives." *Linnean* 24:26–35.

Slaven, Anthony. 2004. "Pearce, Sir William, First Baronet (1833–1888)." *Oxford Dictionary of National Biography*. Oxford: Oxford University Press. Available online at http://www.oxforddnb.com/view/article/21692 (accessed 13 February 2007).

Sleigh, George. 1867. *Strictures on the Lectures Lately Delivered in Queen Street Hall*. Edinburgh: MacLachlan and Stewart.

Smith, Crosbie. 1998. *The Science of Energy: A Cultural History of Energy Physics in Victorian Britain*. Chicago: University of Chicago Press.

Smith, Crosbie, and Anne Scott. 2007. "'Trust in Providence': Building Confidence into the Cunard Line of Steamers." *Technology and Culture* 48:471–96.

Smith, Crosbie, and Jon Agar. 1998. "Introduction: Making Space for Science." In Smith and Agar, *Making Space for Science*, 1–23.

———, eds. 1998. *Making Space for Science: Territorial Themes in the Shaping of Knowledge*. Basingstoke: Macmillan.

Smith, Crosbie, Ian Higginson, and Phillip Wolstenholme. 2003. "'Avoiding Equally Extravagance and Parsimony': The Moral Economy of the Ocean Steamship." *Technology and Culture* 44:443–69.

———. 2003. "'Imitations of God's Own Works:' Making Trustworthy the Ocean Steamship." *History of Science* 41:379–426.

S[mith], G. E. 1922–23. "Alexander Macalister, 1844–1919." *Proceedings of the Royal Society of London, Series B* 94:xxxii–xxxix.

Smith, Henry Nash, 1978. *Virgin Land: The American West as Symbol and Myth*. Cambridge, MA: Harvard University Press. (Orig. pub. 1950.)

Smith, James Edward. 1800–1804. *Flora Britannica*. 3 vols. London: J. White.

Smith, Jean Chandler. 1993. *Georges Cuvier: An Annotated Bibliography of His Published Works*. Washington, DC: Smithsonian Institution Press.

Smith, Laurel. 1995. "In the Museum Case of Otis T. Mason: Natural History, Anthropology, and the Nature of Display in the United States National Museum." MA thesis: University of Oklahoma.

Smith, Samuel Stanhope. *An Essay on the Causes of the Variety of Complexion and Figure in the Human Species*. Cambridge, MA: Belknap Press of Harvard University Press, 1965 [1810 ed.].

Smith, T. T. Vernon. 1857. *The Past, Present and Future of Atlantic Ocean Steam Navigation*. Fredericton: Fredericton Athenaeum.

Smyth, Thomas. 1850. *The Unity of the Human Races Proved to Be the Doctrine of Scripture, Reason and Science with a Review of the Present Position and Theory of Professor Agassiz*. New York: Putnam.

Snobelen, Stephen David. 2001. "Of Stones, Men and Angels: The Competing Myth of Isabelle Duncan's *Pre-Adamite Man*." *Studies in History and Philosophy of Biological and Biomedical Sciences* 32:59–104.

"Some Account of the Commerce between England and France during the Existence of the Continental System; Particularly with Respect to the Book Trade." 1820. *New Monthly Magazine* 14:50–54.

Somerville, Mary. 1851. *Physische Geographie*. 2 vols. Leipzig: Weber.

Sorrenson, Richard. 1996. "The Ship as a Scientific Instrument in the Eighteenth Century." *Osiris* 11:221–36.

Spary, Emma C. 2000. *Utopia's Garden: French Natural History from Old Regime to Revolution*. Chicago: University of Chicago Press.

Speedy, Thomas. 1884. *Sport in the Highlands and Lowlands of Scotland with Rod and Gun*. Edinburgh: William Blackwood and Sons.

Speke, John Hanning. 1859. "Captain J. H. Speke's Discovery of the Victoria Nyanza Lake, the Supposed Source of the Nile: From His Journal—Part 2." *Blackwood's Edinburgh Magazine* 86:391–419.

———. 1859. "Captain J. H. Speke's Discovery of the Victoria Nyanza Lake, the Supposed Source of the Nile: From His Journal—Part 3." *Blackwood's Edinburgh Magazine* 86:565–82.

———. 1859. "Journal of a Cruise on the Tanganyika Lake, Central Africa (Speke's Journal Part 1)." *Blackwood's Edinburgh Magazine* 86:339–57.

———. 1863. *Journal of the Discovery of the Source of the Nile*. Edinburgh: Blackwood.

———. 1863. "The Upper Basin of the Nile, from Inspection and Information." *Journal of the Royal Geographical Society* 33:322–46.

———. 1864. *What Led to the Discovery of the Source of the Nile*. Edinburgh: Blackwood.

Spencer, J. Frederick. 1860. "Surface Condensation" (letter to the editor). *Engineer* 9:22.

Stafford, Barbara Maria. 1994. *Artful Science: Enlightenment, Entertainment, and the Eclipse of Visual Education*. Cambridge, MA: MIT Press.

Stanton, William. 1960. *The Leopard's Spots: Scientific Attitudes toward Race in America, 1815–1859*. Chicago: University of Chicago Press.

Star, Susan Leigh, and James Griesemer. 1989. "Institutional Ecology, 'Translations,' and Boundary Objects: Amateurs and Professionals in Berkeley's Museum of Comparative Zoology, 1907–1939." *Social Studies of Science* 19:387–420.

Stearn, William Thomas. 1981. *The Natural History Museum at South Kensington: A History of the British Museum (Natural History), 1753–1980.* London: Heinemann.

Stearns, Raymond P. 1970. *Science in the British Colonies of America.* Urbana: University of Illinois Press.

Sterling, Keir B. 1974. *The Last of the Naturalists: The Career of C. Hart Merriam.* New York: Arno Press.

Stewart, Susan. 1993. *On Longing: Narratives of the Miniature, the Gigantic, the Souvenir, the Collection.* Durham, NC: Duke University Press.

Stocking, George W., Jr. 1971. "What's In a Name? The Origins of the Royal Anthropological Institute, 1837–1871." *Man. The Journal of the Royal Anthropological Institute* 6:369–90.

———, ed. 1985. *Objects and Others: Essays on Museums and Material Culture.* Madison: University of Wisconsin Press.

———. 1987. *Victorian Anthropology.* New York: Free Press.

———. 1992. *The Ethnographer's Magic and Other Essays in the History of Anthropology.* Madison: University of Wisconsin Press.

———. 1994. "Footnotes for the History of Anthropology: Dogmatism, Pragmatism, Essentialism, Relativism: The Boas/Mason Museum Debate Revisited." *History of Anthropology Newsletter* 21:3–12.

Strachey, Lt. Gen. Richard. 1876. "Transactions of the Sections: Geography." *Report of the Forty-Fifth Annual Meeting of the British Association for the Advancement of Science*, 180–88. London: John Murray.

Strickland, Hugh Edwin. 1855. *Ornithological Synonyms.* Ed. Mrs. Hugh Edwin Strickland and William Jardine. London: Van Voorst.

Stieler, Adolf, and Christian Gottlieb Reichard. 1827. *Hand-Atlas ueber alle Theilen der Erde nach dem neuesten Zustande und über das Weltgebaeude.* Gotha: Justus Perthes Verlag.

Stott, Rebecca. 2003. *Theatres of Glass: The Woman Who Brought the Sea to the City.* London: Short Books.

Strutt, Robert John (fourth Baron Rayleigh). 1930. *Lord Balfour in Relation to His Science.* Cambridge: Cambridge University Press.

———. 1968. *Life of John William Strutt, Third Baron Rayleigh.* Ed. John N. Howard. Madison: University of Wisconsin Press. Reprint of 1924 original, augmented with an introduction by John N. Howard.

Swinney, Geoffrey N. 1998. "Who Runs the Museum? Curatorial Conflict in a National Collection." *Museum Management and Curatorship* 17:295–301.

Sykes, A. H. 2000. "Foster and Sharpey's Tour of Europe." *Notes and Records of the Royal Society of London* 54:47–52.

Tammita-Delgoda, Sinharaja. 2006. *Ridi Vihare: The Flowering of Kandyan Art.* Pannipitiya, Sri Lanka: Stamford Lake.

Taylor, Charles. 1998. "Language and Society." In *Communicative Action: Essays on Jürgen Habermas's "The Theory of Communicative Action,"* ed. Axel Honneth and Hans Joas, 23–35. Cambridge, MA: MIT Press.

Taylor, Griffith. 1947. *Canada: A Study of Cool Continental Environments.* London: Methuen.

———. 1955. *Australia: A Study of Warm Environments*. London: Methuen.

Teather, J. Lynne. 1990. "The Museum Keepers: The Museums Association and the Growth of Museum Professionalisation." *Museum Management and Curatorship* 9:25–41.

te Heesen, Anke. 2007. "On Scientific Objects and Their Visualization." In *Objects in Transition: An Exhibition at the Max Planck Institute for the History of Science, Berlin, August 16–September 2*, ed. Gianenrico Bernasconi, Anna Maerker, and Susanne Pickert, 34–37. Berlin: Max-Planck-Institut für Wissenschaftsgeschichte.

Teich, Mikuláš, and Roy Porter, eds. 1996. *The Industrial Revolution in National Context*. Cambridge: Cambridge University Press.

Tennent, James Emerson. 1860. *Ceylon: An Account of the Island*. 2 vols. London: Longman, Green, Longman, and Roberts.

Testimonials in Favour of John R. Leebody, M.A. Senior Scholar in Mathematics, In Queen's College, Belfast; and First Honorman and Gold Medallist in Mathematics and Mathematical Physics, in the Queen's University of Ireland. As Candidate for the Chair of Mathematics and Natural Philosophy in the Magee College, Derry. 1865. Belfast: Printed by W. & G. Baird.

"Theatres and the Electric Light." 1882. *Telegraphic Journal and Electrical Review* 10:10.

Thirsk, Joan. 1967. *The Agrarian History of England and Wales*. Vol. 1. Cambridge: Cambridge University Press.

Thomas, Nicholas. 1991. *Entangled Objects: Exchange, Material Culture, and Colonialism in the Pacific*. Cambridge, MA: Harvard University Press.

Thomson, Thomas. 1815. "Sketch of the Latest Improvements in the Physical Sciences." *Annals of Philosophy* 5:1–53.

Thwaite, Ann. 2002. *Glimpses of the Wonderful: The Life of Philip Henry Gosse, 1810–1888*. London: Faber.

Todd, A. C. 1964. "The Royal Geological Society of Cornwall: Its Origins and History." In *Present Views of Some Aspects of the Geology of Cornwall and Devon*, ed. K. Hosking and G. Shrimpton, 1–23. Cornwall: Royal Geological Society of Cornwall.

Tomes, Jason. 1997. *Balfour and Foreign Policy: The International Thought of a Conservative Statesman*. Cambridge: Cambridge University Press.

Topham, Jonathan R. 2000. "Scientific Publishing and the Reading of Science in Nineteenth-Century Britain: An Historiographical Survey and Guide to Sources." *Studies in History and Philosophy of Science* 31A:559–612.

———. 2009. "Introduction." *Isis* 100:310–18.

Toplis, John. 1814. *A Treatise upon Analytical Mechanics, Being the First Book of the Mechanique Celeste of P. S. Laplace*. Nottingham: Printed by H. Barnett. Sold by Longman, Hurst, Rees, Orme, and Brown, and Craddock and Joy, London; and J. Deighton, Cambridge.

Torrens, Hugh. 1995. "Mary Anning (1799–1847) of Lyme: 'The Greatest Fossilist the World Ever Knew.'" *British Journal for the History of Science* 28:257–84.

"Trial Trips." 1882–83. *Marine Engineer* 4:30.

Trimen, Henry. 1894. [Obituary of Seneviratne]. *Journal of Botany* 32:255–56.

"A Trip to America." 1866. *Engineering* 1:337–38.

Trist, S. J. 1818. "Notes on the Limestone Rocks in the Parish of Veryan." *Transactions of the Royal Geological Society of Cornwall* 1:107–13.

Truettner, William H., and Nancy Adams, eds. 1991. *The West as America: Reinterpreting Images of the Frontier, 1820–1920.* Washington, DC: Smithsonian Institution Press.

Tucker, Jennifer. 1997. "Photography as Witness, Detective, and Imposter: Visual Representation in Victorian Science." In *Victorian Science in Context,* ed. Bernard Lightman, 378–408. Chicago: University of Chicago Press.

———. 2005. *Nature Exposed: Photography as Eyewitness in Victorian Science.* Baltimore, MD: Johns Hopkins University Press.

Turnbull, David. 1989. *Maps Are Territories; Science Is an Atlas.* Chicago: University of Chicago Press.

Turner, Dawson, and L. W. Dillwyn. 1805. *The Botanist's Guide through England and Wales.* 2 vols. London: Phillips and Fardon.

Turrill, W. B. 1963. *Joseph Dalton Hooker: Botanist, Explorer, and Administrator.* London: Thomas Nelson and Sons Ltd.

Two Centuries of Shipbuilding by the Scotts at Greenock. 1906. London: Offices of Engineering.

Tyler, David Budlong. 1939. *Steam Conquers the Atlantic.* New York: Appleton-Century.

Uragoda, C. G. 1987. *A History of Medicine in Sri Lanka.* Colombo, Sri Lanka: Sri Lanka Medical Association.

Urry, John. 1996. "How Societies Remember." In *Theorizing Museums: Representing Identity and Diversity in a Changing World,* ed. Sharon Macdonald and Gordon Fyfe, 45–68. Oxford: Blackwell.

Van Lohuizen-de Leeuw, E. 1965. "The Kustaraja Image: An Identification." In *Paranavitana Felicitation Volume on Art and Architecture and Oriental Studies,* ed. N. A. Jayawickrama, 253–61. Colombo, Sri Lanka: M. D. Gunasena.

Van Riper, A. Bowdoin. 1993. *Men among the Mammoths: Victorian Science and the Discovery of Human Prehistory.* Chicago: University of Chicago Press.

Vaughan, John E. 1974. *The English Guide Book, c.1780–1870: An Illustrated History.* Newton Abbot: David & Charles.

Verulam. 1813. "Lamarck's System of Philosophy." *Monthly Magazine* 36: 409–11 and 499–500.

Vetter, Jeremy. 2004. "Science along the Railroad: Expanding Field Work in the United States Central West." *Annals of Science* 61:187–211.

Vevier, Charles. 2004. "American Continentalism: An Idea of Expansion, 1845–1910." In *American Empire in the Pacific: From Trade to Strategic Balance, 1700–1922,* ed. Arthur Power Dudden. Aldershot: Ashgate.

Vincent, M. L. 1850. "Professor Scott's Lectures at the Philosophical Institution." *Scotsman,* 5 January.

Visitor's Guide to the Smithsonian Institution and National Museum, Washington, D.C. 1880. Washington, DC: Judd and Detweiler.

Vleuten, Erik van der, and Arne Kaijser, eds. 2006. *Networking Europe: Transnational Infrastructures and the Shaping of Europe.* Sagamore Beach, MA: Science History Publications.

Walford, Thomas. 1818. *The Scientific Tourist through England, Wales, and Scotland.* 2 vols. London: John Booth.

Walker, Brian. 1981. *Sentry Hill: An Ulster Farm and Family.* Dundonald: Blackstaff Press.

Wallace, Alfred. 1876. *The Geographical Distribution of Animals.* New York: Harper and Brothers. (Orig. pub. 1862).

Wallace, Henry. n.d. "Teachings of the British Association." Lecture manuscript held by the Gamble Library, Union Theological College, Belfast.

Waller, Horace, ed. 1874. *The Last Journals of David Livingstone, in Central Africa, from 1865 to His Death: Continued by a Narrative of His Last Moments and Sufferings.* 2 vols. London: John Murray.

Walters, Alice. 1997. "Conversation Pieces: Science and Polite Society in Eighteenth-Century England." *History of Science* 35:121–54.

Warf, Barney, and Santa Arias. 2009. "Introduction: The Reinsertion of Space in the Humanities and Social Sciences." In *The Spatial Turn: Interdisciplinary Perspectives,* ed. Barney Warf and Santa Arias, 1–10. London: Routledge.

Watts, Robert. 1888. "Natural Law in the Spiritual World." In *The Reign of Causality: A Vindication of the Scientific Principle of Telic Causal Efficiency.* Edinburgh: T. & T. Clark. (Orig. pub. in 1885 in *British and Foreign Evangelical Review.*)

———. c. 1894. *Professor Drummond's "Ascent of Man," and Principal Fairbairn's "Place of Christ in Modern Theology," Examined in the Light of Science and Revelation.* Edinburgh: R. W. Hunter.

Webb, James L. A. 2002. *Tropical Pioneers: Human Agency and Ecological Change in the Highlands of Sri Lanka, 1800–1900.* Delhi: Oxford University Press.

Weiner, Annette B. 1992. *Inalienable Possessions: The Paradox of Keeping-While-Giving.* Berkeley and Los Angeles: University of California Press.

Welsh, Margaret. 1998. *The Book of Nature: Natural History in the United States, 1825–1875.* Boston: Northeastern University Press.

White, Paul. 2003. *Thomas Huxley: Making the "Man of Science."* Cambridge: Cambridge University Press.

"The White Star Line." 1885–86. *Marine Engineer* 7:173–74.

Whitehead, Christopher. 2005. *The Public Art Museum in Nineteenth-Century Britain: The Development of the National Gallery.* Aldershot: Ashgate.

"Whittingehame." 1929. *Transactions of the East Lothian Antiquarian and Field Naturalists' Society* 1:132–33.

Wickramasinghe, Nira. 2006. *Sri Lanka in the Modern Age: A History of Contested Identities.* London: Hurst.

"William Gardiner the Botanist." 1847. *Chambers's Edinburgh Journal,* n.s., 7:248–51.

Williams, Stephen. 1991. *Fantastic Archaeology: The Wild Side of North American Prehistory.* Philadelphia: University of Pennsylvania Press.

Williamson, Tom. 2002. *The Transformation of Rural England: Farming and the Landscape 1700–1870.* Exeter: Exeter University Press.

Williamson, William Crawford. 1896. *Reminiscences of a Yorkshire Naturalist.* London: Redway.

Willis, J. C. 1901. "The Royal Botanic Gardens of Ceylon and Their History." *Annals of the Royal Botanical Gardens, Peradeniya* 1:2–3.

Wilmot, Sarah. 1996. "The Scientific Gaze: Agricultural Improvers and the Topography of South-West England." In *Topographical Writers in South-West England*, ed. Mark Brayshay, 105–35. Exeter: University of Exeter Press.

Wilson, Alexander, and Charles Lucien Bonaparte. 1831. *American Ornithology; or, The Natural History of the Birds of the United States*. Edinburgh: Constable.

Wilson, Jessie Aitken. 1860. *Memoir of George Wilson*. Edinburgh: Edmonston and Douglas.

Wilson, John Marius. 1859. *Hand-Book to the English Lakes*. Edinburgh: T. Nelson & Sons.

Wilson, R. Jackson. 1999. "Emerson as Lecturer." In *Cambridge Companion to Ralph Waldo Emerson*, ed. Joel Porte and Saundra Morris, 76–96. Cambridge: Cambridge University Press.

Wilson, T., ed. [n.d.] *Wilson's Hand-Book for Visitors to the English Lakes, with an Introduction by the Late W. Wordsworth, Esq., and a New Map of the Lake District, to Which Is Appended a Copious List of Plants Found in the Adjacent Country*. Kendal: Wilson; London: Longman; and London: Whittaker.

Winsor, Mary P. 2000. "Agassiz's Notions of a Museum: The Vision and the Myth." In *Cultures and Institutions of Natural History: Essays in the History and Philosophy of Science*, ed. Michael T. Ghiselin and Alan E. Leviton, 249–91. San Francisco: California Academy of Sciences.

W[inchell], A[lexander]. 1877. "Preadamite." In *Cyclopaedia of Biblical, Theological, and Ecclesiastical Literature*, ed. John McClintock and James Strong, 8:484–92. New York: Harper and Brothers.

Winchell, Alexander. 1880. 2nd ed. *Preadamites*. Chicago: S. C. Grigg.

Winchester, Juti Anne. 1999. "All the West's a Stage: Buffalo Bill, Cody, Wyoming, and Western Heritage Presentation, 1846–1907." PhD diss., Northern Arizona University.

Winter, Alison. 1994. "'Compasses All Awry': The Iron Ship and the Ambiguity of Cultural Authority in Victorian Britain." *Victorian Studies* 38:69–98.

Winter, James. 1999. *Secure from Rash Assault: Sustaining the Victorian Environment*. Berkeley and Los Angeles: University of California Press.

Withering, William. 1796. *An Arrangement of British Plants; According to the Latest Improvements of the Linnaean System*. 4 vols. 3rd ed. Birmingham: M. Swinney.

Withers, Charles W. J. 1999. "Reporting, Mapping, Trusting: Making Geographical Knowledge in the Late Seventeenth Century." *Isis* 90:497–521.

———. 2000. "Authorizing Landscape: 'Authority,' Naming and the Ordnance Survey's Mapping of the Scottish Highlands in the Nineteenth Century." *Journal of Historical Geography* 26:532–54.

———. 2001. *Geography, Science and National Identity: Scotland since 1520*. Cambridge: Cambridge University Press.

———. 2002. "The Geography of Scientific Knowledge." In *Göttingen and the Development of the Natural Sciences*, ed. Nicolaas A. Rupke, 9–18. Göttingen: Wallstein.

———. 2004. "Mapping the Niger, 1798–1832: Trust, Testimony and 'Ocular Demonstration' in the Late Enlightenment." *Imago Mundi* 56:170–93.

———. 2009. "Place and the 'Spatial Turn' in Geography and in History." *Journal for the History of Ideas* 70:637–58.

———. 2010. *Geography and Science in Britain 1831–1939: A Study of the British Asso-*
ciation for the Advancement of Science. Manchester: Manchester University Press.

———. 2010. "Geographies of Science and Public Understanding? Exploring the Recep-
tion of the British Association for the Advancement of Science in Britain and in
Ireland, c.1845–1939." In *Geographies of Science,* ed. Peter Meusburger, David
Livingstone, and Heike Jons, 185–97. Heidelberg: Springer Science.

Withers, Charles W. J., and Diarmid A. Finnegan. 2003. "Natural History Societies, Fieldwork and Local Knowledge in Nineteenth-Century Scotland: Towards a Historical
Geography of Civic Science," *Cultural Geographies* 10: 334–53.

Withers, Charles W. J., Diarmid A. Finnegan, and Rebekah Higgitt. 2006. "Geography's
Other Histories? Geography and Science in the British Association for the Advancement of Science, 1831–c.1933." *Transactions of the Institute of British Geographers*
31:433–51.

Withers, Charles W. J., Rebekah Higgitt, and Diarmid Finnegan. 2008. "Historical Ge-
ographies of Provincial Science: Themes in the Setting and Reception of the British
Association for the Advancement of Science in Britain and Ireland, 1831– c.1939."
British Journal for the History of Science 43:1–31.

Wittlin, Alma Stephanie. 1970. *Museums: In Search of a Usable Future.* Cambridge, MA:
MIT Press.

Wollaston, A. F. R. 1921. *Life of Alfred Newton.* London: John Murray.

Wonders, Karen. 1993. *Habitat Dioramas: Illusions of Wilderness in Museums of Natural*
History. Uppsala: Almqvist and Wiksell.

Wood, Denis. 1992. *The Power of Maps.* New York: Guilford Press.

Wood, Denis, and John Fels. 2008. *The Natures of Maps: Cartographic Constructions of*
the Natural World. Chicago: University of Chicago Press.

Wood, John George. 1857. *The Common Objects of the Sea Shore: Including Hints for an*
Aquarium. London: Routledge.

Wood, Theodore. 1890. *The Rev. J. G. Wood: His Life and Work.* London: Cassell.

[Woodhouse, Robert]. 1799. "Laplace on the System of the World." *Monthly Review,* 2nd
ser. 29:499– 506.

[———]. 1800. "La Croix on the Differential and Integral Calculus." *Monthly Review,* 2nd
ser., 31:493–505; 32: 485–95.

[———]. 1800. "La Place on Celestial Mechanics." *Monthly Review,* 2nd ser. 31: 464– 79;
32:478–85.

[———]. 1803. "Hassenfratz's Course of Celestial Physics." *Monthly Review* 42:542–43.

[———]. 1803. "La Place on Celestial Mechanics, Vol. 3." *Monthly Review,* 2nd ser.,
40:491–96.

[———]. 1805. "La Place on Celestial Mechanics, Vol. 4." *Monthly Review,* 2nd ser.,
48:477–79.

[———]. 1807. "La Place's Supplement to His Tenth Book." *Monthly Review,* 2nd ser.,
53:483–89.

[———]. 1809. "La Place's Supplement to his 3d Vol. of Mécanique Céleste." *Monthly*
Review, 2nd ser., 60:529–31.

[———]. 1809. "Theory of Capillary Action." *Monthly Review,* 2nd ser., 60:526–29.

Worboys, Michael. 1981. "The British Association and Empire: Science and Social Imperialism." In MacLeod and Collins, *Parliament of Science: the British Association for the Advancement of Science,* 170–87.

Wordsworth, William. 1842. *Guide to the Lakes.* Kendal: Hudson and Nicholson.

Worster, Donald. 1977. *Nature's Economy: The Roots of Ecology.* San Francisco: Sierra Club.

Wosk, Julie. 2001. *Women and the Machine: Representations from the Spinning Wheel to the Electronic Age.* Baltimore, MD: Johns Hopkins University Press.

Wrobel, David M. 2002. *Promised Lands: Promotion, Memory, and the Creation of the American West.* Lawrence: University of Kansas Press.

Wyatt, Mary. 1834–40. *Algae Danmonienses or Dried Specimens of Marine Plants, Principally Collected in Devonshire; Carefully Named According to Dr. Hooker's British Flora.* 5 vols. Torquay: Printed by E. Cockrem.

Wyse Jackson, P. N., ed. 2007. *Four Centuries of Geological Travel: The Search for Knowledge on Foot, Bicycle, Sledge and Camel.* London: Geological Society.

Yanni, Carla. 2005. *Nature's Museums: Victorian Science and the Architecture of Display.* New York: Princeton Architectural Press. (Orig. pub. London: Athlone, 1999.)

Yelloly, John. 1812. "History of a Case of Anæsthesia." *Medico-Chirurgical Transactions* 3:90–101.

Young, Kenneth. 1963. *Arthur James Balfour: The Happy Life of the Politician, Prime Minister, Statesman, and Philosopher, 1848–1930.* London: G. Bell and Sons.

[Young, Thomas]. 1809. "Laplace's Supplement to the Mécanique Céleste." *Quarterly Review* 1:107–12.

[———]. 1821. *Elementary Illustrations of the Celestial Mechanics of Laplace. Part the First, Comprehending the First Book.* London: John Murray.

Z. 1796. "Present State of French Literature." *Monthly Magazine* 2:471–74.

CONTRIBUTORS

SAMUEL J. M. M. ALBERTI is director of museums and archives at the Royal College of Surgeons of England. Previously he held a joint appointment at the University of Manchester between the Centre for Museology and the Manchester Museum. He is author of *Nature and Culture: Objects, Disciplines and the Manchester Museum* (2009) and is completing a monograph on nineteenth-century pathological collections.

LAWRENCE DRITSAS is lecturer in the Science Studies Unit at the University of Edinburgh. His research interests include scientific travel, colonial science, and the institutional transitions from colonial to postcolonial scientific research in Africa. He is the author of *Zambesi: David Livingstone and Expeditionary Science in Africa* (2010). He is currently researching scientific institutions in East Central Africa during the late colonial period.

DIARMID A. FINNEGAN is lecturer in human geography at the Queen's University of Belfast. He is the author of *Natural History Societies and Civic Culture in Victorian Scotland* (2009) and has published widely on the historical geographies of Victorian science. He has research interests in the rhetorical geographies of nineteenth-century debates about science and religion and the history of popular geographical knowledge, particularly as produced and disseminated by late-Victorian Christian missions.

AILEEN FYFE is lecturer in modern British history at the University of St. Andrews. She is the author of *Science and Salvation* (2004) and coeditor, with Bernard Lightman, of *Science in the Marketplace* (2007). Her research interests are in the communication of the sciences to nonspecialist audiences.

GRAEME GOODAY is professor of the history of science and technology at the Centre for the History and Philosophy of Science, Department of Philosophy, University of Leeds. After writing on the history of laboratories, his monographs *The Morals of Measurement* (2004) and *Domesticating Electricity* (2008) have addressed the cultural history of electricity. He is currently director of the research project "Owning and Disowning Invention," funded by the United Kingdom Arts and Humanities Research Council.

SALLY GREGORY KOHLSTEDT directs the Program in History of Science and Technology at the University of Minnesota. Her teaching and research focus primarily on the history of science in American culture, with particular attention to museums, public education, and women and gender issues in science. In 2009 she coedited, with Maria Rentetzi, a special issue of *Centaurus* entitled "Gender and Networking in Twentieth-Century Physical Sciences." Her most recent book is *Teaching Children Science: Hands-On Nature Study in North America, 1890–1931* (2010).

BERNARD LIGHTMAN is professor of humanities at York University, Toronto, Canada, where he is currently director of the Institute of Science and Technology Studies. He is editor of the journal *Isis*. He has edited or coedited several collections, including *Science in the Marketplace* (2007), *Figuring It Out* (2007), and *Victorian Science in Context* (1997). His most recent books are *Evolutionary Naturalism in Victorian Britain* (2009) and *Victorian Popularizers of Science* (2007). He is working on a biography of John Tyndall as part of a collaborative international project that aims to publish all of Tyndall's correspondence.

DAVID N. LIVINGSTONE is professor of geography and intellectual history at Queen's University Belfast and a fellow of the British Academy. He is the author of several books, including *Nathaniel Southgate Shaler and the Culture of American Science* (1987), *Darwin's Forgotten Defenders* (1987), *The Geographical Tradition* (1992), *Putting Science in Its Place* (2003), and *Adam's Ancestors: Race, Religion and Politics of Human Origins* (2008). He is currently working on two books, *Locating Darwinism* and *The Empire of Climate*.

SIMON NAYLOR is senior lecturer in historical geography at the University of Exeter. He has conducted research into the historical geographies of science in South America, Britain, Australia, and Antarctica. He is the coeditor (with James Ryan) of *New Spaces of Exploration: Geographies of Discovery in the Twentieth Century* (2010), and author of *Regionalizing Science: Placing Knowledges in Victorian England* (2010).

DONALD L. OPITZ is assistant professor in the School for New Learning at DePaul University. His publications explore the intersection of science, gender, and religion in the scientific avocations of British gentlewomen and gentlemen in the nineteenth and twentieth centuries, and is currently researching British women's formal education in horticultural science at the close of the nineteenth century.

NICOLAAS RUPKE is Lower Saxony Research Professor of the History of Science at Göttingen University. His areas of expertise are the late modern earth and life sciences. Among his books are *Alexander von Humboldt: A Metabiography* (2008) and *Richard Owen: Biology without Darwin* (2009). He is currently working on the non-Darwinian tradition in evolutionary biology. Rupke is a fellow of the German Academy of Sciences Leopoldina and the Göttingen Academy of Sciences, where he directs the project "Blumenbach Online."

ANNE SECORD is an affiliated research scholar in the Department of History and Philosophy of Science, University of Cambridge. The focus of her research and writings has been on popular, particularly working-class, natural history in nineteenth-century Britain, and on horticulture, medicine, and consumption in the eighteenth century. She is completing a book to be published by the University of Chicago Press, focused on social class, observation, and skill in nineteenth-century natural history. She has also been commissioned to produce a new edition of Gilbert White's *Natural History of Selborne* for Oxford World's Classics.

SUJIT SIVASUNDARAM is lecturer in world and imperial history since 1500 at the University of Cambridge, and Fellow of Gonville and Caius College. He is the author of *Nature and the Godly Empire: Science and Evangelical Mission in the Pacific, 1795–1850* (2005). He organized and contributed to a forum in *Isis* (2010), entitled "Global Histories of Science." His research examines themes in the cultural, religious, intellectual, geographical, and environmental history of modern empire.

CROSBIE SMITH is professor of history of science and director of the Centre for the History of Science, Technology, and Medicine at the University of Kent, Canterbury, England. He is author of *The Science of Energy: A Cultural History of Energy Physics in Victorian Britain* (1998), and coauthor (with Norton Wise) of *Energy and Empire: A Biographical Study of Lord Kelvin* (1989), and (with Ben Marsden) of *Engineering Empires: A Cultural History of Technology in Nineteenth-Century Britain* (2005). In 2009, he and Anne Scott were awarded SHOT's Usher Prize for work on the early history of the Cunard Line.

JONATHAN R. TOPHAM is senior lecturer in the history of science at the University of Leeds, United Kingdom. Among his copublications are *Science in the Nineteenth-Century Periodical: Reading the Magazine of Nature* (2004), *Culture and Science in the Nineteenth-Century Media* (2004), and *Science in the Nineteenth-Century Periodical: An Electronic Index* (2005). He is currently preparing a monograph on science and print culture in Britain, 1789–1832.

CHARLES W. J. WITHERS is professor of historical geography at the University of Edinburgh and a fellow of the British Academy. With David Livingstone, he has coedited *Geography and Enlightenment* (1999) and *Geography and Revolution* (2005). Recent publications include *Placing the Enlightenment: Thinking Geographically about the Age of Reason* (2007), *Geography and Science in Britain, 1831–1939: A Study of the British Association for the Advancement of Science* (2010), and *Geographies of the Book* (2010), coedited with Miles Ogborn.

INDEX

Page numbers in italics refer to figures in the text.